BUSINESS DATA COMMUNICATIONS

BASIC CONCEPTS, SECURITY, AND DESIGN

SECOND EDITION

Jerry FitzGerald
Jerry FitzGerald & Associates

WILEY

John Wiley & Sons
New York   Chichester   Brisbane   Toronto   Singapore

Wiley Series in Computers and Information Processing Systems for Business

Cover Art by Roy Weimann.
Cover Design by Kevin Murphy.

Library of Congress Cataloging in Publication Data:

FitzGerald, Jerry.
  Business data communications.

  (Wiley series in computers and information processing systems for business)
  Includes index.
  1. Data transmission systems.   2. Computer networks.
3. Office practice--Automation.   I. Title.
II. Series: Wiley series in computers and information processing systems in business.

TK5105.F577 1988   005.7'1   87-31627
ISBN 0-471-83727-X

Printed in the United States of America

10   9   8   7   6   5   4   3   2

ABOUT THE AUTHOR

Dr. Jerry FitzGerald is the principal in Jerry FitzGerald and Associates, a management consulting firm. He has extensive experience in computer security, audit and control of computerized systems, data communication security, and systems analysis. In addition to consulting, he conducts training courses and seminars in these subjects.

Prior to starting his own firm, Dr. FitzGerald was a senior management consultant with an international consulting firm, an associate professor in data processing for a state university system, and a senior systems analyst at both a major medical center and a computer manufacturer.

As a consultant, Dr. FitzGerald has been active in numerous system design projects, EDP audit reviews, new system development control reviews, EDP audit training, internal control reviews of online systems, and control/security of data communication networks. This work has included development of the computer security administration function within organizations, redesign of the system development life cycle process, development of data communication networks for organizations, review of the internal EDP audit function on behalf of management, and developing requests for proposals related to the selection and purchase of computer systems.

In addition to numerous articles, Dr. FitzGerald is the author of six books, some of which have been translated into Spanish or Japanese. These books are

Business Data Communications: Basic Concepts, Security, and Design (second edition, 1988); *Fundamentals of Systems Analysis: Using Structured Analysis and Design Techniques* (third edition, 1987); *Online Auditing Using Microcomputers* (1987); *Designing Controls into Computerized Systems* (1981, English and Japanese); *Internal Controls for Computerized Systems* (1978, English and Spanish); and *Fundamentals of Data Communications* (1978, English and Spanish).

Dr. FitzGerald's education includes a Ph.D. in Business Administration from the Claremont Graduate School, an M.B.A. from the University of Santa Clara, and a Bachelor's Degree in Industrial Engineering from Michigan State University. He is also a Certified Information Systems Auditor (CISA) and has a Certificate in Data Processing (CDP).

PREFACE

We live in an information-based society. Everyone who works with computers needs to have a basic understanding of data communications. This book is intended to teach students, educators, and business professionals everything they need to know to be fluent in data communications, especially in a world that increasingly is networking microcomputers.

This book assumes no prior data communications knowledge. It blends technical aspects with practical applications without resorting to a mathematical approach. In addition to the elementary explanations of the technical concepts, there is in-depth coverage of network management, security, and design.

This second edition has been expanded extensively. In addition to three entirely new chapters, over 50 new sections have been integrated throughout the other nine chapters, numerous figures have been added, technical descriptions have been simplified, and the chapters have been reorganized. The most notable changes are

- Three new chapters.
  Chapter 6: Microcomputers and Communications.
  Chapter 7: Local Area Networks (LANs).
  Chapter 12: Computerized Network Design.

- PC-based *software* for design and management.

  A computerized network design software package is available on six diskettes (see Chapter 12).

  A software demonstration disk on the management of LANs is available (see Chapter 7).

- A network design *template* is included with the book.

- Over 100 pages of updated material are incorporated, including new sections on A Brief History of Communications, ISDN, T-1 Circuits, Telecommunication Standards, The Chief Information Officer, Deregulation, and so forth.

- More questions have been added to the end of each chapter, and design problems have been included where the subject matter permits.

- Expanded, in-depth network design coverage has been added to Chapter 11.

Each chapter has its own list of *Selected References.* There is a comprehensive *Glossary* at the back of the book, and of course there is a very detailed *Index*! Three times each year John Wiley and Sons distributes a *newsletter* which is keyed to specific topics in this book. The newsletter is designed to update this book continuously with news of changes in the field of data communications. Finally, the design template, the two software packages, and a comprehensive *Instructor's Manual* provide the book with a superior instructor support package. Contact your Wiley representative for more information on the instructor support materials.

This book is intended for college and university courses, as well as for the working professional who must learn about data communications. After reading this book, you probably will be surprised at the amount of knowledge you have gained and by how well all the pieces have come together to provide a complete end-to-end picture of the data communication environment!

To visualize both the scope and depth of this book, please examine the Contents. It details the specific subjects covered in each chapter.

Jerry FitzGerald
Redwood City, California

ACKNOWLEDGMENTS

The author is grateful to the many people who contributed to the preparation of this second edition. First, my thanks go to the reviewers who made many excellent suggestions and contributions. I used 98 percent of their ideas. The reviewers are Marvin Albin at Western Kentucky University; Ardean A. Anvik, Network Manager, State of Washington; Robert Chew at the University of New Hampshire; Ruth Hudson Lankford at Purdue University; Bruce Sophie at North Harris Community College; Don Stengel at California State University at Fresno; and Bob Verkle at California State University at Los Angeles.

Next, my thanks go to the many people at John Wiley and Sons who made this book possible. These people include Joe Dougherty, college textbook acquisitions editor; Kevin Murphy, designer; Christopher Cosentino, production supervisor; Betty Pessagno, book editor; and Priscilla Todd, supervising editor.

Finally, my thanks go to CONTEL Business Networks of Great Neck, New York, and EXCELAN of San Jose, California. Both of them supplied real-life, PC-based software packages for instructional purposes. Dave Rubin, the vice president and general manager of the Network Analysis Center of CONTEL Business Networks, contributed to Chapter 11, "Network Design Fundamentals," and Chapter 12, "Computerized Network Design." CONTEL Business Networks designs, engineers, implements, and maintains complex communication systems nationwide for industry and government.

Jerry FitzGerald

CONTENTS

PART ONE
FUNDAMENTALS OF DATA COMMUNICATIONS   /1

1. **INTRODUCTION AND NETWORK APPLICATIONS   /3**
    Why Study Data Communications   /4
    A Brief History of Communications in the United States   /5
    Purpose and Scope of This Book   /7
    Definition of Data Communications   /9
    Uses of Data Communications   /10
    Basic Components of a Communication System   /10
    System Progression   /14
    Types of Networks   /16
    Today's and Tomorrow's Networks   /18
    Applications for Data Communications   /26
    The Automated Office   /26
    Voice Mail   /34
    Electronic Mail (E-Mail)   /38
    Communications in Banking/Finance   /40
    Communications for the Airlines   /44
    Communications in the Rental Car Industry   /44

Teleconferencing   /**44**
Electronic Shopping   /**45**
Telecom Careers   /**45**
Selected References   /**48**
Computerized Literature Resources   /**49**
Questions/Problems   /**51**

2.   **FUNDAMENTAL COMMUNICATION CONCEPTS**   /**53**
Introduction   /**54**
Microcomputer/Terminal   /**55**
Modes of Transmission   /**55**
    Parallel Mode   /**55**
    Serial Mode   /**56**
    Asynchronous Transmission   /**57**
    Synchronous Transmission   /**58**
    Isochronous Transmission   /**59**
Coding Terminology and Structure   /**61**
Throughput (TRIB)   /**69**
Baseband/Broadband   /**71**
Connector Cables   /**73**
RS232C/RS449   /**73**
Data Signaling/Synchronization   /**77**
Analog Modulation   /**78**
Modem   /**82**
Digital to Analog   /**83**
Digital/Bipolar   /**85**
Digital Modulation   /**86**
Bits/Baud   /**90**
Station Terminals   /**94**
Local Loops   /**94**
Two-/Four-Wire Circuits   /**96**
Full Duplex/Half Duplex (FDX/HDX)   /**96**
Amplifiers   /**97**
Telephone Company Central Office   /**99**
Switching   /**99**
Interexchange Channels (IXC)   /**99**
Circuits/Channels   /**100**
    Open Wire Pairs   /**100**
    Wire Cables   /**100**
    Coaxial Cable   /**100**
    Microwave Transmission   /**103**

Satellite /104
Fiber Optics /111
Cellular Radio /117
Miscellaneous Circuit Types /119
T-1 Carrier /120
Front End Communication Processor (FEP) /122
Central Control Versus Interrupt /122
Polling/Selecting /123
Response Time /124
Error Control in Data Communications /127
Data Communication Errors /128
Line Noise and Distortion /129
Approaches to Error Control /131
Loop or Echo Checking /132
Error Detection with Retransmisson /132
Forward Error Correction /135
Logging /137
Data Channel /137
Host Computer /137
Selected References /138
Questions/Problems /138

3. **DATA COMMUNICATION HARDWARE** /141
Introduction /142
Host Mainframe Central Computer /142
Front End Communication Processor (FEP) /146
Communication Line Control /148
Protocol/Code Conversion /149
Assembly of Characters/Messages /149
Data and Message Editing /149
Message Queuing/Buffering /150
Error Control /150
Message Recording /150
Statistical Recording /150
Other Functions /151
Line Adapters /151
Line Interface Module /151
Port Sharing Device /152
Intelligent Port Selector /153
Line Splitter /153
Digital Line Expander /154

Port/Line Security Device   /155
Data Compression/Compaction   /156
Line Protectors   /156
Intelligent Controllers   /157
Modems   /159
Optical Modems   /160
Short Haul Modems   /160
Acoustic Couplers   /161
Null Modem Cables   /161
Dumb Modems   /161
Smart Modems   /162
Digital Modems   /164
Features of Modems   /164
Multiplexers   /167
Frequency Division Multiplexing (FDM)   /167
Time Division Multiplexing (TDM)   /169
Statistical Time Division Multiplexing (STDM)   /170
Fiber Optic Multiplexing   /173
T-1 Multiplexing   /173
Multiport Modems   /174
Concentrators   /174
Biplexers   /175
AT&T Multiplexing   /175
Protocol Converters   /177
Hardware Protocol Converter Boxes   /177
Add-on Circuit Boards   /178
Software Protocol Conversion Packages   /178
Hardware Encryption   /180
Terminals/Microcomputers   /181
Microcomputer Workstations   /182
Video Terminals   /183
Teleprinter Terminals   /185
Remote Job Entry Terminals   /185
Transaction Terminals   /185
Facsimile (FAX) Terminals   /186
Dumb/Intelligent Terminals   /188
Attributes of Terminals   /188
Switches   /189
Circuit Switching   /191
Store and Forward   /191
Digital Data Switches   /191
Network Switches   /192

Selected References   /194
Questions/Problems   /194

4.   NETWORK CONFIGURATIONS   /197
Introduction   /197
Voice Communication Network   /199
Voice Grade Leased Circuits   /202
Capacity of a Voice Grade Circuit   /203
Signaling on a Dial-up Circuit   /207
Dial-up Circuits   /208
Echo Suppressors   /210
TASI (Voice Calls)   /211
Voice Call Multiplexing   /212
The Basic Configurations   /212
Point-to-Point Network   /216
Local Intelligent Device   /217
Multidrop Configuration   /217
Multiplex Configuration   /218
Packet Switching   /220
Public Timesharing Networks   /224
ISDN (Integrated Services Digital Network)   /226
PBX (Switchboards)   /228
DTS (Digital Termination System)   /233
Selected References   /235
Questions/Problems   /235

5.   PROTOCOLS AND SOFTWARE   /238
The Basic Software Concepts   /238
Software Design Precepts   /242
Software Testing Precepts   /244
Protocol/Software/Architecture   /245
    Protocol   /246
    Software   /247
    Architecture   /247
Telecommunication Access Method   /248
Teleprocessing Monitor   /252
The OSI Seven-Layer Model   /253
    Layer 1: Physical Layer   /258
    Layer 2: Data Link Layer   /258
    Layer 3: Network Layer   /259
    Layer 4: Transport Layer   /260
    Layer 5: Session Layer   /262

Layer 6: Presentation Layer   /263
Layer 7: Application Layer   /263
ARQ (Automatic Repeat Request)   /264
X.25 Packet Protocol   /264
Binary Synchronous Communications (BSC)   /267
Systems Network Architecture (SNA)   /268
Systems Application Architecture (SAA)   /278
Other Network Architectures   /279
Manufacturing Automation Protocol (MAP)   /281
Digital Network Architecture (DNA)   /281
Distributed Systems Environment (DSE)   /282
Distributed Communications Architecture (DCA)   /282
Burroughs Network Architecture (BNA)   /282
Distributed Network Architecture (DNA)   /283
Distributed Systems (DS)   /283
Transmission Control Protocol/Internet Protocol (TCP/IP)   /284
Open Network Architecture (ONA)   /284
Xerox Network Systems (XNS)   /284
UNIX   /285
Telecommunication Standards   /286
International Organization for Standardization (ISO)   /287
American National Standards Institute (ANSI)   /287
Consultative Committee on International Telephone and
Telegraph (CCITT)   /288
Institute of Electrical and Electronics Engineers (IEEE)   /288
Electronic Industries Association (EIA)   /289
National Bureau of Standards (NBS)   /289
Exchange Carrier Standards Association   /289
Corporation for Open Systems (COS)   /289
Legally Enforceable Standards   /290
CCITT X.nn and V.nn Standards   /291
Selected References   /294
Questions/Problems   /295

6. MICROCOMPUTERS AND COMMUNICATIONS   /297
Microcomputers   /297
Communication Software for Microcomputers   /306
DOS/NETBIOS   /306
Other Communication Software   /309
Microcomputer Protocols   /312
X-ON/X-OFF   /312
XMODEM   /313
KERMIT   /313

X.PC   /315
BLAST   /315
Modems for Microcomputers   /316
Internal Modems   /317
External Modems   /317
Micro/Modem Connectors   /317
Null Modem Cables   /319
Electrical Protection for Micros and LANs   /320
Communication Circuits   /320
Surge/Sag Device   /321
Power Line Conditioner (PLC)   /322
Uninterruptible Power Supply (UPS)   /323
Static Electricity   /324
Micro-to-Mainframe Connections   /324
Selected References   /331
Questions/Problems   /332

7.   LOCAL AREA NETWORKS (LANS)   /334
Local Area Network   /334
Introduction to LANs   /335
Topology   /337
Ring   /337
Bus   /337
Star   /338
Configurations/Standards   /338
Baseband Versus Broadband   /341
Protocols   /342
CSMA/CD   /343
Token Access Method   /346
Bridges and Gateways   /347
LAN Software   /350
Cabling   /352
IBM Cabling   /353
AT&T Cabling   /353
DEC Cabling   /354
LAN Cabling Media   /354
Installing Cables   /358
Single/Dual Cable   /359
LAN Costs   /362
Implementing a LAN   /363
Managing a LAN   /365
Hardware/Software Tools   /365
LAN Analyzer Software   /366

Selecting a LAN   /367
Selected References   /372
Questions/Problems   /373

PART TWO
NETWORK MANAGEMENT AND SECURITY   /375

8.   **NETWORK MANAGEMENT**   /377
The Data Communication Function   /377
Network Organization   /379
    Combining Voice and Data   /380
    The Chief Information Officer   /384
Network Management   /386
    Design and Analysis   /390
    Network Operations   /392
    Failure Control   /392
    Testing/Problem Management   /395
Network Reporting   /397
Network Documentation   /398
Network Status   /400
Test Equipment   /401
    Breakout Box   /404
    Bit-Error Rate Tester (BERT)   /404
    Block-Error Rate Tester (BKER)   /405
    Self-Testing Modems   /405
    Response Time Analyzer   /406
    Data Line Monitor   /406
    Automated Test Equipment   /407
Selected References   /408
Questions/Problems   /409

9.   **SECURITY AND CONTROL**   /410
Why We Need Security   /411
Network Security   /414
Network Control Points   /416
Encryption   /419
Hardware Controls   /427
    Front End Processors   /427
    Packet Switching Controllers   /428
    Modems   /428
    Multiplexers   /429

Remote Intelligent Controllers   /**429**
Terminals—Human Error Prevention   /**430**
Terminals—Security Controls   /**430**
Voice Telephone Security   /**432**
Circuit Controls   /**434**
Microcomputer Controls   /**435**
Database Controls   /**437**
Protocol Controls   /**438**
Layer 1—Physical Link Control   /**438**
Layer 2—Data Link Control   /**438**
Layer 3—Network Control   /**439**
Layer 4—Transport Control   /**439**
Layer 5—Session Control   /**439**
Layer 6—Presentation Control   /**440**
Layer 7—Application Control   /**440**
Network Architecture/Software Controls   /**440**
Management Controls   /**442**
Recovery/Backup/Disaster Controls   /**444**
Matrix of Controls   /**446**
Lists of Data Communication Controls   /**454**
Selected References   /**455**
Questions/Problems   /**456**

**PART THREE
DESIGNING COMMUNICATION NETWORKS   /463**

10.   **COMMUNICATION SERVICES   /465**
Communication Facilities   /**465**
Common Carriers and Tariffs   /**466**
Deregulation   /**469**
Communication Services Offered   /**474**
Private Circuit (Lease) Services   /**474**
Voice Grade Channels   /**475**
Wideband Services   /**476**
Digital Services   /**477**
Satellite Services   /**477**
ISDN   /**477**
T-1 Circuits   /**478**
Software Defined Networks   /**478**
Measured Use Services   /**478**
Direct Distance Dialing (DDD)   /**479**

Wide Area Telecommunications Service (WATS)   **/479**
AT&T Megacom™   **/481**
AT&T Megacom 800   **/481**
Public Packet Switched Services   **/481**
DIAL-IT 900 Service™   **/482**
Discount Voice Services   **/483**
Telex/TWX   **/484**
Other Special Services   **/484**
Foreign Exchange Service (FX)   **/484**
Common Control Switching Arrangement (CCSA)   **/485**
Hotline   **/485**
Selected References   **/485**
Questions/Problems   **/486**

11.   **NETWORK DESIGN FUNDAMENTALS   /488**
Introduction   **/488**
The Systems Approach to Design   **/489**
Thirteen Steps for Network Design   **/490**
1. Conduct a Feasibility Study   **/490**
2. Prepare a Plan   **/492**
3. Understand the Current System   **/494**
4. Design the Network   **/496**
5. Identify the Geographic Scope   **/497**
6. Analyze the Messages   **/499**
7. Calculate Traffic/Circuit Loading   **/504**
8. Develop a Control Matrix   **/507**
9. Determine Network Configurations   **/508**
10. Software Considerations   **/511**
11. Hardware Considerations   **/515**
12. Network (Circuit) Costs   **/516**
Network Cost Analyzer   **/517**
Cost/Benefit Categories   **/517**
Voice Grade Circuit Costs   **/517**
Dial-up Circuit Costs   **/525**
Wideband Circuit Costs   **/527**
Packet Switching Costs   **/528**
Satellite Circuit Costs   **/528**
Digital Circuit Costs   **/530**
Hardware Costs   **/530**
Further Design Ideas   **/530**
13. Sell and Implement the Network   **/532**

Selected References   /**534**
Questions/Problems   /**534**

12.   COMPUTERIZED NETWORK DESIGN   /538
An Overview   /**539**
Modeling Networks   /**541**
Network Optimizer (MINDS^{SM}-Data/PC)   /**546**
Selected References   /**552**
Questions/Problems   /**552**

GLOSSARY   /555

APPENDICES

1.   **DATA COMMUNICATION CONTROL MATRIX   /581**

2.   **CONTROL LISTS FOR DATA COMMUNICATION NETWORKS   /597**

3.   **VERTICAL AND HORIZONTAL COORDINATES   /637**

4.   **HOW TO USE THE LANALYZER™ DEMO DISK   /643**

INDEX   /I-1

PART ONE

FUNDAMENTALS OF DATA COMMUNICATIONS

Part One of this book is devoted to the basic fundamentals and introductory concepts of modern-day data communication networks. There are chapters on . . .

INTRODUCTION AND NETWORK APPLICATIONS

FUNDAMENTAL COMMUNICATION CONCEPTS

DATA COMMUNICATION HARDWARE

NETWORK CONFIGURATIONS

PROTOCOLS AND SOFTWARE

MICROCOMPUTERS AND COMMUNICATIONS

LOCAL AREA NETWORKS (LANs)

1

INTRODUCTION AND NETWORK APPLICATIONS

The purpose of this chapter is to introduce you to the concepts of data communications and to show how systems have progressed toward today's networks. A description of the purpose and scope of this book is included to help you adapt it to your specific needs. The uses of data communications, basic components, types of networks, and applications for data communications are discussed. The chapter introduces the automated office, voice mail, electronic mail, and communications in banking/finance, airlines, and the rental car industry. The chapter closes with a section on telecommunication careers.

WHY STUDY DATA COMMUNICATIONS

It all started around 3300 B.C. with Sumerian clay tablets. They were the ideal way to communicate. Provided you didn't drop them! Next came Greek messengers. Your scroll would always get there—if your runner didn't collapse first. Today you can send and receive vital business information . . . in writing . . . in seconds.

The reasons for studying data communications can be summed up in the occupational history of the United States. In the 1800s we were an agricultural society dominated by farmers. By the 1900s we had moved into an industrial society dominated by labor and management. Now, as we approach the twenty-first century, we clearly have moved into the information society which is dominated by computers, data communications, and highly skilled individuals who use brain power instead of physical power. The industrial society has reached its zenith, and the communication/computer era, which is dubbed the information society, is advancing rapidly.

In an industrial society, the strategic resource is capital. In an information society, the strategic resource is knowledge which creates information that must flow on communication networks. This information society started in the mid-1950s.

Knowledge of data communications is even more important when you realize that satellites are transforming the earth into a global city. In other words, the compression of time that is achieved through satellite communications allows us to be in immediate contact with all other companies or people and to utilize business information in a timely manner.

In an information society dominated by computers and communications, value is increased by knowledge as well as by the speed of movement of that knowledge. This new information economy will completely destroy Ricardo's labor theory of value because in such a society what increases value is not the labor of individuals, but information. Knowledge/information can be created, it can be destroyed, and it is synergetic in that the whole usually is greater than the sum of the parts. In fact, the whole may be many times greater than the sum of the parts if you have the proper communications network to transmit the information. Knowledge that cannot be disseminated (transmitted) may be of zero value.

The main stream of the information age is communications. The value of a high speed data communication network that transmits knowledge/information is that it brings the message sender and the message receiver closer together in time. As a result, we have collapsed the information lag, which is the time it takes for information to be disseminated throughout the world. For example, in the 1800s it might have taken several weeks for specific information to reach the United States from England. By the 1900s it could be transmitted within the hour. Today, with modern data communication systems, it can be transmitted within seconds. Collapsing the information lag speeds up the incorporation of new technology into our daily lives. In fact, today's problem may be that we are unable to handle the quantities of information that we already have.

Finally, the transition from an industrial to an information society means that you will have to learn many new technologically based skills. Instead of becoming a specialist in a certain subject and planning on working in that area for the rest of your life, you will have to adapt and possibly retrain yourself several times during your lifetime. For that reason, the study of data communications will become a basic tool that can be used throughout your lifetime. You will incorporate your knowledge of data communications into several careers such as circuit designer, programmer, business system application developer, communication specialist, and business manager. Even the basic physical job tasks of our society now require technical knowledge in the use of data communications such as citizens band radios in trucks, microcomputers in your home that are connected to national or international communication networks, and personal communication devices such as mobile telephones (cellular radio).

In summary, we forecast that the collapsing of the information lag may be the single most important point in your current study of communications. This is so because new communication technology is being incorporated into the fabric of the information society as fast as people are able to learn how to maintain and use this technology. Once you have learned the basics from this textbook, you will need to "keep up with the communication technology" for the remainder of your life.[1]

A BRIEF HISTORY OF COMMUNICATIONS IN THE UNITED STATES

Today we take data communications for granted, but it was early pioneers like Samuel Morse, Alexander Graham Bell, and Thomas Edison who developed the basic electrical and electronic systems that ultimately became today's voice and data communication networks. In 1837 Samuel Morse exhibited a working telegraph system; today we might consider it the first electronic data communication system. Then in 1876 Alexander Graham Bell invented the first telephone capable of practical use, and it became the basis for our voice communication systems. The telephone was a remarkable improvement, for Morse's system required the operators at each end to use Morse code. This code, which became an auditory signal at the receiving end, could be interpreted into letters, which then became words, sentences, and paragraphs as the operator wrote them down on

[1] Many of the ideas in this section have been extrapolated from the first chapter of the book *Megatrends* by John Naisbitt (New York: Warner Books, 1982).

paper. Obviously, the telegraph was not going to be the most widely used method of communication.

When the telephone arrived, it became the accepted communication device that everyone wanted. In 1879 the first private manual telephone switchboard (PBX—private branch exchange) was installed. By 1880 the first pay telephone was available for use, and the telephone became a way of life for Americans. Now they could call from a public telephone.

In 1885 the certificate of incorporation was registered for the American Telephone and Telegraph Company, and by 1889 AT&T had a recognized logo. This logo took the shape of Philadelphia's Liberty Bell and had the words Long Distance Telephone written on it. By 1910 the Interstate Commerce Commission (ICC) had the authority to regulate interstate telephone business.

The first transcontinental telephone service and the first transatlantic voice connections were both established in 1915. By 1930 the cost of a telephone call from New York to London was reduced from $45 to $30 for the first three minutes. In 1934 President Roosevelt approved the Communication Act, which transferred the regulation of interstate telephone traffic from the Interstate Commerce Commission to the Federal Communications Commission (FCC).

Although the transistor would seem to be more related to computers than to communications, it was invented at Bell Laboratories in 1947. In 1951 the first direct long distance customer dialing began. The first international satellite telephone call was sent over the Telstar satellite in 1962. In 1963 touchtone telephones began to be marketed. Their push buttons were easier to use than rotary dials, and they became quite popular. By 1965 there was widespread introduction of commercial international telephone service by satellite.

The famous Carterfone court decision in 1968 stated that non-Bell equipment could be connected to the Bell System network. This was an important milestone because for the first time independent modem manufacturers could connect their equipment to the Bell networks. Such connections were illegal prior to this decision.

Picturefone service, which allows users to see as well as talk with one another, began operating in 1969. All through the 1970s there were many arguments and court cases regarding the monopolistic position that AT&T held over other companies that wanted to offer communication services. The litigation led to the divestiture of AT&T on January 1, 1984. (Figure 10-1 in Chapter 10 shows the result of this divestiture.)

During 1983–84 traditional radio telephone-type calls were supplanted by the newer cellular telephone networks. By 1985 the Bell Laboratories had invented another new transistor called the ballistic transistor. Ballistic transistors operate 1,000 times faster than the original transistor invented by Bell Laboratories in 1947. Integrated Services Digital Networks (ISDN) began serving the public in

1986. These networks allow the simultaneous transmission of voice, data, and video images.

By 1987 there was considerable competition in both the voice and data communication markets as a number of independent companies began to sell communication services in a manner similar to that of automobile marketing. By the 1990s we can expect to have smaller and less expensive portable telephones to carry around with us, just as we carry a calculator today. We also can expect the charges for long distance calls to be less.

PURPOSE AND SCOPE OF THIS BOOK

Data communications is a complex subject, but this is not a complex book. Most books in this field are complex, and rightly so, because they are intended as reference sources for the experienced systems designer. This book requires no prior experience in data communications, voice communications, or electronic engineering. Rather, this book assumes a basic grounding in data processing and a desire to complement this background with a general knowledge of data communications. After completing a course of study based on this book, you should be able to

- Understand the alternatives available to you in hardware, software, and transmission facilities
- Put that understanding to work by making informed decisions among these alternatives
- Integrate these decisions into a cohesive data communication system design, and carry it forward into reality
- Understand today's technology, as well as how to remain informed
- Perform the above activities for systems of increasingly greater scope and complexity as you build experience, judgment, and confidence

As you read this book, you will encounter many new terms because the world of data communications has its own language. A *comprehensive glossary* is included at the end of the book because it is not always possible to interrupt the presentation of complex subjects with a thorough definition of a new word. The author has evaluated the text carefully to ensure that all the jargon of data communications has been included in the Glossary and that the miscellaneous technical details are described thoroughly.

This book is divided into three parts. Part One introduces the fundamentals of data communications and all of its technical aspects. Part Two defines network

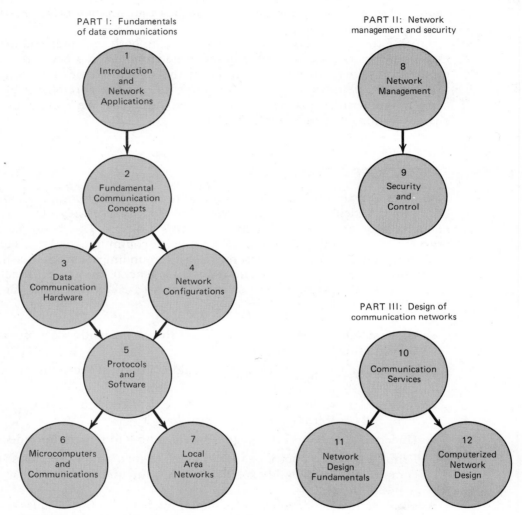

FIGURE **1-1** How this book is organized.

management, security, and control of networks. Part Three describes network design fundamentals and offers a microcomputer-based network design software package. Your learning experience may be taken in three separate parts as shown in Figure 1-1: fundamentals of data communications, network management and security, and design of communication networks.

The *first* five chapters in Part One take you from the introduction, through the applications of communications, and on to the fundamental concepts of communications, hardware, network configurations, and protocols/software for communication networks. The last two chapters of this part examine microcomputers and communications and then present a thorough discussion of local area networks. The *second* segment of your learning experience (Part Two of this book) involves two chapters that discuss the management of networks, and the security and control of networks. Finally, the *third* part of your learning experience (Part Three of this book) covers the design of networks. In it you will learn about the various communication services that are available, along with how to lay out and design a communication network. Both manual network design and computerized network design (using a microcomputer-based software package) are covered.

There are many good books and periodicals on data communications. The bibliographies at the end of each chapter provide a selected sample of current books, serial publications, and computerized literature resources. Before continuing, review both the Glossary at the end of this book and the Selected References at the end of each chapter.

DEFINITION OF DATA COMMUNICATIONS

Data communications is the movement of encoded information from one point to another by means of electrical or optical transmission systems. Such systems often are called *data communication networks.* In general, these networks are established to collect data from remote points (usually terminals or microcomputers) and transmit that data to a central point equipped with a computer or another terminal, or to perform the reverse process, or some combination of the two. Data communication networks facilitate more efficient use of central computers. They improve the day-to-day control of a business by providing faster information flow. They provide message switching services to allow terminals to talk to one another. In general, they offer better and more timely interchange of data among their users and bring the power of computers closer to more users. The objectives of most data communication networks are to

- Reduce the time and effort required to perform various business tasks
- Capture business data at its source
- Centralize control over business data
- Effect rapid dissemination of information
- Reduce current and future costs of doing business
- Support expansion of business capacity at reasonable incremental cost as the organization grows

- Support organizational objectives in centralizing or decentralizing computer systems
- Support improved management control of the organization

USES OF DATA COMMUNICATIONS

While data communications might be used in many different situations, business operations that exhibit some of the following characteristics usually can benefit from the use of a data communication network.

- Widespread use of microcomputers
- Decentralized operations
- A high volume of organizational mail, messenger service, or telephone calls between the various organizational locations (the voice communication corridors, that is, telephone calls, may become or be replaced by the data transfer corridors)
- Repetitive paperwork operations, such as the re-creation or copying of information
- Inefficient and time-consuming retrieval of current business information
- Slow or untimely handling of the organization's business functions
- Inadequate control of the organization's assets
- Inadequate planning and forecasting

Figure 1-2 lists seven types of data communication systems. This figure also summarizes many important characteristics of these typical uses of data communications, giving specific application examples and typical transactions for each application. This tabulation is the framework within which we will work for the remainder of this book. Study this figure carefully. Using your knowledge of the functional characteristics of the specific application examples, consider the information given in the Typical Characteristics of Transactions column. Observe how these characteristics change from one usage to the next.

BASIC COMPONENTS OF A COMMUNICATION SYSTEM

The three basic components of a data communication system are the source, the medium, and the sink. The *source* is the originator of the information; the *medium* is the path through which the information flows; and the *sink* is the mechanism that accepts the information. In this definition, a terminal or mi-

Data Communications Usage Modes	Examples of Applications	Typical Characteristics of Transactions
Source data entry and collection	Sales status data Inventory control Payroll data gathering	Transactions collected several times per day or week, direct response message not issued for every transaction
	Point-of-sale system Airline reservations	Transactions arrive frequently (every few seconds) and demand response within a few seconds
Remote job entry (RJE)	Remote high-speed reading and printing Local access to distant computer power	Transactions usually bunched and require processing times ranging from minutes to hours. Input and output for each transaction may take seconds or minutes.
Information retrieval	Credit checking Bank account status Insurance policy status Law enforcement Government social services Hospital info. systems	Relatively low character volume per input transaction, response required within seconds. Output message lengths usually short but might vary widely with some types of applications
Conversational timesharing	General problem solving Engineering design calculations Text editing	Conversational response required, within a few seconds
Message switching	Company mail delivery and memo distribution	Delivery time requirements range from minutes to hours
Real-time data acquisition and process control	Numerical control of machine tools Remote meter and gauge reading	Remote sensors continuously sampled and monitored at widely varying time intervals
Interprocessor data exchange	Processor, program, and file-sharing applications of all types	Infrequent burst arrivals consisting of large data blocks requiring transmission to another CPU, usually within milliseconds

FIGURE **1-2** Data communication usage modes.

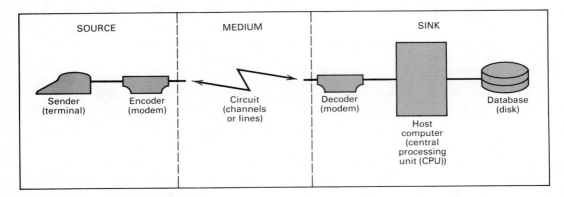

FIGURE **1-3**   Basic components.

crocomputer often alternates as both a source and a sink. The medium is nothing more than the communication line (or circuit) over which the information travels. Usually, the lines are leased from a "common carrier" such as the Bell Operating Companies, AT&T, MCI, or US SPRINT, although an organization can install its own lines. A *common carrier* is a company recognized by the Federal Communications Commission (FCC) or an appropriate state licensing agency as having the right to furnish communication services to individual subscribers or business organizations.

Telecommunications and teleprocessing are other terms used to describe data transmission between a computing system and remotely located devices. The terms *data communications, telecommunications,* and *teleprocessing* are used interchangeably by many writers in this field. We will use data communications in this book because of the combination of computing and communications, although the term *telecommunications* may be used for the integration of data communications, voice communications, and imaging systems, along with the use of host computers and microcomputers.

Figure 1-3 shows the basic components of a data communication system. You must have a sender, which is usually a terminal. It also might be a microcomputer, a video terminal, or some other type. Once the user has entered a message, it goes to the encoder, which usually is called the *modem*. In this example, the modem coverts the signal from its direct electrical pulses (baseband) into a series of varying frequency tones (broadband). Many times the terms *analog* or *digital* are used. Analog is a broadband signal and digital is a baseband signal. An analogy is that an analog (broadband) signal can continually change like the volume control on a radio. On the other hand, a digital (baseband) signal is discrete (off or on) because it operates like a radio's off/on switch. The purpose of this encoding process is to put the transmission into a mode that is compatible with the various

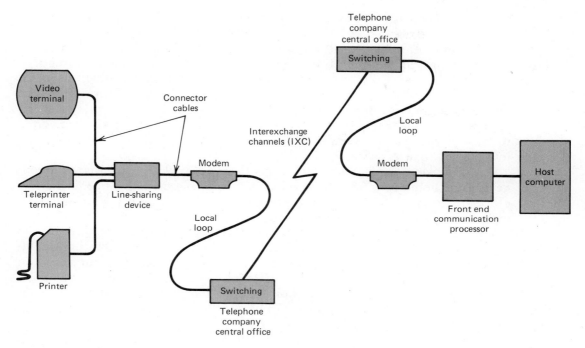

FIGURE **1-4** Basic system.

transmission media such as copper wire, microwave, satellite, fiber optic, or other facilities.

In our figure these transmission media are referred to as a *circuit*. These are the telephone company circuits over which your message moves. Finally, when your message reaches the distant host computer, it first passes through the decoder, which is another modem. This modem converts the signal from broadband (frequency tones) back to baseband (electrical voltages). Finally, your signal (in reality your message request) is passed on to the host computer for whatever processing might be required.

Figure 1-4 depicts a basic data communication system. This system includes terminals, connector cables, a line-sharing device, modems, local loops, telephone company switching offices, interexchange channel (IXC) facilities, a front end communication processor, and a host computer.

The *terminals* or *microcomputers* involve a human-to-machine interface device where people can enter and receive data or information. This type of device might have a video screen, a printing mechanism, and a keyboard. In the future this device may be voice actuated.

The *connector cables* in Figure 1-4 are special cables containing many wires that interconnect the terminal to the modem.

The *line-sharing device* allows multiple terminals to share a single modem. Each terminal sequentially has its turn to transmit and receive data/information.

The *modem* is a solid state electronic device that converts direct electrical signals (+ and − voltages of electricity) to modulated signals that can be sent over data communication circuits. The most common form of modulated signal is a frequency modulated signal where the direct electrical voltages are converted to frequency tones. For example, a high pitched tone might equal a binary 1 and a low pitched tone a binary 0.

The *local loops* in the figure are the connections or "last mile" that interconnects your home or office to the telephone company central office (switching office), or to the special common carrier network if you are using a connection other than the telephone company.

The *central office* (sometimes called *end office* or *exchange office*) contains the various switching and control facilities that are operated by the telephone company or other special common carrier. When you use dial-up communication circuits, your data transmission goes through these switching facilities. When you have a private leased circuit, however, the telephone company wires your circuit path around the switching facilities to provide a clear unbroken path from one modem to the other.

The *interexchange channels/circuits* (sometimes called *IXC circuits*) are the circuits that go from one telephone company central office to another central office. These circuits can be microwave circuits, but they also may be copper wire pairs, coaxial cables, satellite circuits, optical fibers, or other transmission medium.

The *front end communication processor* is a specialized minicomputer with very special software programs. These software programs, along with the front end hardware, control the entire data communication network. For example, a powerful front end communication processor may have 100 or more modems attached to it through its ports (circuit connect points).

Finally, the *host computer* is the central processing unit (CPU) that processes your request, performs database lookups, and carries out the data processing activities required for the business organization.

SYSTEM PROGRESSION

The natural evolution of business systems, governmental systems, and personal systems has forced the widespread use of data communication networks to interconnect these various systems.

In the 1950s we had batch systems with discrete files, and users carried their

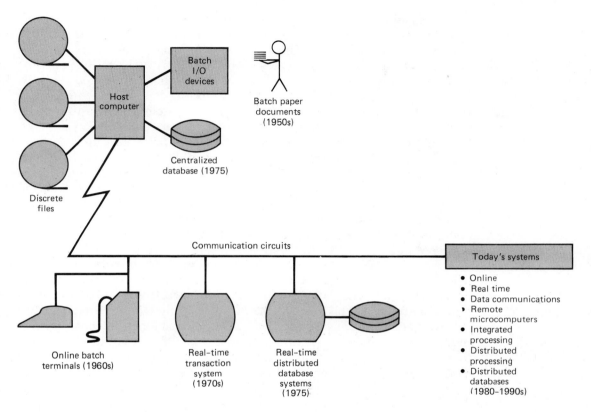

FIGURE **1-5** System progression.

paper documents to the computer for processing. The data communications of that era involved human beings physically carrying paper documents (see Figure 1-5).

During the 1960s we added communication circuits (telephone lines) and gave the users online batch terminals. At this point the users entered their own batches of data for processing. The data communication aspect involved the transmission of signals (messages) from these online batch terminals to the computer and back to the user.

During the late 1960s and into the 1970s we started developing online real-time systems that moved the users from batch processing to single transaction-oriented processing where the response back to the user department was required within approximately three seconds or less. It was during this time that data communications became a necessity.

As the 1970s progressed, we added database management systems that replaced the older discrete files. We also started developing integrated systems whereby one business system might automatically create and pass transactions to some other business system. With integrated systems, the entry of an online real-time transaction could automatically trigger two or three other transactions. For example, when an online terminal user from a purchasing department entered data indicating the purchase of 100 executive desks, the system might initiate three other related transactions. Transaction 1 would go to the Accounts Payable System. There it would set up the original matching file where the purchase order would be matched to the invoice, which in turn would be matched to the receiving dock ticket showing that the goods were received. Transaction 2 might go to the receiving dock to prenotify the personnel that they should expect 100 executive desks in two months. Transaction 3 might go to the Cash Flow Accounting System so preparations could be made to pay for (cash availability) these executive desks. As you can see, both data communications and data processing are interconnected in online real-time systems.

As we enter the 1990s we are fast approaching completely distributed systems where user departments will be given their own computers (probably microcomputers with disks) and the data communication network will have to be even larger and more reliable. The office of the future that interconnects typewriters, word processing machines, facsimile machines, copy machines, teleconferencing equipment, microcomputers, mainframe host computers, and other equipment will put tremendous demands on data communication networks. In addition, local area networks (loop circuits within buildings) will offer greater reliability and speed.

Finally, the ultimate in reliability of the network must be achieved before we move to distributed databases. High reliability is necessary because if one distributed site uses another distributed site's database and the communication circuit fails, it might lock up the database so no one can use it until the communication of the transaction is completed.

TYPES OF NETWORKS

The following is a reasonable scheme for classifying networks, although you may choose to add other applications.

- Single application
- Multiple application
- Organization-wide
- Multiorganization

- Value added
- Common carriers
- International

The *single application* network is built within a single corporation or government agency, and it is used for one specific purpose. For example, in banking you might have a network for the bank balance inquiries for the automated teller machines, checking accounts, or passbook savings.

The *multiple application* network is designed to handle many different applications which can share the network and the common database and/or processing facilities. A multiple application network might be seen in a manufacturing organization. This type of network might handle business systems involving raw material inventory, production planning, the manufacturing process, finished goods inventory, sales and distribution, general ledger accounting, cash flow, accounts payable, accounts receivable, and so on.

Organization-wide networks are developed by large corporations and government agencies that have many computer centers. The networks are designed to interconnect the multiple computer centers. For example, a large government agency could have multiple computer centers in order to keep track of agricultural data, farming patterns, crops, and acreage records. They might place the computer centers in various locations around the country. This organization-wide network would serve its users by allowing local and remote access to any of the data centers and transmission between these data centers.

Multiorganization networks have been constructed to serve groups of similar corporations such as airlines or universities. When you make airline reservations, if any leg of your trip is to be on another airline, the multiorganization reservations network handles transmission of the data to the airline so proper reservations can be guaranteed.

A *value added* network provides a network constructed with leased lines (circuits), and it serves many customers in different geographical areas. It is usually a general purpose computer network like the ones developed by public companies such as Telenet, Tymnet, US Sprint, or MCI. These value added carriers may transmit either data or voice. Their objective is to allow many different users to use their network for a fee, which is dependent on the amount of time the user is using the network (voice calls) or the volume of data being transmitted (data calls). In other words, they lease circuits from the telephone company, build a network, and by doing so, add value to the raw communication circuits because these circuits are now reliable functioning systems.

Common carriers, such as telephone companies, provide nationwide data networks which can be used for a set fee. You also can lease communication circuits from them to build your own network for single or multiple applications that may be either organization-wide or multiorganizational. These common carriers are now offering value added networks. In addition, *special common carriers* lease communication circuits in competition with the telephone companies; therefore, your circuits may or may not be provided by the telephone company.

International networks may be single/multiple application or organization-wide/multiorganizational, and they span the globe. In other words, an international network passes over the borders between countries. Special limitations may be imposed on these international networks with regard to the flow of information (transborder data flow controls). These controls are enacted by the government of each country.

TODAY'S AND TOMORROW'S NETWORKS

By 1990 data communications/teleprocessing will have grown faster and become more important than computer processing itself. Both go hand in hand, but recently we have moved from the computer era to the communications era.

Systems based on teleprocessing can be found in virtually every segment of industry. For example:

- Online passenger reservation systems like American Airlines' SABRE and United Airlines' APOLLO have revolutionized the travel industry. They have helped increase these carriers' market shares by as much as 20 percent. Large car rental and hotel chains could not function effectively without their reservation systems.

- Overnight delivery industry leader Federal Express Corporation considers its COSMOS parcel tracking system as unique. COSMOS enables online inquiry of parcel status from remote locations, locates delayed shipments, and sends invoices to customers automatically. The Federal Express delivery vans even carry on-board terminals.

- The American Hospital Supply Corporation became a networking pioneer in 1974 when it installed the health care industry's first order-entry terminals in hospitals. Today, the ASAP system's success is legendary, and competitors still are trying to catch up.

- The Cirrus banking network, which was organized in 1982, now covers 46 states. Its 1,425 member banks process some 200 million transactions annually and provide such services as cash withdrawals and balance inquiries from checking, savings, and credit accounts. Future services include direct debit retail point-of-sale transactions and international currency conversions from Cirrus' more than 6,500 automated teller machines.

Technological developments are primarily responsible for the enormous increase in the use of communication networks. The two primary technological factors are size and speed. First, the size of electronic components (microprocessor circuits) is decreasing dramatically as circuit density increases. For example, in 1959 one megabyte of memory required a space equal to the size of a room 7 feet square by 8 feet high. Today the same amount of memory requires only about one-half cubic inch of space.

Second, the speed of microprocessor chips used in data communications has increased by many magnitudes. We are on the threshold of a totally new type of transistor, called a *ballistic transistor*, which was first demonstrated by AT&T's Bell Laboratories. The ballistic transistor switches 1,000 times faster than the transistors used in today's communication switches and microcomputers.

IBM's advanced 1 million bits chip technology has resulted in dramatic performance enhancements and increased storage capacity. The high speed chip's dynamic random access memory (DRAM) takes only 80 nanoseconds (billionths of a second) to access stored data. At this speed, coupled with its fastest rate of data flow, this single chip can "read" a 2,200-page document in only one second.

With regard to speed, it was only about 15 years ago that we were first able to transmit at 9600 bits per second (bps) on a standard telephone circuit. Using various digital technologies, we now can transmit at 64,000 bits per second on the same telephone circuit, and with the use of fiber optics it is quite easy to transmit at millions of bits per second. Actually, in the future we will use fiber optics to transmit at billions of bits per second.

There are many exciting prospects ahead with regard to communications. Some of these prospects will be discussed in the following paragraphs.

Videotex will be connected to private homes. *Videotex* involves the two-way transmission between a television set in your home and many other organizations outside your home. It allows you to carry on a two-way dialogue with a doctor, take courses in your home, provide security services for your house, review information retrieval databases, perform teleshopping from your local store, conduct teleconferences (picture and voice) from your home, play video games, have community access to political meetings, view first-run movies, interconnect with satellite television programming, utilize networks for delivery of electronic mail messages, connect with your bank, review the most current news stories, and utilize voice store and forward message systems. Canada already has videotex (called Telidon there), as do England and France.

With regard to voice and data, equipment already is available that combines both voice transmission and data transmission over a single communication circuit. Combining these would be very cost effective in most governmental and private business organizations because much higher circuit usage could be achieved at reduced costs. Approximately three-quarters of today's communication costs are for voice, and one-quarter are for data transmission.

You will have *home satellite TV*. There might be a satellite dish antenna, located on the roof of your house, which will enable you to communicate directly with other people via the satellite. This satellite dish antenna might lead you to transmit either voice or data directly from your house in the United States to someone else's house in Canada or England (see Figure 1-6).

Such use of satellites by individuals leads to widespread possibilities with regard to the freedom and flow of information and ideas. Because the borders of a country may no longer be able to be closed to the free flow of data, information, and ideas, people around the world may become more politically aware. For this reason *transborder data flow* will become more significant.

Many countries today restrict the flow or movement of data across their national boundaries. The United States is probably the most open with regard to the flow of information into and out of the country. Even the United States, however, limits data flow by restricting the sale or delivery of some technological equipment or information to countries that are viewed as less than friendly.

Canada requires that the initial processing of all bank transactions be done in Canada and that foreign networks cross the border only at one crossing point. Canada's 1980 Banking Act prevents the processing of bank transactions outside Canadian boundaries unless some processing also is done within the country. Transmitting financial data outside of the country, or subsequent manipulation of that data, requires government approval. West Germany requires significant local processing of all data transmitted over communication circuits (private or public telephone facilities). Brazil requires corporations to maintain copies of most computer databases inside Brazil, rather than connecting with existing databases outside the country. Most offshore processing of Brazilian data is prohibited. When possible, companies must purchase Brazilian computer equipment and software rather than import it into the country. France is investigating the possibility of taxing data. Sweden has a data inspection board that must approve the export of data files or the transmission of personal data out of Sweden. In England your secret encryption key (encryption will be discussed later in this section) must be shared with the postal and telegraph service. Belgium and France have imposed up to a $400,000 fine for transmitting data defined as sensitive. In Spain you must deposit money in an escrow account before data files can be transmitted out of the country. Such data protection laws are a type of tariff or duty on the free flow of information.

A preliminary agreement between the United States and the European countries represented by the Organization for Economic Cooperation and Development (OECD) has led to a policy declaration on transborder data flow. This declaration states that it is the intent of the 24 OECD member countries and the United States to "promote access to data and information and related services, avoid the creation of unjustified barriers to the international exchange of data and information, and seek transparency in regulations and policies related to informa-

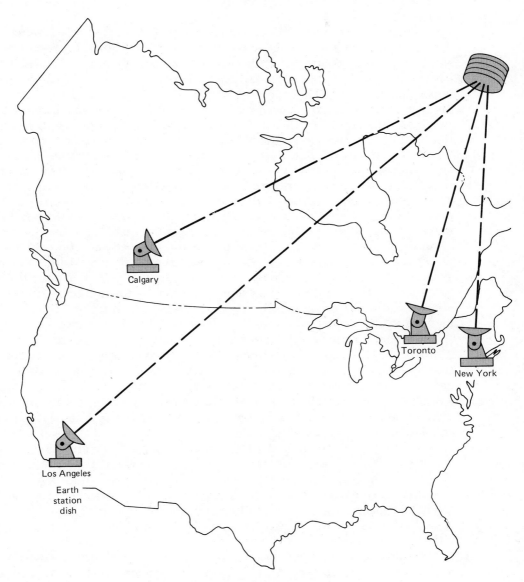

FIGURE 1-6 International information flow.

tion, computer, and communication services affecting transborder data flow." This policy may help ease the free transfer of information between countries, although there still are political and economic barriers, and the licensed common carriers selling communication services sometimes look out for their own special interests first.

In addition to transborder data flow, it also is possible to have an intracompany data flow problem within a single country. In this situation the question is that of a *data monopoly* within a specific country. For example, the U.S. Department of Justice monitors computerized airline reservation systems as a new form of monopoly (called a data monopoly). The Department has stated that it may seek divestiture of the highly successful airline reservation systems from the airlines that own them. It claims that the United Airlines and American Airlines reservation systems are used to monopolize the sales of airline flights. It contends that these two air carriers tend to get most of the reservations because most travel agencies use one of these two reservation systems. The Justice Department further alleges that the computerized reservation systems provide travel agents with a "primary display" that favors their own flights. If travel agents have to scroll or page through many different screens to locate competing carrier flights, then they may be more likely to select those flights that are easy to access.

On the international scene, three U.S. airlines have filed a complaint with the Department of Transportation charging that the Lufthansa Airlines reservation system is biased. The complaint alleges that the flights of U.S. carriers are relegated to the third and fourth screens, if they are shown at all. The three U.S. airlines claim to be losing millions of dollars in annual revenues because of the bias built into the Lufthansa reservation system.

These may be the world's first legal issues involving a data monopoly in our information-based society. A monopoly usually involves goods or services, but in this situation it is claimed that basic information is monopolized in order to enhance one business to the detriment of others. This issue should be watched as a trend for the future. How many other industries or businesses can enhance their own economics through a data monopoly? Do we have the proper laws to address this issue?

As we move into the next decade, you will note that we are moving from the manufacturing/management era to the information era. In other words, information is the single most valuable resource of an enterprise. Information may become more important than management structure, manufacturing ability, or financial capabilities. Thus, a country that restricts information will most likely slow down its economic growth, thereby lowering the standard of living for all its citizens.

The normal voice telephone systems have *store and forward capability* for voice information. In other words, if you call someone who is not at home, the telephone system will accept your voice message and forward it later to the person when he or she calls in. The system can even try repeatedly to contact the person until the telephone is answered and your stored message is delivered.

More encryption will be used on data communication circuits. *Encryption* is the method of encoding data to make it secret during transmission. Encryption on public networks will become standard, as will Forward Error Correction (FEC). *Forward error correction* is the process of automatically correcting most circuit-originated transmission errors without retransmitting the message that contains the error.

More public networks will utilize packet switching. These public networks will have "standard interfaces" that will connect almost any terminal to anything. In other words, any terminal or microcomputer will be able to communicate with any other terminal or microcomputer on the public packet switching network.

Along with more satellites to make better use of television/news, we may even have citizens band radio via satellite. The laws will be changed to encourage satellites, and governments (for economic reasons) will create a good data communication environment within their countries.

More companies will purchase their own computer-based private telephone systems for use within their company. They will interconnect with the rest of the world through rooftop antennas that will transmit via satellite.

Another way these companies will transmit to the rest of the world may be through cellular radio local loops. With cellular radio, radio transmission towers are placed in strategic locations throughout a city. The messages from companies or private homes are transmitted over the airways as radio frequency transmissions to these towers. The towers then connect to land-based communication circuits, microwave circuits, or satellite transmission systems for the long-haul (interexchange channel—IXC) transmission of voice and data. Cellular radio local loops would replace the copper wire local loops that you see throughout the city. In other words, cellular radio towers would both augment and replace telephone lines/telephone poles or underground copper wire cables.

One of the major requirements for adequate data communications in business systems is a large document storage and retrieval system for information that is contained in business documents and for archival storage of documents, market data, books, television shows, movies, local news, and the like. We are on the verge of getting mass data storage on *optical disks*, of the same type that you see today for recorded movies. It is forecasted that optical disks in the future will cost about $10 each.

If you stacked optical disks (similar to the way records are stacked in juke boxes), you could store the entire contents of the U.S. National Archives on only 1,000 12-inch disks. These disks are a "write-once" medium so you have security in the knowledge that no one can modify your data. Optical disks have approximately a ten-year life.

Today there are prototype optical disks that can be erased and rewritten. Today's prototypes are very costly, but when the manufacturing bugs are worked out, they should become quite popular. One prototype makes it possible to squeeze reams of data, sound, pictures, or computer information into a 5¼-inch optical disk. This erasable optical disk can hold as much data as 250 5¼-inch floppy diskettes. For recording, a high power writing laser quickly warms a tiny area on the disk's magnetic surface, causing the magnetized molecules to align with the magnet below. On playback, a specifically filtered, low power laser beam strikes these molecules and is altered slightly as it bounces off the disk. The reflected light then is converted into an electronic signal that can be displayed on a computer screen or fed into a network. For erasing, the laser reheats the area more slowly so it has time to recrystallize gently and take on a neutral characteristic.

One of the new concepts that is emerging is the development of *teleports*. These are large, flexible satellite communication dish antennas that use data transmission frequencies, which are not used in terrestrial communications. Obviously, this is a promising development because teleports can operate in microwave-congested metropolitan areas. These earth stations will be connected with multiple fiber optic cables, coaxial cables, and microwave links to locations throughout the metropolitan area they serve. Teleports usually are constructed by companies that are independent of the telephone companies. The earth station dishes shown in Figure 1-6 are examples of teleports.

A teleport is an urban communication gateway that provides efficient and economical communication services to long distance users. By having these earth stations closer to urban areas, telecommunication customers receive reliable and economic service. Today's teleport earth station (major hub) ranges in size from 20 to 40 feet in diameter. Smaller antennas may be used, and each year new technology provides smaller antennas with the same reliability as older, larger antennas. The transmission rates from a teleport range from 56,000 bits per second to 2.048 megabits per second (mps) or higher. Furthermore, the all-digital bit stream avoids the need for costly analog-to-digital conversion and, consequently, ensures an almost error-free signal. One of the concerns of the local Bell Operating Companies (BOCs) is that the increasing use of teleports allows more of their customers to bypass the local loop, causing them to lose revenues. Currently, there are about two dozen teleports in the United States. Operating teleports are located in Atlanta, Chicago, Dallas/Ft. Worth, Houston, Washington, D.C., Ocala (Florida), Carteret (New Jersey), Raleigh (North Carolina), Seattle, Los Angeles, San Francisco, and New York City. In larger cities the distribution area from the teleport is sometimes as far as 250 miles. This means that there are high speed, terrestrial-based communication links from the teleport out to customers who may be as far as 250 miles away from the teleport itself. At the New York teleport, businesses are linked to the facility via a fiber optic cable network that has been placed throughout the metropolitan New York area. Communications Satellite Corporation (COMSAT) plans to operate international earth station teleports in New York, Chicago, San Francisco, Houston, and Washington, D.C. These international earth stations offer business customers access to international satellites. Numerous other countries also are setting up international teleport earth stations.

Another emerging trend is that of *telecommuting*. This is a trend in which an employee performs some or all of his or her work within the home instead of going to the office each day. If you were the employee doing the telecommuting, some of the benefits might be less time wasted in commuting to work, improved quality of life, ability to optimize the scheduling of both your work and personal life, monetary savings on both clothes and travel, less stress, greater time flexibility, and possibly fewer distractions (a higher level of concentration) at home.

One key area in which telecommuting helps is for "at-home" diagnoses of data processing or data communication problems. This is true especially for after-hours work. In an emergency in which a technician has to communicate with the mainframe, telecommuting is much quicker than a long drive by automobile.

The cable television companies will be increasing their role for two-way communications into and out of each person's home. They are in direct competition with rooftop satellite antennas and the major television networks. These cable TV companies and the common carrier business of data communications may merge into a business cable for the private, commercial, and government markets. This communication pipeline into your home, via both the telephone system and cable TV, is a critical issue because of privacy and security considerations. As an individual consumer, you will need to secure extra privacy for your data and your life. For example, if you use cable TV, someone might determine which television shows you watch and build a profile of your personality. In addition, the TV company could keep track of your purchases, your financial transactions, and anything else that is received or transmitted on the cable to your personal television set. Personal privacy may be a grave concern here.

Did you know that every day 20 million meetings are held in the United States and that more than three-quarters of these meetings last less than 30 minutes? Over one-half of all meetings can be handled by voice communications only, one third of all meetings are for the exchange of information only, and almost 90 percent of air travel in the United States is business travel. For these reasons you will see a distinct increase in *video teleconferencing*. With teleconferencing, people who want to attend a business meeting but are in diverse geographic locations can get together in both voice and picture format. In fact, even documents can be shown and copied at any of the remote locations.

Communications will be enhanced further because integrated circuit chips containing 1 million components per chip will be available. These chips will allow you to develop hardware solutions (firmware) for your current software problems. This capability definitely will increase both the speed and reliability of data communication networks. The chips, which currently allow you to have a video screen, will soon give you an entire wall as your picture screen, and some day an entire holographic wall for your data pictures. Early in the twenty-first century virtually no paper will be used in business communications. Everything will be stored with microchips and large-scale memory devices or optical disks.

The competition for *Digital Termination Systems* (DTS) will become very strong. DTS is a way to bypass the telephone company local loop bottleneck or "last mile" between a user application and long-haul digital transmission (IXC) facilities. DTS is a set of technology and service options that might include microwave, cable TV, telephone company wire pairs, infrared, digital radio broadcast (cellular radio), optical fiber, and rooftop antenna satellite services.

Both electronic mail and voice mail will grow quite rapidly. *Electronic mail* will grow primarily in the business sector and between the private homes of individuals who have microcomputers. *Voice mail* will grow as more special common carriers and telephone companies offer voice store and forward systems.

We are living in an era that is controlled, and soon will be dominated, by data communications. If you think the computer has had an impact on your life or your lifestyle, then you might be surprised, when you look back, to try to determine the changes that were brought about because of data communications. The ultimate in data communications has been a standard part of the television series "Star Trek," where they use a "transporter" to beam people down from space ships to various planets. While this is science fiction today, it might not be science fiction in the twenty-first century.

APPLICATIONS FOR DATA COMMUNICATIONS

The applications for data communications are so wide and so diverse that we could not possibly cover even 10 percent of them in this book. In fact, an entire textbook could be devoted to applications for data communications. Earlier in this chapter some applications were mentioned. In the following sections we would like to discuss selected applications such as the automated office, electronic mail/voice mail, banking/finance communication networks, airlines, and rental cars. The purpose is to provide a frame of reference for the uses of data communications and the way they link our business systems, government agencies, and personal lives.

THE AUTOMATED OFFICE

The automated office is a concept that will have far-reaching effects on the world as we know it today. The term *paperless office* has been used for about 15 years, but society has a way to go before it is ready for this envisioned paperless world.

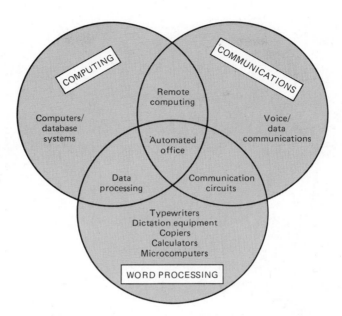

FIGURE **1-7** The three components of the automated office.

However, the technological means to support this capability exists today, and data communication is the key component (see Figure 1-7).

During the 1970s the world began to realize that it was being buried in paper. Most people do not recognize that more than half of the workforce in advanced countries work in offices. Little was done to improve productivity in this area until recently. In the last ten years productivity of manufacturing operations has been improved by at least 90 percent on the average, but during the same time frame the productivity of the office function has been improved by less than half that amount. On the average, it is estimated that one-half of the total costs of U.S. corporations are for office-related work. In some specialized industries such as banking, insurance, and government the figure is much higher, ranging up to 70 percent. The United States alone is spending billions of dollars each year on these office functions supported by managers, professionals, and clerical staff.

Office automation is primarily in the hands of the data communication user; the computer technology already exists. All that is needed is data communications to link together all of the diverse business equipment that is required. Currently available hardware devices allow users to perform their tasks more easily and more efficiently. The part that is still missing is resource sharing, which requires fast and reliable data communications. With the new office net-

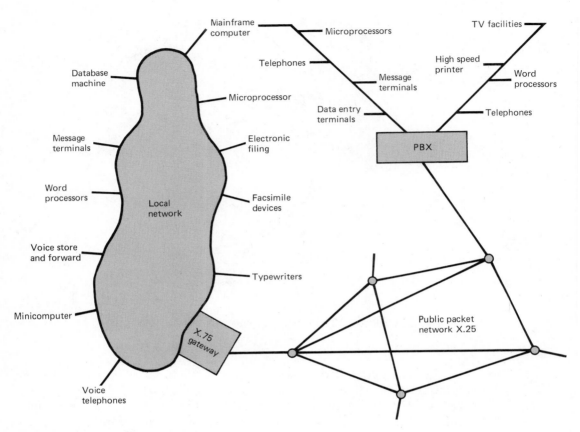

FIGURE **1-8** Automated office networks.

works, especially local area networks and digital PBXs (private branch exchanges are the same as automated telephone switchboards), users will soon be able to achieve this resource sharing and to interconnect diverse operations in a manner that could not be achieved in the past.

Figure 1-8 shows how the business office can be tied together, probably with a local network or PBX, to act as a single integrated company business office. The local networks probably will be tied with other local networks or PBXs across the country or around the world by use of a public switching network interconnected through an X.75 gateway switch to another network. A gateway connects two dissimilar networks.

Probably the one component that will affect office automation the most is the *workstation.* Even though almost everyone can recognize a workstation, people have trouble defining it because of its broad scope. Some would say that workstations are full 32-bit microcomputer systems, while others are quite satisfied to call a dumb terminal (no microcomputer chips) a workstation. Basically, most workstations today are microcomputers. Many of them are customized designs based on microprocessor chips. Workstations are distinguished from minicomputers, which also are 32-bit machines, by number of users and graphic interactions.

Unlike most minicomputers, which generally run multiuser operating software, workstations usually are built to interact with one user at a time. This is changing very fast, however, with the new series of microcomputer chips (Motorola's 68030 and Intel's 80386). Even though most workstations are not multiuser, they are multitasking. If an application is running that does not require frequent screen and keyboard access, it might be possible to tell the system to put the first program away in background but leave it running while working on another program in foreground.

The ideal type of terminal workstation is one with the following features.

- Ability to communicate with all other terminal devices or workstations
- Easy access to multiple databases
- Easy attachment of various types of peripheral devices
- Standalone data processing capability (microcomputer)
- Both voice and data capabilities

As noted above, graphics also have been important to workstation users. Workstations generally count on using high performance pictorial displays. Most now use windowing software and can display several active projects on their screens at the same time. On the software side, various vendors have bundled workstations with specialized software packages. Basically, the line of distinction between workstations and microcomputers (PCs) is fading at a steady rate. Today a number of firms make add-on boards that put a workstation inside a microcomputer or add-on processors that can put a microcomputer inside an already existing workstation. When it comes right down to it, workstations used to be specialized intelligent terminals, but today they are microcomputers and even more expensive customized microcomputers, such as executive workstations. Microcomputers, with their floppy diskettes, probably will become the working tool for one out of every three persons who work in the office. Because these devices are individually programmable, they will become the workhorse of the automated office.

A typical automated office might be interconnected just like the one shown in Figure 1-8 and might contain the types of equipment shown in this figure.

By having such a wide diversity of equipment with the ability to share resources and talk to each other, the office of the future easily can carry out such basic functions as

- **Preparation:** Create business letters, schedules, interoffice memoranda, and reports.
- **Computation:** Compile data to complete forms, generate graphics, develop management charts, and calculate various figures.
- **Dissemination:** Perform local or remote distribution of all of the above in an electronic fashion.
- **Storage:** Store and retrieve documents, maintain mailing lists, and utilize any databases that are available.
- **Voice/Electronic Mail:** Utilize the communication capability for voice communications and/or electronic mail movement.

The digital PBX automated switchboard may become the central focal point or central switch for the automated office. It would be easy to integrate both voice and data communications within the automated office and externally to other organizations with whom you have to communicate or transmit data. The communication decision faced today is, Which system will become the central network, the local area network or the PBX?

People are a major concern in office automation development. Clerical workers, middle management, and upper management currently have a methodology for working *without* a totally automated, paperless type of terminal workstation. Many people see no reason to change their current work methods to what they view as unproven methods.

The two biggest drawbacks to the automated office are fear and change. First is the fear of automation. As a general rule, people fear the unknown, and this is true especially with regard to technology. Second is the massive change required in how one performs one's daily work. How can a supervisor effectively oversee and control an employee's work when it is stored in a machine? In addition, employees tend to resist changing their current comfortable work routines. Today's technology is far ahead of the average person's ability to accept it. For this reason you probably will move into the twenty-first century before the paperless office is achieved.

The first thing that comes to mind for most individuals when they hear the term *office automation* is word processing. It is important to recognize, however, that word processing is more important for increasing the productivity of secretaries than of other staff. Additional capabilities are needed to satisfy the objectives of the automated executive office. The most important objectives in making managers more productive include

- Minimizing the amount of time spent on paperwork
- Improving decision-making capabilities
- Improving local and remote voice and data communications

With these objectives in mind, researchers have concluded that automated office technology can be applied to a number of executive functions including

- Management of the executive workload
- Information access and retrieval
- Word processing
- Sending, receiving, reviewing, and disposing of mail
- Performing computations, and voice and data communications

In managing the workload, the executive will be able to work with an electronic in-basket where paperwork can be viewed and worked with via a desktop video terminal. Items will be placed directly in the work queues of subordinates and automatically be followed up to ensure timely completion. In addition, the executive will have the capability to maintain a diary/action list (to-do list).

In the area of information access and retrieval, the executive will have immediate access to data needed to process paperwork. This includes access to such resources as

- Database systems
- Information search and retrieval systems
- Mail files
- Document files
- External information sources including public networks

The executive will need some access to word processing capabilities to support such activities as

- Checking and modifying materials typed by the secretary
- Retrieving text for composing semistandard letters, contracts, proposals, and other documents
- Simultaneous viewing and changing of text between the executive and the typing pool

The executive will work extensively with the electronic mail and message functions. (Electronic mail is covered in more depth in a later section of this chapter.) As a minimum, the automated office will provide capabilities for receiving and viewing mail/messages as soon as they arrive or when time is available, as well as for sending mail messages.

Almost every executive needs some support in the area of performing computations. The automated office will make available personal computation and calculation capabilities, access to computer modeling software tailored to individual decision-making situations, access to timesharing systems, and ability to use graphics.

The automated office will provide the executive with an opportunity to dramatically improve voice and data communications by supporting electronic mail, voice mail, conference calling, and teleconferencing.

Some people say that office automation is dying in that it does not have a discrete marketplace to sell its products. What really is happening is that office automation is being reborn as the basic framework for business systems. Just as typewriters and multipart carbon copy forms reformed the business office 50 years ago, workstations, microcomputers, and especially data communication networks are reforming the business practices of today. The basic functions supported by office automation systems are shown in Figure 1-9. Notice that these are the functions required by business firms and that the key factor among all of them is the ability to network. As was stated earlier in this section, the communication function is overwhelming and is taking over the office automation function. In essence, companies need an effective strategy to manage the continuing explosion of technology with regard to business office automation. Policy should focus on the right level of decentralization to gain control of assets without constraining the use of those assets.

Six basic benefits are available to a company that automates its office functions and properly networks these functions throughout the entire organization. *First*, the networks provide shared resources that offer a fast, secure, and convenient way for intradepartmental and interdepartmental data sharing. Network users also can share high performance peripherals, such as laser printers and high capacity hard disks. *Second*, the benefit of faster communications through electronic mail and voice mail ensures that users communicate their ideas and information more rapidly and more easily. *Third*, improved data accuracy has the potential of eliminating redundant data entry and lost data, as well as drastically reducing errors upon data entry, transmission, or processing. Having the ability to retrieve information from a single, accurate database might turn out to be the most important benefit of the automated office. *Fourth*, by transmitting data electronically the staff can reach greater levels of efficiency via voice and data transmission methodologies. *Fifth*, the networking together of all the office workstations reduces the possibility of equipment obsolescence because of shared capabilities. This creates the opportunity to use older equipment by providing access to the total resources of the network. *Sixth*, office automation and

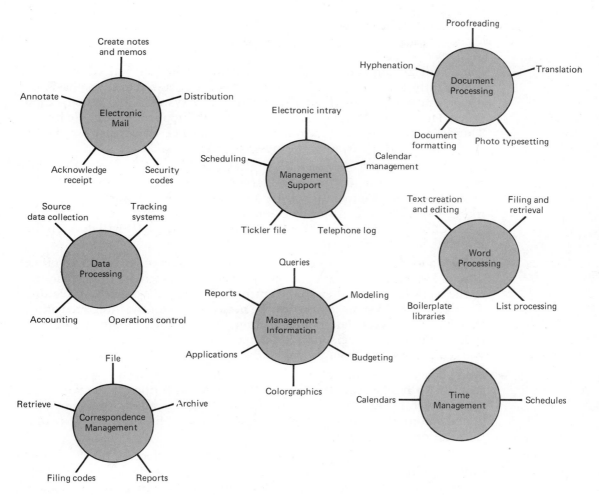

FIGURE 1-9 Functions supported by office automation.

networks make corporate or governmental computing more manageable. In other words, there can be an improved level of management. Traditionally, corporate computing has been planned and managed by controlling the firm's information system resources. Microcomputers and workstations have distributed the computing system, making it more difficult to manage information in a cohesive fashion. Office automation, through networking, now permits the centralized management of a system that actually is distributed.

There probably are more than 40 vendors of hardware and software for the automated office. The following describes the general offerings of some of the larger vendors in this area.

- Data General Corporation built its Comprehensive Electronic Office (CEO) as a tightly linked set of applications. Data General places emphasis on integrating these applications within the file systems and among each other. These systems are highly integrated at the application and file levels. They offer integrated voice/data terminals and voice store and forward software, along with their other business application products.

- Digital Equipment Corporation's approach has been to build an open-ended system architecture. Its All-In-One system is an office station that can be customized to encourage third parties to write the software that can be integrated under the All-In-One umbrella. Consequently, Digital Equipment Corporation has focused on flexibility, hoping others will develop the integrated set of applications.

- Xerox's and Hewlett-Packard's approach to the design of an integrated office system has been to start at the desktop level and later extend these concepts to the work group and then to the entire departmental system architecture. They base their systems on getting the network hub for the integrated office first and then work on the applications. Both are developing integrated office solutions for personal computers and workstations that are networked together in an automated office.

- Wang Laboratory's products were based on different operating systems, thus requiring a bridge/gateway to close the communication gap between the different workstations in the company. Wang now is busy developing integrated communication systems after its purchase of Intecom, a major communication switch manufacturer.

- IBM had a number of its existing office products to select for the integrated office system. It has been marketing its System/36 as the automated office departmental system. Because of IBM's many products, it also is heavily marketing its communication networks, especially those that interface with the IBM host mainframe computers.

VOICE MAIL

The public telephone system that handles voice messages has been around for over 100 years. Because this system is primarily a circuit switching system, it has one tremendous disadvantage. When the remote telephone is already in use or no one is present, the telephone call (the message) cannot be completed. Both voice mail and electronic mail overcome this disadvantage. During the next few years two of the major applications for data communications will be voice mail and electronic mail.

With regard to *voice mail*, great technical advances have occurred to make the telephone more accessible, easier to use, more attractive, and a true message switching system. Voice mail is a flexible means of sending a spoken message to someone even when the person is not at the remote terminal (the telephone). The sender speaks into the telephone, and the message is stored for later forwarding to its recipient. In effect, this turns the telephone system both into a message switching system (instead of circuit switching) and into a store and forward system.

Actually, voice mail is the transmission of a voice message to a recipient voice mailbox. Using a touchtone telephone with its standard 12-key dialing pad, the caller can record a message, listen to the message before transmitting it, and even change it if necessary. The message then can be sent to one or more recipients or even to a predefined group such as a department within a corporation or government agency. When it is convenient, the recipients check their voice mailboxes, scan to see who sent the incoming messages, and choose to listen to some now while saving others for later. Recipients can listen to the message, stop playing it if they are interrupted, skip ahead or back, or replay the message at will. After hearing the message, the recipient can generate a voice reply immediately and send it back to the person who sent the original message. Other options might be to forward the message to a third party and, of course, to discard the original message.

Voice mail has five major advantages over the traditional telephone. With voice mail it is no longer necessary to

- Place several calls to a person to find him or her near the telephone
- Move meeting schedules and match time zone differences around the world
- Place a number of calls to get a similar message to many different people
- Know where a person is located geographically to complete a call to the person (all you need to know is the voice mailbox number which he or she checks each day)
- Type your messages

With voice mail, the sender can place a call without interrupting the recipient, without having to know if the recipient is in the office or traveling around the world, and without regard to the time of day or night.

If a two-way information exchange is necessary because complex points must be discussed, then voice mail has a nice feature that can be used to set up a precise time, date, and telephone number to ensure that the connection can be made the first time. It is here that the newer telephones will have the most advantage because of their ability to transmit both voice and picture simultaneously. Video display telephones allow users to see the person with whom they are speaking (possibly for security reasons) and actually to show documents over the system.

Voice mail will be offered by the major telephone companies and special common carriers such as US Sprint and MCI.

Voice mail will be accepted more readily by general public telephone users than will electronic mail. This is because electronic mail requires, first, the ability to type and, second, access to a keyboard in order to type the text of a message. Psychologically, human beings were built to accept and transmit voice messages, whereas the ability to enter text messages must be learned as a special skill. More people can use their voice than can write a complete sentence correctly.

Basically, voice mail will be the overall glue that holds together and controls the automated office of the future because different operators will need to converse with or leave messages for various other operators. Whereas typists will have no problem with electronic mail, nontypist middle managers and executives will rely more on voice mail because the spoken word is faster than the typed word.

Incidentally, voice mail is achieved by digitizing a voice signal and breaking it into a stream of digital bits. Some systems are now capable of converting a voice signal into a digital bit stream of 4800 bits per second or even 2400 bits per second. This 2400 bits per second digital stream enables you to have four simultaneous voice conversations on a 9600 bits per second data communication circuit.

Some important application features of voice mail that should be considered are

- Closed systems, which subscribers can use and which entail controlled passwords.
- Open systems, used by nonsubscribers or anyone else.
- Type of hardware to be used such as 12-key touchtone telephones or video telephones.
- Training aids such as booklets, audio prompting, or, if you have a video display telephone, a help key that provides pictorial representations of how to use the system.
- Key templates that might be placed over the 12-key pad to enhance explanations of how to use the system.
- System designed to direct messages to any telephone, to only subscriber telephones, or directly to a user's address message queue without regard to the telephone.

- Other message addressing schemes such as to individuals, to a unique telephone, to groups of individuals, to preorganized numeric codes, to system directories, or custom methods used by an individual organization.
- Type of system data that might accompany the message, such as date of call and time, address of sender and receiver, time of message, or other system data.
- Provision of priorities and/or different message categories as might be utilized within a large corporation.
- Ability of users to give answers to requests or to request information. Someone may call in wanting information about a publisher's books or software packages. By asking callers which product they want, and telling them to press 1 for books or 2 for software information, you can disseminate information on your products without operator assistance.
- Depending on size and cost considerations, systems starting with one port and going up to 256 ports. Ports are used for incoming telephone circuits. These systems range from $1,000 (see the description on WATSON in Chapter 6) and may go as high as $500,000.
- Capability of interfacing with PBX (switchboards) systems using either analog or digital communication technology.
- Generation of valuable reports, including accounting functions and various other business-related reports, which can be used by a system administrator for maintenance and billing.
- Message desk option, with which an operator attendant can transfer a voice mail message from a nonuser, or an outside user can establish a voice mail message in the system.
- Outdialing, or the ability to dial an outside call (local or long distance) automatically in order to deliver a user's voice mail message to another user or nonuser of the system. This function is helpful for contacting someone during nonbusiness hours or weekends when people are less likely to call into their voice mailboxes.
- Future delivery, which lets a user create a message and send it to another user at a predetermined future time, either by dropping it into a voice mailbox or by outdialing it.
- Edit functions, including stop, start, skip forward, skip backward, delete, reply, add, and subtract. The edit function is important for users who want to replay and edit their messages.
- Broadcast abilities, allowing one message to be directed to multiple users.
- Time stamp and date stamp, allowing users to embed a time and date,

which can be called up by a user to determine when a message was directed to him or her.

- Receipt acknowledgment, whereby the system notifies a sender that a user has responded to a sent message and offers proof of delivery. It may show how long messages are in the system.

- Message forwarding, whereby a user who has received a message can send it to another system user and also can add dialogue to the first message. This is similar to memo passing.

- Help keys or audio prompting.

- Speed control, so the user can slow down or speed up the received message, and also adjustable volume.

- Various security levels which restrict access to the system by requiring an access code and a follow-up security code.

- Messages ranging from 30 seconds to almost indefinite lengths, depending on disk storage capabilities.

Finally, the number one advantage for voice mail is that everyone who currently has a touchtone telephone already has the terminal required to utilize this system and basically understands how this terminal (the telephone instrument) works.

ELECTRONIC MAIL (E-MAIL)

Electronic mail is similar to voice mail except that the user must key in (or type) the message so the recipient receives a textual style printed message. Electronic mail will be very important in the office of the future, although voice mail will be used more. It also will become a major factor in the private home networks that were mentioned earlier in this chapter.

One of the advantages of electronic mail over voice mail is that people feel more comfortable if they have delivered a lengthy or complex message in writing rather than leaving a verbal message. One of the primary problems with electronic mail, however, is that there is no standard protocol (check protocol in the Glossary) to install as there is with the current switched public telephone network (dial-up).

Electronic mail, like voice mail, can be used to send messages to people when a two-way dialogue is not an immediate necessity. Thus, you can schedule future meetings, send messages to individuals or groups of people, overcome the problem of different time zones, and not worry if your recipient is unavailable when an electronic mail message is sent to an electronic mailbox.

Any user who has a portable terminal or microcomputer can connect to an electronic mail network. In order to review the features of an electronic mail system, reread some of the features that were listed for voice mail; they are the same.

As more competition arises in the public telephone networks, you will be able to transmit both electronic mail and voice mail over your home telephone. All that is required for the electronic mail portion is a small interactive terminal and a modem.

Public electronic mail subscriber services are available from numerous sources that sell mailboxes connected to, and controlled by, that vendor's computer system. Some of the common ones are

Western Union Data Services	EasyLink
MCI Telecommunications	MCIMail
General Electric Information Services	Quik-Comm
US Sprint Communications	Telemail
International Telephone & Telegraph	Dialcom
AT&T Information Systems	AT&T Mail
CompuServe	EasyPlex/InfoPlex
Computer Sciences	Infonet
McDonnell Douglas Electronic Data Interchanges Systems	ONTyme
RCA American Communications	RCA Mail
The Source Telecomputing	The Source

These mailboxes enable users to communicate with other subscribers at both local and remote sites. Although these public electronic mail services are less costly initially than an in-house corporate system, their capabilities are not always well suited to companies' needs. In addition, the levels of security required by a corporate user may not be met.

Most electronic mail is interoffice and intracompany; therefore, large corporations may do better by developing their own in-house electronic mail system. This system might be superimposed on the corporate local area networks or on any other networks that are in existence. You even can put both electronic mail and voice mail on the same intracompany network. Several organizations also market PC-to-PC electronic mail packages. Market observers note that public E-mail firms will have to learn to connect with in-house E-mail or office automation packages such as IBM's PROFS, Wang's OAS, or DEC's All-In-One. The biggest challenge now is interconnecting public and private E-mail systems.

Implementation of an electronic mail system can be a shot of adrenaline to the corporate culture. Users find they can enter messages and get responses within hours, and they can check their mailboxes frequently, regardless of where they are within the corporate facilities, or even around the world. This system usually changes the communication patterns within the organization. For example, upper-level executives tend to talk directly to employees several levels below them when using their electronic mail system. The corporate power structure, with its folkways and mores, usually precludes face-to-face or voice messages over these several levels of management, but E-mail with appropriate carbon copies creates a more direct path down to the first level of employees.

Another impact of electronic mail on the corporate culture is that it allows employees to see the flow of information within the corporation, and they become aware that the company is using a system that is perceived as being attuned to the times.

Probably the most advanced and useful communication systems have a combination of electronic mail and voice mail. Both are needed because many messages are better suited for a voice, while other messages are more appropriate for the written documentation offered by an electronic mail system.

In the business office it might be too expensive and too time consuming to give every employee an electronic mail terminal, but most employees today have a telephone and can use voice mail. Electronic mail also requires more training. Furthermore, as observed earlier, human resistance to major change, such as going from voice conversations to typewritten conversations, makes implementation more difficult.

In summary, both electronic mail and voice mail will be major parts of the future automated office. Voice mail will become a major feature of the world's public telephone systems; electronic mail will become part of these telephone systems only when the user has an interactive terminal. In less developed countries it may be difficult enough to teach a person how to use a voice telephone, much less an interactive keyboard-driven terminal.

COMMUNICATIONS IN BANKING/FINANCE

One of the major applications for data communications is in the banking industry. The increased pace and mobility of our society have placed a tremendous demand on our financial institutions. Customers are now demanding around-the-clock service, multiple banking locations, and virtually instantaneous response to their transactions. To meet these demands, financial institutions have implemented massive data communication networks, automated teller machines (ATM), and automated funds management systems for business organizations and government agencies.

The three main applications in banking are demand deposits, loans, and savings. One customer may have several savings, checking, and loan accounts scattered across several bank branches. All activities of this customer are cross-referenced and transmitted through the data communication network. This cross referencing provides standardization of procedures and customer service across all branches, as well as the ability to extend banking services for easier customer access at any time of the day or night. Such networks also increase marketing opportunities for the bank to sell additional services to customers, such as stock sales, and to provide more timely and comprehensive customer information for management analysis.

Once a bank or group of banks develops a data communication network and interconnects it with one or more host computer processing locations, then any or all of these services can be offered. Some of the financial services that are offered over such networks include

- Automated teller machines where a user can deposit or withdraw money, make transfers between different bank accounts such as savings to checking, pay various bills, and the like.
- At-home banking and telephone bill paying. These are being tested for use with the public telephone system and cable television.
- Point-of-sale terminals (POS) tied directly to a customer's bank account from the store in which the POS terminal is located.
- Automated clearing houses (ACH) that facilitate the paperless dispensing and collection of thousands of financial transactions on a real-time basis.
- National and international funds transfer networks, check verification, check guarantees, credit authorizations, and the like.

Many of the networks developed by banks are private networks that combine point-to-point leased circuits, dial-up circuits, multidrop, multiplex, and packet switching and, as we move into the private home, they will include any of the combinations of private home networks. It should be noted that banking networks which used to be located within a city are now between various states and countries.

Many banks now participate in shared networks: a bank network system in one area of the country interconnects to another bank network in another area.

It should be noted that banks also have to talk between themselves. Some networks available for this type of communication are:

- SWIFT   (Society for Worldwide Interbank Financial Telecommunications)
- CHIPS   (Clearing House Interbank Payments System)

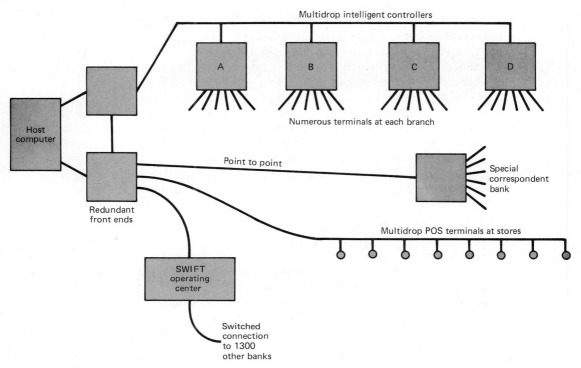

The terminals at branches A–D may be video terminals for tellers and ATMs for customers.

FIGURE **1-10** Bank networks.

- MINTS (Mutual Institution National Transfer System)
- FED WIRE (Federal Reserve Bank Telecommunications Network)
- BANK WIRE (Interbank Network)

These networks are used for transferring information between banks. Banks also create private networks and interconnect them for use with the general public. You might view the preceding list of networks as involving *wholesale* electronic funds transfer between banks. By contrast, private bank networks involve *retail* electronic funds transfer for use by the general public and/or customers of each individual bank. A bank's network for its checking/savings branch offices, ATMs, POS terminals, home banking, and the like would be their retail network. Figure 1-10 shows a typical bank's retail network and its interconnection into the SWIFT wholesale network. If you would like to trace the flow of a

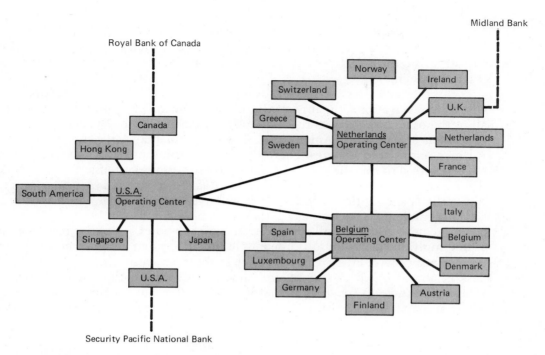

FIGURE **1-11** SWIFT banking network.

message from the SWIFT "USA" Operating Center around the world, look above at Figure 1-11.

One of the most complex networks that has been developed on a worldwide basis is the *S*ociety for *W*orldwide *I*nterbank *F*inancial *T*elecommunications (SWIFT) network. This network interconnects over 1,200 banks in 50 different countries. When the network started in 1977, it had an average volume of 60,000 messages per day. By now the average volume is over 400,000 messages per day.

Figure 1-11 shows the basic diagram of the SWIFT communications network. There are three operating switching centers. From the switching centers, communication circuits go to a regional concentrator located in each country. The banks in the individual country interconnect with their regional concentrator. A regional concentrator in any country has a connection to its normal switching center and a backup connection to an alternate switching center.

If the Royal Bank of Canada had a funds transfer for either Midland Bank (England) or Security Pacific National Bank (United States), it would take the most direct path in Figure 1-11.

COMMUNICATIONS FOR THE AIRLINES

Almost everyone is familiar with the use of data communications in the airline industry. American Airlines' SABRE system was the first automated airline reservation system.

Today the major air carriers have developed individual communication networks for reservations, flight planning, inventory, control, flight scheduling, and all of the other applications required to run a major airline. The major systems have been developed by American Airlines, Continental Airlines, Trans World Airlines, and United Airlines. There also is an interlinking network system (used by all airlines) to transfer messages among these various systems. The interlinking network is maintained by Aeronautical Radio Incorporated (ARINC). American Airlines and United Airlines have the largest base of terminals installed in travel agents' offices as well as for their own use.

Airline networks are built using various point-to-point lease circuits and combining multiplexing and multidropping in order to get an efficient network layout. In today's world the application of data communications to the airline business is not so much in the network itself, but more in the installed base of terminals located in travel agents' offices. This is because there is a correlation between the number of airline booking terminals in travel agents' offices and the number of flights booked on that airline.

COMMUNICATIONS IN THE RENTAL CAR INDUSTRY

Another major application for large national and international communication networks is in the rental of automobiles. Avis Rent-A-Car was the first major company to develop an online real-time rental car network. This network interconnects the major locations where automobiles might be rented, picked up, and returned.

The networks used by rental car companies involve multiplexed circuits. These circuits are leased from the major telephone companies and/or special common carriers which sell communication circuits.

A rental car network is quite similar to an airline network in that it is used to keep track of dates when cars have to be returned to their lessor, make rental agreements with people who rent the cars, calculate the cost and mileage utilized, and perform other general accounting functions.

TELECONFERENCING

One of the growing uses for communications is in teleconferencing between the employees of a single company or those of several companies. *Teleconferencing* allows users to sit in a conference room and talk to another group of users in another conference room thousands of miles away. Participants are able to see and hear the other people when they speak, and the system can focus in on a specific document and allow everyone to read it at the same time. The document can even be copied at either end of the teleconferencing link. Teleconferencing also can be set up as a multipoint network in which several conference rooms in various cities are interconnected.

The primary reason for using teleconferencing is to save time and get a widely dispersed group of people together when they need to discuss a common problem. It also can save considerable amounts of money by reducing travel costs. Teleconferencing has the advantage of being able to transmit documents, pictures, charts, visuals, graphs, or anything else that can be sent over a standard television set.

ELECTRONIC SHOPPING

A recent communication growth area is electronic shopping. A company selling consumer products holds an auction on television, and the viewers at home can bid on the various products. The customers see the goods for sale on their television set and bid on or purchase them through a two-way data transmission. Although it remains to be seen whether this method of selling products over television will be satisfactory to customers in the long run, it has become quite popular, and the stock of companies providing this service has risen rapidly in value.

TELECOM CAREERS

In an isolated village far above the Arctic Circle, a young Eskimo intently studies a computer screen filled with data, generated as part of a University of Alaska business course delivered entirely by data communications. When she finishes the lesson, the student takes a test that is scored within seconds by a computer that is 1,000 miles away in Anchorage. When she needs to do library research, she turns to an electronic card catalog to find the books she needs.

In the remote Outback of Australia, a student has difficulty tuning out the static during a two-way radio transmission that is used for the Australian "Schools of the Air." The two Northern Territory schools, Alice Springs and Katherine, provide correspondence course radio lessons with weekly problem solving and communications with the teacher. Once a year the teacher actually visits the students who live in these remote Outback cattle or sheep stations.

You have seen how communications have affected your own educational career, but these are two other examples of the widespread use of communications in the world of education. Beyond the use of communications as an educational tool, numerous career opportunities are available in both corporations and government agencies. At a time when doctors, lawyers, dentists, accountants, and MBAs are in good supply, jobs for telecommunication managers abound. Today the need for qualified managers in the telecommunication industry is urgent and growing. It is estimated that the industry will double in size by 1990, thereby creating some 100,000 new management jobs. The demand is acute for two reasons. First, technological innovations have created rapidly changing new products, and, second, deregulation has paved the way for new suppliers of communication hardware and software.

Some of the universities that offer telecommunication degrees are Ohio University, the University of Colorado, Golden Gate University, Roosevelt University, and George Washington University. Some universities that sponsor telecommunication research centers and programs are

Harvard University	Program on Information Resources Policy: explores change and develops policy options in information resources
Massachusetts Institute of Technology	MIT Research Program on Communications Policy: focuses on communication issues on an interdisciplinary basis
Stanford University	Information Policy, Engineering, Economic Systems Program: analyzes national and international policies concerning information sources and products
University of Pennsylvania	Communications Program: studies telecommunication policy and offers graduate degrees
University of Southern California	Communications Management/Communications Theory and Research: maintains ongoing communication research programs and offers graduate degrees

For a comprehensive list of programs, contact the Manager of Education for the International Communications Association at (214) 233-3889.

If you choose to direct your own career toward telecommunications, remember that this function is a major support service for the way a company conducts its business operations. In other words, communications are changing the way both private companies and government agencies conduct their basic business functions. For this reason, it is desirable that you have a basic background in general business subjects, such as finance, marketing, economics, and accounting, to go along with your data processing and especially your communications knowledge.

A career in telecommunications will be a very salable item well into the 1990s. Today people who have three or four years of experience make annual salaries of $45,000 to $50,000. Telecommunications is a stronger job market than general data processing.

The following is a typical list of course offerings available at universities and colleges that offer a degree program in data communications/telecommunications

- Telecommunications Technology I: An introductory course covering the various telecommunication technologies; basic electrical theory and the evolution of electrical components; operation of the telephone; sound propagation, transmission systems (microwave, cable, and satellites), switching, and networks.

- Telecommunications Technology II (continuation of Telecommunications Technology I):   the development of data communications; in-depth study of modulation, network system design, multiplexing, network management, and system performance. The role of the telecommunication manager in a business environment also is discussed.

- Introduction to Office Automation: An intermediate course with specific concentration in the areas of telecommunication policy and corporate telecommunication planning; coverage of the present and future state of the art telecommunication systems, and the systems trend toward integration of the "office of the future" technologies with telecommunications. Specific topics might include satellite technology and video, laser communication links, teleconferencing, and cellular radio; security and control systems; introduction to data network protocols and architectures.

- Telephone Interconnect Systems: An advanced course in the design, selection, installation, and management of the voice telephone systems that are available for purchase or lease in today's marketplace; buy versus lease decision checklists; analysis of vendors, equipment types, service requirements, and other market variables of this rapidly changing technology; investigation of competitive system features and case studies of a variety of digital and analog PBX telephone systems.

- Networks—Common Carrier Systems I: An intermediate course covering the latest service offerings of the common carriers in the United States, especially AT&T and the Bell Operating Companies. The history of networks, their social impact, organization, technology, engineering, call rating, maintenance, and network offerings are covered.

- Networks—Common Carrier Systems II (continuation of Networks I):   A thorough analysis of technology, engineering, call routing, and network offerings; investigation of the latest offerings of Western Union, ITT, MCI, AMSAT, SPCC, and other common carrier organizations.

- Video Systems: An advanced course that investigates and analyzes the uses of video as a communication medium in business. Broadcast technologies are discussed, with particular emphasis placed on teleconferencing and in-house video networks; analysis includes energy tradeoffs, writing, production, direction, distribution hardware types, and video network management.

- Integrating Voice, Data, and Video Communication Systems: An advanced course that covers the overall planning, design, and management of a fully integrated corporate communication system, from the selection of various system components (both equipment and services) to the actual layout of facilities; discusses tradeoffs with regard to voice and data facilities and the telecommunication network of the future; case studies illustrate a series of communication networks.

- Microcomputer Communications: The impact of microcomputers on telecommunications including an in-depth study of local area networks (LANs), micro-to-mainframe links, and connection of microcomputers to the various public networks (especially dial-up). Topics include hardware, microcomputer-based communication software, modems, security, and disk/file servers.

- Network management: An advanced course depicting the basic management skills required to be a successful communication network manager. Topics covered include departmental management functions: how to manage the department; network management and control; required reports and documentation; error testing and test equipment; network control and troubleshooting. This course integrates the current communication technology with the best management approaches.

SELECTED REFERENCES

1. Clemmensen, Jane M. *Telecommunications Education in the U.S.: An Informational Guide and Assessment*. Dallas, Tex.: International Communications Association, 1983.

2. *Data Communications*. Published monthly by McGraw-Hill Publications Co., 1221 Avenue of the Americas, New York, N.Y. 10020, 1972– .

3. Halsall, Fred. *Introduction to Data Communications and Computer Networks*. Reading, Mass.: Addison-Wesley Publishing Co., 1986.

4. Held, Gilbert, and Ray Sarch. *Data Communications: A Comprehensive Approach*. New York: McGraw-Hill Book Co., 1983.

5. Hirschheim, R. A. *Office Automation: A Social and Organizational Perspective.* New York: John Wiley & Sons, 1986.

6. Purser, Michael. *Data Communications for Programmers.* Reading, Mass.: Addison-Wesley Publishing Co., 1986.

7. Rosenberg, Jerry M. *Dictionary of Computers, Data Processing, and Telecommunications.* New York: John Wiley & Sons, 1983.

8. Sippl, Charles J. *Dictionary of Data Communications,* 2nd ed. New York: Halstead Press, 1985.

9. *Telecommunications Dictionary and Fact Book.* Ramsey, N.J.: Center for Communications Management, 1984.

10. *Teleconnect: A Monthly Telecommunications Magazine.* Published monthly by Telecom Library, 12 West 21st Street, New York, N.Y. 10010, 1983– .

The following six books are *very* good references, even though some may be a little dated. They contain clear and concise explanations of basic concepts, which remain relatively stable despite continued changes in the telecommunication industry. You will find them useful when searching for alternative descriptions of basic concepts or when researching technical aspects of telecommunications.

11. Davenport, William P. *Modern Data Communication: Concepts, Language, and Media.* New York: Hayden Book Co., 1971.

12. Doll, Dixon R. *Data Communications: Facilities, Networks, and Systems Design.* New York: John Wiley & Sons, 1978.

13. FitzGerald, Jerry, and Tom S. Eason. *Fundamentals of Data Communications.* New York: John Wiley & Sons, 1978.

14. Martin, James. *Telecommunications and the Computer,* 2nd ed. Englewood Cliffs, N.J.: Prentice-Hall, 1976.

15. Sherman, Kenneth. *Data Communications: A Users Guide.* Reston, Va.: Reston Publishing Co., 1981.

16. Techo, Robert. *Data Communications: An Introduction to Concepts and Design.* New York: Plenum Press, 1980.

COMPUTERIZED LITERATURE RESOURCES

Computers have had a major impact on libraries and methods of locating printed information. One approach to locating printed information is to use subject-oriented indexes, but many people feel that using these indexes is too time consuming and tedious. Because of advances in computerized database technology, many of the indexes that can be found in libraries or information centers can be accessed via terminals. In addition, some indexes are available *only* in a computerized format.

The following indexes are available in a number of systems, among them DIALOG (Lockheed Retrieval Services), ORBIT Information Technologies, and BRS (Bibliographic Retrieval Services). These systems may have just a few indexes, or they may have hundreds of indexes with millions of references. Examples of online databases that may be of interest to data communication educators and professionals are outlined below. Contact your librarian for assistance in using these valuable resources.

Abstracted Business Information.   Available only in computerized form. References articles related to the business aspects of data communications such as managing in a data communication environment. Provides bibliographic information and abstracts. 1971– .

Books in Print.   Lists books, symposia, and other monographs sold by U.S. publishers. Excellent for learning what books are available in a particular field. The paper version is indexed by author, title, and subject. The computerized version is useful because terms can be used for which adequate subject indexing is unavailable, as when a subject is either very narrow or an emerging topic. 1948– .

COMPENDEX (COMPuterized ENgineering InDEX).   Emphasizes the engineering aspects of data communications. It is useful for determining how others have applied data communications in an industrial or factory-type situation such as an industrial control application. Includes international journals, technical symposia, reports, government documents, and so on. Provides bibliographic citations and abstracts. 1970– .

INSPEC.   The printed counterparts of this computerized index are *Physics Abstracts, Electrical and Electronics Abstracts,* and *Computer and Control Abstracts.* Because computers are an integral part of data communications, the *Computer and Control Abstracts* portion of this database is an excellent resource. Includes the technical aspects of applications, techniques, hardware, software, technological developments, architectures, economics, and the practical aspects of implementing such systems. References are international in scope with abstracts. 1969– .

PROMT and *Funk and Scott Index.*   These complementary indexes share the same database. *PROMT* abstracts marketing-oriented articles that discuss products, processes, and services for sale. Indexing is by product, country or state, and "event" (for example, sales, new product/process, demand, profits, cost per unit, industry structure/members, regulatory actions). Citations are brief, but abstracts are informative. The *Funk and Scott Index* has no abstracts; it is an index to the same references cited in *PROMT,* plus others that are too short to abstract (for example, a one-line announcement that one firm has contracted with another for a specific product and a specified dollar amount). Items are international in scope and include journals, trade literature, government documents, and so forth. They focus on data communication equipment, networks, office automation, and teletext. 1972– .

Ulrich's International Periodicals Directory.   A guide to journals published in all countries. Provides the name of the journal, publisher information, publication frequency, price, and whether the journal includes such items as advertisements or illustrations. Arranged and indexed by subject. An excellent way to locate journals in a specific field such as computers. The computerized version enhances retrieval because one can obtain the titles of all journals with a specific word in the title. 1932– .

QUESTIONS/PROBLEMS

1. In an information-based society, the _____ resource is knowledge which creates information.

2. Write your definition of the term *data communications* and compare it with the definition given in the Glossary.

3. Define four or five uses of data communications.

4. If the typical characteristics of a transaction are "relatively low character volume per input transaction, response required within seconds, output message lengths usually short but might vary widely with some types of applications," then identify some examples of business applications that might use an information retrieval type of network.

5. What is the difference between data communications and telecommunications/teleprocessing?

6. What is a circuit?

7. Compare the definition of local loop in Chapter 1 with that in the Glossary. Next, look at Figure 1-4 and be sure that you understand the concepts of local loop and central office. Write down these definitions because you will use them repeatedly. Notice how the Glossary can help.

8. Define the progression of systems from the 1950s to the present.

9. Is it possible for a large business organization to have a combination of all seven categories of networks that were described in this chapter?

10. What types of companies can be classified as common carriers?

11. How might transborder data flow restrictions affect a university?

12. Present the issue of data monopoly to the instructor of an economics class and relate his or her comments to your data communication class.

13. What is the difference between voice mail and electronic mail (E-mail)?

14. Identify the closest college or university to you that offers a degree in telecommunications or data communications.

15. Why would transborder data flow restrictions hamper business?

16. Describe the most recent data communication development that you have read about in a newspaper or other periodical.

17. In the real world there are many examples of the source/medium/sink relationship shown in Figure 1-3. Take at least one entry from the Examples of Applications column of Figure 1-2 for each of the first three usage modes and identify some of the source/medium/sink relationships that apply.

18. The section Uses of Data Communications in this chapter lists the characteristics of organizations that can benefit from data communication systems. Examine these characteristics and identify who (or what function) in the organization will benefit and describe how.

19. The section Definition of Data Communications cites eight objectives common to most data communication networks. With reference to these objectives, of what value is it to capture business data at its source? How can a data communication network centralize control over business data?

20. After completing a course of study based on this textbook, what should the student be able to do?

2

FUNDAMENTAL COMMUNICATION CONCEPTS

This chapter defines and discusses the basic concepts of data communications/teleprocessing. The technique used is to examine *technically* the flow of data/information from a remote terminal through the connector cable, the modem, the local loop, the telephone company central office, the circuits, and into the front end/host computer. As the message moves, each technical aspect is defined and described. You do not have to possess an engineering-level understanding of the topics to be an effective analyst of data communication applications. It is important, however, that you understand these concepts and make intelligent use of them. The content of this chapter is basic; knowledge of the content is prerequisite to your competence in this field.

INTRODUCTION

This chapter begins the study of data communications or telecommunications technology by concentrating on what happens when data is moved from point to point. While it requires covering some topics that may verge on electrical engineering, the discussions are phrased in layperson's terms. To begin our discussion of the basic technical concepts of data communications, examine Figure 2-1 which depicts a typical point-to-point network between a microcomputer or remote terminal and a central host mainframe computer. For simplicity, this configuration omits the more complex design configurations such as multidrop,

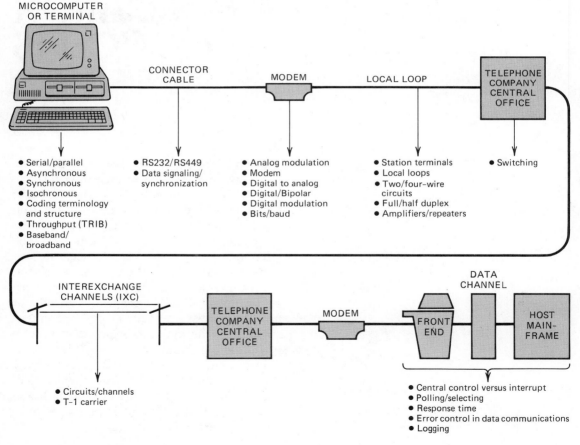

FIGURE **2-1** The basic technical concepts.

multiplex, and local area networks because such complex design methodologies would obscure the basic explanation of how data is transmitted. Instead we will present a simple, straightforward flow approach of the data transmission concepts from the microcomputer to the host mainframe computer.

As you can see in Figure 2-1, each component of the network is identified across the top of the figure, starting with the microcomputer or terminal and going through the connector cable, modem, local loop, telephone company central office, transmission lines or circuits (interexchange channels), and ultimately to the host mainframe computer. The related technical concepts involved in moving data from a remote microcomputer to a host mainframe are identified below each component and are described thoroughly in this chapter.

As the discussion moves through the components shown in Figure 2-1, you can follow the movement of your data signal (message) and learn the basic technical concepts that allow the message to move across the communication network.

MICROCOMPUTER/TERMINAL

The first component in our network is the basic input/output device, which is shown as a microcomputer or terminal in Figure 2-1. Obviously, besides a microcomputer or terminal, this device could be a teleprinter, minicomputer, facsimile device, or any other type of terminal that allows for input and output of data. At this point we will not define the microcomputer itself because we are looking at the flow of a message over a network. Microcomputers, their internal architecture, operations, and communications are covered thoroughly in Chapter 6, Microcomputers and Communications.

MODES OF TRANSMISSION

Systems that transmit data must have consistent methods of transmission over the communication circuits or channels. All systems in our discussion transmit binary data, or data forms that are intrinsically binary. Binary data can be sent over communication circuits in either serial or parallel modes by use of asynchronous, synchronous, or isochronous transmission methodologies.

Parallel Mode   Parallel mode describes the way the internal transfer of binary data is performed within a computer. In other words, if the internal structure of the computer uses an 8-bit element, then all eight bits of the element are transferred between the main memory and any operational register within the same computer cycle. The same is true of the more powerful computers that use a 32-bit element or word length; all 32 bits are transferred between the main memory

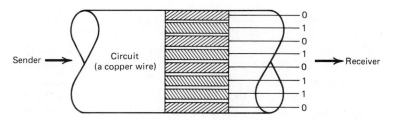

FIGURE **2-2**   Parallel transmission of an 8-bit code.

and any operational register in the same computer cycle. Although parallel transmission is seldom used outside computers, it is illustrated in Figure 2-2. This figure shows how all eight bits of the USASCII[1] code travel down a channel simultaneously, followed a short time later by eight more bits. Normally, this parallel type of transfer is not used in data communications.

Inside computers, data typically travels around in parallel. The bits that form a character move at extremely high speeds between computer components along parallel solder paths, and arrive "together" at the different chips. Some printers use parallel communication, but this is possible only when the printer is very close to the computer, as parallel cables require many wires and do not work reliably in lengths over about 25 feet.

Serial Mode   Serial mode is used predominantly for the transfer of information in data communications. Serial transmission implies that a stream of data is sent over a communication circuit in a bit-by-bit fashion. For example, the bits of information in an 8-bit USASCII code may be sent down a transmission circuit (channel) as shown in Figure 2-3.[2] The USASCII code is used as an example in the following figures; it will be discussed later in this chapter, under Coding Terminology and Structure.

Serial transmission is distinguished from parallel in that the transmitting device sends a single bit, followed by a time interval, then a second bit, another time interval, and so on, until all the bits are transmitted. It takes n time cycles[3] to transmit n bits (eight in the case of USASCII). In parallel transmission, n bits of a character are sent, followed by a time interval, then n more bits are sent, and so on. In other words, with parallel transmission, the n bits of a character are sent in

---

[1] United States of America Standard Code for Information Interchange.

[2] Figure 2-2 and several subsequent figures used to depict forms of transmission are conceptual pictures only; they are not representative of the electrical engineering facts of data transmission. Use them only to help in your understanding of the concepts.

[3] This concept will be discussed in more detail later in this chapter, under Bits/Baud.

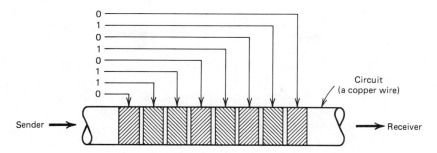

FIGURE **2-3** Serial transmission of an 8-bit code.

one time cycle, whereas with serial transmission, the same n bits of a character require n time cycles.

Thus, most data communication functions are performed by serial transmission. Three types of transmission are commonly used: asynchronous, synchronous, and isochronous. Note that all three of these transmission types are serial in nature.

Asynchronous Transmission Asynchronous transmission often is referred to as start-stop transmission because the transmitting device can transmit a character at any time that it is convenient, and the receiving device will accept that character.

With asynchronous transmission, each character is transmitted independently of all other characters. In order to separate the characters and synchronize transmission, a start bit and a stop bit are put on each end of the individual 8-bit character, so the total is 10 bits per character in transmission as shown in Figure 2-4. There is no fixed distance between characters because, if the terminal is one

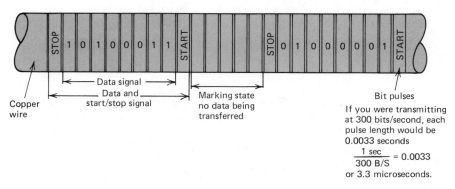

FIGURE **2-4** Asynchronous transmission.

that transmits the character as soon as it is typed, then the distance between characters varies with the speed of the typist. If the asynchronous terminal holds the entire message in buffers and transmits the message when the operator hits the SEND key, then there will be a fixed distance between characters because the terminal unloads its buffer on a character-by-character basis in a fixed timing sequence. Asynchronous transmission has a start bit and a stop bit to separate characters from each other and to allow for orderly reception of the message. Some terminals may have multiple stop bits.

In summary, the synchronization (timing) takes place for an individual character because the start bit is a signal that tells the receiving terminal to start sampling the incoming bits of a character at a fixed rate so the eight data bits can be interpreted into their proper character structure. A stop bit informs the receiving terminal that the character has been received and resets the terminal for recognition of the next start bit. Synchronization of the character (timing between bits) is reestablished upon reception of each character.

Synchronous Transmission   Synchronous transmission (see Figure 2-5) is used for the high speed transmission of a block of characters, sometimes called a *frame* or *packet*. In this method of transmission, both the sending and receiving devices are operated simultaneously, and they are resynchronized for each block of data. Start and stop bits for each character are *not* required. In other words, if you had 100 characters using an 8-bit ASCII code structure, the message part of the block of data would be 800 bits long.

Synchronization is established and maintained either when the line is idle (no data signals being transmitted) or just prior to the transmission of a data signal. This synchronization is established by passing a predetermined group of "sync" characters between the sending and the receiving devices. Figure 2-5 shows how the data signals are continuous and how one long stream of data bits is transmitted from the sending to the receiving device. In other words, the sending device

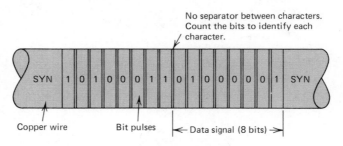

FIGURE **2-5**   Synchronous transmission.

sends some "sync" characters to the receiving device so the receiving device can determine the time frame between each of the bits. Sync usually is written SYN.

The sending device sends a long stream of data bits that may have thousands of bits. The receiving device, knowing what code is being used, counts off the appropriate number of bits, assumes this is the first character, and passes it to the computer. It then counts off the second character and so on. If USASCII code is used, the receiving device counts off the first eight bits and sends them to the computer as a character; it then counts off the second eight bits and sends them as a character, and so on.

Synchronous transmission is more efficient in that there are fewer control bits in proportion to the total number of bits transmitted. The synchronization may take only 16 to 32 bits, while the stream of bits for the data signal may be several thousand bits long.

In asynchronous transmission there are at least one start bit and one stop bit for every character of data. If an error occurs during asynchronous transmission, that error may destroy only one character of data because each character is synchronized with its own start and stop bits. On the other hand, that same error in synchronous transmission probably would destroy the entire message block by breaking the synchronization.

The modems and related equipment for synchronous transmission are more expensive than those used for asynchronous transmission because they must be able to synchronize between themselves.

In summary, synchronous and asynchronous data transmissions are differentiated by the fact that in asynchronous data transmission each character is transmitted as a totally independent entity with its own start and stop bits to inform the receiving device that the character is beginning and ending. By contrast, in synchronous transmission whole blocks of data are transmitted as units after the transmitter and the receiver have been synchronized.

Isochronous Transmission   A third technique, isochronous transmission, combines the elements of both synchronous and asynchronous data transmission. In isochronous transmission, as in asynchronous, each character is required to have both a start bit and a stop bit. However, as in synchronous data transmission, the transmitter and the receiver are synchronized. The synchronization time interval between successive bits is specified to be an integral multiple of the length of one code bit. That is, all periods of no transmission consist of one or more 1-character time intervals. This common timing allows higher precision between the transmitting and receiving equipment than can be achieved with asynchronous techniques only.

Figure 2-6 illustrates the relationships and differences between asynchronous, synchronous, and isochronous transmission. In asynchronous transmission, there is no determination of the spacing between individual characters (indefinite

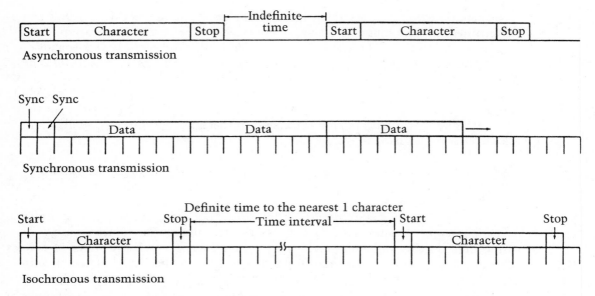

FIGURE 2-6 Comparison of transmission methods.

time). Thus, both the sending and receiving equipment must have clocks to determine the length of a bit, and the receiver must have special recognition circuitry to determine the beginning and end of a character. With synchronous transmission, the clocking signal is used to synchronize the receiver to the sender before a long, multicharacter block of data is transmitted. In isochronous transmission, the clocking is supplied by the sending modem, and the receiving modem synchronizes to it for short periods. Each character begins on some multiple of the length of the bit element.

The primary reason for using isochronous transmission in preference to asynchronous transmission is speed. In practice, asynchronous data transmission generally is limited to 2400 bits per second by the timing precision of the sending and receiving equipment. By contrast, isochronous data transmission can achieve data transmission rates as high as 9600 bits per second. Synchronous data communications may be even more rapid than isochronous.

In summary, the modes of transmission are serial (bit by bit down the line) and parallel (all bits of a character sent simultaneously). Furthermore, serial transmission (which is most prevalent) can be divided into three types.

- Asynchronous transmission, in which the data bits of a character are sent independent of the timing of any other character and are preceded by a start bit and followed by a stop bit

- Synchronous transmission, in which the sending and receiving units are synchronized, and then a stream of many thousands of data signal bits is sent

- Isochronous transmission, in which each character still has a start and stop bit, but the sending and receiving equipment are synchronized for the length of each timing unit

CODING TERMINOLOGY AND STRUCTURE

A *character* is a symbol that has a common, constant meaning for some group of people. A character might be the letter A or B, or it might be a number such as 1 or 2. Characters may also be special symbols such as & or ?. Characters in data communications, as in computer systems, are represented by groups of bits. The various groups of bits that represent the set of characters that are the "alphabet" of any given system are called a *coding system*, or simply a *code*. This section will discuss some of the codes used in data communications.

A *byte* is a group of consecutive bits that are treated as a unit or character. One byte normally is comprised of 8 bits and usually represents one character. However, in data communications some codes in regular use utilize 5, 6, 7, 8, or 9 bits to represent a character. These differences in the number of bits per character arise because the codes have different numbers of characters to represent and different provisions for error checking.

Coding is the representation of one set of symbols by another set of symbols. For example, representation of the character A by a group of 7 bits (say, 1000001) is an example of coding.

As we have seen, information in data communications is normally transmitted serially over a transmission line or channel. Codes for representing the information vary both in the number of bits used to define a single character and in the assignment of bit patterns to each particular character. For example, the bit group 1000001 may represent the character A in one coding scheme (USASCII), but the bit group 11000 may represent the character A in some other code configuration.

United States of America Standard Code for Information Interchange (USASCII), or more commonly ASCII, is the most popular code for data communications. It is the basic standard code and is available on most terminals. This is an 8-bit code that has 128 valid character combinations. The number of combinations can be determined by taking the number 2 and raising it to the power equal to the number of information bits in the code. In this case $2^7 = 128$ characters. The eighth bit is the *parity bit* for error checking on individual characters. (Parity is discussed later in this chapter in the section Error Detection with Retransmission.)

The USASCII code is used widely on both asynchronous and synchronous data communication equipment. Figure 2-7 shows the code structure for this code and an explanation of each of the control characters.

Extended Binary Coded Decimal Interchange Code (EBCDIC) is IBM's standard information code. This code has 256 valid character combinations because there are 8 information bits and parity is carried as a ninth bit. You probably have heard people talk about a 9-channel tape drive. The 9-channel tape drive was developed originally to hold the IBM code with its 8 data bits and 1 parity bit. If used in asynchronous transmission, this code has 11 bits per character because there is 1 start, 8 data, 1 parity, and 1 stop bit. In synchronous transmission it has only 9 bits per character unless the parity bit is stripped off prior to transmission. Figure 2-8 shows the standard EBCDIC code configuration chart.

It should be noted that the bit positional numbering system is different in EBCDIC and ASCII. ASCII numbers its bit positions from 8 to 1 (left to right), while EBCDIC numbers its bit positions from 0 to 7 (also left to right). In fact, IBM addresses everything left to right, such as memory, records, and bits in a byte. Figure 2-9 depicts the positional numbering system differences between EBCDIC and ASCII.

Bit positional difference is mentioned because if you were converting a master file from an EBCDIC system to an ASCII system and the file documentation said that bit number 3 in a status byte reflected an overdraft status of this account, what would you do? To someone who learned on an ASCII system, bit number 3 would be in a different location than it would be to the person who wrote the documentation for the EBCDIC file description.

Binary Coded Decimal (BCD) code is a 6-bit code that has 64 valid character combinations ($2^6 = 64$). The BCD code was the logical extension from the earlier tab card-oriented Hollerith code. Depending on the specific hardware, this code can have 1 or 2 parity bits; therefore, it can be either 6 bits (if the parity bit is not transmitted), 7 bits (1 parity bit), or 8 bits (2 parity bits) during synchronous transmission. A start bit and a stop bit would have to be added for asynchronous transmission. Figure 2-10 shows the BCD code.

One of the oldest codes of data communications is called the *Baudot* code. It is a 5-bit code that has only 32 possible character combinations; however, there are also two functions, called *letters* and *figures*. When one of these two functions is used, it sets the equipment so all characters typed after that point are in a different configuration than they were previously. In effect, this raises the number of valid character combinations to 58, just barely enough for simplified data communications. A version called *International Baudot* has a sixth bit added for parity purposes. Baudot code is used on the earlier teletype equipment and on very slow communication circuits (150 bits per second or less). Figure 2-11 shows the Baudot code configuration.

Self-checking or *"M"-of-"N"* codes also are available. These codes can indicate if one of the bits was changed during transmission. For example, with a 2-of-5

b7 → / b6 → / b5 →					0 0 0	0 0 1	0 1 0	0 1 1	1 0 0	1 0 1	1 1 0	1 1 1
Bits b4	b3	b2	b1	column row ▼ ►	0	1	2	3	4	5	6	7
0	0	0	0	0	NUL	DLE	SP	0	@	P		p
0	0	0	1	1	SOH	DC1	!	1	A	Q	a	q
0	0	1	0	2	STX	DC2	"	2	B	R	b	r
0	0	1	1	3	ETX	DC3	#	3	C	S	c	s
0	1	0	0	4	EOT	DC4	$	4	D	T	d	t
0	1	0	1	5	ENQ	NAK	%	5	E	U	e	u
0	1	1	0	6	ACK	SYN	&	6	F	V	f	v
0	1	1	1	7	BEL	ETB	/	7	G	W	g	w
1	0	0	0	8	BS	CAN	(	8	H	X	h	x
1	0	0	1	9	HT	EM	)	9	I	Y	i	y
1	0	1	0	10	LF	SUB	*	:	J	Z	j	z
1	0	1	1	11	VT	ESC	+	;	K	[	k	{
1	1	0	0	12	FF	FX	,	<	L		l		
1	1	0	1	13	CR	GS	–	=	M	]	m	}
1	1	1	0	14	SO	RS	.	>	N	>	n	~
1	1	1	1	15	SI	US	/	?	O	—	o	DEL

Character Codes

1. Standardized groupings of bits (1s and 0s) to represent alphanumeric and control information.
2. American Standard Code for Information Interchange (ASCII)—ANSI X3.4.
   a. A 7-bit code which yields 128 possible combinations or character assignments.
   b. Ninety-six graphic, i.e., printable or displayable, characters.
   c. Thirty-two control characters, including
      (1) Device-control characters such as Line Feed, Carriage Return, Bell, etc.
      (2) Information-transfer control characters such as ACK, NAK, etc.

Mnemonic and Meaning		Mnemonic and Meaning	
NUL	Null	DLE	Data Link Escape (CC)
SOH	Start of Heading (CC)	DC1	Device Control 1
STX	Start of Text (CC)	DC2	Device Control 2
ETX	End of Text (CC)	DC3	Device Control 3
EOT	End of Transmission (CC)	DC4	Device Control 4
ENQ	Enquiry (CC)	NAK	Negative Acknowledge (CC)
ACK	Acknowledge (CC)	SYN	Synchronous Idle (CC)
BEL	Bell	ETB	End of Transmission Block (CC)
BS	Backspace (FE)	CAN	Cancel
HT	Horizontal Tabulation (FE)	EM	End of Medium
LF	Line Feed (FE)	SUB	Substitute
VT	Vertical Tabulation (FE)	ESC	Escape
FF	Form Feed (FE)	FS	File Separator (IS)
CR	Carriage Return (FE)	GS	Group Separator (IS)
SO	Shift Out	RS	Record Separator (IS)
SI	Shift In	US	Unit Separator (IS)
		DEL	Delete

FIGURE 2-7 ASCII code structure.

Bit grouping for the chart below (Bits 0,1 / Bits 2,3):
- Bits 0,1 = 00 → hex columns 0–3, 01 → hex columns 4–7, 10 → hex columns 8–B, 11 → hex columns C–F
- Bits 2,3 = 00, 01, 10, 11 within each group

Bits 4567	Hex	0	1	2	3	4	5	6	7	8	9	A	B	C	D	E	F
0000	0	NUL	DLE			SP	&	–									0
0001	1	SOH	SBA				/			a	j			A	J		1
0010	2	STX	EUA		SYN					b	k	s		B	K	S	2
0011	3	ETX	IC							c	l	t		C	L	T	3
0100	4									d	m	u		D	M	U	4
0101	5	PT	NL							e	n	v		E	N	V	5
0110	6			ETB						f	o	w		F	O	W	6
0111	7			ESC	EOT					g	p	x		G	P	X	7
1000	8									h	q	y		H	Q	Y	8
1001	9		EM							i	r	z		I	R	Z	9
1010	A					¢	!	|	:								
1011	B					.	$	,	#								
1100	C		DUP		RA	<	*	%	@								
1101	D		SF	ENQ	NAK	(	)	_	'								
1110	E		FM	ACK		+	;	>	=								
1111	F		ITB		SUB	|	¬	?	"								

EBCDIC Code as Implemented for the IBM 3270 Information Display System

Extended Binary Coded Decimal Interchange Code (IBM's EBCDIC).

a. An 8-bit code yielding 256 possible combinations or character assignments.

b. A representative subset, that of the IBM 3270 product family.
   Absence of certain functions not usable by 3270 products (e.g., paper feed, vertical tab, back space) which would show up in EBCDIC code charts for other products that might make use of them.

FIGURE 2-8   EBCDIC code structure.

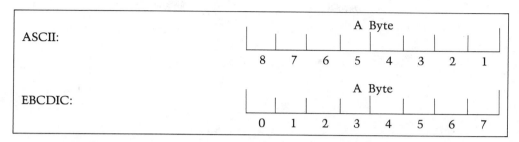

FIGURE **2-9** Bit positional differences between ASCII and EBCDIC.

Zone $\{b_6, b_5\}$

				Digit			
0	0	I	I				
0	I	0	I	b_4	b_3	b_2	b_1
SP	♭	—	&+	0	0	0	0
1	/	J	A	0	0	0	I
2	S	K	B	0	0	I	0
3	T	L	C	0	0	I	I
4	U	M	D	0	I	0	0
5	V	N	E	0	I	0	I
6	W	O	F	0	I	I	0
7	X	P	G	0	I	I	I
8	Y	Q	H	I	0	0	0
9	Z	R	I	I	0	0	I
φ	ǂ	!	?	I	0	I	0
# =	,	$	.	I	0	I	I
@ '	% (	*	□)	I	I	0	0
:	γ	]	[	I	I	0	I
⟩	\	;	⟨	I	I	I	0
√	⧺	△	≢	I	I	I	I

CHARACTER

6	5	4	3	2	1
ZONE			DIGIT		

CODE   DESCRIPTION

CODE	DESCRIPTION
SP	Space
♭	Space (Even Parity)
ǂ	Record Mark
!	Minus Zero
?	Plus Zero
γ	Word Separator
√	Tape Mark
⧺	Tape Segment Mark
△	Delta (Mode Change)
≢	Group Mark

FIGURE **2-10** BCD code structure.

Lower Case	Upper Case	1	2	3	4	5
A	—	●	●			
B	?	●			●	●
C	:		●	●	●	
D	$	●			●	
E	3	●				
F	!	●		●	●	
G	&		●		●	●
H	£			●		●
I	8		●	●		
J	'	●	●		●	
K	(	●	●	●	●	
L	)		●			●
M	.			●	●	●
N	,			●	●	
O	9				●	●
P	0		●	●		●
Q	1	●	●	●		●
R	4		●		●	
S	bell	●		●		
T	5					●
U	7	●	●	●		
V	;		●	●	●	●
W	2	●	●			●
X	/	●		●	●	●
Y	6	●		●		●
Z	"	●				●
Letters (shift to lower case)		●	●	●	●	●
Figures (shift to upper case)		●	●		●	●
Space				●		
Carriage return					●	
Line feed			●			
Blank						
● represents a 1						
Blank represents a 0						

FIGURE **2-11**   BAUDOT code structure.

code, there are always two 1s and three 0s representing a character. If this precise ratio (two 1s and three 0s) is not present in each received character, then an error has occurred during transmission of the data. One disadvantage of this system is that it makes many of the possible combinations unusable. For example,

$$C = \frac{N!}{M! \, (N - M)!}$$

shows how to calculate the number of usable code configurations when using one of these self-checking codes. C (the number of combinations) is determined by taking the various factorials of N and M, where N equals the total number of bits in the code and M equals the number of 1 bits in the code. With the example of a 2-of-5 code we have: C = 5!/2! (5 − 2) ! or only 10 legal combinations. Obviously, this is far too few combinations; therefore, a 2-of-5 code is totally unusable except for transmitting numbers only.

Some years ago IBM developed a 4-of-8 code, where 4 of the 8 bits were 1s and 4 were 0s. Any other combination was an error. This code was of limited value because instead of 256 valid combinations, there are only 70. Also, it does not have the option of stripping off the parity bit before a synchronous transmission and putting it back on upon receipt. On the positive side, this 4-of-8 code detects errors better than a single parity bit. It has the same error checking capabilities as the double parity bit in a BCD code structure.

Now that we have discussed codes, we must review efficiency with regard to a specific code set. One objective of a data communication network is to achieve the highest possible volume of accurate information through the system. The higher the volume, the greater the resulting system's efficiency and the lower the cost. System efficiency is affected by such characteristics of the circuits as distortion and transmission speed, as well as by turnaround time, the coding scheme utilized, the speed of the transmitting and receiving equipment, the error detection and control methodologies, and the mode of transmission. In this section we will focus on the coding scheme.

Transmission efficiency is defined as the total number of information bits divided by the total bits.

Each communication code structure has both information bits and redundant bits. *Information bits* are those used to convey the meaning of the specific character being transmitted, such as the character "A." *Redundant bits* are used for other purposes such as error checking. Therefore, a parity bit used for error checking is a redundant bit because it is not used to identify the specific character, even though it may be necessary. As you can see, if you did not care about errors, the redundant error checking bit could be omitted. Also, message control characters, such as an END OF BLOCK character, are considered redundant bits. These bits are redundant by definition only because they are needed for accurate data com-

$$E_C = \frac{B_I}{B_T}$$

E_C: efficiency of the code
B_I: information bits
B_T: total bits

FIGURE **2-12**   Efficiency of codes.

munications, just as the periods and commas are required when you write a letter.

Figure 2-12 shows that the efficiency of the code (E_C) equals the bits of information (B_I) divided by the bits in total (B_T). This means that if you have an 8-level code with 7 of the bits used to represent the character and 1 representing parity, then you have a code efficiency of 87.5 percent. This is calculated by taking the bits of information (B_I) and dividing by the bits in total (B_T): for example, $\frac{7}{8}$ = 0.875.

If you have a 6-bit code with 2 parity bits, then the efficiency of the code (E_C) is $\frac{6}{8}$ or 75 percent. As is evident by this comparison, some codes are more efficient than others. The addition of parity bits lessens efficiency because parity bits are redundant, not being needed to convey meaningful information but only to check for errors.

The same formula (as is shown in Figure 2-12) can be used to estimate the efficiency of an asynchronous transmission system. As an example, assume that there is an 8-bit code structure where 7 bits represent the data and 1 bit is for parity. In asynchronous transmission there are usually 1 start bit and 1 stop bit. Therefore, the bits of information (B_I) are 7, but the bits in total are 10 (B_T). The efficiency of the asynchronous transmission system, at its maximum, is 7 bits of information divided by 10 total bits for an efficiency of 70 percent. It should be noted that if any other control characters are involved, such as message character counts or other control characters sent at the end of your transmission, the efficiency drops below 70 percent.

The same basic formula can be used if an estimate is needed for the efficiency of a synchronous transmission system. In this case the bits of information (B_I) are calculated by determining how many "information" characters are in the message block. If the message portion of the frame or packet contains 100 information characters, using our 8-bit code, there are 7 bits times 100 characters, or 700 bits of information. Next, the bits in total (B_T) are the 700 bits of information, plus all the redundant bits that are inserted for control and security purposes. These other bits include the parity bit (eighth bit) that is appended to each character, the control bits in the flag at each end of the frame, the bits in the control field, the address field, the frame check sequence, and any internal control characters from the packet frame such as format identifiers, logical channel numbers, sequence numbers, and a couple of synchronization (SYN) characters.

For this example assume that there are a total of 11 control characters for your message. Therefore, the number of redundant bits is 11 control characters times 8 bits per character, plus 100 bits (the parity bit for each of the characters in your message) for a total of 188 redundant bits. This shows that the efficiency of this synchronous system is 700 bits of information (B_I) divided by 888 bits in total (B_T) for an efficiency of 79 percent.

This example shows that synchronous systems are more efficient than asynchronous systems and some codes are more efficient than others. To extend our example further, assume we have compression/compaction software or a hardware scheme that strips off the individual parity bits for each character prior to transmission and puts them back on at the destination. The number of redundant bits (no parity bits) is reduced by 100; therefore, efficiency rises to 89 percent (700 bits of information divided by 788 bits in total). If the frame check sequences and other error checking techniques for synchronous transmission are good enough, you might consider stripping off the individual parity bits for each character prior to transmission in order to gain a 10 percent increase in throughput efficiency of data bits. The next section will describe a more accurate method of measuring throughput efficiency.

THROUGHPUT (TRIB)

Many factors affect the throughput of a data communication system. Probably the most important is the transmission rate of information bits and the communication circuit bandwidth because bandwidth denotes the absolute upper limit of speed. If terminals are multidropped, that is another factor because the circuit must be shared. If terminals are multiplexed, that becomes a factor because each terminal uses a reduced bandwidth (a subset of the total bandwidth on the circuit). Another factor is the capability of the front end communication processor to handle multiple incoming and outgoing communication circuits. If the front end cannot handle circuits or messages simultaneously, the capacity of the system is degraded. Software design is also a factor because it determines which protocol (see Glossary) is used and whether transmission is in full duplex or half duplex. Propagation time, especially on satellite circuits, affects throughput. The time required for the host computer to process a request and/or perform a lookup or update in a database also is a factor in throughput. Error rates in hardware, in software, and on the communication circuit affect throughput be-

$$\text{TRIB} = \frac{\text{Number of information bits accepted}}{\text{Total time required to get the bits accepted}}$$

$$\text{TRIB} = \frac{K(M - C)(1 - P)}{M/R + T}$$

K = Information bits per character

M = Block length in characters

R = Modem transmission rate in characters per second

C = Average number of noninformation characters per block (control characters)

P = Probability that a block will require retransmission because of error

T = Time between blocks in seconds such as modem delay/turnaround time on half duplex, echo suppressor delay on dial-up, and propagation delay on satellite transmission. This is the time required to reverse the direction of transmission from send to receive or receive to send on a half duplex (HDX) circuit. It can be obtained from the modem specification book and sometimes is referred to as *reclocking time.*

FIGURE 2-13   TRIB equation.

cause of possible retransmissions of the same message. The polling scheme (central control) or whether the system is an interrupt system affects throughput. Obviously, many items affect throughput. It is appropriate at this point to examine one of the major parameters of importance: how many usable characters of information can be transmitted per second.

Transmission Rate of Information Bits (TRIB) is the term normally used to describe the effective rate of data transfer. It is a measure of the effective quantity of information that is transmitted over a communication circuit per unit of time.

The American National Standards Institute (ANSI) provides definitions for calculating the transfer rate of information bits. TRIB calculations may vary with the type of protocol used because of different numbers of control characters required and different time between blocks. The basic TRIB equation is shown in Figure 2-13. Figure 2-14 shows the calculation of throughput assuming a 4800 bits per second half duplex circuit. If all factors in the calculation remain constant except for the circuit, which is changed to full duplex (no turnaround time delays, T = 0), then the TRIB increases to 4054 bits per second.

Look at Figure 2-14, where the turnaround value (T) is 0.025. If there is a further propagation delay time of 475 milliseconds (0.475), that figure changes to 0.500. To demonstrate how a satellite channel affects TRIB, the total delay time is now

$$\text{TRIB} = \frac{7\,(400-10)\,(1-0.01)}{(400/600) + 0.025} = 3908 \text{ BPS}$$

K = 7 bits per character (information)

M = 400 characters per block

R = 600 characters per second (derived from 4800 bps divided by 8 bits/character)

C = 10 control characters per block

P = 0.01 (10^{-2}) or one retransmission out of one hundred blocks transmitted—1%

T = 25 milliseconds (0.025) turnaround time

FIGURE **2-14** TRIB calculation.

500 milliseconds. Still using the figures in Figure 2-14 (except for the new 0.500 delay time), the TRIB for our half duplex, satellite link reduces to 2317 bits per second, which is almost one-half of the full duplex (no turnaround time) 4054 bits per second.

BASEBAND/BROADBAND

There are two general categories of electrical current: direct current and alternating current. *Direct current* (dc) travels in only one direction in a circuit, whereas *alternating current* (ac) travels first in one direction (+) and then in the other direction (−). The frequency of a continuous ac wave is the number of times per second that the wave makes a complete cycle from 0 to its maximum positive value, then through to its maximum negative value, and back to 0. Figure 2-15 shows three different *sine waves*. The top configuration shows one complete cycle, the second configuration two complete cycles, and the third configuration three complete cycles. Almost all transmission over telephone lines utilizes the ac waveform known as a sine wave.

Some transmission is performed using direct current where the signals are represented as voltages of dc electricity being transmitted down the wire. These dc signals usually are sent over wire pairs of no more than a few thousand feet in length and are referred to as *baseband signaling* (digital).

Basically, baseband signaling is the digital transmission of electrical pulses. This digital information is binary in nature in that it only has two possible states, a 1 or a 0 (sometimes called *mark* and *space*). The most commonly encountered plus and minus voltage levels range from a low of 3 to a high of 25 volts. These baseband digital signals must be modulated onto the telephone company's interexchange channels by the modem.

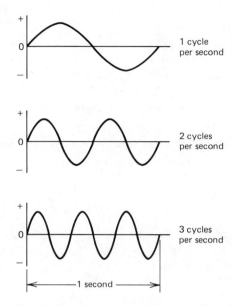

FIGURE **2-15** Sine waves (ac waveforms).

The binary 0s and 1s that are transmitted (groupings of these 0s and 1s make up a character) are represented by different levels of voltage such as +5 volts for binary 1 and −5 volts for a binary 0. The terminal outputs these plus and minus voltages (baseband signal), and they pass over the connector cable and into the modem. This is called a digital signal. The modem then converts it to a *broadband signal* for transmission. Both modems and analog modulation are defined later in this chapter. Broadband signals are an analog waveform.

Leakage of the electrical current either at the terminal connector plug or as it passes over the connector cable causes distortion of your signal, and this in turn causes errors. Capacitance and inductance also cause distortion that can result in errors.

With *capacitance,* as we raise the voltage at one end of a communication circuit there is some delay before the voltage at the other end rises by an equal amount. The copper wire acts rather like a water hose in that it needs to be filled to capacity before the electricity applied at one end is received at the other end. The problem here is that if the pulses (+ and − voltages) are too short in duration, or if there are too many pulses sent per second, then they become indistinguishable when they are received. The faster the pulse rate (baud), the more difficult it becomes to interpret the received signal.

Inductance in a circuit is a force that resists the sudden buildup of electric current. This resistance can be a cause of distortion and, therefore, of errors during transmission.

CONNECTOR CABLES

Go back to Figure 2-1 which illustrates the basic technical concepts of transmitting data. The next component to review is the connector cable, the standards (such as RS232/RS449/X.21), and the data signaling or synchronization for the movement of synchronous data. The synchronization is achieved by having a start bit and a stop bit on each character or sending SYN characters (01101001) to the remote modem. Our discussion will center more on the synchronization/data signaling between the terminal and its local modem and the remote modem.

RS232C/RS449

The RS232C is the connector cable that is the standard interface for connecting data terminal equipment (DTE) to data circuit terminating equipment (DCE). *Data terminal equipment* comprises the data source, the data sink, or both. In reality, it is any piece of equipment at which a data communication path begins or ends, such as a terminal. *Data circuit terminating equipment* provides all the functions required to establish, maintain, and terminate a connection. This includes signal conversion and coding between the DTE and the common carrier's circuit, including the modem.

Figure 2-16 shows a picture of the RS232C interface plug and provides a description of each of its 25 protruding pins. It is the standard connector cable (25 wires/pins) that passes control signals and data between the terminal (DTE) and the modem (DCE). This standard has been supplied by the Electronic Industries Association (EIA). Outside the United States, this RS232C connector cable is known as the V.24 and V.28. The V.24 and V.28 standards have been accepted by the standards group known as the Consultative Committee on International Telephone and Telegraph (CCITT) and are described in the last section of Chapter 5. These standards provide a common description of what the signal coming out of, and going into, the serial port of a computer or terminal looks like electrically. Specifically, RS232C provides for a signal changing from a nominal $+12$ volts to a nominal -12 volts. The standard also defines the cables and connectors used to link data communication devices.

The RS232C has a maximum 50-foot cable length, but it can be increased to 100 feet or more by means of a special low capacitance, extended distance cable.

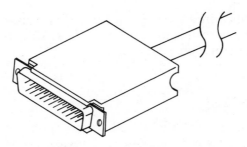

Pin	Function	Pin	Function	Pin	Function
1	Frame Ground	10	Negative dc Test Voltage	19	Sec. Request To Send
2	Transmitted Data	11	Unassigned	20	Data Terminal Ready
3	Received Data	12	Sec. Data Carrier Detect	21	Signal Quality Detect
4	Request to Send	13	Sec. Clear to Send	22	Ring Indicator
5	Clear to Send	14	Sec. Transmitted Data	23	Data Rate Select
6	Data Set Ready	15	Transmitter Clock	24	Ext. Transmitter Clock
7	Signal Ground	16	Sec. Received Data	25	Busy
8	Data Carrier Detect	17	Receiver Clock		
9	Positive dc Test Voltage	18	Receiver Dibit Clock		

The terminal connection to the modem is defined by the Electronic Industries Association (EIA) specification RS232C. RS232C specifies the use of a 25-pin connector and the pin on which each signal is placed.

FIGURE **2-16** RS232C interface.

This is not advised, however, because some vendors may not honor maintenance agreements if the cable is lengthened beyond the 50-foot standard.

As an illustration, let us present the cable distances for Texas Instruments' products. The cable length of the RS232C varies according to the speed at which you transmit. For Texas Instruments, the connector cable length can be up to 914 meters (1 meter = 39.37 inches) when transmitting at 1200 bits per second, 549 meters when transmitting at 2400 bits per second, 244 meters when transmitting at 4800 bits per second, and 122 meters when transmitting at 9600 bits per second. When end users operate at maximum distances, it is important for them to remember that they must meet the restrictions on all types of equipment used, including the electrical environment, cable construction, and cable wiring. This means that when you want to operate at a maximum cable distance, it is necessary to contact the terminal and/or modem vendors to obtain their maximum cable distance before proceeding.

The newer standard is the RS449 which has been adopted as U.S. Federal Standard 1031. The RS449, shown in Figure 2-17, has many advantages over its predecessor, the RS232. A 4,000-foot cable length can be used, there are 37 pins instead of 25 (useful for digital transmission), and various other circuit functions have been added, such as diagnostic circuits and digital circuits. In addition, secondary channel circuits (reverse channel) have been put into a separate 9-pin connector. This same 9-pin connector is used on the AT (Advanced Technology) microcomputers and is known as a DB-9. For some of the new features, look at pin 32 (SELECT STANDBY). With this pin, the terminal can instruct the modem to use an alternate standby network such as changing from a private leased line to a public packet network, either for backup or simply to access another database not normally used. In other words, a terminal can be connected to two different networks, and the operator can enter a keyboard command to switch the connection from one network to another. With regard to LOOPBACK (pins 10 and 14), the terminal can allow basic tests without special test equipment or the manual swapping of equipment or cables.

With microcomputers, the RS232 and the RS449 also are referred to as D-type connectors. The RS232 may be called a DB-25, and the 9-pin RS449 may be called a DB-9. You may wish to look ahead at Figure 6-3 in Chapter 6 to examine the microcomputer pin configurations.

There are also X.20 and X.21 interface cables. The X.20 interface is for asynchronous communications, and the X.21 for synchronous communications. Each is based on only 15 pins (wires) connecting the DTE and the DCE. The lower number of pins requires an increased intelligence in both the DTE and the DCE. X.20 and X.21 are international standards that are intended to interface with the X.25 packet switching networks discussed later in this book.

Another option that may become available in the near future is fiber optic cables in place of the standard RS232 or RS449 electrical cables. Currently, using fiber optic cable, a terminal can be located 1,000 meters (3,280 feet) from a host mainframe computer without the need for a modem. With a 1,000-meter fiber optic cable, these products can communicate at speeds ranging from 19,200 bits per second up to double that speed. Therefore, you get not only greater distance (1,000 meters) but also greater speed. This may be another example in which fiber optics will replace electronics in the future.

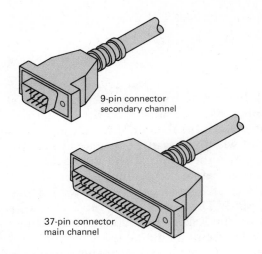

9-pin connector
secondary channel

37-pin connector
main channel

37-PIN CONNECTOR		9-PIN CONNECTOR

First Segment Assignment		Second Segment Assignment			
Pin	Function	Pin	Function	Pin	Function
1	Shield	20	Receive Common	1	Shield
2	Signaling Rate Indicator	21	Unassigned	2	Sec. Receiver Ready
		22	Send Data	3	Sec. Send Data
3	Unassigned	23	Send Timing	4	Sec. Receive Data
4	Send Data	24	Receive Data	5	Signal Ground
5	Send Timing	25	Request to Send	6	Receive Common
6	Receive Data	26	Receive Timing	7	Sec. Request to Send
7	Request to Send	27	Clear to Send		
8	Receive Timing	28	Terminal in Service	8	Sec. Clear to Send
9	Clear to Send	29	Data Mode	9	Send Common
10	Local Loopback	30	Terminal Ready		
11	Data Mode	31	Receiver Ready		
12	Terminal Ready	32	Select Standby		
13	Receiver Ready	33	Signal Quality		
14	Remote Loopback	34	New Signal		
15	Incoming Call	35	Terminal Timing		
16	Select Frequency/ Signaling Rate Selector	36	Standby Indicator		
		37	Send Common		
17	Terminal Timing				
18	Test Mode				
19	Signal Ground				

RS449 is a new EIA specification replacing RS232C. This specification calls for the use of a 37-pin connector. For those devices using a side, forward, reverse, or secondary channel, a second 9-pin connector is specified. RS449 provides for additional control and signaling.

FIGURE 2-17 RS449 interface.

```
Microcomputer                                        Modem
     DTE side     Name and pin number                DCE side
     1 ─────────────── Frame ground ─────────────── 1
     2 ─────────────── Transmit   data  ───────────→ 2
     3 ←─────────────── Receive data ─────────────── 3
     4 ─────────────── Request to send ────────────→ 4
     5 ←─────────────── Clear to send ─────────────── 5
     6 ←─────────────── Data set ready ─────────────── 6
     7 ←─────────────── Signal ground ──────────────→ 7
     8 ←─────────────── Carrier detect ─────────────── 8
    15 ←─────────────── Receive clock ────────────── 15
    17 ─────────────── Transmit clock ──────────────→ 17
    20 ─────────────── Data terminal ready ─────────→ 20
    22 ←─────────────── Ring indicator ────────────── 22
    25 ─────────────── Test/busy ───────────────────→ 25
```

FIGURE **2-18**   RS232/Modem control.

DATA SIGNALING/SYNCHRONIZATION

As you learned earlier in this chapter, synchronous transmission involves the movement of an entire message block at a time. Let us look at data signaling/synchronization as it occurs on an RS232 connector cable. Figure 2-18 shows the 13 most used pins of the 25-pin RS232 connector cable. On the left side is a microcomputer and on the right side is a modem.

Did you ever wonder what happens when you press the "send" key with regard to synchronous data transmission? When a synchronous block of data is sent, the microcomputer and the modem raise and lower electrical signals (plus and minus voltages of electricity) between themselves. This usually is a nominal $+12$ or -12 volts. For example, a modem with an RS232 interface (Figure 2-16) might indicate that it is powered on and ready to operate by raising the signal on pin 6, DATA SET READY. When a call comes in, the modem shows the microcomputer that the telephone line is ringing by raising a signal on pin 22, the RING INDICATOR. The microcomputer may then tell the modem to answer the call by raising a signal on pin 20, DATA TERMINAL READY. After the modems connect, the modem may indicate the connection status to the microcomputer by raising a signal on pin 8, CARRIER DETECT. At the end of the session, the microcomputer may tell the modem to drop the telephone call (release the circuit) by lowering the signal on pin 20, DATA TERMINAL READY. REQUEST TO SEND and CLEAR TO SEND signals go over pins 4 and 5, which are used in half duplex modems to manage control of the communication channel. Incidentally, some of these basic procedures may vary slightly from one manufacturer to another.

Let us discuss an example that handles the flow of a block of synchronous data. When the microcomputer operator presses the "send" key to transmit a block of data, pin 4, REQUEST TO SEND, transmits the signal from the microcomputer to the modem. This informs the modem that a block of data is ready to be sent. The modem then sends a CLEAR TO SEND signal back to the microcomputer by using pin 5, thus telling the microcomputer that it can send a synchronous block of data.

The microcomputer now out-pulses a serial stream of bits that contain two 8-bit SYN (sync) characters in front of the message block. A SYN character is 011101001. This bit stream passes over the connector cable to the modem using pin 2, TRANSMITTED DATA. The modem then modulates this data block to convert it from the digital baseband signal (plus and minus voltages of electricity) to an analog broadband signal (discussed in the next section). From the modem this block of data goes out on to the local loop circuit that goes from your business premise to the telephone company central office. From there it goes on to the long distance interexchange channels (IXC) and the receiving end telephone company central office. Then it moves to the local loop, into the modem, across the connector cable, and into the host mainframe computer. You can follow this movement by reviewing Figure 2-1.

This process is repeated for each synchronous message block in half duplex transmission. The data signaling that takes place between the microcomputer and the modem involves the REQUEST TO SEND, CLEAR TO SEND, and TRANSMITTED DATA pins. Accurate timing between blocks of data is critical in data signaling and synchronization. If this timing is lost, the entire block of data is destroyed and must be retransmitted.

ANALOG MODULATION

Modulation is the technique that modifies the form of an electrical signal (a carrier wave) so the signal can carry intelligent information on some form of communication medium. The modulated signal often is referred to as a *broadband signal*. The signal that does the carrying is called a *carrier wave*, and modulation changes the shape or form of the carrier wave to transmit 0s and 1s.

A continuous oscillating voltage of arbitrary amplitude and frequency carries no intelligence. However, if it can be interrupted or the amplitude altered so it becomes somewhat like a series of pulses that correspond to some known code (such as USASCII), then the oscillating signal can carry some intelligence. In data communications this continuous oscillating voltage is called a *carrier signal* or simply a *carrier*. The carrier signal can be altered in many ways. The most common methods of modulation are amplitude modulation, frequency modulation, and phase modulation.

The equipment in which modulation is performed is called a *modulator*. If the modulator causes the amplitude of the carrier signal to vary, the result is called *amplitude modulation*, and so on for frequency and phase modulation. These modulators are, in fact, the modems that are connected to each end of the transmission lines. In other words, the process of modifying a carrier so it carries a signal that can be interpreted is referred to as modulation, and the process of converting it back again so the original intelligence is recovered is called demodulation. The effect is to take the binary signal (digital signal) from a computer or business machine and modulate it so it becomes a continuous signal (analog signal) that can be transmitted over telephone lines or microwave towers, and so forth. In the demodulation process, the receiving equipment interprets the modulated carrier signal (analog signal) and converts it to a binary signal (digital signal) that is meaningful to the computer or other business machine.

In *amplitude modulation* (AM), the peak-to-peak voltage of the carrier signal is varied with the intelligence that is to be transmitted. The amplitude modulation shown in the top of Figure 2-19 depicts the peaks at one amplitude representing binary 1s, and the peaks of another amplitude representing binary 0s. Amplitude modulation is suitable for data transmission, and it allows efficient use of the available bandwidth of a voice grade line. However, it is more susceptible to noise during transmission than is frequency modulation.

Frequency modulation, through the use of *frequency shift keying* (FSK), is a modulation technique whereby each 0 or 1 is represented by a different frequency. In this case, the amplitude does not vary. This type of modulation is represented by different tones. A high pitched tone (higher frequency) equals a binary 1, and a low pitched tone (lower frequency) equals a binary 0. The bottom of Figure 2-19 shows frequency modulation of a bit stream of 0s and 1s. Modems that use FSK may use three basic frequencies, such as 2125 hertz, 2225 hertz, and 2025 hertz. In this case, when no data is being transmitted, the carrier wave signal is 2125 hertz. When a stream of 0s and 1s is transmitted, the carrier wave switches between 2025 hertz and 2225 hertz, depending on whether it is a 0 or a 1 that is transmitted.

Phase modulation is the most difficult to understand because there is two-phase (0° and 180°), four-phase (0°, 90°, 180°, and 270°), and eight-phase (0°, 45°, 90°, 135°, 180°, 225°, 270°, and 315°) modulation. Figure 2-20 shows *phase shift keying* (PSK) for a stream of 0 and 1 bits. Notice that every time there is a change in state (0 or 1), there is a 180° change in the phase. A 180° phase change can be seen easily, but it is more difficult to see a 45° change in phase.

The other common type of phase modulation is *differential phase shift keying* (DPSK). In DPSK there is a phase change every time a 1 bit is transmitted; otherwise the phase remains the same (see Figure 2-21).

Different modulation techniques are used to obtain data rates of 2400 bits per second and above. At 2400 bits per second, a common technique (used in the Bell

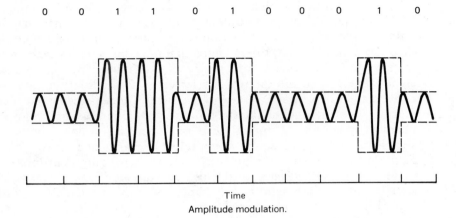

Time

Amplitude modulation.

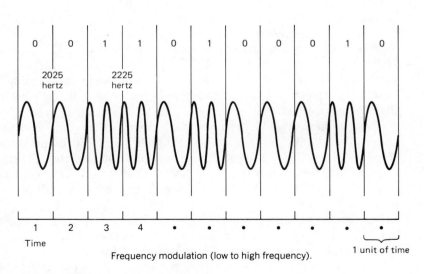

Frequency modulation (low to high frequency).

FIGURE **2-19** Amplitude and frequency modulation.

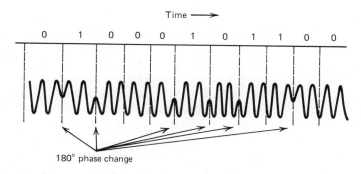

FIGURE **2-20** Phase shift keying (PSK).

201) is *quadrature phase shift keying* (QPSK). QPSK involves splitting the signal into four phases, so that a single frequency tone can take one of four values: 90°, 180°, 270°, or 360° of phase shift. Since a single tone can take any of four values, that tone can represent two bits of information. Note that this is more efficient than simple FSK, where two tones are required to represent one bit of information. For speeds above 2400 bits per second, other techniques are used. For example, most 9600 bits per second modems use a technique called *quadrature amplitude modulation* (QAM). QAM involves splitting the signal into four different phases and two different amplitudes, for a total of eight different possible values. Thus, a signal tone in QAM can represent three bits (two to the third power). The problem with all high speed modulation techniques, however, is that they are more sensitive to imperfections in the communication channel.

Leased communication circuits, as obtained from common carriers, unfortunately tend to exhibit several types of imperfections, which limit the amount of information that can be transmitted across them. The two most common types of

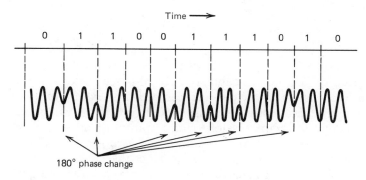

FIGURE **2-21** Differential phase shift keying (DPSK).

imperfections are called envelope delay distortion and amplitude deviation distortion. *Envelope delay distortion* is caused by a phenomenon known as *dispersion*, in which different frequency components of a waveform propagate at different velocities. A complex waveform, such as that produced by a modem, typically is composed of many different frequency components. When the various components of such a waveform begin to propagate at different speeds, they cause the waveform to "spread out" in time. If the waveform spreads out too much, the receiving modem may not be able to interpret it correctly, thus causing line errors. *Amplitude deviation distortion* is an undue amount of variation in the attenuation (loss of signal strength) versus frequency characteristics on the channel. Again, the result of excessive amplitude distortion is an increased error rate because of misinterpretation of the signal by the receiving modem.

In order to transmit data reliably at data rates of 2400 bits per second and above, modem manufacturers developed a technique called *equalization*. Basically, equalization means that the modem tries to compensate for imperfections in the communication channel. Two types of equalization are commonly used today: compromise and adaptive equalization. *Compromise equalization*, which is used in medium speed modems such as the Bell 201, involves an equalizer that is set for average or compromise values based on observed communication line behavior over a large number of lines. Compromise equalization is used at lower speeds (2400 bits per second) because at those speeds the equalization does not have to be as precise, and the circuitry for compromise equalizers is relatively simple and inexpensive. At higher data rates (4800 and 9600 bits per second), the results produced by compromise equalizers are not sufficient, so a more sophisticated technique, *adaptive equalization*, is used. An adaptive equalizer is a device that continuously monitors the signal and adjusts the equalization to obtain the best transmission quality at any given time. In most modern high speed modems, such as the Codex CS 4800 and 9600 or the IBM 3865 (or equivalents), this adaptive equalization is done by a microprocessor with a sophisticated equalization program.

MODEM

From the connector cable, our signal passes next into the modem. Remember, a *modem* takes binary pulses received from a computer, terminal, or other business machine, and converts those signals into a continuous analog signal that is acceptable for transmission over a communication circuit. The name modem is made up of the words *mo*dulator and *dem*odulator. The modem that is transmitting the signal is the modulator because it modulates, or puts some form of intelligence on the carrier wave, whereas the receiving equipment is the demodulator because it demodulates or interprets that signal upon its receipt. A modem has both a digital side and an analog side. The *digital side* is the one toward the business machine (terminal or computer), whereas the *analog side* is the one toward the communication circuit.

As you have seen, to modulate is to adapt a signal so that it can be transmitted over the type of transmission media in use. The modem is the place in which this adaptation is peformed. In addition, even though some communication circuits may have a theoretical maximum transmission capacity in bits per second, you do not always use this full capacity. It is the modem that drives the line. In other words, it is the modem that modulates the signal at whatever speed you wish to transmit. Modems can operate at speeds such as 300, 1200, 2400, 4800, 9600, and much higher in bits per second.

To demonstrate how a modem operates, our examples use frequency modulation. More specifically, they use the frequency shift keying (FSK) technique. To our ears, the change in frequency results in a change in the pitch of the tone. A high pitched tone is a higher frequency and therefore a greater number of cycles per second (hertz), whereas a low pitched tone has a fewer number of cycles per second. You already have seen frequency modulation in Figure 2-19.

To learn about the operation of a modem, examine Figure 2-22. This figure shows the data stream of 0s and 1s that might be coming from a terminal, going into a modem, and going on to the communication circuit. Notice how the 0s and 1s from the terminal are represented as +5 volts of electricity for a 1 and −5 volts of electricity for a 0. Next, the modem modulates or changes that into two frequency tones which are represented as 2225 cycles per second for 1s and 2025 cycles per second for 0s. In other words, when the modem receives −5 volts of electricity, it converts that to a signal of 2025 cycles per second.

If the modem operates at a speed of 1200 bits per second, then the timing between changes from the higher to the lower amplitude is 833 microseconds (1/1200 = 833 microseconds). On the other hand, if there are two 1 bits in a sequence, then the higher amplitude is held on the circuit for a period equal to 1666 microseconds (2 × 833 microseconds).

In reality, a much more complicated sine waveform is sent down the communication circuit; it contains the combination of all the frequencies (2025, 2125, and 2225). The oversimplified description above suits our purpose because we are trying to show how a modem converts electrical signals (baseband) to frequency tones (broadband). Frequency tones (*frequency shift keying*, or FSK) are the most popular method of transmitting data across telephone lines, although other modulation techniques are used at high bits per second rates.

DIGITAL TO ANALOG

On the digital side of the modem (the side connected to the terminal), we have just received a synchronous block of data. It is represented as a sequence of direct electrical pulses (+ and − voltages) which, when counted into their individual 8-

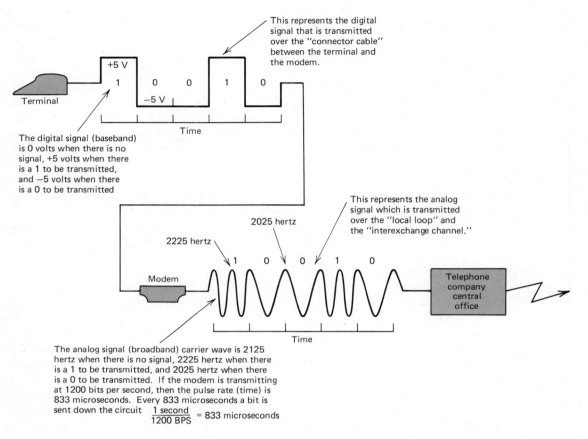

This represents the digital signal that is transmitted over the "connector cable" between the terminal and the modem.

+5 V

1   0   0   1   0

−5 V

Terminal

Time

The digital signal (baseband) is 0 volts when there is no signal, +5 volts when there is a 1 to be transmitted, and −5 volts when there is a 0 to be transmitted

This represents the analog signal which is transmitted over the "local loop" and the "interexchange channel."

2025 hertz

2225 hertz

1   0   0   1   0

Modem

Time

Telephone company central office

The analog signal (broadband) carrier wave is 2125 hertz when there is no signal, 2225 hertz when there is a 1 to be transmitted, and 2025 hertz when there is a 0 to be transmitted. If the modem is transmitting at 1200 bits per second, then the pulse rate (time) is 833 microseconds. Every 833 microseconds a bit is sent down the circuit $\dfrac{1 \text{ second}}{1200 \text{ BPS}} = 833$ microseconds

FIGURE **2-22** Operation of a modem.

bit ASCII code structure, represent individual characters. The modem now proceeds to take each of these plus and minus voltages of electricity and converts them to an appropriate frequency (carrier wave).

Let us develop another example, this time using a different set of frequencies but still using frequency shift keying (FSK), which involves modulating the carrier wave between two different frequency levels. Assume our modem is transmitting at a speed of 1200 bits per second, which means it can change between either of two different frequencies 1200 times each second of time. Therefore, every time the digital side of the modem receives +5 volts of electricity, it transmits a frequency tone signal of 2200 hertz. (Hertz means cycles per second.) Conversely, every time the digital side of the modem receives −5 volts of elec-

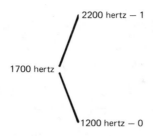

1. When there is no signal to transmit, the modem goes to its carrier wave (1700 hertz). Some modems use either the 2200 or the 1200 hertz as the carrier wave, instead of a third frequency.
2. When a + 5 volt signal is received, the modem out-pulses a 2200 hertz signal, which is a binary 1.
3. When a − 5 volt signal is received, the modem out-pulses a 1200 hertz signal, which is a binary 0.

FIGURE **2-23**   Modem (digital to analog).

tricity, it transmits a signal of 1200 hertz. When no signals are being received by the terminal, the modem falls into a middle frequency (1700 hertz), called its carrier wave (see Figure 2-23).

In order to transmit at 1200 bits per second, our modem has to attain a frequency of 2200 hertz and then hold that frequency pulse on the communication circuit for a time equal to 833 microseconds. The 833 microseconds timing factor is achieved by taking one second and dividing it by 1200 pulses per second; $1/1200 = 833$ microseconds. Also note that, as the modem is out-pulsing these different frequencies, it is doing it on the analog side of the modem. In other words, the modem has carried out its basic function of converting data from digital to analog or baseband to broadband.

DIGITAL/BIPOLAR

Digital signals are discrete on and off signals as contrasted with the continuous form of analog signal. Figure 2-24 shows various digital signals (baseband) starting with the older unipolar and moving through two types of bipolar signals. In the figure, digital signals are drawn as square signals instead of the smooth curve sine wave analog signal. Notice that when there is no signal in unipolar signaling, the voltage level is 0. In bipolar, the 1s and 0s vary from a plus voltage to a minus

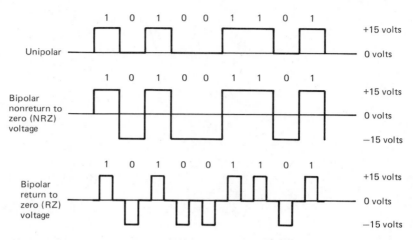

FIGURE 2-24   Unipolar and bipolar signals (digital).

voltage. The voltage ranges from 3 to 24 volts, depending on the equipment used. The AT&T/Bell DDS (Digital Data System) system uses bipolar signals. So does the RS232C connector cable. Figure 2-25 shows the RS232C voltage levels. In Europe bipolar signaling sometimes is called *double current* signaling because you are going between a positive and negative voltage potential. Bipolar is used because it permits faster baud rates than is possible with unipolar signals.

DIGITAL MODULATION

One digital technique of modulation is called *pulse modulation.* It is used for transmitting digital information directly. In other words, an electrical digital signal is not converted to an analog signal. One form of a digital signal might be converted to another form, such as NRZ to RZ, as shown in Figure 2-24, but it is still digital pulses that are being sent down the circuit. Pulse modulation requires a different set of modems.

Another type of digital modulation is *pulse amplitude modulation* (PAM), which gives a different height digital pulse for each different plus or minus voltage (see Figure 2-26). A second type is *pulse duration modulation* (PDM), which gives a longer timed pulse relative to the signal it is measuring. A third type, called *pulse position modulation* (PPM), has more pulses per unit time, depending on the power of the signal that it is measuring. These types of digital modulation normally are used to convert analog voice signals to a digital form for transmission on local loops and long distance interexchange channels/circuits (IXC).

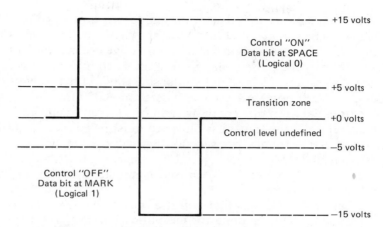

FIGURE **2-25** RS232C voltage levels. RS232C also defines the level and polarity of the signals going to and from the modem. A logical "1" is referred to as a "MARK" and a logical "0" is called a "SPACE."

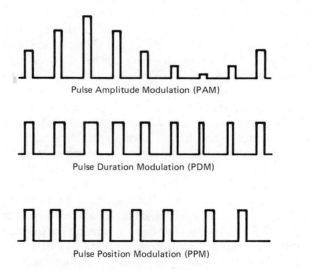

FIGURE **2-26** PAM/PDM/PPM.

Let us explore digital modulation to see how a voice conversation (analog signal) can be converted into a binary pulse stream (digital signal). Figure 2-27 illustrates how the original analog sine wave is converted into digital pulses of 0s and 1s by using pulse code modulation. Each individual sine wave (the top sine wave of Figure 2-15 is an individual sine wave) of the conversation is sampled at a rate of 8,000 times per second. Therefore, 8,000 frames per second travel down the line. If each frame contains eight bits (seven to encode the frame and one for control), then the transmission speed must be equal to 64,000 bits per second (8 bits × 8,000 frames per second). Basically, what you are doing is taking the original sine wave (top of Figure 2-27), encoding it with digital pulses, and redrawing it (bottom of Figure 2-27) at the distant site to which you are transmitting the voice conversation.

Figure 2-27 assumes that you use only three bits (instead of eight) to encode each frame. This is done for simplification to make it easier to understand; actually, eight bits are used in pulse code modulation. As demonstrated in Figure 2-27, a set of three binary numbers (such as 011) codes each one of the eight pulse amplitude heights. These bits are sent down the communication circuit to the distant station or terminal for reconstruction into the original sine wave. This serial transmission of binary pulses is referred to as *pulse code modulation* (PCM). Study this figure to see how each sample (the vertical bars are titled pulse amplitudes) is coded to PAM levels 1 through 8, but remember that in real life there are 128 levels.

After these voice conversations are digitized, 24 of them are combined into one high speed channel called a T-1 carrier. The T-1 channel data rate is 1.544 million bits per second, which includes 1.536 million bits per second of data and 8000 bits per second of synchronization. T-1 channels are then combined into T-2 channels, and the hierarchy goes up to a T-4 channel, all of which are covered later in this chapter.

A pulse code modulation technique such as this can be used to convert an analog voice signal to a digital stream of pulses for digital voice transmission. As might be expected, it is much easier to transmit data bits digitally because they do not have to be converted from an analog sine wave to a digital pulse. Some of the most recent equipment that is being marketed allows you to digitize an analog voice signal and then send the digital transmission over a single voice grade communication circuit with a bandwidth of 4000 hertz.

Incidentally, as you may have guessed by now, other 0s and 1s can be interleaved between the 0 and 1 digital signals from the transmission that we digitized in Figure 2-27. This means that a common carrier can convert from analog transmission to digital transmission and send two or more separate transmissions over a circuit that previously may have handled only a single voice call. In other words, bit multiplexing is used.

The signal (original sine wave) is quantitized
into 128 pulse amplitudes (PAM). In this
example we have used only eight pulse amplitudes
for simplicity. These eight amplitudes can be
depicted by using only a 3-bit code instead
of the 8-bit code normally used to encode
each pulse amplitude.

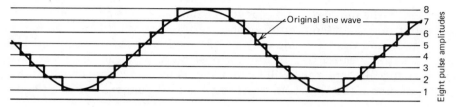

After quantitizing, samples are taken at
specific points to produce amplitude
modulated pulses. These pulses are then
coded. Because we used eight pulse
levels, we only need three binary
positions to code each pulse.[1] If we
had used 128 pulse amplitudes, then a
7-bit code plus one parity bit would
be required.

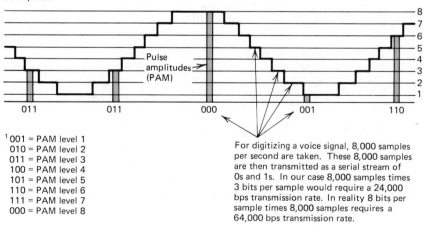

[1] 001 = PAM level 1
010 = PAM level 2
011 = PAM level 3
100 = PAM level 4
101 = PAM level 5
110 = PAM level 6
111 = PAM level 7
000 = PAM level 8

For digitizing a voice signal, 8,000 samples
per second are taken. These 8,000 samples
are then transmitted as a serial stream of
0s and 1s. In our case 8,000 samples times
3 bits per sample would require a 24,000
bps transmission rate. In reality 8 bits per
sample times 8,000 samples requires a
64,000 bps transmission rate.

FIGURE **2-27** Pulse code modulation (PCM).

BITS/BAUD

Bit and baud are terms that are used incorrectly much of the time. They quite often are used interchangeably, but there is a technical difference between them. In reality, the network designer or network user is interested in bits per second because it is the bits that are assembled into characters, characters into words, and thus business information.

A *bit* is a unit of information. A *baud* is a unit of signaling speed that is found by taking the reciprocal of the length (in seconds) of the shortest pulse used to create a character. The length of a pulse of Baudot code used with a 60-word-per-minute teletypewriter is 0.022 second; therefore, the baud rate is 1/0.022 = 45.45 baud (the plural of baud is baud). "Baud" and "bits per second" are not synonyms, but most data communication practitioners use them interchangeably. Bit rate and baud rate coincide only when a code is used in which all bits are of equal length. Because this is true in most cases, one can use these two words interchangeably and they will be understood by data communication people. In transmission technology, one pulse is generally equal to a single-bit state, for example, a 1200-baud circuit generally implies a transfer rate of 1200 bits per second. Because the use of the word "baud" has proved confusing, we do not use it in this book. We use "bits per second" exclusively.

Baud actually measures the signaling rate (pulse rate). The signaling rate is the number of times per second that the signal on the communication circuit changes. In other words, one simple method of increasing the rate of bits transmitted through a line is to combine pairs of adjacent bits into *dibits* ("di" = two). The object is to send each dibit as a separate signal element. Because the laws of information theory cannot be repealed, the process of transmitting two bits at the same time involves four (2^2) different signal states. One way to get four states is depicted in Figure 2-28. One signal state represents 00, another 01, another 10, last one 11. Figure 2-29 shows how, by varying the voltage between 0 and 3 volts, the transmitting device can send two bits simultaneously. This figure depicts the original bit stream of 0s and 1s which were represented by −5 volts for a 0 and +5 volts for a 1. Below that original bit stream is the bit stream of paired bits (dibits) and each of the four different voltage levels that represents each dibit. In other words, when the receiving device receives a +3 volts signal, it interprets

0	0	equals	0 volts
0	1	equals	1 volt
1	0	equals	2 volts
1	1	equals	3 volts

FIGURE **2-28**   Four states with 0 or 1 taken two at a time (baseband signal).

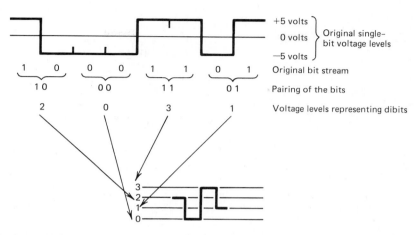

FIGURE **2-29** Dibits represented by different voltages.

that it has received the pair of bits "11." If the receiving device receives the +1 volt signal, it interprets that it has received a pair of bits "01."

In another example, if a modem is able to change between any two frequencies 1,200 times per second, then the signaling rate is 1200 baud.

$$\frac{1 \text{ second}}{1200 \text{ baud}} = 833 \text{ microseconds}$$

Figure 2-30 shows a modem that has a carrier wave of 1700 hertz. It goes up to 2200 hertz when transmitting 1s and down to 1200 hertz when transmitting 0s. (Our example here uses different frequencies.) If this modem operates at 1200 baud, it means the modem is able to switch between 2200 hertz and 1200 hertz at a speed of 1,200 times each second. This in turn means that the modem in Figure 2-30 has a switching rate of 1200 baud and also a bits per second rate of 1200.

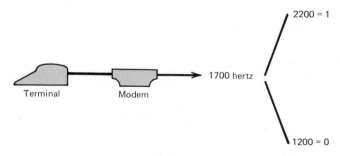

FIGURE **2-30** Single-bit transmission.

Dibits	Tribits	Quabits
00	000	0000
01	001	0001
10	010	0010
11	011	0011
4 Combinations	100	0100
	101	0101
	110	0110
	111	0111
	8 Combinations	1000
		1001
		1010
		1011
		1100
		1101
		1110
		1111
		16 Combinations

FIGURE **2-31**　Dibits/tribits/quabits.

Assume we want to transmit 2 bits at a time instead of 1 bit at a time. Figure 2-31 shows that if transmission is 2 bits at a time (dibits), you must have 4 different combinations. Transmission of 3 bits (tribits) at a time has 8 combinations, and 4 bits (quabits) at a time has 16 combinations.

Now we can determine where bits per second and baud are different from each other. Looking at Figure 2-32, notice that the modem now has four different frequencies that it can switch between 2200, 2000, 1400, and 1200 hertz). Also notice that each frequency now corresponds to a pair of bits rather than a single bit. This is still a 1200 baud modem, which means that it can switch between any of the four frequencies at a speed rate of 1,200 times per second. The difference now is that transmission is two bits at a time.

When the modem operates in the fashion of transmitting dibits, a 1200 baud modem is capable of transmitting 2400 bits per second. Other dibit modulation techniques are shown in Figure 2-33.

You might even imagine a modem that has 16 different levels of frequency. If it is able to switch among any of the 16 different levels at a 1200 baud rate, then it is transmitting at 4800 bits per second. This is because you need 16 different discrete identifiable signals to transmit 4 bits at a time. It should be noted that modem makers never use 16 different frequencies for this type of transmission; they probably use a combination of two or more types of modulation such as combining phase with amplitude modulation.

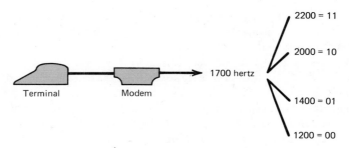

FIGURE **2-32** Dibit transmission using frequency shift keying (FSK).

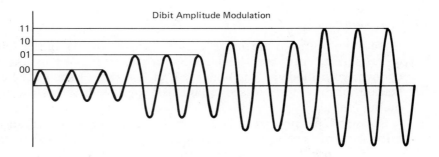

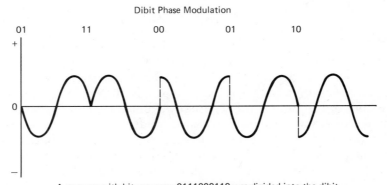

A message with bit sequence 0111000110 was divided into the dibit sequence 01 11 00 01 10, which modulated a carrier wave once per cycle and produced the phase modulated signal.

FIGURE **2-33** Dibit modulation schemes.

In summary, a baud is the signaling rate that tells you how many times per second the signal changes. By transmitting two, three, or four bits with each change of signal, the modem designer is able to transmit more bits per second than there are baud on the circuit.

The goal of a modem designer is to keep the baud rate as low as possible while making the bit rate as high as possible. The reasoning behind this is that if the baud rate is 1200, then the time available to identify the signal at the receiving modem is 833 microseconds (1/1200 = 833). If the baud rate is increased to 2400, then the time available to identify the incoming signal at the receiving modem is only 416.5 microseconds (1/2400 = 416.5). As a result, the receiving modem has only one-half the time to identify the signal when the baud rate is 2400 compared to when it is 1200. Finally, most vendor literature misuses the word "baud," so it has become common practice to think of baud as equal to bits per second, even though this is not correct technically.

STATION TERMINALS

A station terminal is the *terminal block* (leased circuits) or the voice jack (dial-up circuits) that terminates the local loop at your home or business. It is where you connect your modem. The upper half of Figure 2-34 shows the connection to a station terminal for a leased circuit; the lower half shows the connection for a dial-up circuit.

Our signal has now passed over a major boundary point with regard to responsibility for errors and the quality of circuits. It has left our modem and is now using the public local loop circuit of the telephone company or other common carrier.

LOCAL LOOPS

Local loops are sometimes called *subscriber loops.* These are the circuits that go between your organization and the common carrier facility. The newest digital termination system (DTS) local loops entail rooftop antennas that transmit directly from a home or office, microwave radio, cable TV handling digital signals, or radio frequencies such as those used in cellular radio local loops. In other words, the millions of miles of wire pair local loops that are strung along telephone poles or buried underground will be augmented or replaced over the next 20 years or less. One advantage these newer digital termination systems have over wires is that the cost of installation and maintenance is much lower.

Now our signal is on the local loop (see Figure 2-1) and on its way to the telephone company central office. At this point we have to learn about two/four-wire circuits, full duplex/half duplex, amplifiers, and station terminals as they

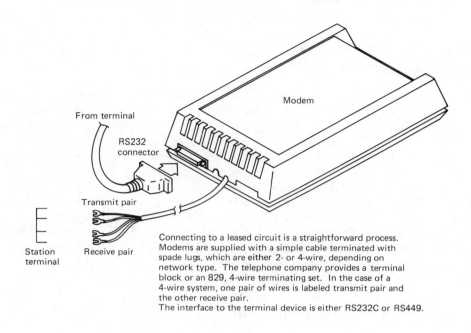

From terminal

RS232
connector

Transmit pair

Station
terminal

Receive pair

Connecting to a leased circuit is a straightforward process.
Modems are supplied with a simple cable terminated with
spade lugs, which are either 2- or 4-wire, depending on
network type. The telephone company provides a terminal
block or an 829, 4-wire terminating set. In the case of a
4-wire system, one pair of wires is labeled transmit pair and
the other receive pair.
The interface to the terminal device is either RS232C or RS449.

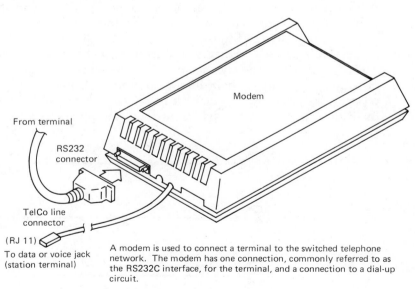

From terminal

RS232
connector

TelCo line
connector

(RJ 11)
To data or voice jack
(station terminal)

A modem is used to connect a terminal to the switched telephone
network. The modem has one connection, commonly referred to as
the RS232C interface, for the terminal, and a connection to a dial-up
circuit.

FIGURE **2-34** Connecting the modem to the local loop.

relate to the movement of a signal between a terminal and the central office, which might be several blocks or miles away.

TWO/FOUR-WIRE CIRCUITS

Your data can travel either on a two-wire circuit, where there are only two wires from modem to modem, or on a four-wire circuit, where there are four wires from modem to modem. The normal dial-up circuits are two-wire circuits, whereas private dedicated lease lines are four-wire circuits.

Two-wire circuits have a problem of *echoes*. When people talk on a two-wire circuit, echoes may occur under some conditions. Echoes arise in telephone circuits for the same reason that acoustic echoes occur: there is a reflection of the electrical wave from the far end of the circuit. The telephone company provides echo suppression circuits to stop echoes during voice conversation. An echo suppressor permits transmission in only one, fixed, direction. These echo suppressors open and close on two-wire lines during data transmission. Echo suppressors are described in Chapter 4.

FULL DUPLEX/HALF DUPLEX (FDX/HDX)

At this point your message can be traveling over the circuit in either full duplex or half duplex communications. Actually, there are three ways to transmit: simplex, half duplex, and full duplex (see Figure 2-35).

Simplex is one-way transmission such as you have in radio or TV transmission.

Half duplex is two-way transmission; however, you can transmit in only one direction at a time. A half duplex communication link is similar to a walkie-talkie link; only one "system" can talk at a time and the other must listen. Instead of using talk and listen buttons, computers use special *control signals*

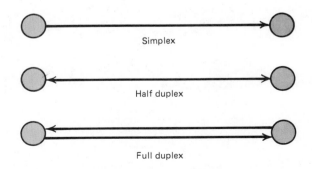

FIGURE **2-35**   Methods of transmission.

built into most serial interfaces, to negotiate which system will send data and which will receive data. The most commonly used control signals for this type of communication are REQUEST TO SEND (RTS) and CLEAR TO SEND (CTS). The amount of time it takes computers using half duplex communications to switch between sending and receiving is called *turnaround time.* For example, assume a message is sent from your modem to a distant modem. At completion of that message transmission, there is a certain amount of turnaround time while the modem at the receiving end changes from receive to transmit, as well as REQUEST TO SEND and CLEAR TO SEND delay.

With *full duplex* transmission, you can transmit in both directions simultaneously. Full duplex transmission does not require turnaround time because the transmit and receive are simultaneous.

Full duplex and half duplex are different from four-wire and two-wire. Full duplex and half duplex are communication methods; two-wire and four-wire are circuits. Full duplex requires a front end processor that has the proper software for simultaneous two-way communication. Also, the local loop and interexchange channels must be a four-wire circuit. Finally, with full duplex the terminal at your node must have the proper devices and functions so it can simultaneously receive and transmit data.

On the other hand, a two-wire or four-wire circuit is nothing more than a configuration that is supplied to you by the common carrier. All voice grade leased circuits are four-wire circuits. All dial-up circuits are two-wire circuits.

On a four-wire circuit it is easy to perform full duplex transmission (simultaneous transmission in both directions). If you want full duplex transmission on a two-wire circuit, you must use a special modem that creates two different frequency channels on the two wires, thus simulating a four-wire circuit (see Figure 2-36).

If you have a four-wire circuit, you also can keep both the send and receive signals from a modem operating simultaneously. This avoids the normal turnaround time for the modem's circuits because the REQUEST TO SEND signal leaves the modem on two of the four wires, while the receive side of the same modem is connected to the other two wires. This is *not* full duplex transmission; it is only a technique that is used to reduce the modem circuitry turnaround time (sometimes called *retrain* or *clocking time*) to zero. Your messages still are sent in half duplex with this technique.

AMPLIFIERS

Now that our message has left the modem and is in the local loop, the signal suffers attenuation. *Attenuation* is the weakening in strength of a signal as it passes down a wire. It is caused by resistance. For example, copper wire pairs can experience an attenuation or loss of signal strength because of the weather. This

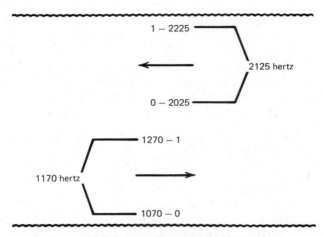

In one direction there is a carrier wave of 2125 hertz, using 2225 for 1s and 2025 for 0s. In the other direction there is a carrier wave of 1170 hertz, using 1270 for 1s and 1070 for 0s.

FIGURE 2-36   Full duplex on a two-wire circuit.

occurs because electrical resistance of wires rises with the temperature. Wet/humid conditions likewise increase attenuation because signal leakage occurs at insulators when they are wet. It should be noted, however, that there is an inherent resistance in any communication media, whether it is a copper wire pair, coaxial cable, microwave link, or the like.

The telephone company places repeater/amplifiers at 1- to 10-mile intervals in order to increase signal strength lost to attenuation. The amplifier and its associated circuits are referred to as a *repeater* because it repeats the signal while increasing the signal's strength. Technically, repeaters do not filter out unwanted noise, but amplifiers do. The distance between the repeaters/amplifiers depends on the degree of attenuation because the signal strength cannot be allowed to fall too low. If too much attenuation occurs, it is increasingly difficult to distinguish it from other noise or distortion that always is present on communication circuits.

On analog circuits it is important to recognize that the noise and distortion that always are present on the circuit also are amplified, along with the increase in signal strength. This means that some noise from a previous circuit link is regenerated and amplified each time the signal is amplified. This amplification has a definite effect on errors, creating the need to retransmit messages. Digital circuits are cleaner and more error free than analog circuits. A digital circuit with its associated digital amplifier re-creates a new signal at each amplifier station.

For this reason, the noise and distortion that are on the previous link of the network are not amplified each time the digital signal goes through an amplifier. This gives a much cleaner signal and results in a lower error rate for digital circuits. The benefit to users is that fewer messages have to be retransmitted because of errors.

TELEPHONE COMPANY CENTRAL OFFICE

The telephone company central office (see Figure 2-1) is the switching center of the telephone company or other special common carrier. At the central office, you can have the common carrier perform circuit switching, or you can have your circuits wired around the electronic switching system. Dial-up circuits are switched, and leased circuits are wired around the switching equipment. The central office also might be called *end office* or *exchange office*.

SWITCHING

Most common carriers today can offer only circuit switching such as we have in the dial-up telephone system. By this method you cannot complete your call if the other telephone or circuit is busy. On the other hand, if you use a public packet switching common carrier, it can perform packet or message switching because it can hold your message in a store and forward area and forward it when the other terminal is available. If you are using *leased circuits* (also called *private* or *dedicated*), your communication circuit is wired around this switching equipment. In other words, there is a direct circuit path from station terminal to station terminal. This method offers a much cleaner circuit with less noise and distortion, and with fewer retransmissions of messages because of errors. Our signal probably travels over wire pairs, coaxial cables, and microwave as it moves across the country or around the world. You know if it goes on a satellite circuit because of *propagation delay,* which is the time necessary for the signal to travel from the earth to the satellite and back (approximately 0.50 second for both directions).

INTEREXCHANGE CHANNELS (IXC)

Our signal has now moved from the telephone company central office to the interexchange channels or long haul communication circuits. At this point, we are using microwave circuits, satellite circuits, wire pairs, coaxial cables, or any other type of circuit media that is available.

CIRCUITS/CHANNELS

A *circuit* is nothing more than the path over which data moves. Many people use the word line interchangeably with the word circuit, although line implies a physical wire or glass fiber connection. Circuit is more appropriate when you are speaking of satellite and/or microwave transmission. Sometimes the terms *circuit*, *channel*, and *line* are used interchangeably.

Often an individual communication circuit is subdivided into separate transmission subchannels, as is done with multiplexing. Some users refer to a channel when they are speaking of a single transmission facility as well as when they are speaking of a circuit that has been subdivided into numerous channels (more correctly, *subchannels*).

Also, there is the term *link*. There can be many links in a cross-country communication circuit because a link is any two-point segment of a communication circuit. A *two-point segment* is a circuit that goes between point A and point B (any two terminals or microcomputers). Therefore, a multidrop circuit has many links as it traverses the country. Voice telephone lines are sometimes referred to as *trunks* or *trunk lines*.

Many types of transmission media are in use today. There are also very exciting aspects in store for the future as the various transmission media enter our homes for direct communication through such techniques as videotex, electronic mail, voice mail, and two-way cable TV. The discussion below defines each type of transmission media or circuits used to transmit data.

Open Wire Pairs   These are copper wires suspended by glass insulators on telephone poles (see Figure 2-37). They are spaced approximately 1 foot apart. While they are still familiar in many areas, they are being replaced by cables and other more modern transmission media. Open wire pairs are fast becoming a part of the past.

Wire Cables   These are insulated pairs of wires that can be packed quite close together (see Figure 2-38). Bundles of several thousand wire pairs are placed under city streets and in large buildings throughout the country. In fact, your own house probably has a set of four wires connecting your telephone to the telephone company central switching office. Wire cables usually are twisted (twisted wire pairs) in order to minimize the electromagnetic interference between one pair and any other pair in the bundle. Wire cables also are being replaced by more efficient transmission media such as coaxial cable, microwave, satellite, and optical fibers.

Coaxial Cable   Figure 2-39 shows a single coaxial cable and a bundle of coaxial cables. Each individual coaxial cable consists of copper in the middle with an

FIGURE **2-37** Open wire pairs.

FIGURE **2-38** Wire cables.

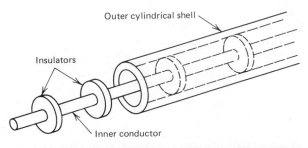

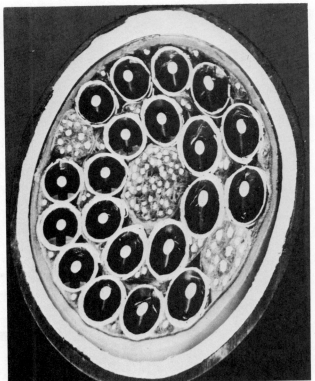

FIGURE **2-39** Coaxial cables. Exterior view (top) and cross-sectional view (bottom) of a bundle of 20 coaxial cables.

FIGURE **2-40** Microwave tower.

outer cylindrical shell for insulation. This type of circuit can transmit at substantially higher frequencies than a wire pair. Therefore, it is far more efficient to use a coaxial cable because it can contain many telephone conversations. A 2-inch-diameter bundle of coaxial cables can handle approximately 20,000 voice or data telephone calls simultaneously. Coaxial cables have very little distortion, cross talk, or signal loss; therefore, they are a better transmission medium than either open wire pairs or bundles of wire cables.

Microwave Transmission This transmission medium is the one most used for long distance data or voice transmission. It does not require the laying of any cable because long distance dish or horn antennas with microwave repeater stations are placed approximately 25 to 30 miles apart (see Figure 2-40). It is a line-of-sight transmission medium. A typical long distance antenna might be 10 feet across, although over shorter distances in the inner cities these antennas might

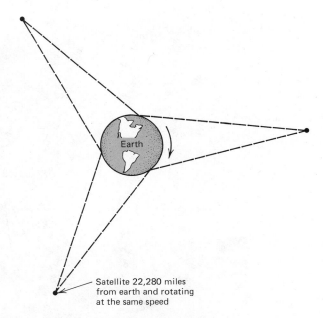

Satellite 22,280 miles
from earth and rotating
at the same speed

FIGURE 2-41 Geosynchronous satellites in space.

get down to 3 feet in diameter. In larger cities we now have microwave congestion; so many microwave dish antennas have been installed that they interfere
with each other, and the air waves are saturated. This problem will force future
users to seek alternative transmission media such as satellite or optical fiber
links.

Satellite This transmission medium is similar to microwave transmission except, instead of transmitting to another nearby microwave dish antenna, it transmits to a satellite 22,300 miles out in space. Figure 2-41 depicts a geosynchronous
satellite in space. Figure 2-42 shows the satellite in operation.

One of the disadvantages of satellite transmission is that a delay occurs because the signal has to travel far out into space and back to earth (propagation
delay). A typical signal propagation time is approximately 0.5 to 0.6 second for
the delay in both directions. In addition, there may be a delay for going through
ground-based switching equipment. This delay is controlled by the common
carriers to avoid disruptive voice telephone conversation. In data communications, however, this type of delay is disastrous if you are using a protocol with the
stop and wait ARQ. This is so because when an individual message block is sent,
you have to wait for a positive or negative acknowledgment before sending the

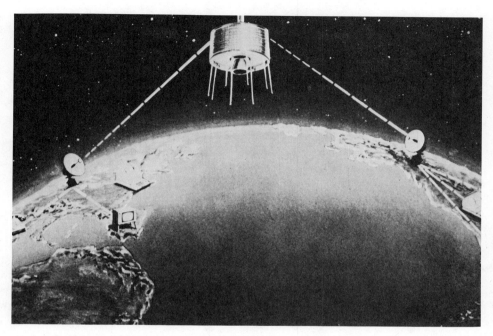

FIGURE **2-42**   Satellite in operation transmitting to its ground station antennas.

next message block. As you will see in Chapter 5, the newer software allows you to send a burst or group of messages, and then the called party responds to this group or burst (continuous ARQ). (You might go to Chapter 5 and read the section titled ARQ.)

This message delay, or path delay, with stop and wait ARQ must be taken into account when half duplex, stop and wait transmission protocols are utilized. For example, modems provide a variable timer to set the CLEAR TO SEND delay. The value of this timer usually is set to be greater than the round-trip delay (propagation time) of the overall path. In the case of a satellite, this path length is approximately 22,300 miles from earth to the satellite and 22,300 miles return, plus a possible several thousand miles more on terrestrial links between the satellite earth station and wherever your message is to be delivered. This propagation time easily can *exceed* 0.5 second. For this reason, a common practice is to set the CLEAR TO SEND delay on the modem to 700 milliseconds for satellite transmission. On a terrestrial-based circuit, this CLEAR TO SEND delay might be set for only 50 milliseconds.

At this point you might wonder, "What is the CLEAR TO SEND delay?" The *CLEAR TO SEND delay* in a modem is the time from when it stops transmitting

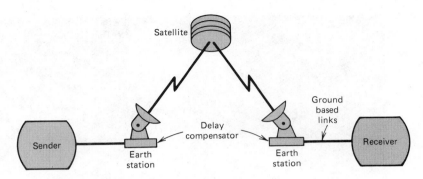

FIGURE **2-43**   Satellite delay compensator.

and waits for return of the acknowledgment that tells the modem whether the message was received correctly (ACK) or incorrectly (NAK) at the distant end of the telecommunication path. Remember, a message is sent, then transmission stops and waits until either an ACK or a NAK is received. Only then can the next message be sent. This wait time is set by the CLEAR TO SEND timer on the modem. Even on a terrestrial-based circuit it might vary because there is a shorter propagation time for a message that is sent 10 miles than for one that is sent 3,000 miles.

With that in mind, it is reasonable to ask how you can reduce the delay time of 700 milliseconds without going to full duplex transmission or changing your software protocols. The answer is that some networks have *delay compensators* in which the acknowledgment is returned by the local delay compensator before the message is sent on the satellite link. Suppose the sender in our example (see Figure 2-43) sends Message 1. The delay compensator at the sender side of the satellite circuit sends an acknowledgment back to that sender. The sending terminal then is able to send Message 2 immediately, for which an acknowledgment is received immediately. After that, Message 3 can be sent immediately, and it also receives another immediate acknowledgment from the delay compensator at its end of the circuit. By the time the third message has been sent, the receiver (the other end of the satellite circuit) may be ready to send back the actual acknowledgment to the message stating whether it had been received correctly or incorrectly. Because the delay compensator provides the sending station with an acknowledgment before the data takes the satellite hop, the sender can immediately follow the first transmission with the second, and so forth.

Satellites use different frequencies for receiving and transmitting. These frequencies are in the ranges of 4 to 6 gigahertz (GHz), 12 to 14 GHz, and 20 to 30

GHz.[4] One gigahertz is equal to 1 billion cycles per second. The older *C-band* transmits in the 4 to 6 GHz and therefore requires a larger dish antenna.

Ku-band is the transmission spectrum between 12 and 14 GHz. In addition to data transmission, the Ku-band is used to transmit television programs between various networks and individual television stations. One of the problems encountered when transmitting data or television using Ku-band is *raindrop attenuation*. This problem occurs because waves that high in the spectrum are so short that they can be absorbed by raindrops. It is not really a major problem but just something that engineers need to work around. Currently, COMSAT (the satellite company) guarantees 99.99 percent reliability. One advantage of Ku-band is that the waves are so short they can be caught and concentrated in much smaller dish antennas, such as those householders put on their roofs, thus permitting direct transmission between a satellite and the home. It is estimated that a dish antenna ranging from 30 to 48 inches in diameter is all that would be required for direct Ku-band transmission to your home.

Unlike the C-band, which has been the predominant commercial satellite bandwidth, the Ku-band is a high enough frequency that it does not interfere with ground-level transmissions, such as microwave and radar. This means a Ku-band dish can be installed without a Federal Communications Commission license. In addition, very small aperture terminals (VSAT) can be utilized with the Ku-band. VSATs are small, inexpensive satellite dishes that significantly lower the cost for an organization that wants to enter a satellite network. A typical Ku-band satellite network provides either one-way or full duplex communications among VSATs installed at a large number of remote branch offices and a larger earth dish or "hub" installed at a central site. A receive only VSAT dish may cost several thousand dollars, while an interactive (full duplex) dish antenna may cost about $9,000. Major earth dish "hubs" can cost as much as several hundred thousand dollars. Companies such as Tymnet, McDonnell Douglas Network Systems, and AT&T Communications offer shared hub services to customers. This means that users only have to buy their own inexpensive VSATs and not a significantly more expensive central hub dish. To put this cost in perspective, one-tenth of a satellite transponder can be leased for about $13,000 per month. This one-tenth of a transponder can support a 120-node full duplex network. (Transponders receive incoming signals, and retransmit them back down to earth.)

Current satellite networks are used primarily to provide information services through data that is transmitted or received. Among the largest satellite users are

---

[4] 1 Hz (hertz) equals 1 cycle per second.
1 KHz (kilohertz) equals 1,000 cycles per second.
1 MHz (megahertz) equals 1 million cycles per second.
1 GHz (gigahertz) equals 1 thousand million cycles per second.

news wire services, television networks, stock quotation services, and similar business data functions in which there is a download system. Through satellites, central television networks download news stories to subscribing television stations around the country or around the world. Similarly, stock quotations might be downloaded through a satellite network from the New York Stock Exchange to the ultimate user of that information.

Security poses a serious problem for satellite communications because it is easy to intercept the transmission as it travels through the air. Laser communication systems, which have begun to revolutionize long distance data transmission on earth, now are moving into the satellite arena. Whereas terrestrial-based fiber optic networks send pulses of laser light down hair-thin fibers of glass, future satellites will exchange information by transmitting modulated laser beams across thousands of miles of empty space. Such laser intersatellite link (ISL) systems currently are under development in West Germany, France, and the United States. These systems could be deployed on satellites by the early 1990s.

Laser ISL technology is a cousin of microwave communications. Information can be carried on a beam of laser light in much the same way that it is modulated onto a microwave signal. Whereas microwaves use a parabolic antenna, a laser system uses an optical telescope to concentrate and aim the beam of light. At the target satellite, a receiving telescope captures and diffuses the laser signal, concentrates it, and focuses it onto a detector. One major application of laser ISL technology is to cross-link military satellites in geostationary orbits at 22,300 miles altitude so that they can communicate between themselves without having to rely on a vulnerable ground station back on earth.

Another issue is satellite crowding in space. There is now both orbit and frequency congestion among the various satellites. Figure 2-44 depicts a band of satellites around the earth, first as if you were looking from space, and then from earth. You can place only so many synchronous satellites around the band. Notice that this figure shows a 4° separation between each satellite. If you brought them too close together, they would touch. Because of the frequency congestion—most of the available frequencies are in use today—we will have to launch larger satellites (fewer required) in the future, while also expanding available frequencies to meet future needs for satellite data, voice, and image transmission. Looking at the top half of Figure 2-44, find your location on the globe. Then chart an imaginary line from where you are to one of the satellites. Notice how this line tends to go directly south, or possibly a little southeast or southwest, if you are in the United States. It is for this reason that satellite dishes in the United States usually face in a southerly direction and possibly a little southeast or southwest, depending on which satellite is being used for signal reception.

Another example of the use of satellites would be interconnecting two geographically separated local area networks (LANs) over a satellite link (see Figure 2-45). There would be a satellite earth station, comprised of a 3.7-meter antenna,

from Space . . .

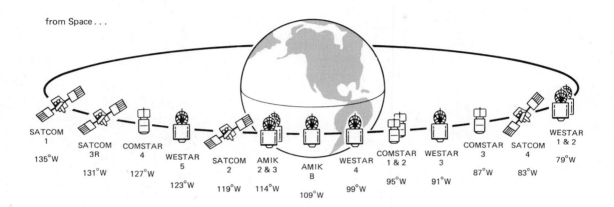

from Earth . . .

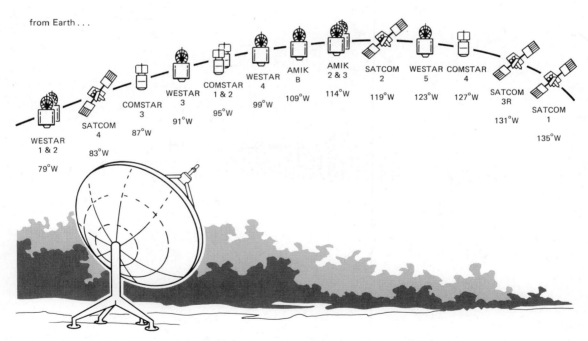

FIGURE **2-44** Geosynchronous satellite band around the equator.

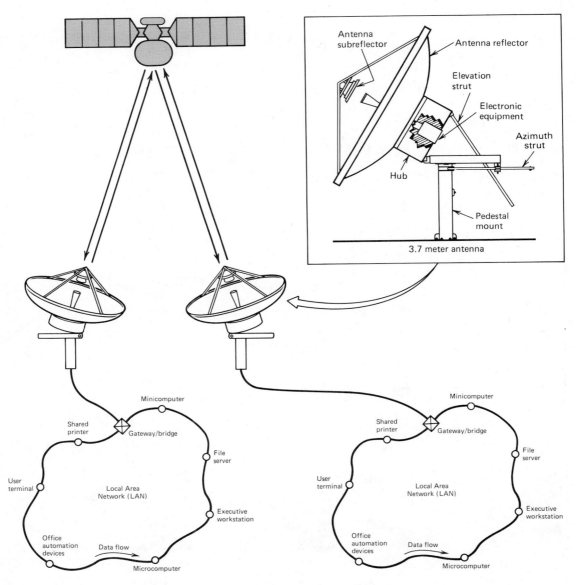

FIGURE **2-45**   Satellite connecting two local area networks (LANs).

at each LAN location. For a large organization it would be easy enough to interconnect more than 100 different geographically separated sites using this earth station technique.

Fiber Optics Fiber optics is the newest of our technologies for the transmission of data, voice, and images over a continuous line. Instead of carrying telecommunication signals in the traditional electrical form, this technology uses high speed streams of light pulses that carry encoded information inside hair-thin strands of glass called optical fibers. At the end of their journey, solid state electronics reconverts these pulses of light back into electrical signals so they can be processed by conventional microcomputers, terminals, and host mainframe computers. Fiber optic technology is a revolutionary departure from the traditional message-carrying systems of copper wires and microwave radio signals. One of the main advantages of fiber optics is its large bandwidth which enables it to carry huge amounts of information. This capacity makes it an ideal system for the simultaneous transmission of voice, data, and video (image) signals.

As noted above, the optical fibers carrying the pulses of light are made of hair-thin strands of plastic or glass. Plastic is falling into disuse, however, because glass can be made into a purer product, enabling the signal to be transmitted over a longer distance before an amplifier/repeater station is needed to increase the signal strength. A new, even purer type of glass, called halide glass, will bring improvements in the distance between amplifiers for fiber optic communications. Halide glasses are made from thorium, lanthium, and lutecium, whereas conventional glasses are made from silicon dioxide (sand). Because halide glass is purer, light can travel farther before it attenuates enough to require amplification. The purer the glass, the less often the signal has to be amplified. For example, a windowpane made of halide glass could be 125 miles thick and still be translucent. Not having to amplify the signals as often would be an extremely important development because of the current problems in amplification of light signals.

Other uses for halide glass are in the military. The Navy could use these glass fibers to transmit information from undersea sensors, such as those used to detect submarines; the enemy submarine would be unaware of the transmission because there is no electromagnetic radiation from optical transmissions. The Air Force would be interested because conventional glass turns dark when exposed to radiation, meaning that optical communication could be lost following a nuclear attack. Halide glass shows promise of being able to withstand radiation because it does not turn dark.

When light signals are amplified today, they must first be converted back to electronic signals, amplified in power, and then reconverted to a light pulse for transmission down the next link of the fiber optic communication circuit. Not only does this process slow down the signal somewhat, but it also entails costly amplification equipment. Light amplifiers are being developed in various organizational research and development departments that can take a pulse of light, amplify its power without its having to be converted to an electrical signal, and pass it down the next link of the fiber optic circuit. Furthermore, having several thousand miles between amplifiers would significantly decrease the cost and complexity of undersea cables.

British Telecom (England's major telephone company) recently demonstrated the first all-optical light amplifier. This amplifier, which still is experimental, amplifies and retimes light pulses directly, so they need not be converted back to electricity first as do conventional amplifiers. It operates at 140 million bits per second.

One of the current technological barriers with regard to optical fibers involves the inability of a pulse of light to switch between different communication circuits. Although fast by most standards, electrical impulses move through computer chips at only about two-thirds the speed of light. There is no problem in transmitting light signals down an optical fiber, but if they could be switched as optical light pulses (rather than electrical pulses), we could gain a 30 to 50 percent increase in speed. Currently, an optical switch actually converts the signal back to an electrical pulse, switches it, and reconverts it back to an optical pulse for further transmission.

The process of switching without converting the light signal back to electricity is called *photonic switching*. Photonic switching diverts the light signal from one fiber optic line to another by changing a field of current between the incoming and outgoing circuit ports. These photonic switches are fashioned from materials such as gallium arsenide through which paths called *waveguides* are carved. The light beams travel down these waveguides. These waveguides are placed 5 to 10 microns apart, or about one-tenth the width of a human hair. Problems arise when researchers try to build switches capable of handling more than 8 to 10 light beams. The photonic switches of today are still in the research laboratory, and they do not operate as fast as their electronic switch counterparts, although this limitation should be overcome soon. Even though a photonic switch may not be able to match the speed of an electronic switch, it does increase system reliability by doing away with the electronic conversion equipment.

AT&T and 21 other telecommunication companies are constructing the first undersea cable to span the Pacific Ocean with a laser-powered, digital light wave communication system. The Hawaii 4/Transpac 3 fiber optic system is tentatively scheduled to begin service in January 1989. This light wave system will stretch nearly 7,200 nautical miles across the Pacific and will require nearly 250 undersea amplifiers located approximately 30 miles apart. This fiber optic cable system is designed to handle approximately 37,800 simultaneous telephone calls.

Manufacturers all seem to claim different fiber optic transmission speeds, but the Bell Laboratories has reported a speed of 420,000 bits per second, without any amplification, over a 203-kilometer-long fiber optic cable. Bell Laboratories is using a new type of low loss optical fiber cable to achieve this speed. In a second experiment, a 2-gigabit signal was sent 130 kilometers without any amplification. AT&T's new low loss optical fiber uses a new cladding around its core, which reduces light loss to such a degree that, after traveling 200 kilometers, the emerging signal is ten times stronger than with traditional optical fiber cables.

Figure 2-46 shows a fiber optic cable and depicts the idea of the optical core, the cladding, and how light rays travel in optical fibers. The optical core is very pure glass over which the pulses of laser light or light emanating from LEDs (light-emitting diodes) passes. The cladding is heat fused around the optical core and consists of glass that has a different refractive index, thereby making the light signals bounce down the optical core. Earlier cables, called *step index cables*, had mirrorlike coatings around the optical core or a coating of plastic to cause this reflection.

An average cable bundle contains about 72 fibers, and a large cable contains about 144 fibers. The earliest fiber optic cable systems were multimode, meaning they carried several light waves down the fiber simultaneously. But multimode cables were plagued by excessive signal attenuation (reduced optical intensity) and dispersion (spreading of the optical pulse). Single mode optical fiber cables transmit a single direct beam of light. It achieves higher performance, in part because the core diameter has been reduced from 50 microns to about 8 to 10 microns. This smaller diameter core allows the fiber to send a concentrated light beam farther than multimode because the light strikes the core/cladding boundary at a much smaller angle, causing less attenuation and dispersion.

The construction of an optical cable includes more than the optical core and cladding. Typically, the optical core is surrounded by its cladding, then protected by a layer of Kevlar (bulletproof vest material), and finally jacketed with some type of thermoplastic covering.

Optical transmitters convert electrical energy from the computer or terminal into light pulses that are coupled into, and transmitted through, the optical fiber. The light sources commonly used are either light-emitting diodes (LEDs) or injection laser diodes (ILDs). LEDs come in two types, edge and surface emitting, with the former generally considered the more reliable and powerful. Laser diodes are used for longer distances. ILDs are more expensive than LEDs, but their service life generally is less by a factor of ten. Edge-emitting LEDs are estimated to be capable of lasting over 1,000 years. Both LEDs and ILDs are available in 830-nanometer and 1,300-nanometer wavelengths. LEDs can deliver coupled power into a fiber of up to 1,000 microwatts, while ILDs typically supply power levels of 1,000 to 3,000 microwatts with a resulting ability to send the signal over a greater distance before any amplification is required.

Fiber optic cable is an attractive security measure because it is almost totally immune to unauthorized access by tapping. Taps can be made only by someone

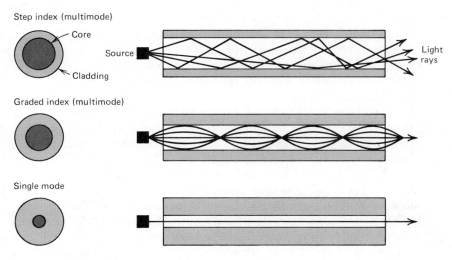

FIGURE **2-46** Fiber optic cable.

breaking the cable, polishing it off, and inserting a splice, or by nicking into the core to detect the light. The first method, which uses a T-splice adapter, gives a detectable power loss in fiber optic systems, so it can be detected. The second method, nicking, might be possible with a step index cable, which has a silica core and plastic cladding around it. The plastic cladding can be nicked so the light leaks out, although if too much light leaks out, the signal is lost. The nicking technique is almost impossible to accomplish in a graded index fiber because the optical core and the cladding are heat fused together. They are both glass, but of different refractive indexes.

Splicing optical fibers is one of the technical problem areas that still needs a lot of work. Although techniques are improving, until very recently an oscillo-scopelike device and/or a microscope was needed to tell the installer when the two optical fiber cable ends were aligned. Splicing an optical fiber requires that a technician align the two fiber optic cores perfectly; otherwise some light leaks out. The optical cores must be joined precisely along six different dimensions: not only the x, y, and z dimensions, but also three angular dimensions. If the installer is not precise enough, light is lost and the signal degrades. In addition, the connections have to be strong enough to last thousands of hours while with-standing vibration and constant changes in temperature.

The basic technique for splicing, called *fusion splicing*, involves cutting the cable, polishing the ends of the splice, then fusing them with a high voltage electricity torch. Given the thinness of the fiber, this process usually is done under a microscope by a well-trained technician. To lower the cost of this proce-dure, a number of companies have developed simpler methods. For example, General Telephone & Electronics has a kit that uses a diamond cleaver to cut the fiber, thus eliminating the need to polish the ends of the splice. The technician fits the cut fibers together, places a plastic collar over them, and pulls the fiber ends together in a device that aligns them automatically along all six axes. Other splicing kits also are on the market, and as these kits become simpler, the use of optical fiber cables will increase further.

Optical fibers are immune to electrically generated noise and therefore have a very low error rate. For example, a fiber optic cable might have a bit error rate of 10^{-9} (1 bit out of 1 billion) as compared with a 10^{-6} (1 bit out of 1 million) error rate found in metallic lines. Optical fibers also provide complete isolation be-tween transmitters and receivers, thus eliminating the need for a common ground. This provides electrical isolation from hardware and eliminates problems such as ground loop within an installation. For communications in a dangerous atmosphere, such as a petroleum refinery or a paint factory, it has another advan-tage because static spark is eliminated. For this reason, optical fiber cables may be the ideal solution in the robotic-controlled factory of the future. Communica-tions within a factory, where many machines are operating, can be degraded severely by electrical noise, static electricity, and voltage fluctuations.

The small size and light weight of fiber cables offer users better opportunities to secure this medium physically. Because fiber optic cable is nonconductive, it is free from electromagnetic noise radiation and therefore is completely resistant to conventional passive tapping techniques. With passive techniques you do not actually cut into the communication medium but only get in close proximity to it. Finally, in most cases fiber optic cable is less restricted under harsh environmental conditions than its metallic counterparts. It is not as fragile or brittle as might be expected, and it is more resistant to corrosion than copper. The only chemical that affects optical fiber is hydrofluoric acid. Also, in case of fire, an optical fiber can withstand greater temperatures than copper wire. Even when the outside jacket surrounding the optical fiber has melted, a graded index fiber optic system still can be operational in an emergency signaling system. One word of warning is in order, however. Care must be taken when these cables are pulled through a building so that the cable is not separated; its tensile strength may not be as great as the tensile strength of some other cables, such as coaxial.

One of the biggest advantages of fiber optic cable is the bandwidth because this provides the potential of transmitting data at speeds up to 10^{14} bits per second. These bandwidth frequencies are 10,000 times greater than the upper ranges of radio frequency bands (microwave and satellite).

Optical fibers require extremely precise manufacturing techniques because a single finished fiber is less than half as thick as a human hair. Moreover, mere traces of impurities can cause severe signal attenuation. There are two main types of optical fiber fabrication: modified chemical vapor deposition (MCVD) and outside vapor deposition (OVD). MCVD is an inside method of fabrication in that it deposits the core material (typically in the initial form of metal halide vapors) on the inside of a cylindrical silica substrate tube. The vapors deposited inside the tube (which has a $\frac{1}{2}$-inch center hole and a $\frac{3}{4}$-inch outer diameter) eventually form the core of the fiber, and the tube itself forms the cladding. Once the chemical deposition is complete, the tube is heated to a molten state and stretched by a computer-controlled fiber drawing process into its final form. A 2-foot tube can yield more than 30 miles of optical fiber. MCVD is the preferred method of fabrication for both multimode and single mode fiber because of its consistency in high volume production.

In the OVD process, the metal halide vapors react in an open environment, producing a fine-grained white soot. The soot is blown against a long ceramic rod that rotates about its axis. Once the proper diameter is achieved, the blank is cooled and separated from the rod. The porous blank then passes through an oven in which cleansing gases remove any impurities. The blank is then heated and stretched through the standard drawing process.

Fiber optics will not be confined to the cable market. IBM has developed a new experimental fiber optic computer chip which is capable of receiving data from input/output devices, such as printers and terminals, at a speed of 4 million bits per second. At this rate, the entire text of a 20-volume encyclopedia could be transmitted via fiber optic cables in less than three seconds. The new chip uses hair-thin fiber optic lines to receive laser light pulses that convey computer data signals between the main computer and its peripherals. The chip serves as part of the connection between the optical light pulses and a computer's memory. A separate photodetector converts the optical pulse to an electrical signal. While this experimental fiber optic chip operates four times as fast as its predecessor and receives optical pulses directly, it is nowhere near where we will be going. In the future, you will see microprocessor chips (computers) that operate completely on light pulses, not electricity.

Cellular Radio   Traditional mobile telephones win no awards for being there when you need them. The reason is that 20 channels typically are being shared by 2,000 subscribers, and there are times when you cannot get a connection or even a dial tone. This situation is changing because of cellular radio, a form of high frequency radio in which antennas are spaced strategically throughout a metropolitan area (see Figure 2-47). A service area or city is divided into many cells, each with its own antenna. This arrangement generally provides subscribers with reliable mobile telephone service of a quality equal to that of a hardwired telephone system. Users (voice or data transmission) dial or log into the system, and their voices or data are transmitted directly from their automobile, home, or place of business to one of these antennas. In this way, the cellular radio system is a replacement for the hardwired local loop.

This system has intelligence. For example, as you drive your automobile across the service area or city, you move away from one antenna and closer to another. As the signal weakens at the first antenna, the system automatically begins picking up your signal at the second antenna. With cellular radio, therefore, transmission is switched automatically to the closest antenna without communication being lost.

Today cellular radio is used widely by radio paging companies and users of mobile telephones. Future applications might include remote troubleshooting. Your automobile mechanic, for example, could use cellular diagnostic equipment to poll remotely your car's onboard computers and identify the cause of a problem. Most people think of a car telephone when anyone mentions cellular equipment, but in the very near future you will see many more applications. Cellular pay telephones will appear in taxicabs, buses, and trains (they already are in airplanes). Burglar alarm systems will use cellular radio because when burglars defeat alarm systems, 90 percent of the time they do it by cutting the telephone wires. Paramedic rescue units already use cellular radio.

The real future of cellular radio is based on the philosophy of dividing the

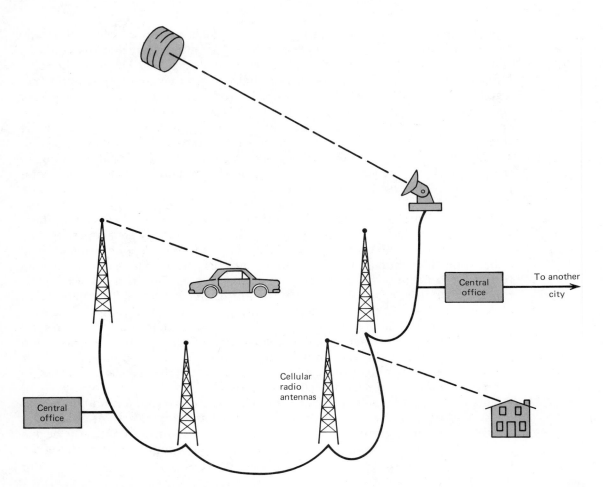

FIGURE **2-47** Cellular radio system.

entire United States, or maybe the world, into various cells so that a person can call from anywhere to anywhere. It would be easy to interconnect the cells of the entire United States with the cells of, let us say, France by using either satellite communications or fiber optic undersea cables. Just as most of us today think of a telephone as a fixed-location device (located in our home, school, place of business, or specific pay telephone booth), in another ten years the telephone will be a personal item. You will carry it with you, just as you might carry a calculator or a microcomputer today. Thirty years ago no one anticipated that we all would have electronic calculators small enough to carry with us. In the future, you will have small portable telephones, with the number assigned to you, and the telephone unit will be carried with you rather than connected to the wall of your home.

Miscellaneous Circuit Types   Other types of circuits are waveguides, tropospheric scatter circuits, short distance radio, and submarine cables.

A *waveguide* is a conductive tube down which radio waves of very high frequency travel. These tubes may be from 2 to 15 inches across. Waveguides usually are placed in the back of microwave towers to transmit the signal from the repeater/amplifier of electronic equipment to the dish antenna. They are used over very short distances, 10 feet or less. The American Telephone & Telegraph WT4 waveguide system can carry 230,000 two-way telephone calls simultaneously.

The *troposphere*, which extends upward from the earth about 6 miles, scatters radio waves and can be used for communication links of up to about 600 miles. *Tropospheric scatter* is especially useful in the South Pacific, where islands are widely scattered and separated by long distances. One problem with this method of transmission is that it requires very large antennas, on the order of 60 × 120 feet. Tropospheric scatter circuits can be used to transmit television over shorter distances, and up to several hundred voice communications can be transmitted over a link of approximately 100 miles. Tropospheric scatter circuits are not advised for data transmission because of the high error rate encountered. Obviously, microwave links, or even satellite circuits, are preferable.

Short distance radio is the type of transmission circuit that is used by walkie-talkies, police radio, taxis, and other community services. Depending on the distance involved, short distance radio might be used on the cellular radio circuits or it might be a type of independent high frequency radio.

Submarine cables are transmission circuits that go between the various continents. The most popular form in use today is a bundle of coaxial cables. Currently, Bell Laboratories is developing a high speed underwater optical fiber cable. This cable probably will come into use by 1988, and it may be able to handle transmission speeds of 274 million bits per second, which would accommodate 4,000 voice channels.

The *FM band* is two 10-kilohertz-wide subchannels on the commercial FM broadcast band that normally is used by FM radio stations. These two bands can be used for broadcast transmission of business data at rates of from 9600 to 19,200 bits per second. One advantage of this subcarrier technology is price. The receiving terminals cost about $100, and, unlike receive only stations used in satellite broadcasting systems, subcarrier terminals can be handheld devices that travel with the users. Of course, a satellite network has a far greater geographic range, but a single FM subcarrier can cover a 20- to 40-mile radius. One of the uses currently available is the broadcast of stock market price quotations to users with the handheld FM receiving device.

T Carrier	Number of Voice Channels		Speed (bits/second)
	PCM	ADPCM	
T-1	24	48	1,544,000
T-1C	48	96	3,152,000
T-2	96	192	6,312,000
T-3	672	1,344	44,376,000
T-4	4,032	8,064	274,176,000

FIGURE 2-48   T carrier system.

T-1 CARRIER

The *T carrier system* is the North American telephone industry standard for interconnecting digital communication systems. It is a hierarchy of digital transmission and multiplexing standards ranging from T-1, which operates at a data rate of 1.544 megabits per second, up to T-4, which operates at a data rate of 274,176 megabits per second.

Figure 2-48 shows the T carrier system and its capacities. Notice the second column, PCM. Earlier in this chapter we showed the traditional method of digitizing voice (see Figure 2-27). Digitized voice usually is transmitted over T carriers, and new methods of digitizing voice are appearing each year. One of the newest methods is called *adaptive differential pulse code modulation* (ADPCM). This method digitizes voice at 32,000 bits per second instead of the traditional 64,000 bits per second of standard pulse code modulation (PCM). As you can see, the use of ADPCM makes the T carrier twice as efficient because it can handle 48 voice channels (telephone circuits), rather than 24 as in pulse code modulation.

GTE Laboratories, Inc. has developed a digital speech encoding technique that, when combined with other compression techniques, may eventually enable T-1 digital communication facilities to support 8 to 12 times the normally supported voice channels. Called *adaptive subbands excited transform* (ASET), the speech sampling technique digitally encodes speech at 16,000 bits per second, compared with the industry standard of 64,000 bits per second pulse code modulation and the newer 32,000 bits per second adaptive differential pulse code modulation techniques.

A 1.544 bits per second digital T-1 link could support 96 voice channels encoded by using the 16,000 bits per second. A silence detection compression technique that fills gaps in speech with parts of other conversations could increase the 96-channel capacity by a factor of two or three. Assuming the optimum, the technology would enable a single T-1 link to support up to 288 voice channels, compared with the 24 PCM or 48 ADPCM channels per T-1 channel that can be achieved now.

AT&T Communications has T carrier capability between hundreds of central office locations throughout the United States. Organizations with large-capacity transmission needs (either voice or data) can make use of this service. For example, as you can see in the table (Figure 2-48), a T carrier that uses pulse code modulation can be subdivided into 24 voice grade communication circuits. Use of ADPCM doubles the amount of voice grade circuits to 48. These can be used for voice transmission, data transmission, or both. Large-capacity data communication users lease these T carrier circuits directly from the telephone companies or other special common carriers.

T-1 links transport data according to a signaling format called DS-1. DS-1 stipulates that data be transmitted in 192-bit frames, followed by a framing bit. The framing bit is used to synchronize the clocks in both user and telephone company equipment. The frames are transmitted at a rate of 8,000 per second. Take 193 bits per frame, multiply it by 8,000 frames per second, and you get 1.544 megabits per second.

DS-1 also specifies that the signal maintain a minimum density of 1s. At least one 1 must be included in every 24-bit sequence. The density specification is used to preserve synchronization throughout the network.

AT&T has specified another signaling standard, DS-0, which divides frames into virtual channels. The DS-0 standard evolved from requirements for transmitting voice over T-1 links. Since T-1s are intrinsically digital, voice transmitted over a T-1 link must be digitized. The most common method of digitizing voice is pulse code modulation. With PCM, voice is digitized by sampling voice signals at the rate of twice the effective bandwidth, or 8,000 times per second. Each analog sample is digitized to a resolution of eight bits. Because there are 8,000 samples per second and each sample is eight bits, each voice signal requires a data rate of 64,000 bits per second.

Twenty-four voice channels are multiplexed on a T-1 link (see the definition for *multiplexer* in the Glossary). The equipment that multiplexes these channels together onto a single T-1 link is called a *digital channel bank*.

Because individual data applications seldom operate at the maximum T-1 rate of 1.544 megabits per second, users can acquire multiplexers to aggregate lower speed traffic economically for transmission over T-1 facilities.

For users of T-1 channels, all-digital devices are available to reconfigure these channels. These devices, referred to as *digital cross-connect systems*, are used to reconfigure the channels within the T-1 digital communication links. Remember that T-1 channels operate at 1.544 million bits per second and typically are segmented into 24 individual channels, each of which operates at 64,000 bits per second. Traditionally, if any one of these 24 different circuits within a T-1 link had to be routed to a different location, then the digital signal had to be broken into 24 individual analog signals, put through a type of distribution patch panel, and returned to the digital signal state. This technique requires that a multiplexing device, called *channel banks*, be configured back to back.

Digital cross-connects are used to switch multiplexed channels within T-1 1.544 megabits per second digital facilities, without demultiplexing the signal as you would with channel banks. For example, a cross-connect that is connected to three T-1 links could route one of the 24 channels operating at 64,000 bits per second within an incoming T-1 to any of the other outgoing T-1 links.

Digital cross-connects are replacing channel banks, which provided the same switching function when configured back to back. One channel bank would demultiplex the digital T-1 signal into 24 channels, which could then be patched to a port on the other channel bank and remultiplexed. Digital cross-connects support multiple inbound and outbound T-1 links, enabling any one 64,000 bits per second circuit within an incoming T-1 to be routed to any other outgoing T-1 channel. These digital cross-connect switches are not new, but they are becoming more cost effective so that large corporations or government agencies using T-1 facilities might find them viable.

AT&T makes extensive use of pulse code modulation internally and transmits quite a bit of its information in pulse code modulated digital format over T carriers. The T-1 carrier uses pairs of wires with digital amplifiers spaced approximately 6,000 feet apart to carry its 1.544 million bits per second (24 voice/data circuits). The T-1 carrier is used for short haul transmission over distances up to approximately 50 miles. As a rule of thumb, a T-1 carrier with a capacity equivalent of 24 voice grade lease lines costs about the same per month as 12 to 14 individual lease lines might cost if leased separately.

FRONT END COMMUNICATION PROCESSOR (FEP)

Our signal has now passed through similar facilities/equipment at the other end of the IXC (telephone company central office, local loop, modem, and RS232C connector cable) on its way to the front end communication processor (see Figure 2-1). The front end communication processor is the basic network control point, controlling such functions as central control versus interrupt, polling/selecting, response time, error control, logging, and many others. An in-depth description of front ends appears in Chapter 3.

CENTRAL CONTROL VERSUS INTERRUPT

In our example, the front end processor probably is a central control system. This means that, when the "send" key on the terminal is hit, the data signaling/

synchronization cannot commence until the terminal is polled: the front end sends a signal asking, "Do you have anything to send?" Because the "send" key has been hit, the REQUEST TO SEND (pin 4 of the RS232C connector) can begin the data signaling/synchronization process. If this were an interrupt system, the terminal would interrupt the front end immediately when the "send" key was hit.

POLLING/SELECTING

Polling and selecting take place in a centrally controlled system. *Polling* is the process of sending a signal to a terminal in order to give it permission to send messages that it might have ready. *Selecting* is the process of sending a signal to a terminal in order to determine if it is in a current status that will allow it to accept a message from the central computer site. Most people tend to refer to both of these conditions as polling.

Polling is performed by the front end communication processor, although it can be performed easily by the host computer or a remote intelligent terminal controller. There are several types of polling such as roll call, fast select, and hub go-ahead.

With *roll call polling*, the front end communication processor consecutively works through a list of terminals, first polling terminal 1, then terminal 2, then terminal 3, and so on, until all are polled. Roll call polling can be modified to prioritize terminals in the following sequence: 1, 2, 3, 1, 4, 5, 1, 6, 7, 1, 8, 9, and so on. Terminal 1 may have priority because of extremely heavy usage.

Typically, roll call polling involves excessive "wait time." The front end has to poll a terminal and then wait for a response. The response that the front end is waiting for might be an incoming message that was waiting to be sent or a negative response indicating that nothing is to be sent, or the full "time-out period" may expire because the terminal is temporarily out of service. Usually, there is a timer that "times out" the terminal after waiting, say, one-tenth of a second without getting a response. If some sort of failsafe time-out is not used, the system poll might lock up on an out-of-service terminal indefinitely. Incidentally, more sophisticated systems totally remove an individual terminal from the polling list after getting three consecutive time-outs.

Fast select polling schemes were developed in order to eliminate the time of waiting for a response when a terminal does not have a message to send, and to eliminate the time-out wait when a terminal is not operating correctly. In a fast select type of poll, the front end polls the terminals until the first incoming message is received. Assume we have 20 terminals and use fast select polling. The front end sends a poll character to terminals 1, 2, 3, 4, 5, and 6, and then the first incoming message is received. If the first incoming message is from terminal

2, the front end stops polling and waits long enough to receive any incoming messages from terminal 6 (the last terminal polled). Next, the front end resumes polling at terminal 7 and polls terminal 7, 8, 9, and so on, until the next message is received. After waiting an appropriate time for receipt of any message from the last terminal polled, the front end once again resumes polling where it left off and continues to terminal 20. After polling all 20 terminals, it may automatically start the polling list again at terminal 1. With this type of polling scheme, the front end does not wait for a negative response from terminals with nothing to send, and, if a terminal is not operating correctly, it does not waste time waiting for a time-out before proceeding to the next terminal.

Hub go-ahead polling is used in multidrop (see Glossary) configurations. The front end passes the poll character to the farthest terminal on the multidrop circuit. That terminal then sends its message and/or passes the polling character to the next inbound terminal. That terminal also passes the poll back to the next inbound terminal, and so on until it reaches the terminal closest to the front end. The closest terminal passes the poll back to the front end, and it restarts by again passing the poll to the farthest terminal. This technique relieves the front end of many polling tasks because the terminals themselves undertake the process of sending messages in a sequential fashion. It should be noted that hub go-ahead polling assumes more intelligence in each of the terminals in order to properly handle the poll. Intelligent terminals also are necessary because there must be a means of bypassing a terminal that is out of service, since it would be a polling break in the multidrop link.

RESPONSE TIME

Response time in its simplest form is the time that elapses between the generation of an inquiry at a remote terminal and the receipt of the first character of the response at the same terminal. Therefore, response time includes transmission time to the computer, processing time at the computer, access time to obtain any needed database records, and transmission time back to the terminal.

The best indicator of response time for a system not yet developed is a set of statistics drawn from another operating system that supports the same application and uses the same protocol. In other words, examining a similar network with similar operating characteristics and applications is the best indicator of how a planned network will perform. The problem is that finding such a duplicate system is almost impossible. If no like network exists from which to draw performance data, then some predictive techniques must be used.

When using these predictive techniques, do not rely on average response times. Instead, always state the question as: "X" percent of all response times must be less than or equal to "Y" seconds. In other words, a typical statement might be

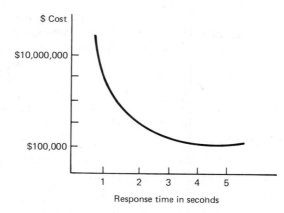

FIGURE **2-49** Cost versus response time.

that 95 percent of all response times must be less than or equal to three seconds. Mean and standard deviation of these figures might be used to identify the reliability of the final response time figure.

A typical cost versus response time curve is shown in Figure 2-49. When the response time is shortened, the cost increases; when it is lengthened, the cost decreases. Factors that affect system cost include speed and capacity of the host computer, speed and size of the front end communication processor, capacity of the communication circuits, remote intelligent control devices, and software programs/protocols.

Queuing theory allows for the definition of such elements as service time, facility utilization, and wait time at the host. A *queue* is a line of items to be handled or serviced. Single server and multiserver queuing relationships must be understood within the environment of system priorities. We cannot begin to explain, in this text, the techniques of statistical and queuing formulation. That should be reserved for a more advanced course, although modeling is introduced in Chapter 12. Estimations can be of the best or worst case. These techniques often yield average results that describe the average operational performance of a network. Statistical views of network performance based on queuing theory can vary from real performance by as much as 20 percent, but they can provide estimates of their own accuracy.

Simulation is a technique to model the behavior of the communication system or network. Response time is viewed as an elapsed time incurred, which is part of the accumulation of the elapsed times of a series of individual events. Sophisticated programs can be written to simulate the action of a series of events, and these programs add up the elapsed times of each event. Simulation programs are run on large machines and can execute several thousand polls within a few seconds in order to generate a statistical view of the projected network. Simulators typically ignore error conditions because error conditions are the exception and not the rule. They can be built into sophisticated simulators, but this vastly increases the complexity of the programs. Queuing analysis can be used to verify the predicted results of simulation. Most vendors offer simulators to assist in this area. These simulators examine the effects of many parameters of a communication system on projected performance such as

- Number of intelligent control units per circuit
- Number of terminals per intelligent control unit
- Printers/printer buffer size/printer speed
- Modem delay
- Propagation delay
- Statistical time division multiplexer delay (if any)
- Line protocol overhead
- Message lengths/occurrences/rates
- Host computer processing delays
- Database access delays
- Multiple queues or single queues
- Polling/selecting

The specific components that contribute to Response Time are Message Input Time, Application Processing Time, and Message Output Time.

$$RT = MIT + APT + MOT$$

The *message input time* is the sum of the polling time, transmission time (including modem turnaround time and time for acknowledgment), and queuing time in the front end communication processor or host computer. The transmission time component usually is stable, but the other factors are determined statistically according to traffic volume. A typical input time might be 0.85 second.

The *application processing time* includes all program processing time and all input/output accesses to the database. As might be expected, these timings are variable, depending on message traffic and the number of transactions being handled by the host computer. An example of a typical application processing time might be 0.75 second.

The *message output time* is the sum of the internal queuing in the host computer and/or front end communication processor and the transmission time (including all modem turnaround, selection, and acknowledgment times). Again, the transmission time component usually is stable, whereas internal queuing is a variable figure depending on the current volume of transactions at the host/front end communication processor. An example of a typical message output time is 0.90 second.

If the sum of the typical average times is approximately 2.5 seconds, imagine what would happen if another half second is added for propagation delay time or other delays. In a typical communication application, the component that becomes the most sensitive to increased volume is the application processing or database handling time in the host computer. Response time on a current network is easy to measure by use of a network analyzer (or even a stopwatch). Predicting it during the design stage, however, requires detailed network analysis involving queuing theory or simulation and a lot of common sense.

Queuing theory/simulation programs break the process into more segments than just message input time, application processing time, and message output time. Simulation takes into account such factors as terminal buffering, effect of an intelligent terminal control device, statistical time division multiplexers, mode of transmission used by the modem, communication circuit speed, error rates on communication circuits, queuing at transmission nodes/front end/host computer, line configurations such as point to point/multidrop/multiplex, message lengths, expected arrival times of messages, propagation delays, any priorities built into the system, average versus peak loads, central control versus interrupt, type of applications, speed of output devices, and intrinsic factors within the host computer, such as its hardware architecture, software, or protocols.

ERROR CONTROL IN DATA COMMUNICATIONS

There are two categories of errors. The first category involves corrupted (changed) data, and the second involves lost data. With regard to selecting an error control system, some of the following factors should be considered.

- The extent and pattern of error-inducing conditions on the type of circuit used
- The effects of no error control, error detection with retransmission, or automatic error detection and correction
- The maximum error rate that can be tolerated
- Comparison of the cost of increased accuracy with the present cost of correcting errors
- Comparison of different application systems as to the overall transmission accuracy currently being achieved
- Cost of errors remaining in the received station to flag them and reenter them

Data Communication Errors   Errors are a fact of life in today's data communication networks. Depending on the type of circuit/line, they may occur every few minutes or every few seconds or even more frequently. They occur because of noise on the lines (types of line noise are discussed in the next section). No data communication system can prevent all these errors from occurring, but most of them can be detected and many corrected by proper design. Common carriers that lease data transmission lines to users provide statistical measures specifying typical error rates and the pattern of errors that can be expected on the different types they lease.

Normally, errors appear in bursts. In a burst error more than one data bit is changed by the error-causing condition. This is another way of saying that 1-bit errors are not uniformly distributed in time. However, common carriers usually list their error rates as the number of bits in error divided by the number of bits transmitted, without reference to their nonuniform distribution. For example, the error rate might be given as 1 in 500,000 when transmitting on a public voice grade telephone circuit at 1200 bits per second.

The fact that errors tend to be clustered in bursts rather than evenly dispersed has both positive and negative aspects. If the errors were not clustered (but instead were evenly distributed throughout the day), with an error rate of 1 bit in 500,000 it would be rare for two erroneous bits to occur in the same character, and consequently some simple character checking scheme would be effective. But this is not the case because bursts of errors are the rule rather than the exception. They sometimes go on for time periods that may obliterate 50 to 100 or more bits. The positive aspect is that, between bursts, there may be rather long periods of error-free transmission. Therefore, no errors at all may occur during data transmission in a large proportion of messages. For exa#ple, when errors are #ore or less evenly distrib#ted, it is not di#ficult to gras# the me#ning even when the error #ate is high, as it is in this #entence (1 charac#er in 20). On the other hand, if errors are concentrated in bursts, it becomes more difficult to recover the meaning and much more reliance must be placed on knowledge of message #######[5] or on special logical/numerical error detection and correction methods.

It is possible to develop data transmission methodologies that give very high error detection and correction performance. The only way to do the detection and correction is to send along extra data. The more extra data that is sent, the more error protection that can be achieved. However, as this protection is increased, the throughput of useful data is reduced. Therefore, the efficiency of data throughput varies inversely as the desired amount of error detection and correction is increased. Errors even have an effect on the length of the block of data to be transmitted when synchronous transmission is used. The shorter the message

---

[5] In case you could not guess, the word is "context."

blocks used, the less likelihood there is of needing retransmission for any one block. But the shorter the message block, the less efficient is the transmission methodology as far as throughput is concerned. If the message blocks are long, a higher proportion may have an error and have to be resent.

Considerable variation in the error rate is found from one time of the day to another in transmissions sent over the dial-up switched network. The error rate usually is higher during periods of high traffic (the normal business day). In some cases the only alternative open to the user of these facilities is to transmit the data at a slower speed because higher transmission speeds are more error prone. Dial-up lines are more prone to errors than private leased lines because they have less stable transmission parameters, and because different calls use different circuits, they usually experience different transmission conditions. Thus, a bad line is not necessarily a serious problem in dial-up transmission; a new call may result in getting a better line. Line conditioning, a service that is not available on dial-up lines but only on private leased lines, consists of special electrical balancing of the circuit to ensure the most error-free transmission.

Line Noise and Distortion Line noise and distortion can cause data communication errors. In this context we define noise as undesirable electrical signals. It is introduced by equipment or natural disturbances, and it degrades the performance of a communication line. If noise occurs, the errors are manifested as extra or missing bits, or bits whose states have been "flipped," with the result that the message content is degraded. Line noise and distortion can be classified into roughly 11 categories: white noise, impulse noise, cross talk, echoes, intermodulation noise, amplitude noise, line outages, attenuation, attenuation distortion, delay distortion, and jitter.

White or Gaussian noise is the familiar background hiss or static on radio and telephones. It is caused by the thermal agitation of electrons and therefore is inescapable. Even if the equipment were perfect and the wires were perfectly insulated from any and all external interference, there still would be some white noise. White noise usually is not a problem unless its level becomes so high that it obliterates the data transmission. Sometimes noise from other sources such as power line induction, cross modulation from adjacent lines, and a conglomeration of random signals resembles white noise and is labeled as such even though it is not caused by thermal electrons.

Impulse noise (sometimes called *spikes*) is the primary source of errors in data communications. An impulse of noise can last as long as 1/100th of a second. An impulse of this duration is heard as a click or a crackling noise during voice communications. This click does not affect voice communications, but it might obliterate a group of data bits, causing a burst error on a data communication line. At 150 bits per second, 1 or 2 bits would be changed by a spike of 1/100th of a second, whereas as 4800 bits per second, 48 bits would be changed. Some of the sources of impulse noise are voltage changes in adjacent lines or circuitry surrounding the data communication line, telephone switching equipment at the telephone exchange branch offices, arcing of the relays at older telephone exchange offices, tones used by network signaling, maintenance equipment during line testing, lightning flashes during thunderstorms, and intermittent electrical connections in the data communication equipment.

Cross-talk occurs when one line picks up some of the signal traveling down another line. It occurs between line pairs that are carrying separate signals, in multiplexed links carrying many discrete signals, in microwave links where one antenna picks up a minute reflected portion of the signal from another antenna on the same tower, and in any hardwired telephone circuits that run parallel to each other, are too close to each other, and are not electrically balanced. You are experiencing cross-talk during voice communication on the public switched network when you hear other conversations in the background. Cross-talk between lines increases with increased communication distance, increased proximity of the two wires, increased signal strength, and higher frequency signals. Cross-talk, like white noise, has such a low signal strength that it normally is not bothersome on data communication networks.

Echoes and echo suppression can be a cause of errors. (Echo suppressors will be discussed in Chapter 4.) An echo suppressor causes a change in the electrical balance of a line. This change causes a signal to be reflected so that it travels back down the line at reduced signal strength. When the echo suppressors are disabled, as in data transmission, this echo returns to the transmitting equipment. If the signal strength of the echo is high enough to be detected by the communication equipment, it causes errors. Echoes, like cross talk and white noise, have such a low signal strength that they normally are not bothersome.

Intermodulation noise is a special type of cross talk. The signals from two independent lines intermodulate and form a product that falls into a frequency band differing from both inputs. This resultant frequency may fall into a frequency band that is reserved for another signal. This type of noise is similar to harmonics in music. On a multiplexed line, many different signals are amplified together, and slight variations in the adjustment of the equipment can cause intermodulation noise. A maladjusted modem may transmit a strong frequency tone when not transmitting data, thus yielding this type of noise.

Amplitude noise involves a sudden change in the level of power. The effect of this noise depends on the type of modulation being used by the modem. For example, amplitude noise does not affect frequency modulation techniques because the transmitting and receiving equipment interprets frequency information and disregards the amplitude information. Some of the causes of amplitude noise may be faulty amplifiers, dirty contacts with variable resistances, sudden added loads by new circuits being switched on during the day, maintenance work in progress, and switching to different transmission lines.

Line outages are a catastrophic cause of errors and incomplete transmission.

Occasionally, a communication circuit fails for a brief period of time. This type of failure may be caused by faulty telephone branch office exchange equipment, storms, loss of the carrier signal, and any other failure that causes an open line or short circuit.

Attenuation is the loss of power that the signal suffers as it travels from the transmitting device to the receiving device. It results from the power that is absorbed by the transmission medium or is lost before it reaches the receiver. As the transmission medium absorbs this power, the signal gets weaker, and the receiving equipment has less and less chance of correctly interpreting the data. To avoid this problem, telephone lines have repeater/amplifiers spaced through their length. The distance between them depends on the amount of power lost per unit length of the transmission line. This power loss is a function of the transmission method and circuit medium. Attenuation increases as frequency increases or as the diameter of the wire decreases.

Attenuation distortion refers to high frequencies losing power more rapidly than low frequencies during transmission. The received signal can thus be distorted by unequal loss of its component frequencies.

Delay distortion occurs when a signal is delayed more at some frequencies than at others. If the method of data transmission involves data transmitted at two different frequencies, then the bits being transmitted at one frequency may travel slightly faster than the bits transmitted at a different frequency. A piece of equipment, called an *equalizer*, compensates for both attenuation distortion and delay distortion.

Jitter may affect the accuracy of the data being transmitted. The generation of a pure carrier signal is impossible. Minute variations in amplitude, phase, and frequency always occur. Signal impairment may be caused by continuously and rapidly changing gain and/or phase changes. This jitter may be random or periodic. Phase jitter during a telephone call causes the voice to go up and down in volume.

APPROACHES TO ERROR CONTROL

Error control implies (1) techniques of design and manufacture of data communication transmission links and equipment to reduce the occurrence of errors (an area that is outside the scope of this book), and (2) methodologies to detect and correct the errors that are introduced during transmission of the data. In the sense of the second meaning of error control, the methodologies fall into three categories, and possibly four if you consider the option of ignoring the errors.

- Loop or echo checking
- Error detection with retransmission
- Forward error correction (FEC)

Loop or Echo Checking Loop or echo checking does not use a special code. Instead, each character or other small unit of the message, as it is received, is transmitted back to the transmitter, which checks to determine whether the character is the same as the one just sent. If it is not correct, then the character is transmitted a second time. This method of error detection is wasteful of transmission capacity because each message (in pieces) is transmitted at least twice and there is no guarantee that some messages might not be transmitted three or four times. Some of this retransmission of characters for a second or third time might not be necessary because the error could have occurred on the return trip of the character, thus requiring the transmitter to retransmit the character even though it was received correctly the first time. Loop or echo checking usually is utilized on hardwire, short lines, with low speed terminals. This type of error checking does give a high degree of protection, but it is not as efficient as other methods. It sometimes is confused with full duplex transmission.

Error Detection with Retransmission Error detection and retransmission schemes are built into data transmitting and receiving devices, front end computers, modems, and software. These schemes include detection of an error and immediate retransmission, detection of an error and retransmission at a later time, or detection of an error and retransmission for up to, say, three tries and then retransmission at a later time, or the like. Error detection and retransmission is the simplest, and if properly handled, the most effective and least expensive method to reduce errors in data transmission. It requires the simplest logic, needs relatively little storage, is best understood by terminal operators, and is most frequently used. Retransmission of the message in error is straightforward. It usually is called for by the failure of the transmitter to receive a positive acknowledgment within a preset time. Various methods are used to determine that the message that has just been received has, in fact, an error imbedded in it. Some of the common error detection methods are parity checking, constant ratio codes, and polynomial checking.

Parity checking: If you examine a character from the ASCII coding structure, it soon becomes apparent that one of the eight bits encoding each character is redundant, that is, its value is determined solely by the values of the other seven and therefore is unnecessary. Since this eighth bit cannot transmit any new information, its purpose is to confirm old information. The logic of its use is shown in Figure 2-50. The most common rule for fixing the value of the redundant bit uses the parity (evenness or oddness) of the number of 1s in the code. Thus, for an even parity code system using ASCII,

- Letter V is encoded 0110101. Because the number of 1s is 4, already an even number, a 0 is added in the parity (eighth) position, yielding V = 01101010.

- Letter W is encoded 0001101, which had an odd number of 1s. Therefore, a 1 is added in the parity position to make the number of 1s even, yielding W = 00011011.

A little thought will convince you that any single error (a switch of a 1 to a 0 or vice versa) will be detected by a parity check but nothing can be deduced about which bit was in error. Moreover, if the states of *two* bits are switched, the parity check may not sense any error. Of course, it may be possible to sense such an error because the resulting code, although correct as far as parity is concerned, is a code that is "forbidden," for example, undefined or inappropriate in its context. Such detection requires more circuitry or software. Many systems today do not use parity because it cannot correct errors and is only about 50 percent accurate in detecting errors. Systems are described as having *odd parity, even parity,* or *no parity.*

Another parity checking technique is the cyclical parity check (sometimes called *interlaced parity*). This method requires two parity bits per character. Assuming a 6-data-bit code structure, the first parity bit provides parity for the first, third, and fifth bits, and the second parity bit provides parity for the second, fourth, and sixth bits. Figure 2-51 shows an even parity cyclical parity check on a 6-bit code.

Constant ratio codes: M-of-N codes are special data communication codes that have a constant ratio of the number of 1 bits to the number of 0 bits. The most common one is IBM's 4-of-8 code that was discussed earlier. M-of-N codes detect an error when the number of 1 bits and 0 bits are not in their proper ratio. For example, in the 4-of-8 code there always are supposed to be four 1 bits and four 0 bits in the received bit configuration of the character. When this ratio is out of balance, the receiving equipment knows that an error has occurred. M-of-N codes are not used widely because they are inefficient. As an example of their ineffi-

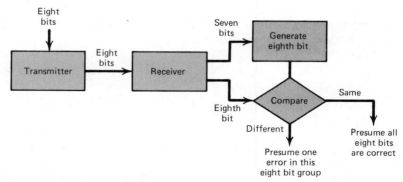

FIGURE **2-50** Parity checking logic.

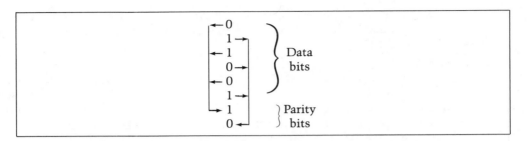

FIGURE 2-51 Cyclical parity check for a 6-bit code.

ciency, consider that the 4-of-8 code has 70 valid character combinations while a 7-bit ASCII code has 128 valid character combinations ($2^7 = 128$).

Polynomial checking: Polynomial checks on blocks of data often are performed for synchronous data transmission. In this type of message checking, all the bits of the message are checked by application of a mathematical algorithm. For example, all the 1 bits in a message are counted and then divided by a prime number (such as 17), and the remainder of that division is transmitted to the receiving equipment. The receiving equipment performs the same mathematical computations and matches the remainder that it calculated against the remainder that was transmitted with the message. If the two are equal, the entire message block is assumed to have been received correctly. In actual practice, much more complex algorithms are utilized.

One of the most popular of the polynomial error checking schemes is *cyclical redundancy check* (CRC). It consists of adding bits (about 10 to 25) to the entire block. A communication protocol, using a 16-bit CRC, calculates a 16-bit number that is a function of all the data in the block being sent. This 16-bit number is added to the end of the message block. The receiver recalculates its own 16-bit CRC as the block is received. If the numbers are the same, everything is acceptable. If they are different, an error has occurred. Once an error is detected, the correction mechanism tells the sender to retransmit the message. The sender goes back to the last successful (prior) message and begins transmitting again. A 16-bit CRC detects every error that is 16 bits or smaller with 100 percent probability. In CRC checking the data block can be thought of as one long binary polynomial, P. Before transmission, equipment in the terminal divides P by a fixed binary polynomial, G, resulting in a whole polynomial, Q, and a remainder, R/G.

$$\frac{P}{G} = Q + \frac{R}{G}$$

The remainder, R, is appended to the block before transmission, as a check sequence k bits long. The receiving hardware divides the received data block by the same G, which generates an R. The receiving hardware checks to ascertain if the received R agrees with the locally generated R. If it does not, the data block is assumed to be in error and retransmission is requested. A 25-bit CRC code added to a 1000-bit block allows only three bits in 100 million to go undetected. That is, for a 2.5 percent redundancy, the error rate is 3×10^{-8}.

Forward Error Correction   This approach uses codes that contain sufficient redundancy to permit errors to be detected and corrected at the receiving equipment without retransmission of the original message. The redundancy, or extra bits required, varies with different schemes. It ranges from a small percentage of extra bits to 100 percent redundancy, with the number of error detecting bits roughly equaling the number of data bits. One of the characteristics of many error correcting codes is that there must be a minimum number of error-free bits between bursts of errors. For example, one such code, called a *Hagelbarger code,* corrects up to six consecutive bit errors provided that the 6-bit error group is followed by at least 19 valid bits before further error bits are encountered. Bell Telephone engineers have developed an error correcting code that uses 12 check bits for each 48 data bits, or 25 percent redundancy. Still another code is the *Bose-Chaudhuri code,* which, in one of its forms, is capable of correcting double errors and can detect up to four errors.

To show how such a code works, consider this example of a forward error checking code, called a *Hamming code,* after its inventor, R. W. Hamming.[6] This code associates even parity bits with unique combinations of data bits. Using a 4-data-bit code as an example, a character might be represented by the data bit configuration 1010. Three parity bits P_1, P_2, and P_4 are added, resulting in a 7-bit code, shown in the upper half of Figure 2-52. Notice that the data bits (D_3, D_5, D_6, D_7) are 1010, and the parity bits (P_1, P_2, P_4) are 101.

As depicted in the upper half of Figure 2-52, parity bit P_1 applies to data bits D_3, D_5, and D_7. Parity bit P_2 applies to data bits D_3, D_6, and D_7. Parity bit P_4 applies to data bits D_5, D_6, and D_7. For the example, in which D_3, D_5, D_6, D_7 = 1010, P_1 must equal 1 since there is but one 1 among D_3, D_5, and D_7 and parity must be even. Similarly P_2 must be 0 since D_3 and D_6 are 1s. P_4 is 1 since D_6 is the only 1 among D_5, D_6, D_7.

Now, assume that during the transmission, data bit D_7 is changed from a 0 to a 1 by line noise. Because this data bit is being checked by P_1, P_2, and P_4, all three parity bits now show odd parity instead of the correct even parity. (D_7 is the only data bit that is monitored by all three parity bits; therefore, when D_7 is in error, all three parity bits show an incorrect parity.) In this way, the receiving equipment can determine which bit was in error and reverse its state, thus correcting the error without retransmission.

The bottom half of Figure 2-52 is a table that determines the location of the bit

[6]William P. Davenport, *Modern Data Communication: Concepts, Language, and Media* (New York: Hayden Book Company, 1971), p. 96.

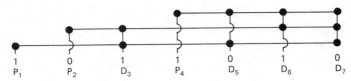

1	0	1	1	0	1	0
P_1	P_2	D_3	P_4	D_5	D_6	D_7

Checking Relations Between Parity Bits (P) and Data Bits (D)

0 = Corresponding parity check is correct 1 = Corresponding parity check fails			Determines in which bit the error occurred
P_4	P_2	P_1	
0	0	0	→ no error
0	0	1	→ P_1
0	1	0	→ P_2
0	1	1	→ D_3
1	0	0	→ P_4
1	0	1	→ D_5
1	1	0	→ D_6
1	1	1	→ D_7

Interpreting Parity Bit Patterns

FIGURE **2-52** Hamming code for forward error correction.

in error. A 1 in the table means that the corresponding parity bit indicates a parity error. Conversely, a 0 means the parity check is correct. These 0s and 1s form a binary number that indicates the numerical location of the erroneous bit. In the example above, P_1, P_2, and P_4 checks all failed, yielding 111, or a decimal 7, the subscript of the erroneous bit.

The Consultative Committee on International Telephone and Telegraph (CCITT) has incorporated a forward error correction technique called Trellis-coding into its V.32 specification of 9600 bits per second dial-up modems. Forward error correction is essential in certain circumstances. In some media like simplex communication links, there is no return communication channel to provide a negative acknowledgment (NAK), retransmission cannot be requested, or a message may be sent to thousands of receivers with no return message capability.

Satellite transmission is one area in which the argument for forward error correction makes sense. A round trip from the earth station to the satellite and back includes a significant delay. Error rates can fluctuate depending on everything from the transmission frequencies to the condition of equipment, sun spots, or the weather. When compared to satellite equipment costs, the additional cost of forward error correction is insignificant.

Forward error correction use in broader markets such as modem-based communications is limited today because of the costs associated with coding forward error correction algorithms. With time, however, board-level forward error correction modules will be replaced by chip-level devices. Forward error correction modules add $500 to the cost of the communication devices, but chip implementations of forward error correction are coming on the market.

LOGGING

When our message reaches the front end communication processor, the first task the front end performs is to double-log the message. Logging messages onto a disk for short-term storage protects them in case of system failure or interrupt. It gives the recovery software instant access to the last few messages received, thus enabling proper recovery and restart.

Messages also are logged onto a magnetic tape for a long-term transaction trail and/or historical purposes. These magnetic tapes might be saved for a few days and possibly be converted to microfilm. They can provide a long-term history of all transactions entering the system and all returned messages after processing in the host computer.

DATA CHANNEL

The data channel actually is part of the host computer. It takes a message that has been stripped of its communication control characters by the front end and moves it into the memory of the host computer for processing.

HOST COMPUTER

Our message now has been passed from the front end communication processor, through the data channel, and into the host computer. The host computer processes the message as requested. It may retrieve information from a database, or it may update databases, depending on the message instructions. Any security restrictions placed on you through individual passwords, terminal-to-host identification codes, or restrictions placed on the front end communication processor through which your message entered will be validated.

SELECTED REFERENCES

1. Arick, Martin R. *Data Communications: Concepts and Solutions.* Wellesley, Mass.: QED Information Sciences, 1987.

2. *Catalog of Technical Publications.* Florham Park, N.J.: AT&T Communications, January 1985 (AT&T Communications Publication 10000).

3. *Computerworld: Newsweekly for the Computer Community.* Published weekly by CW Communications, 375 Cochituate Road, Box 9171, Framingham, Mass. 01701-9171, 1967– .

4. *Computing Canada: The Newspaper for Information Processing Management.* Published biweekly by Plesman Publications, Ltd., 2 Lansing Square, Suite 2, Willowdale, Ontario M2J 5A1, Canada, 1975– .

5. Freeman, Roger L. *Reference Manual for Telecommunications Engineering.* New York: John Wiley & Sons, 1984.

6. Gagliardi, Robert M. *Satellite Communications: An Introduction.* New York: Van Nostrand Reinhold Co., 1984.

7. Gurrie, M. L., and P. J. O'Conner. *Voice/Data Telecommunications Systems: An Introduction to Technology.* Englewood Cliffs, N.J.: Prentice-Hall, 1986.

8. Korn, Israel. *Digital Communications.* New York: Van Nostrand Reinhold Co., 1985.

9. *Lightwave: Journal of Fiber Optics.* Published monthly by Howard Rausch Associates, 235 Bear Hill Road, Waltham, Mass. 02154, 1984– .

10. *Network World.* Published weekly by CW Communications, Box 9171, 375 Cochituate Road, Framingham, Mass. 01701-9171, 1983– .

11. *Proto: AT&T Bell Laboratories Report to Executives on New Technologies.* Published by AT&T Bell Laboratories, 600 Mountain Avenue, Murray Hill, N.J. 07974.

12. *Telecommunications: Covering the Total Spectrum of Communications Worldwide.* Published monthly by Horizon House-Microwave, 610 Washington Street, Dedham, Mass. 02026, 1967– .

13. Weik, Martin H. *Fiber Optics and Lightwave Communications Standard Dictionary.* New York: Van Nostrand Reinhold Co., 1980.

QUESTIONS/PROBLEMS

1. Inside computers, data typically travels around in _____ but on a data communication circuit it travels in a _____ fashion.

2. What is the difference between start-stop transmission and asynchronous transmission?

3. Looking at Figure 2-1, can you define the type of signal (whether it is baseband or broadband) as it passes over or through the following devices: terminal, connector cable, modem, local loop?

4. If there was a 3-of-7 code, it could represent $C_3^7 = 7!(3!4!) = 35$ distinct symbols, out of the $2^7 = 128$ bit combinations. We then could say that the 3-of-7 code is $35/128 = 0.273$ or 27.3 percent "efficient." How efficient would a 5-of-10 code be?

5. If you experiment with a touchtone telephone and listen carefully (and if you are not tone deaf), you will discover that each key generates two tones. One of the tones is the same for all keys in the same row; the other is the same for all keys in the same column. Label the "row tones" a, b, c, d from lowest to highest pitch, and the "column tones" x, y, z, from the lowest to the highest pitch. Make an equivalence table between the digits 0 through 9, * and #, and the tone code letter pairs. In the terminology of bits, dibits, tribits, etc., when you press one key what are you generating?

6. Parallel transmission almost always is used over short distances, as in cables between computer components, whereas serial transmission is used over long distances. Can you think of the main reason why?

7. What is a start bit in asynchronous transmission?

8. How efficient would a 6-bit code be in asynchronous transmission if it had two parity bits, one start bit, and two stop bits (some very old equipment utilizes two stop bits)?

9. Which circuit type has the greatest capacity for carrying voice or data transmissions: satellite or optical fiber?

10. If the signal-to-noise ratio is 10 dB, how much more powerful is your signal than the background noise?

11. Which modulation technique is more error free during transmission: amplitude, frequency, or phase modulation?

12. Draw a digital signal that has a characteristic of returning to 0 between consecutive data bits.

13. Use Figure 2-32 as a model and draw a situation in which frequency division multiplexing is used and the modem transmits 3 bits per baud.

14. Is asynchronous or synchronous transmission used at your organization?

15. What is the difference between RS232C and RS449?

16. Diagram the data signaling/synchronization for transmitting a message.

17. What is the difference between a station terminal and an RS232C connector plug?

18. What is the transmission rate of information bits if you use EBCDIC (eight bits with one parity bit), a 400-character block, 9600 bits per second modem transmission speed, 20 control characters per block, an error rate of 1 percent, and a 30-millisecond turnaround time?

19. What is the TRIB in Question 18 if you add a half-second delay to the turnaround time because of satellite delay?

20. If a modem is able to switch between any two frequencies 2,400 times per second, what is the signaling rate (baud) in microseconds?

21. Suppose you send in a request for data to the central host database and your response gets back in three seconds. If the host database processing time was 1.5 seconds and your message took 0.73 second to go from your terminal to the host, how long did it take for the message to return from the host to your terminal?

22. If the transmission speed is 9600 bits per second and a spike (impulse noise) hits the line for one-tenth of a second, how many bits are destroyed?

23. Why does the 4-of-8 code not use a parity bit?

24. A frequently used form of checking of decimal numbers is the modulo nine check. A check digit is appended to each numerical quantity. The digit is the remainder, upon division by 9, of the sum of the digits in the original number. For example, if the original number is 73842, then the sum of the digits is 24 (7 + 3 + 8 + 4 + 2). The check digit calculation is: 24/9 = 2 + Remainder of 6; therefore, the check digit is 6. The number with the check digit appended would then be 738426. Assume that during transmission there is an error that changes the number 738426 to the number 739426. Let us check the received number (739426). The sum of the digits is 25 (7 + 3 + 9 + 4 + 2); therefore, 25/9 + 2 + Remainder of 7. Since 7 does not equal 6, the number has been detected in error. Experiment with this method to find out under what circumstances it works and when it fails. If you are particularly industrious, look up "casting out 9s" in a mathematical dictionary or investigate this property in a book on number theory.

25. How efficient is the method above if we are transmitting 8-digit numbers? 12-digit numbers?

26. Using the format of Figure 2-52, encode the data character using 1011.

27. Do errors in data communications normally appear in bursts or are they distributed evenly over time?

28. A signal suffers a loss of power as it travels from the transmitting device to the receiving device. What is this called?

29. What is the most frequently used approach to error control?

3
DATA COMMUNICATION HARDWARE

The previous chapter discussed methods of communicating data; this chapter describes equipment to accomplish those methods. A large and ever-growing array of equipment is available to data communication networks for transmitting and receiving data as well as for performing the basic tasks of handling the data as it flows through the network. Hardware selection takes into account the network requirements, which, in turn, are derived from the basic business requirements of the organization that uses the data communication network. The hardware's technical capabilities are described along with its purpose within the network. Numerous figures show the specific location of the hardware within the network environment.

INTRODUCTION

Now that you have completed Chapter 2 on the fundamental communication concepts, you should understand the basic technical concepts, as well as the flow of a message (signal) through a network. This chapter defines each piece of hardware that is used in a network. We will not attempt to configure all these pieces into the different types of network configurations until the next chapter. By the time you have completed Chapters 2, 3, and 4, however, in addition to the basic technical concepts, you should understand the various pieces of data communication hardware and the network configurations that draw the technical concepts and the hardware together.

HOST MAINFRAME CENTRAL COMPUTER

The host computer generally is considered to be the central computer or central processing function for a business or government data communication application processing system. In distributed processing, several host computers may be tied together by the data communication network. While the host computer is not truly a part of the network, it performs many network functions because these functions may be shared by the host computer and front end communication processor.

The suitability of a computer to serve as the central computer for an online, real-time data communication system depends on both its own capabilities and the capabilities of other hardware attached to it. Many computers on the market today can be used for online, real-time data communication networks, provided that the hardware attached can handle the tasks for which the central computer is inefficient. In other words, the characteristics that make a computer suitable for data communications do not necessarily make it good for "number crunching." In particular, data communication work involves many short periods of activity to service a single arriving or departing character or message. A computer whose hardware or software makes this kind of operation clumsy or time consuming does not perform well in the data communication environment. For such a machine to be effective, auxiliary hardware is required.

Data communications falls into three categories, dependent on the interface between the data communication network and the processing functions of the central computer.

The *first* of these categories is a standalone computer configuration. In this configuration, the computer is designed to handle a specific set of communication facilities and terminals. The circuitry to handle data communications is built directly into the computer. In other words, the computer's architecture is designed so that it can interact in a real-time mode. Figure 3-1 shows a standalone

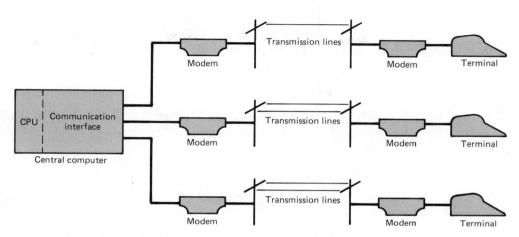

FIGURE **3-1** Standalone communication configuration.

communication configuration in which the central computer is able to handle all communication tasks. This type of computer is a stored program computer with communication as well as computing capabilities. It typically is used in a mode where the emphasis is on communication rather than data processing. It often is seen in the manufacturing environment for process control, and in areas where the user queries a database on the status of a certain product, inventory level, or the like. This field is dominated by minicomputers that have been developed and programmed for special purpose processing and communication functions.

The *second* category is a network that uses microcomputers, micro-to-mainframe connections, and local area networks. While some people may think this configuration involves a small network, many of the larger local area networks can handle several hundred terminals or microcomputers. Usually, this type of network is totally within the corporate/government business office or facility/campus area. It does not require long distance communication circuits, and, when users want to transmit outside of this local network, they must address their messages so they can be transferred to long distance networks. Because these networks are so important in today's business and scientific environment, two other chapters in this book are devoted to them: Chapter 6, which discusses micro-to-mainframe connections, and Chapter 7, which discusses local area networks (LANs).

The *third* category is a general purpose computer. Large general purpose computers are used for both data communications and perhaps some batch processing, but with greater emphasis on the online, real-time data communication

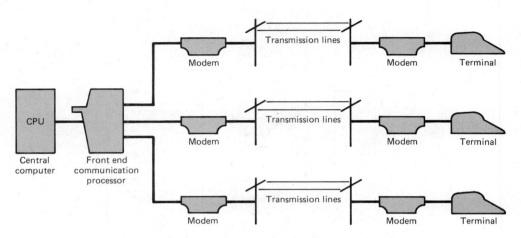

FIGURE **3-2** Front end central computer configuration.

portion of the system. In this configuration, there is a distinct division of labor between the front end and the general purpose computer.

The front end can take two forms. The first is that of a nonprogrammable, hardwired, communication control unit that is designed by the computer manufacturer to adapt specific line and terminal characteristics to the computer. The second form is that of a front end communication processor that is programmable and can handle some or all the input/output activity as well as performing some processing. Figure 3-2 depicts a front end/central computer configuration.

Such a configuration is employed primarily in situations where the input/output and computing processing requirements are very large and where rapid response time is of the essence. This type of configuration is used in large data communication networks.

The general trend today is to remove everything you can from the host computer and move it further out into the network. Figure 3-3 shows the downline movement of some of the functions that can be moved out of the host computer. This movement increases the efficiency of each piece of hardware because it offloads some of its duties to the next piece of hardware.

For example, the data channel (a part of the host computer) handles the movement of all data into the host computer memory and the movement of completed processing out of the host computer memory. The front end communication processor now handles most, if not all, of the control functions related to data communications. Modems perform modulation and specific diagnostic checks. Next, some of these control functions are passed off to other devices such as switches, statistical multiplexers, and remote intelligent controllers that might be located hundreds or even thousands of miles away from the host computer.

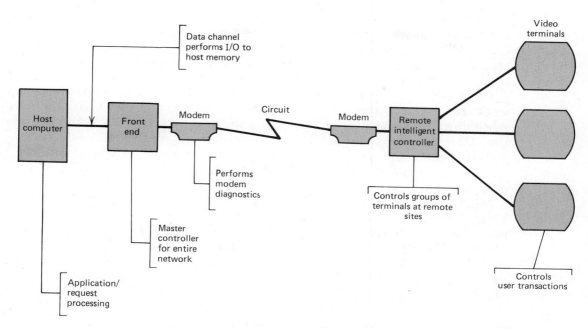

FIGURE **3-3** Downline network control.

This also implies that the massive quantities of software that used to reside solely in the host computer are removed and located downline in other pieces of hardware such as front ends and remote intelligent controllers.

Finally, the ultimate proof of this movement of software and hardware functions out of the host computer is exemplified by distributed data processing where application user departments now have minicomputers or microcomputers to perform their own data processing functions. The result is that user application departments are given their own distributed database so they can store their own files. As this happens, each business or government function has its own data processing host computer, even though it might be a small microcomputer, and its own distributed database files. Therefore, data communication networks become the fabric or glue that holds the business organization together; there are multiple host computers, ranging in size from a mainframe down to a microcomputer.

Networks can operate in a *central control mode* or an *interrupt mode*. The central control mode involves the centralized polling of each station device (terminal or node) on the network. *Polling* is the process of individually giving each of the terminals permission to send data one at a time.

The interrupt philosophy implies that when a terminal sends data, the incoming data stream interrupts the host computer and the host stops processing so it can handle the incoming data. Because this mode of operation is wasteful of processing capacity, it generally is not used on host computers except for some minicomputer systems that have very few terminals and on local area networks.

The host computer contains software to operate the data communication network. For example, teleprocessing monitors are used to control the routing, scheduling, and movement of data within the host computer. Telecommunication access programs also can be used to handle the routing, scheduling, and polling of terminals out in the network, although in today's marketplace these telecommunication access programs usually are offloaded to a front end communication processor.

Finally, as we move toward the twenty-first century, you might find that the term *host computer* is becoming somewhat blurred because some of the new microprocessor chips are 32-bit chips that can operate at 2 million instructions per second when coupled together. As a result, they have the power of today's mainframe host computers. Put another way, the computing power of a host mainframe computer today may be in the microcomputer of the future.

FRONT END COMMUNICATION PROCESSOR (FEP)

A front end communication processor is a computer that has been programmed specifically to perform many different control and/or processing functions required for the proper operation of a data communication network. This computer might be a mainframe, a minicomputer, or, as is becoming the case more often today, a microprocessor-based digital computer.

As mentioned in the preceding section, the primary purpose of this device is to offload some of the processing and control functions of the data communication network from the host computer system to a specially designed and programmed communication processor. These devices are programmable, and they are equipped with extensive software packages. The software defines the architecture of the system; it determines which of the various protocols and software programs are used for communicating with this communication processor device.

The primary application for this type of device is the interface between the central data processing system (host computer) and the data communication network with its hundreds or even thousands of input/output terminals or nodes. Many of the newer and more powerful communication processors can perform message processing because they have enough storage capacity, processing power, and disk units. For example, the processor might receive inquiry messages

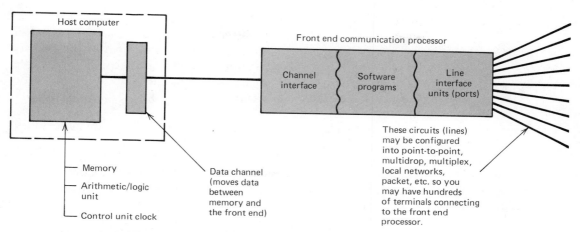

FIGURE 3-4   Front end communication processor.

from remote terminals, process the messages to determine the specific information required, retrieve the information from an online random access storage unit, and send it back to the inquiring terminals without involving the host computer. In systems of this type, application-oriented processing is as important as message receipt and transmission.

Another function related to processing is message switching. This function occurs when the communication processor receives a message, determines that it only needs to be switched to some other terminal or node, and performs the circuit switch or message switch. It also can utilize a store and forward process to hold the message and forward it at a later time.

Some of the basic component parts of a front end communication processor are as follows (see Figure 3-4).

- *Channel interface* is the hardware interface that permits the communication processor to connect directly to the standard data channel of a host computer.

- *Line interface units* (also called *ports*) are hardware devices that link the communication processor with the modems that terminate each communication circuit. Along with the line interface units, there might be a communication multiplexer if multiplexing is built into the processor. A specific front end may have the capacity for 10 or 200 ports (circuits).

- *Software* is the set of stored programs that usually are highly specialized and that define the specific architecture/protocols of the front end. The software determines which standard protocol (such as X.25) should be used for a given front end. Some of this software now is being built into *firmware*. That is, the programs themselves are coded into circuit chips instead of being programmed logic. Some sophisticated communication processors have firmware (circuit cards that have program code built in) that can serve as a protocol converter so you can interconnect the X.25 protocol to the protocols of other host mainframe computers.

There are over two dozen vendors of front end communication processors, including each of the major host mainframe manufacturers.

The general functions of front end communication processors are discussed below. Not all communication processors perform all of these functions; therefore, you must check an individual model to be sure it performs the functions necessary for your network.

Communication Line Control

Polling/selecting of individual terminals, intelligent terminal controllers, or network nodes. *Polling* involves asking each terminal whether it has a message to send, and *selecting* involves asking each terminal if it is in a condition to receive a message. Both imply a network architecture that involves central control.

Automatic answering, acknowledgment, and dial for outgoing calls.

Port selection allowing several circuits to share a single port. The *port* is the plug or connection point where the individual lines enter the front end communication processor. You can control which group of terminals can have access to the host computer. The communication processor has several incoming circuits, each of which is in contention for a single incoming port. These ports also can be put on a priority basis, with certain incoming communication circuits taking priority over others.

Ability to address messages to specific circuits or terminals. Examples are a broadcast address that goes to all terminals of the system, a multiple address that goes to a select group of several terminals, or a single address that goes to a specific communication circuit or specific terminal.

Circuit switching allowing one incoming circuit to be switched directly to another when it is available. This creates a straight-through transmission path for the movement of messages from one terminal location to another. Associated with circuit switching is a store and forward capability similar to that in voice mail or electronic mail systems. When the second half of the circuit path is unavailable, the communication processor records the incoming text and transmits it to the other terminal when the circuit is free.

Automatic routing of messages to a backup terminal when a specific terminal or circuit is out of order. This is a type of switching function, but it is used when there are problems in the network.

Addition or deletion of communication line control codes. Line control codes (the grammar of data communications), such as END OF BLOCK, BEGINNING OF BLOCK, or START OF MESSAGE, must be deleted before the message is passed to the host computer or must be added before the message is passed to the outgoing communication circuits.

Protocol/Code Conversion

Code conversion, that is, the software or hardware conversion from one code format to another such as ASCII to EBCDIC. Code conversion is available from any code to any code.

Conversion from one protocol to another, which allows different machines to talk to each other if they all use different protocols such as HDLC, SDLC, and X.25.

Assembly of Characters/Messages

Assembly and disassembly of bits into characters. Bits are transmitted in serial fashion on a communication circuit. The front end assembles these serial bits into parallel characters.

Assembly/disassembly buffering in order to handle synchronous or asynchronous modes of transmission.

Handling of transmission speed differences where different communication circuits transmit at different bits per second rates, such as 2400 bits per second versus 56,000 bits per second.

Data and Message Editing

Control editing, which involves adding items to a message, rerouting messages, rearranging data for further transmission, or looking for nonexistent addresses.

Message compression or compaction, a methodology for transmitting meaningful data messages but through the transmission of fewer data bits.

Application editing, using some of the expanded processing and storage capabilities of the newest front ends. It provides for editing of either application errors or human factors errors that occur during data entry.

Triggering of special remote alarms if certain parameters are exceeded.

Signaling of abnormal occurrences to the host computer.

Assignment of consecutive serial numbers to each message and, possibly, time stamping and date stamping of each individual message.

Message Queuing/Buffering

Buffering several messages in the main memory of the front end processor before passing them to the host computer or out to the remote terminal station (node.)

Slowing the flow of messages when the host computer or the remote terminal station (node) is overburdened by traffic.

Queuing messages into distinct input queues and output queues between the front end communication processor and the host computer or between the front end communication processor and the outgoing communication circuits.

Giving priorities to different communication circuits or automatically assigning priorities to various types of messages to speed throughput of various message types.

Handling the time-out facility, such as when a specific terminal station does not respond or when a circuit ceases to respond. The system times-it-out and, in the future, skips it so normal operation can be continued.

Error Control

Error detection for parity on single characters.

Error detection and automatic retransmission for parity checks on message blocks (cyclical redundancy checks and others).

Forward error correction techniques to reduce the errors flowing through the communication circuits.

Message Recording

Logging all inbound and outbound messages on magnetic tape for a historical transaction trail.

Logging the most recent 20 minutes on a magnetic disk for immediate restart and recovery purposes.

Monitoring for specific messages to identify trouble or as a security check.

Statistical Recording

Maintaining a continuous record of all data communication traffic such as number of messages processed per circuit, minutes of downtime per circuit, number of errors per circuit per hour or per day, number of errors encountered per program module, terminal stations that appear to have a higher than average error rate, average length of time in the queue for each message, number of busy signals on dial-up circuits, and any other unique statistical data collection required by the organization.

Facilitating online development of various graphs and charts showing the efficiency of the network on an hourly or daily basis.

Keeping track of online diagnostics performed by vendors.

Other Functions

Multiplexing.

Dynamic allocation of task management queues.

Automatic switchover to a backup host computer in the event of a primary host failure.

Circuit concentration where a number of low speed communication circuits might be interfaced to a higher speed communication circuit.

Performance of some functions of the Packet Assembly/Disassembly (PAD), if the front end also is serving as a Switching Node (SN) in a packet switching network. In this situation it contains the multiple databases required for alternate routing and packetizing of messages.

LINE ADAPTERS

Rather than being one specific piece of hardware, line adapters are a class of communication hardware. The line adapter performs a specific task or allows the interconnection of terminals or microcomputers to host mainframe computers in many types of configurations. Examples of this hardware are line interface modules, port sharing devices, intelligent port selectors, line splitters, digital line expanders, port/line security devices, data compressors, line protectors, bridges/gateways, network test equipment, and so forth. All of these devices are discussed in this section, with the exception of bridges/gateways and network test equipment which are presented in later chapters in conjunction with other related network material. Specifically, bridges/gateways are covered in Chapter 7, Local Area Networks, and network test equipment is discussed in Chapter 8, Network Management.

Line Interface Module   Line interface modules enable terminal users to connect to more than one network and switch between them, without plugging or unplugging any of the connector cables. For example, this device enables a terminal user to connect both to a central host computer and a local minicomputer, alternately accessing screens from either computer by using simple keyboard

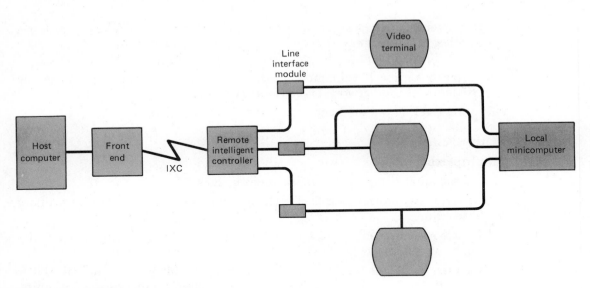

FIGURE **3-5** Line interface module.

instructions (see Figure 3-5). An extremely simplistic line interface module used with microcomputer printers is the two-position switch box that allows users to connect two printers to one microcomputer. In this case, users have to switch manually to whichever printer they want to use for printing a particular job.

Port Sharing Device. A port sharing device allows several incoming communication circuits to use a single port on a front end communication processor. All front end communication processors have a fixed capacity of ports. For example, if such a processor is designed to handle 50 ports, up to 50 incoming circuits can be connected to it. When users want to exceed the design capacity of a front end processor, a port sharing device may be used. Reexamine Figure 3-2 where you will see that three incoming circuits utilize three of the ports on the front end communication processor. Now look at Figure 3-6 where you will observe that four incoming communication circuits employ only one port on the front end communication processor because a port sharing device is used. Figure 3-6 shows a port sharing device with local terminals; no modems are required.

While use of a port sharing device may not be a long-term solution, it can become a short-term holding action until a new network can be configured or new hardware purchased. Telephone companies sometimes refer to these devices as bridges, although current usage generally restricts the term *bridge* to an intelligent device that allows interconnection of two dissimilar networks (defined in Chapter 7).

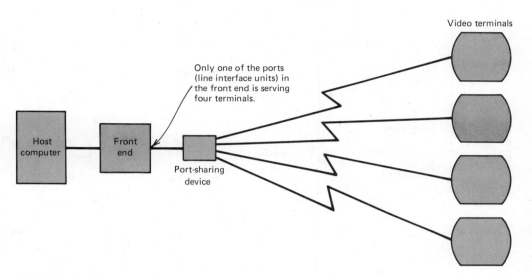

FIGURE 3-6 Port sharing device.

Intelligent Port Selector An intelligent port selector replaces the older telephone rotary switch. It provides the same connection facility as a rotary and gives a response equivalent to a busy signal if no ports are available. This device is utilized when you have many incoming dial-up communication circuits but not enough ports to allow all of the terminals to connect at the same time. It is configured into the network at the same physical position as the port sharing device of Figure 3-6. Intelligent port selectors are nothing more than devices that answer incoming telephone calls and connect to the front end if one of the ports is available. For example, if there were ten ports, it would try port 1, then port 2, port 3, and so on. If any one of the ten was free, you would be connected. If all ten were busy, you would get a busy signal and would have to dial back at a later time. Intelligent port selectors can handle different speeds of transmission (bits per second), and different communication codes (ASCII or EBCDIC); interface between dial-up or dedicated (lease line) ports; offer a busy signal or the opportunity to stay on the line to be placed in a queue for future connection; and collect network statistics on utilization.

Line Splitter Line splitters are similar to port sharing devices except in the matter of location. Line splitters usually are located at the remote end of the communication circuit, whereas port sharing devices are at the central site housing the host mainframe computer. A line splitter is a "switch" that allows several terminals to be connected to a single modem. The three terminals in Figure 3-7

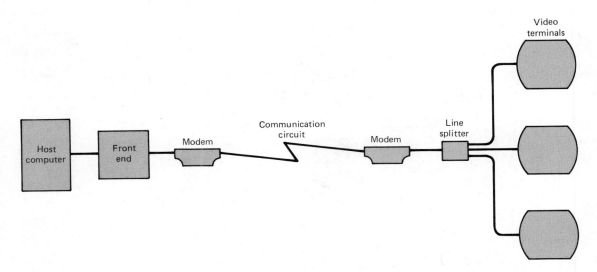

FIGURE 3-7 Line splitter.

share the total circuit capacity of the communication link, but the line splitter ensures that only one terminal at a time is utilizing the communication circuit. In other words, these three terminals are treated as if they were on a multidrop circuit. A *multidrop circuit* is one in which two or more terminals or terminal controllers share the communication circuit, although only one terminal at a time is allowed to transmit its data. With a line splitter, costs are reduced to whatever it costs for a single communication circuit and a single modem pair. Otherwise (see Figure 3-7) your network would require three communication circuits, three modem pairs, and three ports on the front end.

Digital Line Expander Digital line expanders enable customers to squeeze a greater number of voice or data channels into the bandwidth of a given communication facility. Companies with three or more analog leased circuits that run between two points might be able to save money by using a line expander because one such device provides up to eight voice communication circuits over a single 56,000 bits per second digital communication circuit. These eight voice communication circuits also can be intermixed with digital transmissions. An example of intermixing might be if you were to have six voice channels and one 9600 bits per second data channel, or four voice channels and one 19,200 bits per second data channel. Notice that you would be switching from analog to digital communication techniques to accomplish this kind of intermixing.

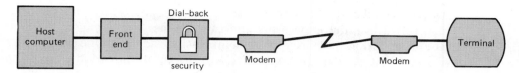

FIGURE **3-8** Port/line dial-back security.

Port/Line Security Device Port/line security devices provide dial-back security and are configured as shown in Figure 3-8. The procedure for using one such device is as follows.

1. The user calls the host mainframe computer from the remote terminal.

2. The dial-back security device at the central site intercepts the call and asks for the user's name and password. Some of these devices have circuitry for automatic ring blocking. This inhibits the modem from answering the caller by detecting a ring as it occurs and absorbing the ringing energy. In this case, the caller must enter an ID/password immediately because the dial-back security device waits only a short time before disconnecting the caller. The purpose of this ring blocking is to prevent a caller from hearing the carrier wave tone. In this case, computer hackers may not even realize that they have called a dial-up modem device connected to a computer.

3. The user at the remote terminal enters his or her name and password.

4. After receiving the name and password, the dial-back security device instructs its modem to disconnect the circuit (hang up).

5. The dial-back security device verifies the user and, through the use of the password, obtains that user's preestablished telephone number from the central system.

6. Assuming the user is a valid one, the dial-back security device then instructs its modem (at the central site) to dial up the modem at the remote site.

7. The remote site modem answers, and the session between the terminal user and the host computer begins.

Port/line devices provide security by allowing computer system access only to those individuals who have proper passwords and by restricting those users to "registered" telephone numbers. Thus, a hacker who is searching for entry into a computer system cannot break through the dial-back mechanism. Even if the hacker has the correct password and name, the system is prevented from calling back to the hacker's telephone.

Data Compression/Compaction Data compression/compaction devices are controlled by microprocessors and can increase the throughput of data over a communication link literally by compressing the data. One simple way to do so is to look at the data being moved over the communication circuit and send an instruction which provides a count of any characters that repeat themselves in sequence. This basic technique is called *run length encoding*, and it normally is used in combination with other techniques. Another technique, *code book compression*, uses specific codes to indicate a pattern of characters and phrases that are stored in each data compressor's memory. Thus, one character can be sent over a link to indicate many other characters.

A more complex method of encoding actually replaces standard ASCII and EBCDIC code. This technique, called *Huffman encoding*, uses tables of typically sent characters within a language and adjusts the number of bits it takes to send each character based on the relative frequency of that character in the language. Different Huffman tables can be used to optimize this technique for different applications or languages. A more flexible method, *adaptive Huffman encoding*, uses a mathematical algorithm to update the tables in real time to optimize the compression capability.

Data compression devices are located in two places: (1) between the terminal and the modem at the remote end of the communication circuit, and (2) between the modem and host mainframe computer at the central site. Data compressors can be used on either dial-up or leased private circuits.

One specific example of a data compressor is a standalone device that compresses data in half duplex mode for synchronous communication. During operation, this device accepts data from the terminal at 14,400 (9600 × 1.5) bits per second, and compresses the data so it can be sent to the modem for transmission at 9600 bits per second. Obviously, the reverse process occurs at the other end of the circuit where the data is decompressed and transferred to the front end processor. This type of unit can accept input speeds of 4800, 9600, and 14,400 bits per second. Another compression unit can compact a data stream of 2400 bits per second from a terminal into a modem's 1200 bits per second operating speed. This unit has a 2:1 compression ratio. Some manufacturers are predicting an 8:1 compression ratio in the not too distant future.

Line Protectors When non-AT&T devices are connected to the standard Direct Distance Dial (DDD) telephone circuits, the devices can be registered in one of three ways: permissive, programmable, and fixed loss loop. The type of registration determines how the terminal, mode, or telephone connects to the telephone line.

A *permissive* connection is used to connect a standard telephone to the telephone line or a data terminal to a switchboard. Permissive devices limit the

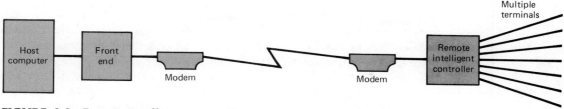

FIGURE **3-9** Remote intelligent controller.

amplitude (power) of the signal presented to the telephone line and use an RJ-11 plug.

A *programmable* device has two main advantages. First, the modem always transmits at its maximum allowable level, and second, the telephone company installs a "data-quality" circuit. It uses an RJ-45 plug (see Glossary).

The *universal* connection or *fixed loss loop* connection has a switch on the jack to prevent the installation of the wrong connector cable. When the fixed loss loop mode is used, the telephone company installs an attenuation pad to ensure that signals arriving at the telephone company central office do not damage their equipment.

These three devices protect the telephone company's circuits from extraneous signals that might damage the switching facilities at the central office.

INTELLIGENT CONTROLLERS

Remote intelligent controllers, sometimes called *intelligent terminal controllers*, usually reside at the distant or far end of a communication circuit (see Figure 3-9). A remote intelligent controller is simply a scaled-down front end communication processor. In fact, you can have a remotely located front end processor for an area of the country or world where hundreds of terminals are located. In this case you connect the remotely located front end to the host computer front end by high speed data circuits.

Remote intelligent controllers control four to sixteen local terminals, although some are much more powerful. There is a unique address for each controller and, therefore, a further address or unique memory space for each terminal connected to the controller. Remote intelligent terminal controllers are used primarily because they allow an organization to have full duplex transmissions between the two electronic devices (front end and remote local controller).

A remote intelligent controller might perform any or all of the functions performed by a front end, but as noted above, usually it is scaled down and not as powerful, although each vendor's terminal controller has its own set of functions. These controllers started out as simple devices to control four or eight video terminals. Their power has been increasing ever since so that today one device might control sixteen multifunctional terminals in a branch bank. The terminals in a branch bank might be entirely different from each other, say, four video terminals, four simple teller inquiry terminals, two high speed printers, one sophisticated wire transfer data entry terminal, one facsimile machine, and so on. This variety makes a remote intelligent controller especially desirable.

Today, microcomputers, especially the powerful AT version, are replacing many of the intelligent control devices. Microcomputers serve as intelligent controllers when interfacing on micro-to-mainframe communication links. The use of microcomputers as terminals is causing a problem with regard to intelligent controllers. For example, the long-standing mainstays, the IBM 3174 and 3274 controllers, are being overwhelmed with transmissions of data from microcomputers. This development can be attributed to our movement from a batch environment, through the single transaction processing environment, to our present entrance into the file transfer environment.

We used to send our batches of data to the data center for key entry. This occurred entirely outside of the communication links. Then we sent our batches of data over communication links, but immediate processing was not required. Today we send most of our business transactions over the communication circuits on a transaction-by-transaction basis. These individual, short transactions are processed and immediately returned to the sending microcomputer.

We also are transmitting entire files of data (thousands of characters) by using microcomputer systems. This file transfer is overloading controllers which were built to handle many short transactions of 100 or less characters each. For example, if one microcomputer, which is connected to a controller that has 16 microcomputers attached to it, starts transmitting a file, then the entire response time might be slowed by one or two seconds. If a second microcomputer starts transmitting a file at the same time, the response time might degrade to one or two minutes. But if a third microcomputer starts transmitting a file simultaneously with the first two, it can bring down (cause a crash) the intelligent control device. Thus, the people at the central site would have to restart the controller, which would delay everyone's data processing needs for 5 to 20 minutes.

Newer versions of the older controllers now possess even more power. For example, some of them can perform operations, such as protocol conversion, allow direct connection to local area networks, handle 32 terminals instead of 16, perform multiplexing, and handle file transfer more easily than can be done through individual transactions.

MODEMS

As you learned in the previous chapter, modem is an acronym for *MOdulator/ DEModulator*. A modem takes digital electrical pulses (baseband signaling) received from a terminal or microcomputer and converts these signals into a continuous analog signal (broadband signal) that is acceptable for transmission over a communication transmission circuit. In this section we will expand our understanding of modems by depicting some of their alternate features and uses. You may want to review Figure 2-22 in Chapter 2 before beginning this discussion.

Modems are classified by the speed at which they operate. The less expensive modems used with microcomputers usually operate at 300, 1200, and 2400 bits per second, although microcomputers easily can handle greater speeds. With larger and more centralized networks, modems traditionally operate at speeds of 4800, 9600, 14,400, 50,000, 56,000, and 1.544 million bits per second. An example of a fast modem that can be used with microcomputers might be Racal-Milgo's RM-1822D modem that can send data over ordinary telephone lines at speeds ranging from 300 to 18,000 bits per second. At line speeds of 300, 1200, and 2400, this modem can talk to any Hayes-compatible modem (that is, it has the ability to communicate with a modem manufactured by Hayes). Above 2400 bits per second it must talk with another RM-1822D modem.

High speed modems do save on line charges. For example, at 1200 bits per second it takes approximately one hour to transmit a 120-page file, but at 18,000 bits per second transmission time drops to approximately four minutes.

To illustrate the dramatic effect of higher communication speeds, here are some theoretical times for sending files of different lengths at different speeds. First, let us define some of the files. A 1,000-byte file (10,000 bits using start-stop asynchronous transmission) is about one-half a video screen in terms of output. A 10,000-byte file might be a four- to five-page document. A 50,000-byte file might be several complex graphic pictures. A 100,000-byte file might be a good-sized spreadsheet or a 45-page document. Figure 3-10 illustrates comparative transfer times for these file sizes.

The mathematics in Figure 3-10 were calculated as described here. First, the number of bytes in the file are multiplied by 10 bits per byte (8 bits per character plus 1 start and 1 stop bit). The result equals the number of bits to be transmitted. Then the number of bits to be transmitted is divided by the bits per second transmission speed of the modem, which gives the theoretical transfer speed in seconds. Remember that this is a general calculation which does not account for any control characters or the possibility of errors that might cause retransmission of the same message. Now look at Figure 3-10 to see approximately how long it takes to transmit a file, and remember it is the modem that governs the transmission speed.

| | Bits per second | | | |
File size	300	1200	2400	9600
1,000 bytes	33 sec	8 sec	4 sec	1 sec
10,000 bytes	5 min 30 sec	1 min 20 sec	40 sec	10 sec
50,000 bytes	27 min 36 sec	6 min 40 sec	3 min 10 sec	1 min 5 sec
100,000 bytes	55 min 33 sec	13 min 53 sec	6 min 56 sec	1 min 35 sec

FIGURE 3-10   File transfer time.

Optical Modems   Transmitting on an optical fiber requires the use of an *optical modem*. This type of modem converts the electrical signals from a terminal or microcomputer to pulses of light that are transmitted down the optical fiber. Optical modems connect to RS232 and RS449 connector cables. They operate using asynchronous or synchronous transmission up to 10 million bits per second. Current models can transmit up to 5 kilometers (1 kilometer is equal to 3,280.8 feet) without using amplifiers/repeaters, but that certainly will be increased in the future, as will the 10 million bits per second speed. Within the optical modem is a laser or light-emitting diode that originates the light pulses, along with the electronics needed to convert the electrical signal prior to transmitting the digital pulses of light.

A specific example of a fiber optic modem might be one that is designed to operate over distances of up to 2.2 miles, operates in full duplex mode, and transmits asynchronously at speeds up to 19,200 bits per second. This fiber optic modem is plug-compatible with RS232C devices and was designed to link asynchronous microcomputers with central host computer systems. Electronic data signals enter the device and are converted to light sources. A light-emitting diode transmits the data over fiber optic cable to the remote modem, which reconverts the light signal to electronic signals. This inexpensive fiber optic modem sells for approximately $100.

Short Haul Modems   Another type of modem is a *short haul modem* in which you use your own wire pair cable to transmit direct electrical baseband signals. Sometimes this is called a 20-milliamp circuit. There are inexpensive (costing as little as $40) interface devices that can interface any RS232 port from a modem to a standard 20-milliamp loop system. Typically, these systems transmit at 19,200 bits per second over a distance of several miles. This type of modem also is called a *line driver*. When you put more electrical power (amperage) onto the communication circuit, you can drive your signal for a distance greater than several miles.

Such short haul modems are used within buildings, a plant, college campus, or university facility. An example of a line driver is one that operates asynchronously over full duplex, four-wire circuits at speeds up to 19,200 bits per second for a distance of more than 1 mile. When this same line driver is used at lower speeds, transmission distance increases to 18 miles at 110 bits per second.

Microcomputer users often need short haul modems. For example, if you want to connect two microcomputers together and they are several thousand feet apart, a small short haul modem may be the answer. One such device consists of a box approximately the size of a package of cigarettes. The standard RS232 connector plug is on one side of the box, and the cable attaches to the other side. You attach these small modems first to each end of your cable and then to the serial ports of the microcomputers. They get their power through the serial ports. Because short haul modems do not require an external power source, they sometimes are referred to as *modem eliminators*. These tiny modems are able to send data down the cable for distances of approximately 3,000 feet at 9600 bits per second, or up to 6 miles at 1200 bits per second. This type of short haul modem is analogous to increasing your RS232 connector cable beyond its 50-foot maximum length.

These modems are ideally suited for interconnecting microcomputers between two different offices in the same building. Typically, a pair of these modems with 500 feet of cable costs approximately $150. You should use a shielded RS232 cable, although two wires (twisted wire pair) or telephone cords with the standard RJ-11 telephone connectors also work. To transmit data between microcomputers using these short haul modems, you need to call up your communication software package and transmit as the package instructs.

Acoustic Couplers An older type of modem is an *acoustic coupler* (see Figure 3-11). This modem is used primarily for dial-up because it can interface with any basic telephone handset. All you do is call the computer and place the telephone handset into the acoustic coupler. The coupler performs the typical modem functions of converting direct electrical signals from the terminal to frequency modulated tones (frequency shift keying) that can be sent over any telephone communication circuit.

Null Modem Cables A *null modem cable* allows transmission between two microcomputers that are next to one another (within 6 to 8 feet) without the use of a modem. This specially configured cable connects the two microcomputers. A complete explanation of how a null modem cable works is presented in Chapter 6, Microcomputers and Communications.

Dumb Modems Modems typically have been differentiated as either smart or dumb. This differentiation is based on their varying abilities to respond to a

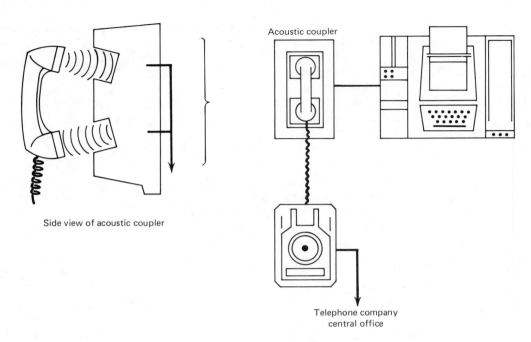

Side view of acoustic coupler

Acoustic coupler

Telephone company
central office

FIGURE **3-11** Acoustic coupler.

command language through which a user's communication software package
instructs the modem to perform various tasks, such as dialing calls, answering
incoming calls, and redialing calls. *Dumb modems* must be set manually for
parameters of speed, originate or answer mode, and so forth. Switches are used to
set these parameters, and then the user dials the call by using a telephone.

Smart Modems　By contrast, *smart modems* are commanded to perform their
functions through the use of a command syntax language. This language is used
to control their functions, such as changing speed and dialing calls. For example,
if a Hayes modem is used with your microcomputer, then you might type the
letters AT. These letters instruct the modem to pay attention to the next set of
letters because they constitute a command or parameter change. Therefore, typing the sequence ATD tells the modem to pay attention, and the D tells it to dial
a number. The complete command might be ATD555-1212, which tells the
modem to pay attention, to dial a number, and to dial 555-1212.

Did you ever wonder what all those modem lights mean? While their
significance might vary from modem to modem, here is an explanation of the
eight lights on a Hayes Smartmodem 1200.

MR—Modem Ready. This light is on when your 110-volt ac electrical power is turned on.

TR—Terminal Ready. This light shows that the modem has received a signal from the microcomputer to which it is connected, telling it that the microcomputer is now ready to do something, such as send data.

SD—Send Data. This light is on when data is being sent from the RS232C serial port of your microcomputer to the modem.

RD—Receive Data. This light is on when the modem is receiving data from the distant computer and the modem in turn is sending it to the RS232C serial port of your microcomputer.

OH—Off Hook. This light is on when the modem is off hook with regard to the telephone line. If the light is on, you probably are connected to the telephone company and using the circuit.

CD—Carrier Detect. This light turns on when the modem detects a carrier wave tone from a distant modem.

AA—Auto Answer. This light is on if the modem is set to answer an incoming call automatically.

HS—High Speed. This light is on when the modem is set to work at 1200 bits per second. It is not on when the modem is operating at 300 bits per second.

Beyond smart modems is a classification that might be called *intelligent* or *advanced modem features*. These more expensive modems contain microprocessor chips and internal read only memory (ROM) coding to provide sophisticated communication protocols and diagnostic checking that run within the modem itself.

This is where you begin moving into a classification of equipment that overlaps with other equipment because some of the newer, more sophisticated pieces of communication hardware cross over the boundaries that define other pieces of hardware. For example, some modems not only perform digital-to-analog conversion but also operate as multiplexers, security restricter devices, encryption devices, error detection and retransmission devices, and so forth. The problem is that there are many types of hybrid equipment on the market. As a result, it is no longer possible to say that a device is solely a modem because it might be a combination modem, multiplexer, and encryption device.

When a microcomputer uses a modem to dial a host mainframe computer, a lot more happens than meets the eye. In an ideal situation, you only have to know how to operate the communication software, which provides the communication capabilities. If you have a modem with a command syntax (language), the software uses the microcomputer's operating system to set up a communication circuit to the modem via the microcomputer's serial port. It sets the requested speed and then issues any other commands required to set up the modem's dialing or answering capabilities. This is explained further in Chapter 6.

Digital Modems If the communication circuits use digital transmission for their entire length, instead of converting to analog transmission as is done with normal telephone circuits, you have the equivalent of a *digital modem*. This modem shapes the digital pulses and performs all auxiliary functions needed, such as loopback testing and checking the circuit diagnostics. Its special function is to convert a digital signal to a more precise and more accurate digital signal. For example, a digital modem can take a weak electrical signal, put very precise timing characteristics between the pulses, put it out at a certain strength, and control its electrical characteristics. This is done to reduce noise, distortion, and errors (digital modulation was described in Chapter 2). Digital modems are much simpler than digital-to-analog modems, as evidenced by their cost. While a 9600 bits per second digital-to-analog modem might lease for $100 per month, its digital counterpart might lease for $15 per month, a factor of seven times.

Digital transmission (discrete on and off pulses) involves sending baseband signals between your premise and the telephone company's central office and then transmitting digital pulses over the long distance communication circuits between cities. When using digital transmission, you use a digital modem which is necessary to shape the digital pulses. The correct name for this modem is either a channel service unit (CSU) or a data service unit (DSU). Sometimes they are referred to as a CSU/DSU.

The *channel service unit* performs transmit and receive filtering, signal shaping, longitudinal balancing, voltage isolation, and equalization functions, and it supports remote loopback testing. The *data service unit* is a channel service unit that, in addition, provides bipolar conversion functions. These bipolar conversion functions ensure proper signal shaping and signal strength for transmission that is almost error free. In the early days of digital transmission, you had to use both a DSU to manage the interface to the communicating computer and a CSU to manage the connection over the digital communication circuit. Today's digital modems incorporate both DSU and CSU into what is now called a DSU. In other words, most DSUs incorporate the CSU, which was used to manage the electrical characteristics on the circuit.

Features of Modems Now that we have seen the types of modems and discussed the basic function of a modem (to modulate a signal), let us discuss some of the features of various modems. The features listed below may not be in all modems; rather, this is a comprehensive list of the types of features that may be built into modems.

- *Loopback* functions for diagnostic purposes probably are the single most important feature built into modems. Automatic loopback allows the user to set a remote modem on loopback and send a message to that modem. The message is looped back to the original sender, where it can be checked for accuracy to help diagnose where a fault might be in the network. Loopback switches allow you to diagnose whether the problem is in the connector cable between the terminal and modem, whether it is on the digital or analog side of the modem, or whether the problem lies in the local loop communication circuit itself. The digital side of the modem is the side that plugs into the terminal; the analog side is the side that plugs into the telephone circuits, unless you are on a complete digital network. In that case you would use a data service unit instead of a standard modem.

- Some modems have the ability to be turned on or off from a remote station. These contain automatic answering and automatic calling capabilities, so that a remote terminal can be started from thousands of miles away.

- Some modems allow the simultaneous transmission of both voice and data. One model allows a voice conversation to go over the circuit while simultaneously transmitting a data stream of 2400 bits per second.

- Some modems allow a *reverse channel* capability for message acknowledgments. The purpose of a reverse channel is to avoid interruption of the ongoing message stream but still provide for a path over which message acknowledgments can be sent. For example, the modem could receive messages at 2400 bits per second but simultaneously send back, in the reverse direction, one character acknowledgment for each received message. This tells the original transmitter whether the message was received correctly. This resembles full duplex transmission, but it is different technically because of the vast differences of speed in each direction. The reverse channel may be transmitting at only 10 to 75 bits per second.

- Multiplexing is built into some modems.

- Microprocessor circuits are built into some modems for automatic equalization to compensate for electronic instabilities on the transmission line. This equalization compensates for attenuation/delay distortion which causes errors that require retransmission of messages.

- Many modems today have built-in diagnostic routines for self-checking of their own circuits in order to determine where a fault might lie.

- Some modems have alternate speed switches so you can switch up or down in speeds, let us say from 9600 to 4800 bits per second.

- Some modems have *split streaming* by which the modem transmits three message streams at different speeds. One message stream is transmitted at 4800 bits per second and the other two at 2400 bits per second.

- Some modems are more efficient because they have a lower clocking or retrain time. *Retrain time* is the turnaround time when the message direction changes. For example, when you are transmitting in half duplex mode (one direction at a time), the modem is in the transmit mode, but when you receive a message, it must switch its electronic circuits to be in a receive mode.

- For efficiency, some modems have a longer flywheel effect for brief drops of the carrier wave (the 2025/2225 hertz signal). When the carrier wave drops for a couple of microseconds, you technically have lost the circuit. With a built-in flywheel effect, however, the modem can remain connected to the circuit and pick up where it left off after the carrier wave returns. This assumes the drop is not for too long a period of time.

- Modems can be plugged into other equipment by means of various standard interface cables. For example, the plug and cable that come with the modem should be specified through one of the standards such as RS232 or RS449. It can be quite embarrassing to order a modem and then learn that it will not plug into your terminal.

- Integrated modems are built into a device. For example, a modem can be built into a terminal, a front end communication processor, or a microcomputer.

- Certification of modems involves AT&T and the Bell Operating Companies. For modems to operate over the dial-up telephone network without data access arrangements (DAA), they must be registered and certified with the Federal Communications Commission. Certification is required only for dial-up and not for modems operating over lease circuits. When using the dial-up network, you will be wise to select certified modems because data access arrangements add extra cost to your circuits. The purpose of a data access arrangement is to limit the amplitude (power) of the signal presented to the telephone circuit. See the section Line Protectors, earlier in this chapter.

- Some modems can perform full network analysis, monitoring such features as the RS232C interface and circuit characteristics (analog), although usually this is performed by other test equipment. Tests are carried out during normal data transmission using out-of-band signaling (unused portions of the bandwidth).

- Some modems provide status indication lights on the modem's front panel, such as CLEAR TO SEND (CTS) or REQUEST TO SEND (RTS). These indicators are helpful but do not constitute comprehensive diagnostics, although at a remote site they may be the only form of diagnostics. When problems develop at a remote site, you look at the modem status lights to

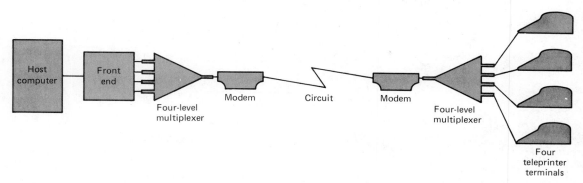

FIGURE **3-12** Multiplexed circuit (TDM).

see whether the terminal is "bringing up" the RTS and whether the modem is "bringing up" the CTS. If not, one of them is working improperly.

• Some vendors emphasize failsafe modems: a spare modem is mounted internally in the same package and tied to the central site control facilities. If the primary modem fails, the spare is switched in automatically.

MULTIPLEXERS

To *multiplex* is to place two or more simultaneous transmissions on a single communication circuit. An important aspect of multiplexing is transparency. *Transparent* means that the multiplexer system does not in any way interrupt the flow of data or the RS232 interface signals. Neither the computer, nor the modem, nor the terminal/microcomputer using the modem knows that the multiplexer system is being used regardless of whether leased or dial-up circuits are being used. When the line is multiplexed at one end and demultiplexed at the other, each user's terminal thinks it has its own connection to the host mainframe computer. Multiplexing usually is done in multiples of 4, 8, 16, and 32 simultaneous transmissions over a single communication circuit. Figure 3-12 shows a typical four-level multiplexed circuit. Multiplexers can be separated into major categories, such as frequency division multiplexers, time division multiplexers, and statistical time division multiplexers.

Frequency Division Multiplexing (FDM) Frequency division multiplexing can be described as having a stack of four or more modems that operate at different frequencies so that their signals can travel down a single communication circuit. Another way of looking at frequency division multiplexing is to imagine a group

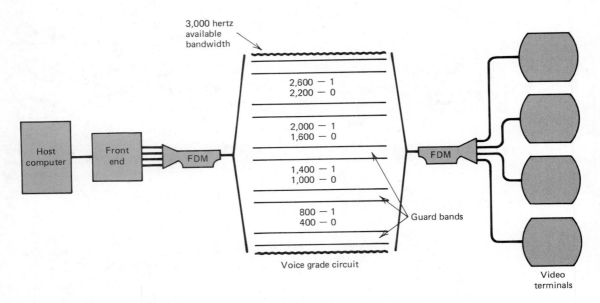

3,000 hertz available bandwidth

2,600 — 1
2,200 — 0

2,000 — 1
1,600 — 0

1,400 — 1
1,000 — 0

800 — 1
400 — 0

Guard bands

Voice grade circuit

Host computer

Front end

FDM

FDM

Video terminals

FIGURE 3-13 Frequency division multiplexed circuit (FDM).

of people singing. There might be a bass, a baritone, an alto, and a soprano. What you hear is the combination of the four people singing, but sometimes you can identify clearly one or more of the individual singers.

With FDM, the frequency division multiplexer and the modem usually are a single piece of hardware. Compare Figures 3-12 and 3-13 to see that the TDM has two separate pieces of hardware (a time division multiplexer and a modem), whereas the FDM has a single piece of hardware (multiplexer/modem).

In FDM, the frequency division multiplexer uses the available bandwidth of a voice grade circuit, dividing it into multiple subchannels. When we discussed modems in Chapter 2 (you might review Figure 2-22), the modem modulated the signal into only one pair of frequencies (two frequencies) which were used to transmit the binary 1s and 0s. In that case the frequencies were 2225 hertz and 2025 hertz. Now, our multiplexer in Figure 3-13 is subdividing the bandwidth of the voice grade circuit into four pairs of frequencies in order to allow four simultaneous tranmissions of 0s and 1s.

The guardbands in Figure 3-13 are the unused portions of bandwidth that separate each pair of frequencies from the others. They keep the signals in each of the four subchannels from interfering with the adjacent subchannels and allow space for frequency drift. The guardband serves the same purpose as does a plastic insulator surrounding a copper wire; it keeps adjacent copper wires or subchannels from interfering with the others' transmission.

Another characteristic of FDM is that the subchannels need not all terminate at the same location; therefore, frequency division multiplexing can be used in a multidrop network where each dropoff operates at a different frequency. This changes the multidrop philosophy of sharing the circuit to one in which all terminals on the multidrop circuit can use the circuit simultaneously.

Frequency division multiplexers are somewhat inflexible in that, once you determine how many subchannels are required, it may be difficult to add more subchannels without purchasing an entirely new frequency division multiplexer that is divided into a greater number of subchannels. Also, the maintenance cost on frequency division multiplexing equipment usually is greater than that required for time division multiplexing equipment.

Time Division Multiplexing (TDM) Time division multiplexing is really a type of time slicing or sharing the use of a communication circuit among two or more terminals. Each terminal takes its turn. In TDM, the multiplexer takes a character from each transmitting terminal and puts them together into a frame. The frames are put onto a high speed data stream for transmission to the other end of the circuit. In Figure 3-14 we show a four-character frame for a four-level multiplexer. This is *pure* multiplexing because it is totally transparent to everyone on the network, including the system programmers. In pure multiplexing your messages are never held back or slowed up by the multiplexer, as in statistical time division multiplexing which is discussed later in this section.

In Figure 3-14 a character is taken from each terminal, placed in its frame, sent down the circuit, and delivered to the appropriate device at the far end of the circuit. If each of the four terminals transmits at 1200 bits per second, then the time division multiplex bit stream has to transmit at 4800 bits per second. Notice that there is no terminal addressing here. Each position of the four-position frame gives its character to the appropriate terminal at the other end, even if that character is a blank. When you start addressing each character position of a frame, then you are moving into the world of statistical time division multiplexing.

Time division multiplexing generally is more efficient than frequency division multiplexing, but it does require a separate modem. (Frequency division multiplexing is in reality a special modem.) It is not uncommon to have time division multiplexers that share a line among 32 different low speed terminals, although these might be replaced with statistical time division multiplexers that can hold 32 higher speed terminals. It is easy to expand a time division mutliplexer from, let us say, 8 to 12 channels. All TDM channels usually originate at one location and all terminate at another location, but this is not always the case. Time

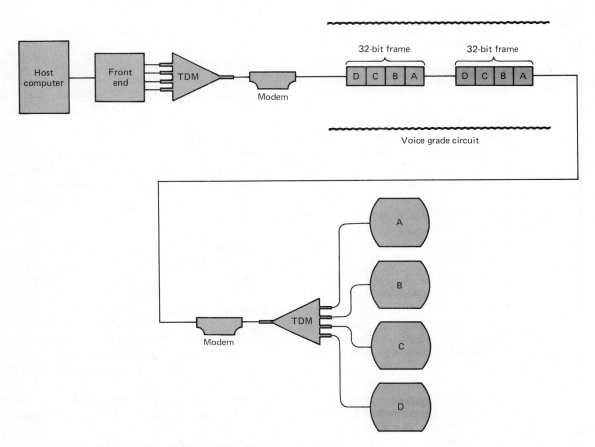

FIGURE **3-14** Time division multiplexed circuit (TDM).

division multiplexers usually are less costly to maintain than frequency division multiplexers.

Statistical Time Division Multiplexing (STDM) Statistical time division multiplexing allows the connection of more terminals to the circuit than the capacity of the circuit. In its simplest context, if you have 12 terminals connected to a statistical time division multiplexer and each terminal can transmit at 1200 bits per second, then your total is 14,400 bits per second transmitted in a given instant of time. However, if the STDM/modem/circuit combination has a maximum speed of only 9600 bits per second, then you would find that there might be a period of time when the system is loaded above its capacity.

The technique of statistical time division multiplexing takes into account the fact that there is some downtime and that all terminals do not transmit at their maximum rated capacity for every possible microsecond that is available. With this in mind, you start by addressing each character in a frame or message and time division multiplex on a statistical basis.

For example, assume we have a statistical time division multiplexer that multiplexes individual characters from 12 terminals. In this case, a terminal address is picked up in addition to the character and is inserted into the frame. Using the same four-character frame that was used in the previous example, look at Figure 3-15. Notice that, in addition to the eight bits for each individual character, we have added five bits of address space (ADD.). These five bits of address allow you to address 32 different terminals using binary counting ($32 = 2^5$). Now the multiplexer takes a character from each terminal only when the terminal has a character to send. The technique used is to scan through the 12 terminals and take characters from, let us say, terminals 1, 4, 5, and 12. These are sent immediately. Then all 12 are scanned again to determine which terminals need servicing. This process is repeated indefinitely. At the other end of the communication circuit, the character is given to the proper device because the 5-bit address that is included with each 8-bit character identifies the terminal device.

Another type of statistical time division multiplexer involves multiplexing entire messages from terminals. With this type of multiplexer you use one of the new bit-oriented protocols and interleave entire messages rather than characters. For example, when terminal 1 has a message to send, the multiplexer picks up the entire message (all of its characters) and passes it to the modem for transmission, after which the multiplexer immediately scans for the next terminal that has an entire message to send. The primary difference is that the first statistical scheme is for asynchronous character-by-character transmission, while the second scheme is for synchronous or block transmission.

Stat muxes, as they are called, are becoming the most commonly used type of multiplexer. Because they use software and a microprocessor chip built into the multiplexer, they can support a number of devices at different speeds, without the modem having to equal the total composite speed of all of the attached devices. An important aspect of statistical multiplexing is *flow control*. Because the statistical multiplexer allocates the usage of the communication link based on statistical demand of each attached device, the multiplexer must have a way to stop and start transmission from any of its attached terminals when one device tries to communicate too much data. Without flow control and internal data buffers, statistical multiplexers would lose almost as much data as they send on a busy link.

While STDM may be very efficient, you should be aware that it can cause time delays. When traffic is particularly heavy, you can have anywhere from a 0- to 30-second delay of your data. Some data is held back by buffers when too many terminals transmit at maximum capacity for too long a period of time. A side

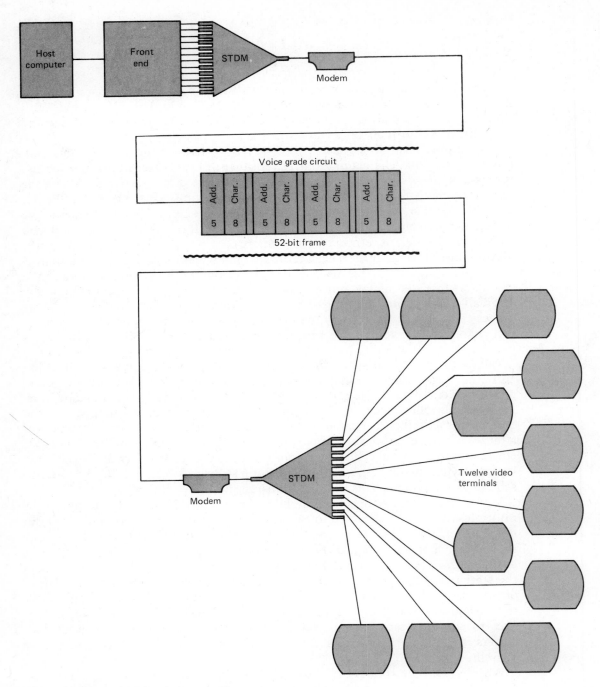

FIGURE 3-15 Statistical time division multiplexing (STDM).

effect of statistical multiplexing is the overhead. While the software provides the benefits of statistical multiplexing, it also reduces the throughput efficiency on a circuit by 8 to 30 percent. One technique that is implemented to improve the throughput of statistical time division multiplexers is to compress the data, thus reducing the number of bits transmitted per character or per message. Again, notice that several pieces of hardware are combined into a single piece of hardware. In this case, statistical time division multiplexing and data compression are combined.

Fiber Optic Multiplexing Fiber optic multiplexers are available, and they interconnect with the RS232, RS449, V.35, and T-1 interfaces. For example, a fiber optic multiplexer box might take up to 16 channels of data, with each channel transmitting at a capacity of 64,000 bits per channel, and multiplex it onto a 14 million bits per second fiber optic link. Fiber optic multiplexers operate similarly to time division multiplexers, but with much higher data transmission capacities. The transmission distances usually are limited to 1 to 3 miles unless amplifiers are used somewhere along the fiber optic cable. Fiber optic multiplexers are ideal for multiplexing T-1 circuits (1.544 million bits per second). Fiber optic T-1 multiplexers can multiplex up to eight T-1 streams of data on a single fiber pair at a composite data rate of 14 million bits per second over 3 kilometers without amplifiers and, therefore, without cable splices. You will recall that we discussed how difficult it is to splice optical fiber cables in Chapter 2.

T-1 Multiplexing T-1 multiplexing involves a special kind of multiplexer combined with a high capacity digital service unit (DSU) made especially for managing the ends of the T-1 link. T-1 multiplexers are expensive and typically are sold based on the "port" capacity or the maximum number of devices that can be connected to the T-1 multiplexer. Because T-1 links can carry digitized voice, data, and image (video) signals, it is important to differentiate T-1 multiplexers by their ability to provide each of these three services.

One typical T-1 multiplexer can take a T-1 communication circuit that operates at 1.544 million bits per second and multiplex it onto 48 voice and data communication circuits. These 48 communication circuits can operate in either synchronous or asynchronous mode. Another T-1 multiplexer can subdivide the 1.544 million bits per second channel into 96 circuits at 9600 bits per second or even 200 channels that support 4800 bits per second. Other T-1 multiplexers can accept very high speeds (up to 768,000 bits per second for compressed video transmission) and, at the same time, multiplex lower speed (such as 4800 bits per second) data transmission paths through the multiplexer.

In addition to T-1 multiplexers, there is a similar device called a *channel bank*. Both T-1 multiplexers and channel banks multiplex high speed communication circuits (T-1 circuits) into lower speed communication channels. The difference, however, is that channel banks are designed to handle analog voice transmissions, while T-1 multiplexers are designed primarily for digital data transmissions. Because a channel bank's standard input rate is 64,000 bits per second, it typically cannot divide a T-1 channel into more than 24 subchannels. Channel banks therefore are used primarily to digitize analog voice transmissions.

Channel banks also have limited data-carrying capabilities. With difficulty, they can handle 9600 bits per second rates transmitted by the appropriate modem, but they cannot directly handle data rates at 56,000 or 64,000 bits per second except via a dataphone digital service connection. (Dataphone digital service is a digital transmission service offered by AT&T.)

Channel banks are more suitable than T-1 multiplexers for moving voice calls between PBXs (switchboards). The fact that channel banks cannot divide a T-1 communication link into more than 24 subchannels also can result in inefficient handling of multiple data circuits. It takes eight channel bank termination devices (four at each end of the link) and four T-1 circuits to transmit 96 data streams at 9600 bits per second from Point A to Point B. Each of the four T-1 circuits carries 24 channels, each of which is capable of transmitting at 64,000 bits per second, but actually carrying only 9600 bits per second. By contrast, it takes only two T-1 multiplexers and one T-1 communication circuit subdivided into 96 fully utilized channels to carry the same 96 data streams at 9600 bits per second.

At the higher capacity end of the scale, AT&T offers a digital multiplexer that can combine up to 28 T-1 communication channels into a single 45 million bits per second T-3 transmission facility. You should review Figure 2-48 in Chapter 2 to put these capacities in perspective.

Multiport Modems Multiport modems usually are high speed synchronous modems in the range of 9600 bits per second. These modems generally encode several bits into a single signal element such as you saw in Chapter 2 in the discussion of bits versus baud. At 2400 bits per second, 2 bits are encoded; at 4800 bits per second 3 or 4 bits are encoded, depending on whether it is 1600 baud or 1200 baud, respectively. At 9600 bits per second, 4 or more bits are encoded. Typical 4800 bits per second modems may be channelized (subdivided) 2 × 2400. Typical multiport 9600 bits per second modems may be channelized using any of these combinations: 2 × 4800, 4 × 2400, or 2 × 2400 with 1 × 4800 (see Figure 3-16). The multiport modem may be looked upon as a bit-oriented time division multiplexer. Multiport modems use synchronous transmission; they cannot handle asynchronous transmission.

Concentrators In today's terminology, concentrators are special forms of multiplexers, or even biplexers. Concentrators are used for the same purposes as multiplexers. In fact, they originally were intelligent multiplexers.

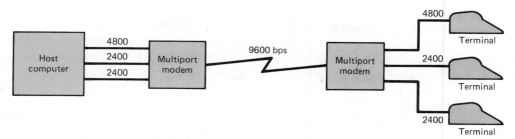

FIGURE **3-16**   Multiport modem.

The primary use for a concentrator is to combine circuits. In this respect you can have 16 low or medium speed circuits that are concentrated into one or two high speed lines. For example, you might concentrate approximately twelve 4800 bits per second communication circuits into one 56,000 bits per second digital communication circuit. Even though this does not work out in a perfectly equal way, the statistical intelligence takes care of the small difference. Like statistical multiplexers, concentrators can buffer or hold back data. Some concentrators even can perform switching functions to switch messages to different communication circuits.

To avoid confusion, you can assume that when you have *pure* multiplexing, there is no basic programmed intelligence in the device, but when you have intelligence and programming capability, it might be a statistical multiplexer or a concentrator. Also, the newer STDM/concentrators perform other functions such as circuit contention and switching.

Biplexers   Biplexers sometimes are called inverse multiplexers (see Figure 3-17) because a biplexer can take a 19,200 bits per second transmission circuit and divide it into two incoming or outgoing 9600 bits per second circuits.

AT&T Multiplexing   The following summaries of eight types of multiplexing used by AT&T will help you have a more complete understanding of multiplexing.

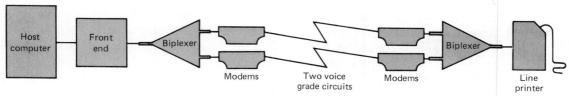

FIGURE **3-17**   Biplexer.

Frequency division multiplexing (FDM) divides the assigned bandwidth of a communication channel (its frequency range) into narrower bands so that the signals can be sandwiched for common transmission. Frequency division multiplexing is used for analog signals.

Time division multiplexing (TDM) divides the communication channel into separate time slots. This allows two or more signals to be assigned to the same circuit path at slightly different points in time.

Bit compression multiplexers (BCM) can be used to reduce from 64,000 to 32,000 the number of bits per second that are required to carry voice transmission. This cuts in half the amount of information needed to represent a voice signal, thus allowing more digitized signals to share one circuit path.

Circuit switched digital capacity (CSDC) uses time division compression multiplexing to send and receive data over an ordinary telephone line. In this case the multiplexing equipment is located at your premise as well as at the telephone company's central office location. This equipment takes an incoming data stream of 56,000 bits per second and partitions it into 3-millisecond segments. The data stream then is speeded up to 144,000 bits per second, compressing each segment in time from 3 to 1.38 milliseconds, which are sent as bursts over the wire pair. Not only does this technique speed data transmission, but it also allows you to use the same communication circuit for simultaneous voice and data communications, or to alternate between the two.

Digital channel banks, such as AT&T's D4 and D5 Digital Terminal Systems, do time division multiplexing. This technique interleaves representative pieces of 24 conversations onto one data stream. Each piece is transmitted in a burst of data that lasts about 5 microseconds. This is similar to the technique called TASI (see Index).

DATAKIT™ virtual circuit switch (AT&T) uses statistical multiplexing techniques to transmit information in packets. Mathematical algorithms sample the bit streams from many sources and break them into data packets for transmission over a high speed communication circuit. When there is a pause in one message stream, the statistical multiplexing technique stuffs a packet or two from another message stream into that pause. In effect, this interleaves packets of information. Again, this same philosophy is used with TASI.

Light wave multiplexers combine streams of electrical pulses that already are multiplexed. For example, AT&T's digital channel banks are light wave multiplexers that convert the streams of electrical pulses into a single stream of light pulses. AT&T's DDM-1000 can multiplex 56 of the electrical pulse streams, each of which represents 24 voice channels. This multiplexer therefore makes it possible to have more than 1,300 telephone calls travel simultaneously through a single glass fiber.

Time compression multiplexing (TCM), also known as *ping-ponging*, uses half duplex transmission to simulate full duplex communications over twisted pair wiring. For example, by transmitting 272 kilobits per second at half duplex, TCM can provide an average transmission rate of 136 kilobits per second in full duplex (see CSDC above).

The principles of multiplexing predate the invention of the telephone. Alexander Graham Bell experimented with a type of multiplexing for telegraph systems, but the technology needed to apply his technique did not exist in his day. Today's multiplexing systems are utilized on voice, data, image, and optical pulses of light. To do this engineers must find a way to divide the circuit or channel so that signals can travel together without interfering with one another.

PROTOCOL CONVERTERS

Protocols are a formal set of conventions governing the format and control of inputs and outputs between two communicating devices. In other words, protocols are the rules that allow two machines to communicate. As you already know, when you use the English language to communicate with someone, there are various rules (protocols) that you must observe. For example, when you write to someone, you must observe the rules of English grammar and use proper sentence construction and punctuation.

Protocol converters are hardware or software utilized to interconnect two dissimilar computer systems or terminals so they can talk to each other. As an analogy, if an American who speaks only English wants to speak with a Frenchman who speaks only French, they need to have an interpreter to carry on a conversation. This language interpreter serves the same purpose as a protocol converter.

Protocol conversion is one of today's hottest topics, particularly when you consider its application to the problem of micro-to-mainframe communications. Because of IBM's dominant position in the computer industry, most protocol conversion allows non-IBM equipment to communciate with IBM equipment. In general, the basic approaches to protocol conversion can be divided into three categories: hardware protocol conversion boxes, add-on circuit boards for microcomputers, and software that resides in host mainframe computers.

Hardware Protocol Converter Boxes   These boxes convert the communication protocol used by one computer vendor to that required for another computer vendor's equipment. For example, some protocol converters allow an asynchronous terminal to communicate with IBM host computers that use synchronous data link control (SDLC) or binary synchronous control (BSC) protocols. Other

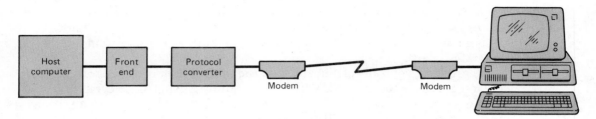

FIGURE 3-18 Protocol converter (host end).

protocol converters allow asynchronous terminals to interconnect with public packet switching protocols, such as X.25. These three protocols (SDLC, BSC, and X.25) are discussed in Chapter 5, Protocols and Software.

In today's networks, protocol converters might be located at either the host end of the communication circuit or the remote terminal end of the circuit. Figure 3-18 shows the location of a protocol converter at the host computer end of a communication link. Figure 3-19 shows a protocol converter at the remote terminal end of the communication link. Some protocol converters, similar to that shown in Figure 3-19, offer 1 to 16 ports for the connection of individual terminals. Incoming information from the terminals have their protocols converted to the host computer's protocol before being transmitted.

Add-on Circuit Boards Add-on circuit boards convert the microcomputers' protocols (transmission methodologies) to the protocol of the host computer to which they are transmitting. For example, there are two distinct types of add-on protocol conversion boards for IBM personal computers. The first converts the personal computer into a 3278 terminal that plugs directly into a 3270 controller via a coaxial cable. The second makes the IBM personal computer function as a 3278 terminal but with an already built-in 3276 controller. These add-on circuit boards are used for protocol conversion, which offers the user a direct micro-to-mainframe link with IBM host mainframe computers. A complete description of micro-to-mainframe communications is in Chapter 6 (Microcomputers and Communications) because we only want to define protocol conversion hardware in this chapter.

Software Protocol Conversion Packages In addition to these first two hardware approaches to protocol conversion, host mainframe software protocol conversion packages are available. These software packages can support almost any ASCII terminal or microcomputer. If an organization has a very large number of terminals or microcomputers to connect to its host mainframe computer, software protocol conversion may be the most cost-effective method because the

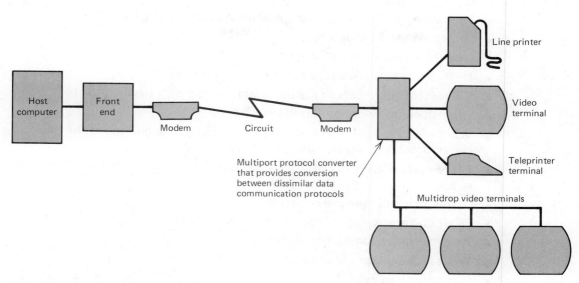

FIGURE **3-19** Protocol converter (remote terminal end).

software package at the host mainframe can support hundreds of remotely located terminals or microcomputers. The user, however, must take into account that the software of the host mainframe computer uses some of the computer cycles, which may slow processing or overload the host mainframe central computer. A hardware protocol converter, such as the one shown in Figure 3-18, does not use or interfere with host mainframe computer cycles.

Hardware vendors sometimes make overlapping and conflicting claims about protocol converters. For example, many protocol converters also can perform such tasks as multiplexing, concentrating, and packet assembly/disassembly for packet switched networks. (Packet networks are described in the next chapter.) Once again you have a piece of hardware that might be a multiplexer, a protocol converter, or both. While this situation may be confusing from a user's viewpoint, you must get used to it because that is what happens in the real world! In the next few years various hardware vendors probably will clash because of overlapping features and functions between front end communication processors, modems, multiplexers/concentrators, and protocol converters. By using microprocessor chips instead of minicomputers, all of the above pieces of hardware could be combined into a single piece of hardware that performs all the same functions now being marketed as separate pieces of hardware.

When comparing protocol converters, the following selection criteria should be considered.

- How many terminals can the product support concurrently?
- Which types of terminals are supported?
- How easy is it to add new types?
- Is hardware or software protocol conversion needed?
- What is the increment of expansion if it is necessary to add more terminals?
- Does the quality of emulation meet user needs?
- What additional features and functions are available on each of the protocol converters being evaluated?
- Is the system adaptable for future enhancement, or will it rapidly become obsolete?
- Is this protocol converter reliable?
- What diagnostic capabilities does the product have?
- Are qualified references from other users available?
- What is the quality of support and maintenance?
- Is it easy to install?
- Is it easy to use?
- Can it be customized?
- Is it cost effective?
- Does its use entail additional or hidden costs?

HARDWARE ENCRYPTION

The description of what encryption is and how it works appears in Chapter 9, Security and Control. In this section we only want to discuss software encryption versus hardware encryption.

Software encryption involves the use of stored programs to do the encryption. This technique is vulnerable to anyone who can copy the contents of the computer's memory. In data communications, software encryption usually is not used. Software encryption is used for the storage of data on disks/tapes or for other types of programming security problems.

In data communications, *hardware encryption* devices are employed. They are located most often as shown in Figure 3-20. Data from the host computer's front end or the terminal enters the encryption device, where it is secretly encoded. Most hardware encryption devices have standard RS232 interfaces. These devices expect digital information as input; their output also is digital. This is one reason why the encryption device usually is placed between the terminal and the modem. After encryption, the data is transformed appropriately by the modem. This is a straightforward "link" encryption setup.

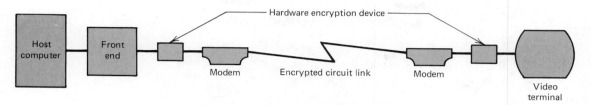

FIGURE **3-20**   Hardware encryption.

Hardware encryption involves a lockable box that is about the size of a modem. When this box is opened, the secret encryption key is destroyed automatically. Most encryption today is done on a link-by-link basis, except for military encryption, which is done on an end-to-end basis. For practical purposes, you can assume that *link-to-link* encryption is encryption of the data from modem to modem, where *end-to-end* encryption is encryption from user to user.

Most hardware encryption devices today employ the data encryption standard (DES) that was verified and authorized by the United States National Bureau of Standards. Most encryption devices also allow the insertion of several secret keys at one time to reduce the cost of key management. *Key management cost* is the cost incurred when you have to go to the remote end of a communication circuit to change the key. By having several keys, you can have a master key that is changed, for example, every two years; by transmitting new keys under the encryption of the master key, you can change the basic key daily if that is desirable.

TERMINALS/MICROCOMPUTERS

The input/output hardware device that is at the remote end of a communication circuit probably is the one piece of equipment with which you are most familiar. Obviously, both terminals and microcomputers can be used as input/output devices. (Chapter 6 contains more in-depth coverage of microcomputers.)

Shopping for a terminal can be a real problem if it is not done correctly. For example, a bargain terminal may have the wrong code format or an incompatible protocol, which requires the additional purchase of a protocol converter. It also must be determined which interface cable can be used with the selected terminal. Some systems use an RS232, others an RS449, and still others fiber optic cable. The selected terminal or microcomputer workstation must be compatible with the communication protocols used by the host mainframe computer, and its general characteristics and uses must meet the user's specific day-to-day business requirements.

There are several major categories of terminals, each of which is discussed in the following sections.

FIGURE 3-21 Microcomputer.

Microcomputer Workstations Microcomputer workstations are either general purpose microcomputers or specially designed input/output workstations that have custom-designed microprocessor chips. Many vendors have developed customized microcomputer workstations for use in office automation, data entry, the automated factory, banking, and a whole host of other special situations. The basic input/output device is a microcomputer. Its functions include editing data, possibly storing data, protocol conversion if required, mathematical calculations, and prompting for information or forms design (handled on the video screen); all can be handled locally without assistance from the host mainframe computer.

Technically, there is a difference between a workstation and a microcomputer. A *workstation* usually provides all of the tools that professionals need in their daily work. Among these tools are specialized applications like mathematical modeling, computer-assisted design (CAD), intensive programming, and networking. Today's workstation usually has more computing power than the average microcomputer. Workstations must offer true multitasking capability so that the user does not spend a large amount of time waiting for the computer to finish one job before starting the next one. On the other hand, a *microcomputer* may not have the ability to handle all of the specialized applications, and its multitasking ability may be significantly less. Most microcomputer users may be satisfied with printer spooling and telecommunications as a replacement for true multitasking capability. The primary use of a microcomputer is for a few functions such as word processing, accounting, and spreadsheet modeling. As microcomputers get more powerful microprocessor chips, they will be as powerful as contemporary workstations. Therefore, today you may be able to see the difference between a workstation and a microcomputer, but tomorrow they may be the same thing.

A microcomputer is shown in Figure 3-21. A complete description of microcomputers and how they operate is presented in Chapter 6, Microcomputers and Communications.

Video Terminals These terminals have a television screen and a typical typewriter keyboard. Sometimes they are called cathode ray tubes (CRT), video display units (VDU), or video display terminals (VDT). Alphanumeric video terminals are used in the business office, whereas graphic video terminals are used by graphic designers. The intelligent version of a video terminal might be used for computer-assisted design.

In addition to a standard keyboard, the video terminal has a marker on the screen called a *cursor*. It moves about the screen to show the terminal operator the next position in which a character will be printed.

When selecting a video terminal, you should consider what the transmitting line speed is, if users can rotate or tilt the screen for easy viewing, if split-screen mode is available so two or more screens can display simultaneously, whether the character matrix size is large enough for easy viewing, if a detachable keyboard would be advantageous, if a self-test mode is available, how many characters can be displayed horizontally and how many lines vertically, whether it can accommodate a separate printer that can be turned on and off as needed, and which editing functions are available such as character insertion/deletion, line insertion/deletion, erasing, and paging.

With regard to the individual terminal operator, some specific items must be taken into account when you are ordering a video terminal. Because eyestrain and fatigue are the most common complaints of video terminals operators, the following arrangements will increase productivity.

- A video tube screen filter should be obtained to protect you from electromagnetic radiation, which can emanate as microwaves, infrared waves, and ultraviolet waves. You want to block all three.
- An antiglare screen should be used to eliminate glare from the front face of the video tube. Glare is the number one complaint of people who use video terminals or microcomputers.
- Cursors should be visible from 8 feet away, and they should be seen easily at

3 feet. Sometimes it is advantageous to have the option of either a blinking cursor or a cursor that is on as a steady light.

- Detachable keyboards or a sloped keyboard may be desirable.

- Reverse video may ease eyestrain. With this feature, the operator can change from a dark background with light characters to a light background with dark characters.

- Multiple colors also may reduce eyestrain.

- Adjustments for tube brightness, focus, and contrast are desirable to accommodate the variety of operators who might use a video terminal.

- Glare and eyestrain can be reduced with the use of window shades, proper lighting, and movable screens that cut reflective glare.

- To prevent muscular aches and fatigue, VDTs should have movable keyboards, document holders, and screens that tilt to a comfortable viewing angle for the user. The height of the video screen and its ability to tilt might be critical to the terminal operator who wears bifocal lenses. Today's computer furniture often requires that video screens be placed above eye level. This consideration should be borne in mind when purchasing such furniture because operators should be able to look *down* onto the screen.

- The characters on the video monitor tube should not jitter or flicker because this movement causes eyestrain. By using a magnifying glass to look at the characters, you should be able to determine whether they jitter or shake. *Jitter* is caused by an insufficient video reflection voltage which makes the electron beam inaccurate when refreshing the character. *Flicker* is caused by a fast decay of the phosphor and a slow refresh rate. Technically, you should accept nothing below 65 hertz (frequency). Many video data terminals are 45, 50, or 55 hertz, which does not provide a fast enough refresh rate to avoid noticeable or even subliminal flicker.

- On the controversial subject of screen color, displays with black characters on white screens provide the most legibility, visual comfort, and work efficiency. Many users, however, feel that yellow/green tube color is the easiest to see and tends to reduce eyestrain. You should seriously consider purchasing video monitors that have multicolor capabilities because color tends to reduce eyestrain and enhances productivity. Color also is used to make it easier to understand an application.

- Users should always obtain the highest possible video monitor tube resolution. For example, some video monitors have resolution up to 800 (horizontal) by 560 (vertical) pixels. The higher the resolution (a greater number of pixels), the clearer the picture and the less eyestrain. (*Pixels* are "picture elements"; each one is a small dot on a video tube.)

- Users should always strive to obtain a dot matrix printer with the greatest density of pins possible. For example, a dot matrix printer with a 24-pin print head is preferred. With laser printers, you should get 300 dots per inch.

Teleprinter Terminals These non-video screen terminals produce a paper print-out and have a typewriterlike keyboard. In this discussion, they have no programmable capacity (intelligent terminals are a separate category). This type of terminal can perform its printing through an impact, a dot matrix, or even an electronic ink jet. Its characters can print at a speed range from 10 characters per second to 300 or more characters per second. This always has been the most common form of terminal, although the video terminal will replace it in the future because of the cost.

The keyboard on a teleprinter terminal is a typical typewriter keyboard, although it may contain several special function keys for data communication networks. Also, a third function might be assigned to a single key such as start, stop, delete, or end of transmission. The first two functions are lower case and upper case for an alphabetic character.

Some of the features to look for in a teleprinter terminal are size of the dot matrix, maximum print speed, bidirectional printing, size of the printable character set, number of character positions across the paper, form feed mechanism or individual sheet insertion, and graphics printing capability. Some teleprinter terminals offer a self-test answer-back mode whereby the terminal circuitry can be tested for problems. Some also have a built-in acoustic coupler/modem.

Remote Job Entry Terminals Usually, these are nodes of a network or terminal stations where there are several types of connected devices. Data often is transmitted from a host computer to a remote job entry terminal that might have a variety of terminal devices such as a video terminal, a high speed printing terminal, several data entry devices (such as disk or tape), and perhaps a microcomputer. Terminals in such an area operate at higher speeds, such as 9600 bits per second or greater, because large quantities of data are transmitted from this terminal station to the central host computer.

Transaction Terminals These generally low cost terminals are used by individuals in their homes or place of business. The most common transaction terminal is the automated teller machine (ATM) which banking institutions use for cash dispensing and related functions. Your telephone is a type of transaction terminal that accepts voice transactions for transmission.

A transaction terminal that may be popular in the future is a display telephone videophone. With it you can transmit both voice and picture from your home or office.

Other transaction terminals are point-of-sale terminals in a supermarket; these

FIGURE **3-22** Omnifax G32 digital desktop facsimile.

enter charges directly from the supermarket to your bank account, or they can be used for checking credit or verifying checks. These terminals can be built into electronic cash registers.

Yet another transaction terminal will be your personal television set through cable television use of two-way teletext/videotex data communications.

Facsimile (FAX) Terminals Transmission of an exact picture of a hard copy document, including legal signatures, is one of the most important features in the automated office. This is especially so in areas such as legal contracts, medical records, and authorizations, and for the control of business records.

Although analog facsimile has been used widely in business, government, and professional areas, modern high speed digital facsimile devices provide many added features such as encryption, security identification codes, and delivery verification. Figure 3-22 shows a picture of a facsimile device. Whereas analog facsimile devices transmit a page in one or two minutes, digital facsimile devices now transmit a page ($8\frac{1}{2} \times 11$ inches) in less than a minute. If you use a standard copy machine that can reduce two $8\frac{1}{2} \times 11$ pages down to one $8\frac{1}{2} \times 11$ page, you can double this transmission speed.

Facsimile transmission can be very threatening to post office systems because you can send a letter from one area to another for anywhere from $0.30 to $0.50, depending on the volume of traffic per month and the cost of your communication circuits. For this reason, facsimile transmission is in direct competition with both electronic mail and voice mail.

Once you have a facsimile device connected to a communication circuit, the basic steps for transmission are

- Establish the call either by dialing manually or by having an automatic call placed (*physical circuit connection*).

- Complete the premessage procedure, or *handshaking,* which consists of identifying the called station (facsimile machine) and any other procedures that might be required to set up the session.

- Transmit the message (the *session*), which involves synchronization between the two devices, any error detection and correction methodologies, and movement of the message from one facsimile device to the other.

- Complete the postmessage procedure, which includes END OF MESSAGE signaling at the conclusion of the page, any signaling that signifies more than one page, end of transmission, and anything else required to end the session.

- Manually or automatically disconnect the call (*physical circuit disconnection*).

Facsimile devices are connected using standard interface cables, such as the RS232C interface. Microcomputer-to-facsimile terminal links allow microcomputer users to create a file on their microcomputer and to send it directly to a remote FAX receiver, without having to go through the process of printing it, physically carrying it to a facsimile machine that scans it, and then transmitting it. Users also can receive a FAX file, display it, manipulate it on a microcomputer screen, and then print it or send it to another FAX terminal.

The international Group 3 facsimile standard provides for transmission of any text or graphics image over ordinary telephone circuits. A scanner in the facsimile device breaks the original document into pixels at 200 dots per inch horizontally and 100 dots per inch vertically. (The "fine" mode has 200 dots per inch in both directions.) Then it converts the pixels into a bit stream to feed into a modem. The receiving device takes the bit stream and prints the image. The most sophisticated devices use laser printers.

Because facsimile images always are pictures, a page can occupy a considerable amount of storage. Typically, it is 30 to 60 kilobytes for a page of text or 120 kilobytes in fine mode. This is true even when using Group 3's data compression scheme. A 20-megabyte hard disk devoted to facsimile images can store only 170 fine-mode pages. As a standard ASCII computer file, a page takes up only 2 to 4 kilobytes; thus, an ASCII page transmitted at 1200 bits per second is faster than a facsimile page transmitted at 9600 bits per second.

Future facsimile developments will concentrate on faster transmission and higher resolution rather than on character or attribute coding because these two features can be agreed on internationally. The proposed Group 4 standard requires several resolution levels up to 400 dots per inch and an improved compression scheme. It also is designed for an all-digital telephone network instead of dial-up telephone circuits. It should make optical character resolution more practical.

Dumb/Intelligent Terminals Whether a terminal is considered dumb or intelligent does not depend on whether it has a video monitor or hard copy printer, or on whether it is used for data entry or as a transaction terminal. Any of these previously mentioned terminals might be either dumb or intelligent. A *dumb terminal* does not participate in control or processing tasks. It usually does not have any internal storage for memory, nor does it have any microprocessor chips; it has the bare minimum required to operate. The older, really dumb terminals transmitted asynchronously, and when a character was typed, the terminal transmitted the character immediately. Thus, central computer systems had to have receiving buffers for each of these terminals so that they could assemble the entire message before acting on it.

An *intelligent terminal* is one that has a built-in microprocessor chip with the capability of being programmed, and it has internal stored memory. The specific market for intelligent terminals has been superseded by microcomputers. In today's world, microcomputers and specialized microcomputer workstations constitute the bulk of intelligent terminals. Many intelligent terminals also have external disk storage capability or their own internal software-driven communication programs. Today, you should regard an intelligent terminal as a microcomputer.

Attributes of Terminals Many attributes might be considered when you are evaluating a terminal. The most popular method of entering data is through the use of a typewriterlike keyboard, but you also should consider other more technologically advanced methods. Some of the more important attributes are discussed below.

- *Light pens* are used to touch the screen. The light activates whatever it is you are selecting from the screen.
- *Touch screen* works the same way as light pens except your finger physically touches a portion of the screen to make a selection.
- A *mouse* is moved around the surface of a table, while at the same time the cursor moves around the video monitor screen in order to make various

selections. When you find whatever it is you want to select (the cursor is at that location), you press a button on the mouse.

- *Direct voice entry* is starting to become feasible, but it still is in the early stages of its technological development.
- *Page scanners* are available that automatically read a printed page and enter it into the computer. Some scanners only have the ability to scan graphics, some can scan only text, and others have special add-on devices that enable the scanner to handle both graphics and text. Special software packages interpret the ASCII characters so they can be recognized. Page scanners that are connected to microcomputers can produce their output on diskettes. The price of page scanners varies tremendously—from a few hundred dollars to many thousands of dollars.
- *Numeric keypads* are used for entering numbers.
- Small cameras are available for the *direct digital entry* of graphics or pictorial data.

The speed at which the terminal operates (characters per second) and its ability to transmit asynchronously or synchronously is an important attribute, depending on the type of data that must be entered.

Output attributes include monochrome or color or the type of hard copy printer utilized. A color display certainly is advantageous over a monochrome display because it enhances productivity and reduces eyestrain. As mentioned previously, the higher the resolution of the video monitor tube, the lower the eyestrain and fatigue. With regard to printers, the higher the resolution of the print mechanism, the more acceptable the printer, although with printers the speed at which the printing is performed might be its most important feature. Speed may be especially important for those who share one printer with several users. For example, a typical dot matrix printer might print 100 to 200 characters per second, but a laser printer might be able to print 8 to 10 full pages per minute.

Other miscellaneous attributes that may be considered are the interface that the terminal will use (RS232 or other), the compatibility of the data transmission protocol with that of the central host computer, the code that is utilized (ASCII or EBCDIC), whether any applications require a portable terminal, the adequacy and availability of maintenance and support from the terminal's vendor, and so forth.

SWITCHES

When data is to be sent over a network with many terminal locations, some arrangement must be made to enable different terminals to communicate with each other. Early communication systems used permanent connections between

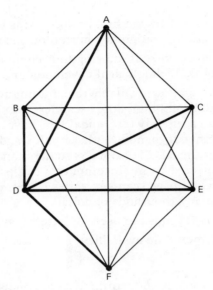

FIGURE **3-23** Direct wire connection for six terminals.

each pair of terminals. When the number of terminals exceeds two or three, however, this system gets increasingly expensive. Figure 3-23 depicts this situation; it shows six terminals with every station individually connected to every other station. Fifteen lines are needed in this case. The number of lines needed for a network of this type may be calculated as

$$\text{Number of lines} = \frac{N^2 - N}{2}$$

where N = the number of terminals or stations.

A more economical and flexible type of interconnection arrangement is to allow for temporary connections between any two stations that wish to communicate with each other. This process is called *switching*, a methodology that eliminates the need for direct wire connection between all station pairs in a network. If Station D has the ability to connect any two stations temporarily, then the 15-line configuration of Figure 3-23 can be reduced to a 5-line configuration, consisting of one line from each station going into the central switching station (D).

The basic functions of a switching system are to

- Interconnect the various stations to provide a communication path between the stations

- Control the establishment and release of the various connections
- Check network equipment to determine whether it is busy or inoperative

Circuit Switching   Circuit switching is the most common type of switched network. Your telephone system uses circuit switched connections. In circuit switching, the central switch site (usually the telephone company's central office) establishes a connection between two stations, and a message goes directly from one terminal to another. You might picture this as a voice message going from one telephone to another after the two communication circuits are switched together at the telephone company central office. If the two telephones cannot be switched together (connected), you get a busy signal and have to try again (redial) at a later time. The major disadvantage of circuit switching is that you cannot get a message through the system if the two circuits cannot be connected together. For this reason, store and forward networks and packet switched networks are more popular. Packet switching, a software type of switching, is described in the next chapter.

Store and Forward   Store and forward switching systems first try circuit switching. If one of the circuits is busy, the central switching site stores the incoming message from the sending terminal by copying it onto a storage medium such as paper tape, magnetic tape, or disk and retransmits that message to the destination terminal at a later time. Typical of the older type of store and forward switching system is the manual "torn tape" switch. In this system Station A transmits a message to the switching center. The message is received on punched paper tape and torn off the paper tape perforator that is connected to Station A. At this point, the message may or may not be stored, depending on whether the station to which it is addressed is busy or free. If the station is free, the punched paper tape is placed in the paper tape reader that is connected to the station to which the message is addressed.

With modern high speed computers, data communication networks are able to combine the store and forward concept with the circuit or line switching concept. This combination offers data communication network users the highest level of throughput. A modern communication switching network offers circuit or line switching so the originator of a message can be connected immediately to the station to which the message is addressed; if that station is busy, the computer center accepts the message, stores it on a magnetic device (generally disk), and automatically transmits the message to the proper addressee as soon as the line is free.

Digital Data Switches   These switches provide users with a way of switching between different host mainframe computers without having to access back through the modems (as shown in Figure 3-24). Digital data switches sometimes

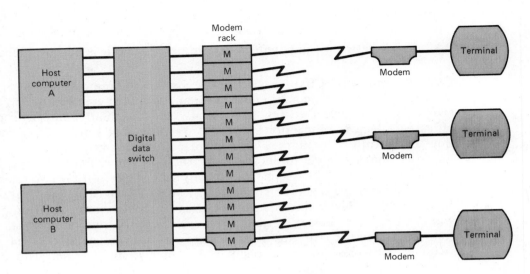

FIGURE 3-24 Digital data switch. A modem rack is a cabinet at the central site that holds many modems. Sometimes only the modem circuit cards are in the rack, and sometimes the entire modem box is in the rack.

are referred to as digital PBXs or digital switchboards. A digital data switch also can manage traffic and set up inbound and outbound queues of messages during periods of peak usage; this maximizes the available resources. Digital data switches offer port contention, which is a sharing of the ports on a front end communication processor. This sharing can be done on a first-come, first-served basis, or some incoming messages can be given priority over others. Some digital data switches provide security restriction through the use of passwords.

Network Switches   Even though switching can be built into the front end communication processor, statistical multiplexer/concentrator, PBX, or host computer, there are many standalone message switching systems. A typical message switching system might be based on the minicomputer or microcomputer. It functions as both a message switch and a store and forward system. Sophisticated message switches utilize stored program techniques and highly reliable third generation solid state circuitry.

A typical standalone switch might have 12 or more asynchronous communication circuits that can operate at different speeds. The system should be able to support various code structures and be expandable to handle more communication circuits. The switch should be able to handle half duplex or full duplex and should provide an interface to the Direct Distance Dial (DDD) network, Telex, TWX, and other international networks.

A standalone switch should be able to dial and receive calls automatically, as well as switch messages between any of its circuits. Another feature should be a delivery/verification/confirmation response, with the message sender knowing when the message was delivered and having a positive delivery acknowledgment. Sophisticated switches allow various addressing schemes such as 3, 5, or 7 addressing characters in order to be able to address a multitude of circuits, individual terminals, and individual devices that might be attached to a microcomputer. Finally, a good switch should provide for in-transit storage of an adequate number of messages (at least 500). It should have the ability to retrieve messages that were sent during the day, possibly during previous days, and message logging for transaction trails/historical purposes. Some switches also provide alarms in case of circuitry failure and self-diagnostics to locate the failure. The telephone company central office is a switch.

Since many people feel that switches are an obscure piece of hardware, we wish to expand on how switches are used. Today most switches are located in the telephone company's central office where they are used for the important function of switching calls (circuit/line switching) between the calling and called parties. For example, when you dial a telephone number, the switch finds the other telephone instrument, whether it is located across town, across the state, or across the country.

Switches also may be used in connection with a "pay-per-view" television service in which you dial in appropriate identification numbers and a special movie or television show is directed to your home with billing at a later time. Cellular telephone companies are major users of switches. Many other types of businesses can use switches for some of the following tasks.

- **Opinion Polling.** Viewers can register opinions on subjects ranging from politics to personalities with results tabulated and made available during the broadcast.

- **Financial Services.** Brokerage houses can offer sales leads, stock tips, and instant stock quotations.

- **Banking.** Banks can provide information on budgeting, investments, loans, insurance, and estate planning. Banks already use switches to route messages through their automated teller machine (ATM) networks.

- **Catalog Shopping.** Department stores or mail order companies can invite customers to purchase goods by telephone without operator assistance.

- **Computer Security.** Businesses that want tighter control over access to their database can furnish a list of approved users and passwords to the telecommunication company (common carrier) for insertion into the switch. Unauthorized attempts to gain access can be routed to a busy signal or a rejection announcement.

- **Cable Trouble Reporting.** Cable television companies can eliminate or reduce staffed switchboards needed to obtain outage reports.
- **Network Services.** Remote meter reading and diagnostics on various home systems can be automated at far less cost than with existing methods.
- **Lottery by Telephone.** In states that already sponsor lotteries, this service could help lottery commissions avoid the expense of furnishing and maintaining lottery terminals operated by humans.

As you can see, switches can perform many functions and may become increasingly important in the future.

SELECTED REFERENCES

1. *Auerbach Data Communications Reports.* Published monthly by Auerbach Publishers, 6560 N. Park Drive, Pennsauken, N.J. 08109, 1965– .
2. *Black Box Catalog of Data Communications and Computer Devices.* Published monthly by the Black Box Corp., P.O. Box 12800, Pittsburgh, Pa. 15241.
3. *Communications Products & Systems.* Published bimonthly by Gordon Publications, 13 Emory Avenue, Randolph, N.J. 07869–1380.
4. Coughlin, Vincent J. *Telecommunications: Equipment Fundamentals and Network Structures.* New York: Van Nostrand Reinhold Co., 1984.
5. *Datacomm Catalog.* Published by Glasgal Communications, 151 Veterans Drive, Northvale, N.J. 07647.
6. *Datapro Reports on Data Communications.* Published monthly by Datapro Research Corp., 1805 Underwood Boulevard, Delran, N.J. 08075.
7. Held, Gilbert. *Data Communications Networking Devices: Characteristics, Operation, Applications.* New York: John Wiley & Sons, 1986.
8. *Telecom Gear: The Market Place to Buy and Sell Telecommunications Equipment.* Published monthly by Telecom Gear/Telecom Jobs, 12 West 21st Street, New York, N.Y. 10010.

QUESTIONS/PROBLEMS

1. What is the difference between a front end communication processor and a modem?

2. What is the term used to describe the combination of placing two or more signals on a single channel?

3. If you were buying a multiplexer, why would you choose either TDM or FDM?

4. For data communication transmissions, which would you use, hardware encryption or software encryption?

5. Three terminals (T_1, T_2, T_3) are to be connected to three computers (C_1, C_2, C_3) so that T_1 is connected to C_1, T_2 to C_2, and T_3 to C_3. All are in different cities. T_1 and C_1 are 1,500 miles apart, as are T_2, C2 and T_3 and C_3. The points T_1, T_2, and T_3 are 25 miles apart, and the points C_1, C_2, and C_3 also are 25 miles apart.

$$
\begin{array}{l}
\text{25 miles} \left\{ \begin{array}{l} T_1 \text{———}1{,}500 \text{ miles———} C_1 \\ T_2 \text{———}1{,}500 \text{ miles———} C_2 \end{array} \right\} \text{25 miles} \\
\text{25 miles} \left\{ \begin{array}{l} \\ T_3 \text{———}1{,}500 \text{ miles———} C_3 \end{array} \right\} \text{25 miles}
\end{array}
$$

If telephone lines cost \$1.00 per mile, what is the line cost for three independent lines? If a multiplexer/demultiplexer pair costs \$2,000, can you can save money by another arrangement of the lines? If so, how much?

6. List some systems applications in which an automatic answer capability is necessary; similarly, list some systems applications in which an automatic calling capability is necessary.

7. Review the list of functions of a front end communication processor. Identify what you consider to be the four most important functions of a front end.

8. Identify and describe the three basic components of a front end.

9. How would you keep "hackers" out of your dial-up system?

10. If your modem transmits at 2400 bits per second and also uses a 4 to 1 compression ratio data compression/compaction device, what is your data transmission in bits per second?

11. If you had to transmit a ten-page term paper of 20,000 bytes to your professor, approximately how long would it take (file transfer time) if your modem speed was 2400 bits per second using asynchronous transmission?

12. Are short haul modems and null modem cables the same thing? Explain your answer. The answer is in the chapter, but use the Glossary too.

13. One of the modem lights is OH. What does that mean?

14. Can you identify the major categories of terminals discussed in this chapter?

15. Can circuit switching and store and forward switching be combined in the same system? If so, describe how it would work.

16. If you have 11 microcomputers to interconnect without switching, how many lines (circuits) are required?

4

NETWORK CONFIGURATIONS

This chapter describes network configurations. The described configurations are the basic building blocks that are interconnected when you develop multiorganizational or international networks. Each configuration can stand alone as a single application network, or they can be interconnected.

INTRODUCTION

This chapter explains network configurations in enough detail, along with diagrams, that you can understand how they operate. By this time you already have mastered the fundamental communication concepts (Chapter 2) and data communication hardware (Chapter 3). This chapter assembles all of these concepts and hardware together so you can see the many ways that a network might be configured.

From the smallest system with just two or three microcomputers to a large system including host mainframe computers and spanning long distances, data communication systems are organized into *networks*. It is the network that provides communications between computers of all types, printers, terminals, and many other pieces of hardware. Computers on a network typically are referred to as host mainframes, minicomputers, or microcomputers. Systems on a network (electronic boxes) which provide usage of other peripherals, such as hard disks and printers, are called *servers*. They are so called because they provide service from the various peripheral devices to all network users. Servers perform this task in an organized fashion, thus keeping the system free of chaos.

Some networks are referred to as *wide area networks* (WAN). They link systems that are too separated, either physically or geographically, to be included in a small in-house network. In other words, wide area networks cover a much larger geographical area than that of a local area network. They usually transmit data over "public thoroughfares" and therefore must utilize the communication circuits of a publicly registered common carrier (telephone company). A *common carrier* is a government-regulated private company that furnishes the general public with communication facilities, primarily circuits. The most noted examples of common carriers are AT&T, the Bell Operating Companies (local telephone companies), Bell Canada, US Sprint, MCI, and so forth. Wide area networks use a broad range of communication media for interconnection and can be located as close together as a few city blocks or as far away as another country. They generally use transmission media, such as microwave or satellite transmission, but they are not limited to these two forms. As a result, wide area network is a general term referring to all networks that cover a wide area because the various terminal stations are geographically separated by longer distances.

Another term that is gaining in popularity is *hybrid networks*. A hybrid network is a flexible or open-ended network concept that takes advantage of a mixture of different technologies or media to achieve a large wide area network or backbone network. Hybrid networks have been in existence for years, but they were never given a label until recently. You can consider a hybrid network as any mixture of two or more network configurations combined in such a way as to optimize whatever it is the business entity wants to achieve. Hybrid networks may use a mixture of different types of transmission media (satellite, microwave, and so forth), dial-up communication circuits, private dedicated leased circuits, local area networks, and many other devices or protocols.

Local area networks (LANs) connect devices within a small local area, usually one building or adjacent buildings. LANs usually do not have their transmission media (circuits) cross any public thoroughfares (roads). In most local area networks, devices with the processing power of a microcomputer are interconnected. This kind of network is not described here because we have devoted a whole chapter to it (Chapter 7, Local Area Networks).

Any network has three key elements: media, devices, and protocols. The transmission media are the various wires, cables, microwave links, satellite links, optical fiber cables, cellular radio, and so on. The *media* provide pathways over which data can travel; they also might be referred to as transmission facilities, channels, circuits, lines, or links. *Devices*, which also are called stations, provide a means for inputting and outputting data over the transmission media. These devices might be modems, multiplexers, gateways, protocol converters, front ends, terminals, microcomputers, and the like. *Protocols* provide the common set of rules for managing the network pathways and the data traffic. They are the software programs that tie the overall network together and make it a functioning entity.

As an example of the above three network elements (media, devices, and protocols), have you ever thought of the group of friends with whom you closely associate as a network? These people are an informal information network who exchange various bits of data (voice communications) among themselves. This informal network transmits its voice communications over a type of media: the air over which the sounds of your voices travel. Furthermore, this informal information network employs several devices for moving the voice communications. These devices are your voice box (larynx), your mouth and tongue to help form sounds, body language, voice inflection, volume, pitch, and so on. Finally, this same informal information group observes various protocols or rules for exchanging information.

Two of the basic rules of this network involve whether this group operates under a philosophy of central control or one of interrupt communications. If the group uses central control, then one person might be the formal leader who gives other people permission to speak, much as the professor in the classroom calls on students and each has a turn to speak. On the other hand, if the group operates under an interrupt philosophy, then each person interrupts or cuts in with whatever it is he or she wants to say when the current speaker finishes a sentence or idea. Protocols also come into play with this informal information group in the form of which words might not be appropriate with which groups of people, the speed with which individuals speak, and so forth. In this situation your brain is the computer, and the basic rules of conversation you have learned over the years become the software programs or protocols by which you operate.

As you continue through this chapter, you will notice that we first discuss voice/telephone networks and then move through the many other data communication networks.

VOICE COMMUNICATION NETWORK

The basic voice communication telephone network is not only the largest, but also one of the oldest of our twentieth-century electromagnetic data communication networks. This system involves the telephone in your home which you use

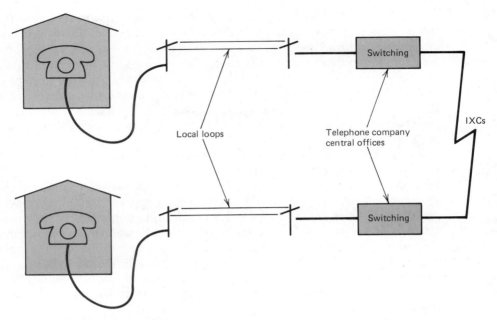

FIGURE **4-1** Telephone interconnection.

to call friends, relatives, or other people. In today's business communication cost structure, the basic voice telephone system commands three-quarters of the business revenues spent for communications, and only one-quarter is spent on true data transfer communications.

This network involves the connection of your telephone to the telephone company central office (also called end office or exchange office) where switching is performed.

Switching includes identifying and connecting independent transmission circuits to form a continuous path from your telephone to the telephone you are calling. Figure 4-1 shows the interconnection of telephones through the central office in our basic voice network. You will note that the local loop on this figure is nothing more than the copper wire pairs that go from your home to the telephone company central office. The central office is where the electromagnetic or the electronic switching system equipment is located. The central office uses the telephone number that you dialed as an "address" and searches out the other telephone so the two can be connected.

Calls made within the same central office or interoffice trunks are known as *local calls* (see Figure 4-2). Calls that use the tandem trunks and the tandem office are known as *unit calls*, for which there may be an extra charge for each

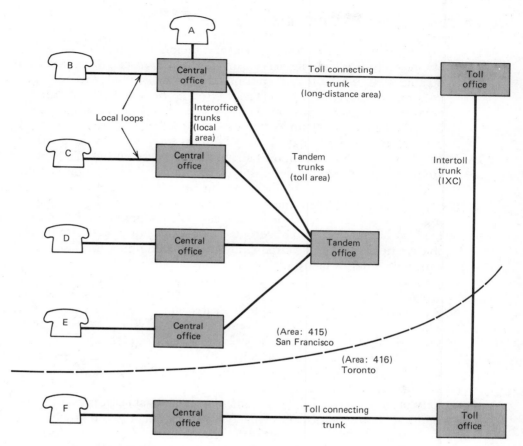

FIGURE 4-2 Central office hierarchy.

minute of time. The calls that use intertoll trunks (those that cross a LATA) usually are known as *long distance calls.*

The worldwide telephone network is so complicated that it would be difficult to present a single drawing showing its configuration. Department of Defense officials have indicated that the network in the United States is so integrated that if offers a unique emergency backup capability, for it would be almost impossible to destroy all communications in the United States. It should be noted that even though Figure 4-2 shows telephones at the end of the local loops, terminals or other data transmission equipment also might be attached to the dial-up telephone network.

VOICE GRADE LEASED CIRCUITS

Voice grade leased circuits are normal telephone circuits that have been taken out of dial-up telephone service by the common carrier and dedicated to one organization. These leased communication facilities sometimes are called *private circuits, private lines, leased circuits,* or *dedicated circuits.* The leasing organization makes them available for use on a 24-hour basis, seven days a week. It might be helpful to understand the distinction between a leased circuit and a dial-up circuit. If you have a leased circuit from New York to Atlanta, it is one continuous, unbroken circuit path between the two cities. This circuit is "wired around" any telephone switching equipment at the telephone company central offices; therefore, switching equipment cannot cause errors or distortion of messages. When not leased, the same dial-up circuit is wired through all the switching equipment that is required to locate the telephone numbers that are dialed. Dial-up circuits do not have one continuous, unbroken circuit path between these cities. Voice grade circuits are the most common form of leased communication channel or facility.

In today's marketplace there are two different types of leased voice grade circuits. The first is a standard leased circuit that is a physical connection or, as with satellite or microwave, a dedicated circuit between two different points. The entire capacity of this circuit, whether it is a physical wire or dedicated satellite/microwave circuit, is yours for your use alone.

Because the maximum capacity of these conventional leased circuits are not used for the entire 24 hours of the day, AT&T has developed *virtual leased circuits,* which also are called *software defined networks* (SDNs). This type of network is a service for voice transmission and data communications that carries up to 9600 bits per second. Once your data transmission is within the software defined network, AT&T routes the circuit according to instructions stored in the memories of its network switches. In this way AT&T allows a user to assemble a private network of leased voice grade circuits by using AT&T's switched facilities. While this operates as a private leased line, you do not have one continuous, unbroken circuit path from Point A to Point B for 24 hours of every day. This path is for your use *only* during the moments that you are transmitting data; at other times the path may be used by another organization. This system works because of our present-day sophisticated transmission techniques and ultra high speed digital switching facilities. These facilities have reliability rates that are as good as physical connections, and in many cases error rates that are better than the old physical wire connections.

The basic concept behind these virtual networks (SDNs) is to provide a leased line service that is less costly than dedicating one physical wire or circuit path to an organization that wants to lease a circuit from Point A to Point B. As more common carriers convert more of their networks to digital switches, they will be able to create private leased line networks without dedicating a physical wire to the customer leasing the circuit. Large communication circuit users can look forward to leased virtual networks as the price of leased "physical" circuits climb. Chapter 10 (see the section titled Private Circuit [Lease] Services), will give you more details on voice grade lease lines and software defined networks so you can use it when designing networks.

CAPACITY OF A VOICE GRADE CIRCUIT

Analog transmission takes place when the signal that is sent over the transmission media continuously varies from one state to another. This is analogous to having a dimmer switch on an electric light that allows the light to vary from very bright to very dim, but it is a continuous varying as contrasted with a light that a switch just turns on and off. (The on/off switch is digital.)

Most telephone circuits utilize analog transmission because they were developed for voice transmission, not data transmission. New systems today are built in a digital fashion.

In order to understand bandwidth in an analog channel, it may be desirable to review a sine wave and frequency or hertz. Figure 2-15 (Chapter 2) illustrated sine waves that had different frequencies (different number of cycles per second). Hertz means cycles per second; therefore, a 2125 cycles per second tone also can be called a 2125 hertz tone.

Figure 4-3 shows that most transmissions over voice grade circuits are within the human hearing range. This is because most of the telephone systems were built for human speech rather than for data. Human hearing is in the range of approximately 20 to 20,000 hertz, although most people cannot hear above 14,000 hertz. Figure 4-3 also shows the frequency range for coaxial cable, microwave, satellite, and laser (optical fibers). Bandwidth refers to a range of frequen-

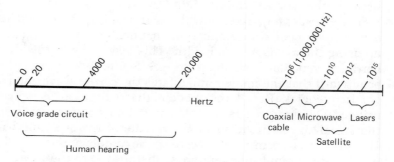

FIGURE 4-3 Frequency spectrum (zero to 10^{15} hertz).

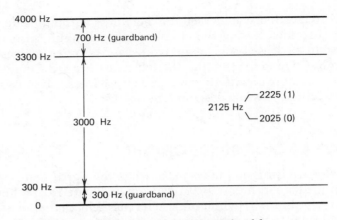

FIGURE **4-4** Voice grade circuit bandwidth.

cies. For example, Figure 4-3 shows that the bandwidth of a voice grade communication circuit is from 0 to 4000 hertz.

A voice grade communication channel or circuit is the most common communication circuit today. This circuit is used with dial telephones, and it typically is the circuit that is installed when organizations lease private or dedicated communication circuits. The bandwidth of this communication circuit is 4000 hertz.

Figure 4-4 shows how this 4000 hertz bandwidth is divided for data transmission use. To start, there is a 300 hertz guardband at the bottom and a 700 hertz guardband at the top. These prevent data transmissions from interfering with other transmissions when these circuits are stacked on a microwave or satellite link; thus, they prevent the frequencies from overlapping between your communications and the communications of others. They can be compared to the plastic insulation that is put around a copper cable to keep it from shortcircuiting with another copper cable.

Figure 4-5 demonstrates how the guardbands provide 1000 hertz of empty space between adjacent communication channels. Each voice grade circuit is stacked up using the available bandwidth. This leaves the bandwidth from 300 to 3300 hertz for your data transmission (see Figure 4-4). It is within this 3000 hertz bandwidth that the modem modulates the carrier wave and transmits the different frequencies. Recall the modem that had a carrier wave of 2125 hertz. It utilized 2225 hertz for 1s and 2025 hertz for 0s. Returning to Figure 4-4, note that the signal travels right down the middle of the available bandwidth. Modems always try to transmit and receive in the middle of the available frequency bandwidth, where amplitude and phase distortion are the lowest (fewer retransmis-

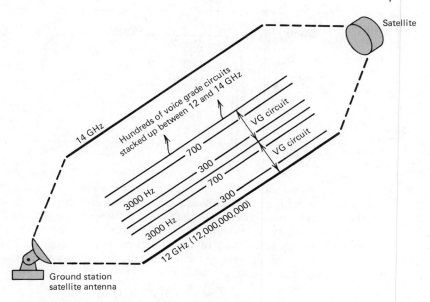

FIGURE 4-5 Guardbands.

sions due to errors). Figure 4-6 shows the frequency assignments for a Bell 202 modem.

It is this 3000 hertz bandwidth that limits the maximum transmission speed. For example, the modem controls the speed at which data bits can be transmitted (300 to 19,200 bits per second on voice grade circuits). The 3000 hertz bandwidth limits the maximum speed.

Let us examine the relationship between bandwidth and bits per second. Assume a situation that uses a dial-up circuit, with 2600 hertz out of the 3000 hertz available to the modem. The remainder might be unusable because of noise and distortion, or it may be used for ringing operators, signaling the return of coins in a pay telephone or other telephone company signaling functions.

In 1928 H. Nyquist worked on time intervals between data samples. He showed that one cycle of a signal can contain a maximum of two baud. His work proved that the maximum theoretical signaling speed in baud is twice its bandwidth, or

$$\text{Baud} = 2 \times \text{Bandwidth}$$

This theoretical signaling speed became known as the *Nyquist rate,* and it is the maximum rate of transmitting pulse signals through a system. This work was

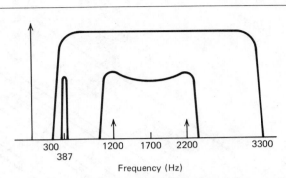

A significant factor to consider when using a 202 modem is line turnaround time required for switching from transmit mode to the receive mode. Echo suppressors in the telephone equipment that are required for voice transmission on long-distance calls must be turned off by the modem to transmit digital data. The modem must provide a 200 ms signal to the line to turn off the echo suppressors every time it goes from transmit to receive mode; hence, if short records are being transmitted, the turnaround time can slow the throughput considerably.

SPECIFICATIONS

Data:

Serial, binary, asynchronous, half duplex on 2-wire lines

Data Transfer Rate:

0 to 1200 bps—switched network

0 to 1800 bps—leased lines with C2 conditioning

Optional 5 bps AM reverse channel transmitter and receiver available for switched-network units (387 Hz).

FIGURE **4-6**   Bell 202 (half duplex/1200 bps).

significant because it showed that if a communication channel is sampled at a rate at least twice the highest baseband frequency at which the message transmission occurs, then the samples should contain all of the information that was in the original message.

In 1948 Bell Laboratories engineer Claude Shannon, who created information theory, wanted to find out how much information could be delivered in a telephone call of a certain duration. He first looked at how well the signals used to convey the information could be distinguished from one another. To avoid garbled messages, the signals or code words used to convey information must keep their "distance" from one another.

First Nyquist, and then Shannon, proved that there is a theoretical maximum capacity, and it is based on bandwidth. A random stream of bits going across the 2600 hertz bandwidth has a maximum capacity of 25,900 bits per second. This is demonstrated by using Shannon's law:

$$\text{Maximum bits per second} = \text{Available bandwidth} \quad \text{LOG}_2 \left(1 + \text{Signal-to-noise ratio} \right)$$

$$\text{Maximum} = 2600 \ \text{LOG}_2 \left(1 + \frac{1000}{1} \right)$$

$$\text{Maximum} = 25{,}900 \text{ bits per second}$$

This calculation uses a 30-decibel signal-to-noise ratio. The *signal-to-noise ratio* is the strength of the transmit signal in decibels (dB) compared with the level of background white noise (Gaussian noise) on the channel/circuit. A signal-to-noise ratio of 30 dB is 1000/1, and a signal-to-noise ratio of 20 dB is 100/1. This means that the signal is 1000 or 100 times more powerful than the background noise on the circuit.

Incidentally, if the signal-to-noise ratio is lowered to 20 decibels in the above example, then the maximum transmission capacity is 17,300 bits per second. Therefore, the higher the signal-to-noise ratio, the greater the maximum capacity in the channel. Also, the greater the bandwidth, the greater the maximum capacity.

SIGNALING ON A DIAL-UP CIRCUIT

There are two basic aspects to consider in the use of a telephone: the information that is conveyed during the conversation (the *session*), and the coded control signals that set up and terminate the call (*control signaling*). Data transmission also has a session and control signaling. Some of these control signals convey information, such as the telephone number that might be passed from end office to end office, or the status of certain equipment can be signaled, such as whether it is busy or whether a person has answered a call.

The control signals that are passed through the telephone network do some of the following.

- The dial tone indicates that dialing can begin.
- A busy signal indicates to the caller that the call cannot be completed. Sixty pulses per minute implies that the called person's telephone is in use or that the central office to which they are connected is overloaded.

- A high speed busy signal (120 pulses per minute) indicates that the long distance trunk lines (IXCs) are busy or overloaded.
- A ringing tone indicates that the called number is ringing.
- A pulsing loud noise indicates that something has gone wrong and the caller should restart the process.
- The tones heard when dialing a number indicate that the numbers are being transmitted from your telephone to the telephone company central office.
- There are many other control signals for the telephone company, such as signals for passing a number between central offices, recording billing information, giving the status of certain equipment, ringing long distance operators, diagnosing and isolating system failures, controlling special equipment such as echo suppressors, and indicating coin return in pay telephones.

DIAL-UP CIRCUITS

Dial-up is a network in which the organization uses the public telephone system, whereas private leased circuits are for your exclusive use. Dial-up networks are usually point to point, but now you can build a three-point dial-up connection which is known as a *conference call*. In other words, the user might dial up two other terminals and simultaneously interconnect these three terminals. The basic configuration for dial-up circuits in a data communication network is the same as the one shown in Figure 4-1, except there is a terminal instead of a telephone.

When using the dial-up telephone system for voice communication only, the basic procedure is to

- Lift up the telephone receiver to place the telephone in an off-hook position. This signals the central office that you want to make a call. The central office responds with a dial tone to indicate that you may dial the number.
- Dial the telephone number. Notice that after you dial the first digit, the dial tone breaks (stops). That is a control signal to tell you that your number is being received at the end office.
- After dialing, you receive one of three signals: a busy signal, a ringing signal, or a loud pulsing signal that identifies a system malfunction.
- Begin your verbal conversation when someone answers the telephone. When two data terminals talk to one another, it is called a *session*.

When the device at the end of the line is not a standard telephone, but a data-type device such as a video terminal, the procedures to sign onto the system and begin working are

- Lift up the telephone receiver to place the telephone in an off-hook position. This signals the central office that you want to make a call. The central office responds with a dial tone to indicate that you may dial the number.
- Dial the telephone number.
- Wait for the tone; when you hear the tone, place the telephone receiver in the acoustic coupler cradle.
- When the online light flashes on, depress the carriage return on your terminal keyboard (this step may vary in different systems).

Note that the above four steps would be done through the keyboard on a microcomputer if you were using a regular modem or nonacoustic coupler.

- The central computer may ask for any of the following information (this is the user sign-on procedure).

  User ID   *Enter your ID and hit the carriage return or enter key.*

  Password   *Enter your password and hit the carriage return or enter key.*

  Account Number   *Enter your account number and hit the carriage return or enter key.*

  *   *You must now tell the computer which programming language, system, or file you want to use.*

- The system responds, and you are now ready to use the timesharing system through the dial-up network. This procedure varies with different systems; there may be fewer or more steps.

Even though the above example may not match perfectly the sign-on procedures for the dial-up network at your organization, it is typical of the procedures used when entering a computer-based dial-up system. In addition, the acoustic coupler now is a very outdated modem type. It is mentioned here because some still are in use.

In some cases a dumb ASCII terminal must connect to a mainframe via a protocol converter. The ASCII terminal does not have the intelligence to dial, but there are intelligent modems on the market. With an intelligent modem, the user merely strikes a preset series of keys (that is, CR-LF-CR), which prompts the command sequence within the modem to request the telephone number from the ASCII terminal keyboard.

Before proceeding to the next section, here are four famous "firsts" related to the dial-up telephone system.

1. The very first telephone directory, a single sheet listing of 50 names, was published in February 1878 in New Haven, Connecticut. There were no telephone numbers or addresses. You had to ring the operator to ask for the person or business by name. This was in the days of the old hand-cranked telephone.

2. The first telephone numbers appeared in the Lowell, Massachusetts, telephone book in 1880. Numbers were listed along with names during the measles epidemic. The purpose was to help inexperienced operators make connections when they did not know all the subscribers.

3. The first yellow pages directory appeared in 1883 in Cheyenne, Wyoming. It was not a business directory, but merely the result of a printer running out of white paper.

4. The first yellow pages directory with business listings and advertisements was printed in Detroit, Michigan, in 1906.

ECHO SUPPRESSORS

Two-wire circuits (dial-up circuits) have a problem with echoes. When people talk on a two-wire circuit, echoes may occur under some conditions. Echoes arise in telephone circuits for the same reason that acoustic echoes occur: there is a reflection of the electrical wave from the far end of the circuit.

Telephone companies install echo suppressors on two-wire circuits to prevent the echoing back of your voice when you are talking to someone. The echo suppressor permits transmission in one direction only. When you talk, your voice closes the echo suppressor. When the other person starts talking, his or her voice closes the echo suppressor in his or her direction; because you have stopped talking, your echo suppressor opens. Obviously, if you are both talking, the power of your voices so overwhelms the echoes that you do not hear them. This is because the echo suppressors close in both directions when you both talk. Figure 4-7 depicts the operation of echo suppressors.

Lease data circuits (private dedicated lines) do not have echo suppressors; therefore, echo suppressors are not present on a four-wire lease circuit used for data transmission. Echo suppressors are present in dial-up circuits.

When a computer is dialed using a modem, there is a loud ringing tone when the computer answers. The reason for this tone is that it disables the echo suppressors. The tone is held on the line for approximately 200 to 400 milliseconds in order to disable or close the echo suppressors in both directions. Immediately after the tone ceases, the carrier wave signal comes up, perhaps a 1700 hertz tone, and it is this carrier wave signal that keeps the echo suppressors closed in both directions. If the carrier wave is lost (perhaps because of an electrical failure) for

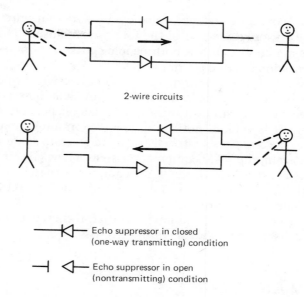

2-wire circuits

Echo suppressor in closed
(one-way transmitting) condition

Echo suppressor in open
(nontransmitting) condition

FIGURE 4-7 Echo suppressors.

approximately 150 to 300 milliseconds, the echo suppressors reopen. This reopening causes the dial-up data transmission to have many garbled or destroyed characters. The only choice at this point may be to redial the call and start over in order to close the echo suppressors in both directions.

Many garbled characters remain after the echo suppressors reopen because echo suppressors take approximately 150 to 200 milliseconds to open and close. To illustrate this effect, assume we are transmitting when such a lapse occurs. The first 150 milliseconds of the next transmission are lost because the echo suppressor is not fully closed and the data bits cannot get through. If we change directions, the first 150 milliseconds of your data transmission also are lost while the echo suppressor closes. It is for this reason that dial-up modems disable the echo suppressors before transmission.

TASI (VOICE CALLS)

The *Time Assignment Speech Interpolation* (TASI) technique is used on some of today's long distance, frequency division multiplexed voice lines. It allows for the packing of extra voice conversation into a fixed number of circuits.

Usually when two people conduct a telephone conversation, both parties do not speak at the very same moment, and for a small portion of the time neither speaks. Most long distance voice circuits are four-wire circuits. When each person speaks, only one pair of the four wires is used. Thus, two of the four wires always are empty, unless both people speak simultaneously, which cannot last for a very long period of time.

TASI electronic switching equipment is designed to detect a user's first word, and within a few milliseconds the equipment assigns a communication circuit to that speaker. Actually, an almost undetectable portion of the first syllable may be lost, but it is seldom noticed in voice communications. When a person ceases talking, the circuit is switched away and given to someone else. When the person speaks again, the TASI equipment assigns a new circuit path. Occasionally, if the circuits are overloaded, the TASI equipment may be unable to find a free path. Even though it is for a very brief period of time, several words might be lost when this occurs (see Figure 4-8).

The benefit of TASI equipment is that, if there are 100 circuits, more than 100 voice calls can be handled simultaneously.

VOICE CALL MULTIPLEXING

Multiplexing of data communication transmission was discussed in Chapter 3 in the section on multiplexers. The telephone company further multiplexes received signals, whether they are voice or data. We refer to the process as *voice call multiplexing*, even though technically it may be data that is being multiplexed.

AT&T Communications has devised a group multiplexing function. The telephone company takes groups of calls that are destined for the same area and multiplexes them together. This permits groups of calls to travel on a single coaxial cable, microwave system, satellite transmission, or optical fiber transmission media. Figure 4-9 shows the names given to these groupings and the number of voice grade circuits involved in each group.

For example, assume that your company already has multiplexed 16 conversations onto one outgoing circuit. The telephone company may further multiplex this one circuit into a group (12 voice grade circuits) for transmission to its destination. In addition, the telephone company may again multiplex it into a jumbo group of circuits. Multiplexing done by the telephone company is *pure* multiplexing (not statistical), and it is totally transparent to users. There is no holdup or delay of messages.

THE BASIC CONFIGURATIONS

The basic *configurations* (also called *topologies*) that are used to lay out networks are called star, ring, bus, and hybrid. They are used to configure an overall net-

FIGURE **4-8**   Before and after TASI.

Name	Number of Voice Grade Circuits
Group	12
Super Group	60
Master Group	600
Jumbo Group	3600
Jumbo Group Multiplex	10,800

FIGURE **4-9** Group multiplexing.

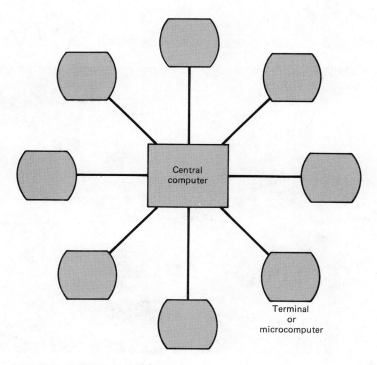

FIGURE **4-10** Star configuration.

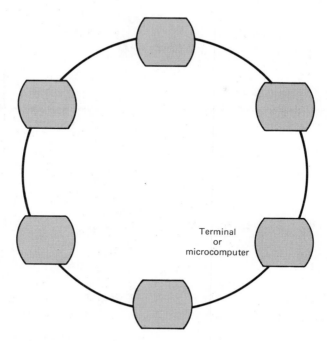

FIGURE **4-11** Ring configuration.

work, but you often only need to configure a single data circuit between two business offices of your organization. This is mentioned so that you have the proper perspective on configuring an entire network as opposed to configuring a couple of data circuits between several business offices.

A *star configuration* (shown in Figure 4-10) connects all of the terminals or microcomputers to the central computer, each with its own incoming data circuit. Data flows back and forth between the central computer and the terminal.

A *ring configuration* (shown in Figure 4-11) connects all of the terminals or microcomputers with one continuous loop. Within this loop the data usually travels in one direction only, making a complete circle around the loop.

A *bus configuration* (sometimes called a *tree configuration*) is shown in Figure 4-12. With this configuration, each of the terminals or microcomputers is connected to a single cable that runs the entire length of the network. Messages travel directly to or from the intended terminal or microcomputer.

With regard to the *hybrid configuration*, it can be an intermixing of the other three. For example, you can have a star configuration with a ring at the end of one of the links where the terminal or microcomputer might have been. Another

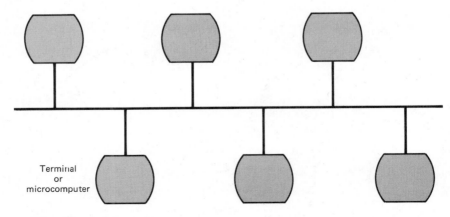

FIGURE **4-12** Bus configuration.

hybrid consists of a star configuration in which one of the links going out from the central computer is a bus configuration.

Chapter 7 (see the section titled Topology) contains a discussion of how these basic configurations are used to configure local area networks.

POINT-TO-POINT NETWORK

Figure 4-13 demonstrates a leased circuit that goes from point to point. A point-to-point circuit or network means that an organization builds a private network and, in doing so, has a communication circuit going from its host computer to a remote terminal. Point-to-point circuits sometimes are called *two-point circuits*. This type of configuration is quite advantageous when the remote terminal has enough tranmission data to fill the entire capacity of the communication circuit. When an organization builds a network using point-to-point circuits, many point-to-point circuits may emanate from the front end communication processor ports to the various remote terminals wherever they are located.

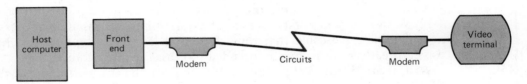

FIGURE **4-13** Point-to-point configuration.

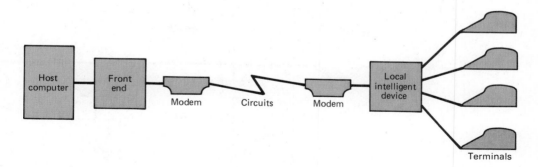

FIGURE **4-14** Local intelligent device.

LOCAL INTELLIGENT DEVICE

Figure 4-14 shows a local intelligent device configuration that allows a point-to-point circuit to be connected to a local intelligent device controlling one or more terminals. Local intelligent controllers frequently control 16 terminals simultaneously. The local intelligent terminal controller loads the point-to-point circuit more efficiently and also serves as a security restrictor. In addition, the device saves costs because some of the intelligence that would have to be built into each terminal can be built into the controller, thereby allowing the use of terminals that are less intelligent and lower in cost. As the next section shows, local intelligent device configurations also can be used in a multidrop network.

MULTIDROP CONFIGURATION

Figure 4-15 shows a multidrop circuit configuration. Notice that the first *dropoff* point (where a terminal is attached to the circuit) has only a single teleprinter terminal, but the second dropoff has a local intelligent terminal controller that is managing a cluster of terminals. Either of these configurations is possible on a multidrop circuit. Organizations that design multidrop configurations do so to load the communication circuit more efficiently, reduce circuit mileage, and thus save money. When there are various branch offices or government agencies throughout a city or country, multidrop configurations can be a very efficient method of interconnecting these various locations.

In a multidrop configuration, each dropoff point shares the line and is serviced or responded to in sequential fashion. In other words, only one dropoff point can use the circuit at a time. It is only because we can switch between the various dropoff points so fast that it appears to each user that he or she has sole use of the

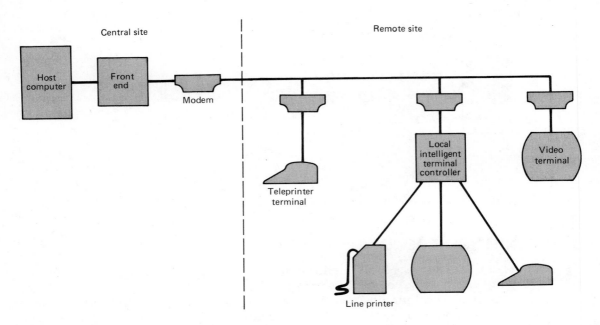

FIGURE **4-15** Multidrop circuit.

entire circuit. It is not uncommon to have 50 or 60 terminals interconnected on a single multidrop circuit. There might be five to seven local intelligent devices (one per dropoff point) and perhaps six to ten terminals connected to each local intelligent device where that circuit has been dropped off at a branch or agency.

The efficiency of a multidrop circuit can be described as: number of drops per line = traffic + error control + expected response time.

MULTIPLEX CONFIGURATION

Figure 4-16 shows a typical multiplexing configuration. To *multiplex* is to place two or more signals on the communication circuit simultaneously. Multiplexing may be performed by using either frequency division multiplexing or time division multiplexing.

The primary benefit of multiplexing is to save communication circuit costs between the host computer or business entity and many far-flung remote sites. Figure 4-17 shows how several levels of multiplexing can be used to save on communication costs. Reading from right to left, the first level of multiplexing is where four terminals are multiplexed onto a single IXC (interexchange channel) for transmission to a distant site. The second level involves multiplexing the

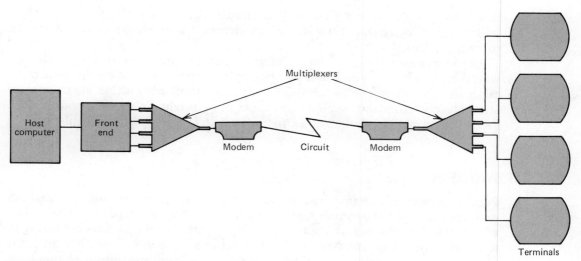

FIGURE **4-16** Multiplexing.

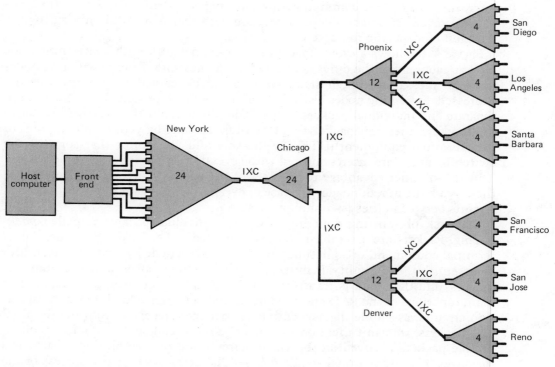

FIGURE **4-17** Levels of multiplexing.

resulting 12 signals over a single IXC circuit. Finally, the third level involves multiplexing 24 signals over a single IXC circuit. The modem pairs were left out to simplify this figure.

As you examine this illustration, think about how much more circuit mileage would be involved if you had a point-to-point circuit going from the host computer site to each of the 24 terminal device locations. Also, think about how an alternate configuration, such as multidropping, might be used to connect the various terminal device locations, in contrast to one that uses multiplexing.

PACKET SWITCHING

Packet switching is a store and forward data transmission technique in which messages are split into small segments called *packets.* Each packet is switched and transmitted through the network, independent of other packets belonging to the same transaction or anyone else's transactions. Packets belonging to different messages or transactions can travel via the same communication channel because no transmission channel is reserved exclusively for any pair of terminals or microcomputers. The communicating terminals are connected via a virtual circuit. A *virtual circuit* is a communication path that is utilized only during the duration of a specific message transmission.

Figure 4-18 shows a typical packet switching network with a *switching node* (SN) at each end of the cities connected to the network. These switching nodes route the messages through the network to whichever city and terminal they are addressed. Switching nodes also may perform the task of splitting messages into separate and individual packets. If the nodes do not split the message into separate packets, then this packetizing task must be done by your own computer or data terminal equipment (DTE). By the way, did you notice that some of the switching nodes are interconnected with several IXCs (communication circuits)? This is a distinct advantage to the user of the packet switched service because there is a built-in redundancy (in case of disaster) with regard to communication circuits between cities (switching nodes).

The task of splitting your message into individual packets is called *packetizing.* Packets are assembled and disassembled either by the customer's data terminal equipment, or sometimes at the switching node by a *packet assembly/disassembly* (PAD) facility. In either case, packetizing is an almost instantaneous process, and then data is transmitted in a virtually uninterrupted stream. The main functions of the PAD are to establish and clear the virtual telecommunication circuits, assemble the asynchronous characters received from the terminal into packets, transmit them on the virtual circuit, and, at the other end, disassemble packets received on the virtual circuit and reassemble them back into messages. An international standard called X.3 describes how the process of the PAD function should operate.

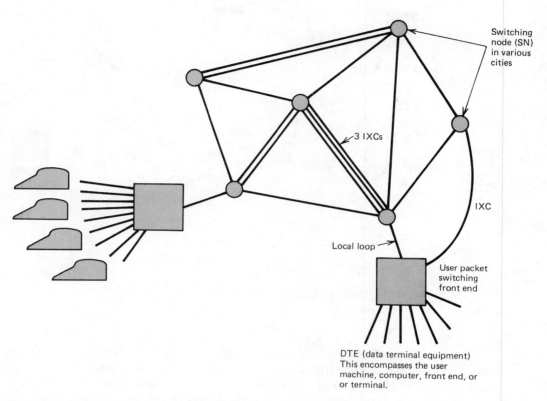

FIGURE **4-18** Packet switching network.

A typical packet might be a 128-character message block. In other words, no matter what the length of your original message, it will be split into one packet, but more likely into several 128-character-long packets. Notice that every message on a packet network is precisely the same size and contains the very same control characters within and/or surrounding the message. Once your message is packetized, either at your data terminal equipment or at the network's switching node, it is ready for transmission. At the top of Figure 4-19, you will see a picture of a typical packet. Notice that there is a header address (the packet's destination), some control characters (if a message is broken into several packets, the system must number these packets so they can be reassembled at the other end of the communication circuit), up to 1024 bits of information (128 characters), and an error check (16 bits). The receiving switching node therefore can ask to have the entire message retransmitted if any bits in the packet are corrupted during its movement over the communication circuits.

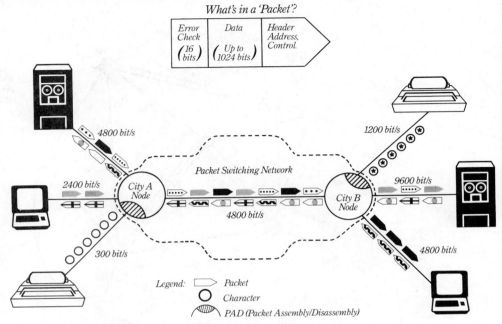

FIGURE **4-19** Packet switching concepts.

The lower part of Figure 4-19 shows a packet switching connection between two different cities. The little boat-shaped figures that are shown on the communication circuits and have different symbols in them represent individual packets of separate messages. Notice how packets from separate messages are interleaved for transmission.

For asynchronous transmission, the PAD at the receiving end reassembles all of the individual packets into their original message format. You might remember that the PAD is used to assemble the characters of asynchronous transmission (character-by-character transmission) into a 128-character packet for transmission. As a result, low speed asynchronous terminals can use a packet switching network.

In the case of synchronous transmission, network users usually are responsible for packetizing their own synchronous messages. They use their own data terminal equipment or computers prior to transmitting the message to the switching node, although the switching node also can packetize the message.

Packet switching is becoming popular because most data communications consist of short bursts of data with intervening spaces that usually are of longer duration than the actual burst of data. Packet switching takes advantage of this characteristic by *interleaving* bursts of data from many users to maximize use of the shared communication network. This is why you use a virtual circuit when transmitting (you share it with others). This interleaving is achieved by assembling the bursts of data into packets that contain your message, along with various addressing, control information, and error checking information. The overall packet switching network connects various cities, or other areas, with multiple circuit paths between them.

The charges for using a public packet switching network are not related to the distance between the various switching nodes, as is true with other communication circuits. You usually are charged a basic price depending on how many packets you transmit or the total usage time.

Many companies and government agencies have set up their own in-house packet switching networks because it is a very efficient way to design private networks. In addition, an organization can use many public packet switching networks on a number-of-packets-transmitted charge basis, as well as on a usage time basis.

Packet switched networks usually are more reliable than other types of networks because there are redundant communication circuits between various cities (you saw this in Figure 4-18). This means that if a switching node or IXC circuit fails, an entirely different path might be utilized. Moreover, because digital switching is used, it is more reliable than older analog switching techniques.

At this point, we should distinguish between circuit switching, store and forward switching, and packet switching. In *circuit switching*, communication circuits are switched between each other. If the entire circuit path is not available, you get a busy signal which prevents completion of your circuit connection (much like a dial-up telephone circuit). In *store and forward switching* (also called *message switching*), the entire transmitted message is accepted by the central switch and forwarded either immediately (the two circuits can be switched together at that moment) or at a later time when the two circuits can be switched together or when the receiving terminal requests its stored messages. In *packet switching* (which basically is a store and forward technique), your message is first split into the appropriate number of packets (normally 128 characters per packet) and then sent on to the receiving terminal. Usually, your message is forwarded immediately because packet switching networks are designed to have enough capacity (multiple circuits) to operate with immediate delivery. But many packet switched networks have the store and forward capability in case of communication circuit overloads, disastrous situations, an inoperable receiving terminal, or simply to offer electronic mail service.

One other point should be made. Any organization that does not want to develop its own in-house network and hire the required technical expertise is a prime candidate for the use of public packet switched networks. Use of public timesharing networks (virtually all of them use packet switching) eliminates many of the technical in-house expertise requirements and lowers overall network costs.

PUBLIC TIMESHARING NETWORKS

Public timesharing networks and remote computing services are available for virtually any data processing application, business function, or information retrieval service that any organization might desire. Various governments and private companies now offer network services to organizations that might wish to subscribe. Almost all of these network services utilize the packet switching concept for the transmission of data/messages over its network.

These public timesharing networks often are called *value added networks* (VANs). The concept of value added networks began when vendors leased circuits between cities, combined them into a packet switching network, started charging on the basis of message volume or amount of time used (rather than the traditional distance method), and sold the service to users. This was a first generation VAN. Second generation VANs came into being when the vendors selling these services identified additional user requirements and added them into the network. In this phase the vendors added services such as electronic mail and other security features. Today the third generation VAN is built around a microprocessor-based digital message switch similar to the fully automated digital PBX switchboards. This provides more flexibility in configuring the packet switching value added network to fit more closely the system definitions required by each individual user.

This digital message switch capability is encouraging large users (corporations and governments) to develop their own in-house third generation packet switched VANs. These VANs allow the user to interconnect an assortment of terminals, computers, and microcomputer workstations. Such packet switched networks can accept various line speeds and different protocols and can be used as a corporate *backbone network* to which all other networks within the corporation are connected. That is, they can accommodate private leased circuits, direct distance dialing, and connections to both public timesharing networks and in-house local area networks. In Figure 4-19 you might visualize the two nodes as being two different corporate offices within the same organization.

Many information retrieval services, such as Dow-Jones News Retrieval Service, The Source, Dialog, and CompuServe, connect their computers to public timesharing networks. Regardless of where the users might be located, they usually can connect through a local dial-up telephone call, or, as you might expect, they sometimes can have a leased communication circuit to the information provider. Within the United States, numerous public timeshare networks utilize packet switching capabilities, including

- GTE/Telenet
- Tymnet/McDonnell Douglas
- Accunet (AT&T)
- IBM Information Network

- GE Information Services
- UNINET
- Boeing Computer Services (BCS)

Packet switched public timeshare networks are available in most countries of the world. Some of these networks are Datapac (Canada), PSS (England), Transpac (France), ARPAC (Argentina), Euronet (connecting major European cities), and Austpac (Australia). The European Economic Community (ECC) is promoting *Euronet*, a database information network for member countries. This provides an opportunity to design and install a truly international service. Euronet is developing a videotex gateway so that the average homeowner may have access to it in the future.

Public timeshare networks and packet switching network architecture mix together very well. Both hardware technology and better software have improved the reliability of these networks, so they are dependable. The only continuing fear might be that of message security. Because messages are flowing over a network shared by many others, you may need to incorporate some type of security. The ultimate security is to encrypt messages prior to transmitting them on the public timeshare network. Encryption may be required for sensitive messages because the various password schemes and other protection devices have been violated in the past. For example, hackers have gained access to several sensitive databases of the United States government. The important topic of data communication control and security is covered in Chapter 9.

According to popular thought, we are becoming a society of home-based computer commuters, and as this trend develops we will connect to a whole myriad of public timesharing networks from our homes. These networks will be packet switched public timeshare networks, and we will connect to them through the dial-up network, a leased circuit, a cellular telephone radio connection, cable television connections, or possibly small rooftop satellite antennas. Only time will tell how popular the connection of the private home to these public timeshare networks will become. The number one issue it creates is the cost associated with connecting to and using the network and its associated databases or computing facilities.

Along with private home networks is another service known as *videotex*. Using the existing telephone network, cable TV, or "through the air" television transmission networks, videotex can provide users with instant access to a wide range of information and services. Any organization with information or services to offer can rent space on a videotex system to become a service provider to all videotex users. The United States has only been conducting experiments with videotex, and to date they have not been very successful. Apparently, the general public in the United States is not willing to pay for the cost of this service.

Outside the United States, government agenices are spearheading videotex networks. For example, France has its Teletel, England its Prestel, Australia Viatel, Canada Telidon, Japan Captain, and West Germany Bildschirmtext. In contrast in the United States videotex development is largely within the domain of private industry rather than government. In countries where videotex is very popular, the government has underwritten some of its cost and charges users far less than the basic cost of providing the service. Probably because of this subsidized low cost, it is more popular in countries outside the United States.

ISDN (INTEGRATED SERVICES DIGITAL NETWORK)

Integrated Services Digital Network (ISDN) refers to an emerging set of communication standards for a global telecommunication network that delivers voice, data, and image (video) in a digitized form. There may be competing ISDN networks within national boundaries (each nation's telephone system wants to develop its own ISDN network), but from the telecommunication user's viewpoint ISDN appears to be a single, uniformly accessible worldwide network.

The communication standards-developing organizations have launched an effort to develop ISDN standards in the hope of thwarting vendor-specific or de facto standards with regard to ISDN. The Consultative Committee on International Telephone and Telegraph has recommended standards on the basic user interface (also known as 2B + D), which details the physical and electrical specifications for an interface that supports two 64,000 bits per second digital channels and one 16,000 bits per second signaling channel. Because of this push by the international standards-making bodies, ISDN may be the first network type to have a true set of standards on a worldwide basis *before* being implemented widely in the many countries of the world.

Basically, ISDN is a communication service (communication channel) that provides direct digital transmission at high speeds and allows you to integrate your data, image (video), and voice transmissions. There are two kinds of ISDN communication channels; basic and primary.

The *basic access service* is referred to as 2B + D and consists of two 64,000 bits per second B channels and one 16,000 bits per second D channel. The letters B and D refer to the function of the channel, not its bandwidth. The B channels are used for all data other than control messages. The D channel is the conduit for "out-of-band" control messages and is used to specify, for example, what devices are connected at each end of the circuit link. The control messages give the network its intelligence. The two 64,000 bits per second B channels can handle digitized voice and data transmission, while any control messages/characters are sent simultaneously down the 16,000 bits per second D channel.

Primary access service is offered to commercial customers of an ISDN who want to hook up their PBXs or local area networks. It consists of a multiple group

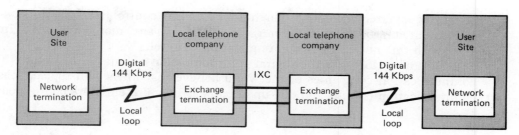

FIGURE **4-20**   ISDN showing network termination and exchange termination boxes.

of 64,000 bits per second B channels. For example, one of the primary access services is called primary rate and consists of 23 B channels (64,000 bits per second each) and one D channel. Obviously, these are for high volume data transmission users. This 24-channel ISDN (23B + 1D) is the standard in North America, but Europe seems to prefer a 32-channel ISDN (31B + 1D). The 24-channel primary rate connection offers a 1.544 million bits per second bandwidth (T-1) divided into 23 B channels at a rate of 64,000 bits per second and one D channel at 16,000 bits per second.

The interconnection for the basic access service to your organization might be as shown in Figure 4-20. It involves a *network termination box* at the customer/user site. Users get a single access communication circuit (local loop) that can carry 144,000 bits per second (2B + D or 64K + 64K + 16K) between the user's premise and the local telephone company end office. The D channel uses a Q.931 message format which contains the calling party number, the type of service requested, any user-to-user information that must be transmitted, and so forth. The D channel is referred to as "out-of-band" signaling because it is outside of the B channel that carries any voice, data, or image transmission. Once the data transmission from the user site reaches the local telephone company central office, it enters an *exchange termination box.*

An exchange termination box separates the type of signal (such as data, voice, and image) for further transmission to the telephone company end office in the city to which the transmission is being sent. At the distant end office, the information enters another exchange termination box. From this point, the information is sent over another single access line (144,000 bits per second local loop) to the receiving user site. At the receiving premise, the information again enters a network termination box, and it is distributed to the appropriate station, node, terminal, or microcomputer device within the user organization.

AT&T now has a chip that can be incorporated into communication boards that interface digital telephones, terminals, and microcomputers with the Integrated Services Digital Network. The chip implements the 2B + D ISDN standard. The chip also has the built-in capacity of formatting data (acting like a protocol converter) according to high level data link control (HDLC) protocols. If implemented into a microcomputer, the chip allows any microcomputer to integrate with this newer ISDN communication channel.

In summary, ISDN is the digital communication circuit of the future. The communication companies of the world believe that ISDN will replace leased communication circuits and dial-up circuits as they presently are configured.

PBX (SWITCHBOARDS)

A *private branch exchange* (PBX) is a switchboard within an organization in which all the telephone lines of the organization terminate. Usually, several telecommunication circuits go from this switchboard to the telephone company's end office. These circuits generally are referred to as *trunk lines* when they are devoted primarily to voice transmissions from a switchboard. In data transmission we refer to these same circuits as *leased circuits, private circuits,* or *dedicated circuits.*

First generation manual PBX systems started in the late 1800s. In those days you had to ring the operator to ask for the business or person by name. There were no telephone numbers. You may recall from an earlier section that the very first telephone directory was published in 1878, and the first telephone numbers did not appear until 1880. These first generation switchboards were totally manual. An operator connected the wires coming from a home or place of business to the other set of wires that led to whomever was being called.

A business organization could have from one to dozens of incoming trunk lines (local loops) going to the PBX. From the PBX there were inexpensive wire pair connections to numerous telephones located throughout a building. When a call arrived at the PBX switchboard, an operator determined the extension of the called party, manually plugged the call into the correct extension, and pushed a toggle switch to ring that telephone. When the called person answered, a busy light turned on at the switchboard, signifying that the line was in use. This busy light turned off as soon as either party hung up, at which time the PBX operator removed the plug, making the extension and trunk available for other calls.

Second generation PBX systems appeared on the scene in 1929 with the Bell System 701 family of switchboards. The major difference between the two generations was that an operator was no longer required to handle outgoing calls or intraoffice calls. These switchboards sometimes were referred to as *private automatic branch exchange* (PABX). We will discuss the automating of switchboards later in this section, but first we wish to introduce the term key system.

The *key system* was one of the first telephone systems that the telephone company installed on a customer premise. It would be more correct technically to call the PBX we described earlier as a key system because this switchboard had clear lighted buttons, thick connecting cables, and a telephone switchboard operator who would plug in a cable and use a toggle switch to connect and ring voice telephone instruments.

With modernization, the original cable-connected key systems/PBXs became more automated. In the next stage outgoing calls did not require operator intervention and an incoming call could be switched automatically to an extension within an organization, also without operator intervention. The term *PABX* was popular for many years, but today people refer to these automated switchboards as PBXs, key systems, and sometimes *CBXs* (computer branch exchanges).

The 1970s saw the development of a category of PBXs that now is termed the second and a half generation. This is not quite a full-generation jump. These systems used a computer to control the switch, and they offered functions such as programmable telephone moves between various locations and least cost routing of telephone calls. *Least cost routing* involves a switchboard that can choose between a telephone trunk line that the organization already leases, different common carriers, WATS, or dial-up. In this case, the lease line would have the lowest cost, with WATS or varying common carriers next, and dial-up the highest cost. Because the earlier second generation PBXs did away with most of the operators, they lost the intelligence necessary to call a person back when a line was free. The stored programs inherent in the generation $2\frac{1}{2}$ systems restored such functions because the program could call the person back when a line was free.

The first of the third generation systems appeared about 1979. Since then many vendors have announced third generation products, such as the American Telephone & Telegraph ATTIS System 85 Release 2 and the Rolm CBX II. These true third generation PBXs have three main characteristics: distributed architecture, nonblocking operation, and integrated voice and data.

Our developing technology is leading us to the all-digital switchboard, sometimes referred to as a *digital PBX*. Digital PBXs are popular because the switching is so fast and so error free that it allows data transmissions to go through the same switchboard as the one used for voice communications. In fact, if you are using a digital switchboard, your voice transmissions must be digitized prior to going through the switchboard. With digital switchboards, therefore, everything is transmitted in a digital format.

Digital PBXs use a *distributed architecture* that is either hierarchical or fully distributed. *Hierarchical systems*, such as the Wang IBX and the ATTIS System 85, distribute routine functions to the switching module, but real control resides in the central processor. In *fully distributed systems*, the switch module processes its calls independent of any other system component. Fully distributed systems are said to be inherently more reliable because failure of a single control module, no matter how catastrophic, cannot produce overall system outages. Those in the hierarchical system camp counter that fully distributed PBXs cannot perform under high load conditions because of internode control and synchronization problems.

Nonblocking operation is when intraoffice calls *always* can be placed between any two telephones within an organization. For example, if the switchboard carries lines for 2,000 telephones within the organization, then 1,000 people always can be talking with people at the other 1,000 extensions. *Integrated voice and data* means that both voice and data can be transmitted over the same communication circuit and through the same PBX.

While some vendors currently are claiming to offer PBXs of the fourth generation, at this point it probably is only generation $3\frac{1}{2}$, just as happened earlier in the mid-1970s with generation $2\frac{1}{2}$. Therefore, today's generation $3\frac{1}{2}$ PBXs have features such as inherent broadband architecture and integrated packet switching. *Inherent broadband architecture* is the ability of a PBX to transmit at extremely high data rates, usually 56,000 bits per second or greater, whereas *integrated packet switching* adds the capability of handling packet switching services.

These newer digital PBXs can use digital transmission and switching techniques because they are designed with 32-bit microprocessor chips that act as the central controller/central intelligence. A typical digital PBX uses a star architecture so that individual telephones or microcomputers can be connected directly to the central digital switchboard (see Figure 4-21). Some features of these new digital switchboards are

- Simultaneous transmission of voice and data.
- Format and protocol conversion that allows the interconnection of different vendors' word processors, host mainframes, microcomputers, and terminals.
- A local area network controlled from within the switchboard. Microcomputer workstations on the local area network are connected to the switchboard in a star configuration, and the PBX digital switch redirects messages among microcomputers connected to the local area network.
- Both voice mail and electronic mail.
- Modemless switching for digital transmission.
- Asynchronous and synchronous data transmission.
- Wire pair cables instead of more expensive coaxial or fiber optic cables.
- Both circuit switching and store and forward capabilities.
- Nonblocking switching.

These digital switchboards may become the central control switching point of all networks in a corporation or government agency. In other words, there might

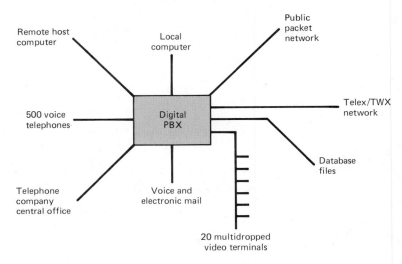

FIGURE **4-21** Digital PBX.

be extensive competition between the vendors of digital switchboards, mainframe and minicomputer vendors, and the common carriers who already offer communication switching. Especially in large organizations, it usually is quite cost effective to have one central switch (digital switchboard) rather than have some of the switching performed by the data processing department, some by the voice communication department, and some by the common carriers. Figure 4-22 shows the digital PBX serving as a central network switch for voice, data, and image (video) transmissions.

Another type of switchboard is a *data PBX*, which is variously called a *digital switch* or a *digital cross-connect switch*. Data PBX is used only when switching data (no voice transmissions), and it handles digital transmission. A typical data PBX might be one that can connect up to 144 asynchronous 9600 bits per second circuits. It automatically switches among all of the circuits, depending on the address to which a message is destined. Data PBXs offer other features, such as

- Allowing terminal users to select different computers or destinations from their keyboard without moving any cables.
- Providing port contention so that users can be connected to the first available port, or they can "camp-on." (To *camp-on* is to be held in a waiting queue until a port is freed.)
- Offering many other features of voice PBXs, such as call forwarding and security restriction.

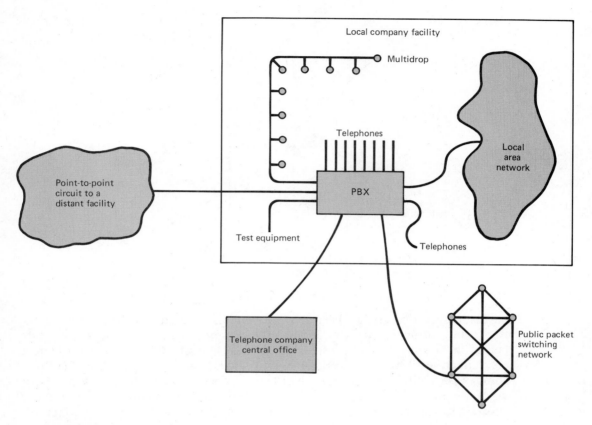

FIGURE **4-22** PBX network.

Although it is nearly 20 years old, another widely used switching technology is *CENTREX*. CENTREX is the Bell Operating Companies' tradename for an automatic system that can route incoming or outgoing calls directly to the person being called without the intervention of a PBX switchboard operator. Each telephone station has its own telephone number, the last four or five digits of which are an internal organization telephone extension number. With CENTREX, incoming telephone calls go directly to the telephone being called. People within the organization can dial inside extensions (four or five digits) or outside telephone calls to other organizations, also bypassing the telephone switchboard operator. Any central telephone numbers still can be routed to an information switchboard operator if the organization so desires. Through CENTREX all of the switching equipment can be housed and maintained at the central office of the local telephone company. This feature makes switching equipment more reliable, allows faster maintenance, and eliminates the need to manage or test the equipment yourself. Modern-day CENTREX offers many of the features of a digital PBX, such as

- Automatic callback where a calling party encountering a busy station can be called back automatically when the called station becomes available
- Call forwarding
- Call holding
- Conference calling
- Ability to change a telephone number through software rather than rewiring to a new telephone instrument
- Ability to connect to dictation equipment
- Paging people throughout a facility
- Speed calling in which often-used numbers are stored in the system.

DTS (DIGITAL TERMINATION SYSTEM)

A *digital termination system* is a communication service designed to provide flexible, low cost digital communications within a community. This configuration is nothing more than a local loop configuration within a community. It is the "last mile" to the final user premise, whether the premise is a private home, a large corporation, or a government agency. These local loop configurations have long been dominated by the telephone companies and their copper wire pairs that are located either underground or on overhead telephone poles.

The pressing issue related to digital termination systems is bypassing. *Bypass* is any communication alternative that does not use the local telephone company's twisted pair local loop. If private homes or companies want to bypass the local telephone company's local loop, they have a variety of options. Some of the competing common carriers might be those that offer cable television, direct satellite transmission, intracity microwave transmissions, infrared transmissions, and cellular radio.

Two of the technologies that are available for voice or data over short distances (less than 15 miles), and that can provide 96 channels of voice or data transmission, are infrared and microwave. Both of them, for example, can be used on college campuses to connect various buildings or two campus facilities that are not contiguous.

Infrared is a short distance transmission based on light. It is a low cost transmission medium and easy to install. Furthermore, there are no regulatory agency restrictions regarding this technology as long as transmission is within a single

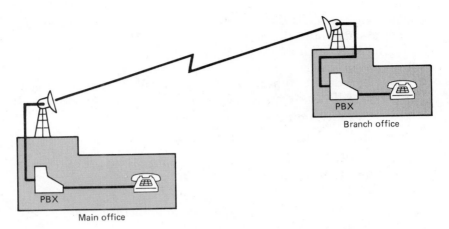

FIGURE **4-23**   Digital termination system bypass.

facility. Elaborate path alignment is not necessary, and it does not require frequency clearance or right-of-way clearance such as that required for microwave. The drawbacks of infrared are its restriction to line of sight and its suceptibility to interference from smoke or fog.

Microwave is based on transmission of a signal at specified frequencies within three frequency groupings (23 gigahertz, 18 gigahertz, and 2 to 10.5 gigahertz). Microwave requires frequency clearance from the regulatory agencies because it can interfere with nearby microwave antennas. An example of digital termination system bypass, using microwave, is shown in Figure 4-23.

Cellular radio units can be purchased or leased from telephone companies or other independent vendors. Small *VSAT satellite antennas* also can be used, especially if a public packet switching vendor sets up a packet network on a satellite link.

As more corporations purchase digital switchboards (allowing them to own the central switch in their organization which controls both voice and data communications), they will be further tempted to use one of the digital termination systems to bypass the local telephone companies. Of course, the logic of bypassing is to save on costs by leasing fewer telephone company circuits. The bottom line on the DTS is that the local telephone companies are losing revenues as more businesses link their offices directly with one another.

The Federal Communications Commission classifies the providers of digital termination systems as either limited or extended common carriers. *Limited common carriers* may operate digitial termination systems in as many as 29 standard metropolitan areas. By contrast, *extended common carriers* may operate digital termination systems in no fewer than 30 standard metropolitan areas.

SELECTED REFERENCES

1. Goeller, Lee, and Jerry Goldstone. *Business Communications Review Manual of PBXs.* New York: Telecom Library, 1987– .

2. Marney-Petix, Victoria. *Networking and Data Communications.* Reston, Va.: Reston Publishing Co., 1985.

3. McDermott, Martin. *Teleconnect Guide to the Business of Interconnect,* 2nd ed. New York: Telecom Library, 1984.

4. Rosner, Roy Daniel. *Distributed Telecommunications Networks: Via Satellites and Packet Switching.* New York: Van Nostrand Reinhold Co., 1982.

5. Rosner, Roy Daniel. *Packet Switching: Tomorrow's Communications Today.* New York: Van Nostrand Reinhold Co., 1982.

6. St. Amand, Joseph V. *Guide to Packet-Switched, Value-Added Networks.* New York: Macmillan Publishing Co., 1986.

QUESTIONS/PROBLEMS

1. Select a set of characteristics that will serve to point out differences, advantages, and disadvantages of point-to-point and multidrop lines.

2. A _____ network is a mixture of two or more network configurations to achieve a large wide area network. Some organizations call this a backbone network.

3. Define the three key elements of a network.

4. What is the difference between a private circuit, leased circuit, and a dedicated circuit?

5. An organization can lease a private network but not have physical wires interconnecting the different nodes. What is this called?

6. When you dial a call, sometimes you get a 60 pulses per minute busy signal and sometimes a 120 pulses per minute busy signal. What is the difference?

7. When you dial into a computer, the first thing you hear is a loud tone. What does that tone do?

8. Assume there are four terminals in each of the following cities: New York, Chicago, and Washington, D.C. Each city has a four-level multiplexer and circuit leading to Atlanta. If Atlanta has four terminals and a 16-level multiplexer, how many circuits are required to transmit all the

data to Miami? *HINT:* Go to Chapter 11 and copy Figure 11-15 so you can draw this configuration.

9. Assume each city (in Question 8) has its four terminals transmitting at the following rates:

New York City—4000 bps

Chicago—1200 bps

Washington, D.C.—9600 bps

Atlanta—9600 bps

How many voice grade circuits are required between Atlanta and Miami if the Atlanta/Miami modems can transmit at 14,400 bits per second?

10. What type of topology does the following illustrate?

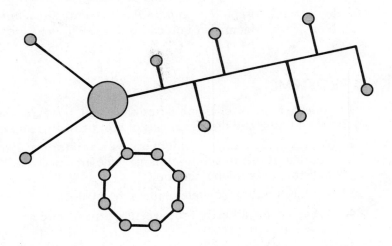

11. If there is a dial-up computer system at your organization, compare the steps for gaining access to this computer system with the steps that are listed in this chapter.

12. Can you configure a system that combines local intelligent devices, multidrop, and multiplex configurations?

13. Define the difference between circuit switching, message switching, and packet switching.

14. Identify several different packet networks.

15. Identify types of terminals and their locations for a network that is utilized by your organization, your department, or at your university or business.

16. Using Figure 4-17 as a guide, draw a multidrop configuration. *HINT:* Draw it on a copy of Figure 11-15 from Chapter 11.

17. If the total bandwidth of a satellite channel is 360,000 hertz, how many voice grade channels are supported? *HINT:* See Figure 4-5.

18. What is a 2B+D? Define it.

19. What is a PBX? Define it.

20. What is DTS? Define it.

5

PROTOCOLS AND SOFTWARE

This chapter traces the flow of a message through the critical software packages and defines the differences between protocols, software, and network architectures. The seven layers of the Open Systems Interconnection model are explained in a simplified manner. Other topics include telecommunication access method/monitor software, ARQ, X.25, BSC, SNA, network architectures, and telecommunication standards.

BASIC SOFTWARE CONCEPTS

As you may recall, in Chapter 2 we traced the flow of a message (data signal) through a communication network and discussed the basic technical concepts. (At this point we suggest that you review Figure 2-1.) In this chapter, we will relate the various software packages to that same message flow as it is received at the front end communication processor and passed on to the host mainframe computer.

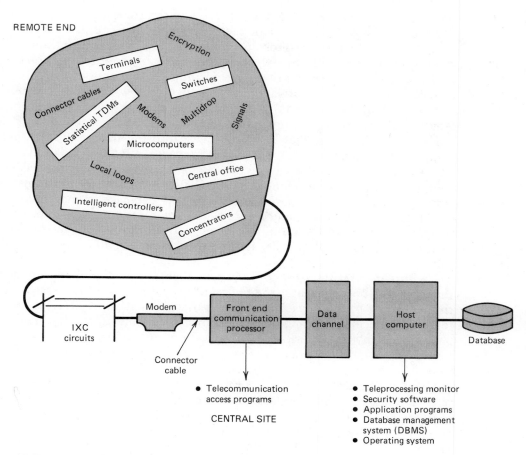

FIGURE 5-1 The basic software concepts.

To determine the location of various software packages, look at Figure 5-1. For example, at the remote end of this data communication network, there may be six or more software packages located in various pieces of hardware, such as switches, statistical time division multiplexers, terminals, intelligent controllers, telephone company central office switches, remote concentrators, and, of course, microcomputers. With the exception of microcomputers, all of these pieces of hardware utilize specialized software programs to perform specific functions required for network control or message movement. Microcomputers can accept more generalized software programs and therefore also can perform business application functions. The software programs located at the remote end of the network often are scaled-down subsets of the software located in the front end communication processor or the host mainframe computer located at the central site in Figure 5-1. From previous chapters, you already know that the front end/host can perform switching, multiplexing, control functions, and any of the application programs that a microcomputer can perform.

Now look at the six software packages listed below the front end/host computer in Figure 5-1. As you go through this chapter, you will learn that the telecommunication access programs, along with the teleprocessing monitor, are the software programs that bring messages into the host computer, get them ready for processing, and move them back out to the remote end of the network. The security software does just what the name implies—restricts access to the computer, programs, and stored database files. Application programs perform required day-to-day business functions. The database management system (DBMS) organizes, stores, and retrieves the organization's data/information. Finally, the operating system is the set of software programs that runs the host mainframe computer, controls, and coordinates among all of the other software packages.

The intent of Figure 5-1 is to show you where software packages are located at the front end/host computer and to describe the flow of your message as it moves through these software packages. Let us assume that your message has left the remote microcomputer and has passed over the connector cables, modem, telephone company end office, and IXC telecommunication circuits, and is now on its way through the modem and connector cables (at the central site) into the front end communication processor.

The message probably would not have been permitted to leave the microcomputer or remote terminal unless the *telecommunication access program*, located in the front end communication processor, polled the remote terminal and gave it permission to send the message. Note that the telecommunication access programs also can be located in the host computer, but only the older systems' architectures use the host. It is more efficient, and the processing loads are distributed better if these programs are located in the front end communication processor.

IBM is a good example of a dual location because it has telecommunication access programs both in the host computer (telecommunications access methods—TCAM) and in the front end communication processor (network control programs—NCP).

After the telecommunication access programs bring the message in from the network, log it, check for errors, and perform any other functions that this software is designed to do, the message is passed from the front end through the data channel and on to the host computer.

The *teleprocessing monitor* software package (located in the host) now takes control of the message. The detailed workings of the telecommunication access programs and the teleprocessing monitor are described later in this chapter.

The teleprocessing monitor software handles functions such as placing messages into input or output processing queues, file management, and sometimes database management, depending on the database management system. Its other functions are handling restart/recovery for problems, recording statistics for accounting purposes, keeping track of statistics for performance evaluation, handling communication tasks within the host computer such as the processing of requests and interfacing with application programs/database management systems, checkpointing, error handling within the host, security checking, and performing various utility functions such as message switching or checking various control tables.

Once the teleprocessing monitor has placed a message into the host computer's input processing queue, the input message may go through a *security software* package for review. A variety of security packages are available to check functions such as the validity of passwords, whether the password is a valid one for the originating terminal, whether the password allows access to the requested data, whether the originating terminal is allowed to make the request that was entered, whether both the password and terminal that sent the request are valid at the specific time of day the message entered the system, whether the terminal or password are restricted to specific application programs, files, records, data items, and the like.

Assuming that a request passes all the security checks, the next step is for a specific *application program* to be called out to process the request that has been made.

At this point the application program begins to process the request. Let us assume that a *database management system* lookup is required to process this request.

Your database lookup proceeds something like this. The database management system software takes the program subschema from the application program and matches it against the database schema. *Subschemas* delineate the specific logical data that the program is allowed to retrieve, while *schemas* delineate the specific logical data layout in the database.

Once the location of your data has been pinpointed on the disk, the database management system asks that the computer *operating system* physically read the disk because many database management systems do not physically read disks. When the data has been read from the disk and placed into a computer memory buffer, the database management system again takes control from the computer operating system and sorts out the specific data that your program subschema allows you to retrieve. The data is sent back to the application program, and processing of the message continues.

When the message processing has been completed, the results are passed back to the teleprocessing monitor and the response is put in the host computer's output queue.

Next, the telecommunication access programs in the front end take the message. They select the remote terminal that originated the request to determine whether it can receive the response. If the answer is yes, the response is transmitted to the terminal. At this time, the round-trip transaction has been completed.

Notice that this simple input transaction might have involved any of the five software packages (host/front end) and possibly an application program in the host computer. Furthermore, on the remote terminal end it may have involved any of the six software packages listed. This simple transaction may have gone through 50,000 lines of program code as it passed through the remote software packages, telecommunication access programs, teleprocessing monitor, security package, application program, database management system, and the host computer operating system. At this point you might review the section on Response Time in Chapter 2, especially the equation $RT = MIT + APT + MOT$.

SOFTWARE DESIGN PRECEPTS

The data communication environment poses some unique problems for the program designer, the most basic of which is lack of control over the time dimension. In conventional batch processing, the designer plans the program so it can refuse to deal with inputs until conditions are just right. If things go wrong, the designer simply stops the "clock" and causes the program to abort. This luxury is not often available in the data communication world because inputs arrive at the computer at a time and sequence beyond the designer's control.

In Chapter 2, we said that communication errors are a fact of life. This fact also impacts heavily on the program designer. The error detection schemes described in Chapter 2 deal mostly at the level of one to a few dozen characters. They are, of course, ineffective for catastrophic errors such as line breaks and interruptions lasting as long as one or more messages. The programs must accept the responsibility for prudent action in these situations. Finally, although it happens infrequently, computer hardware fails, too. When it does, it produces its own set of problems for the program designer.

Three factors make data communication programs different.

- Lack of control over input timing
- Communication errors
- Computer failures

What can the program designer do to deal with these factors? The first step is to ensure that the software provides proper *message accountability*. Basically, message accountability is a recordkeeping function that ensures no inputs or intended outputs "fall through the cracks" and that recovery from a communication or computer interruption can be accomplished with minimum damage to users.

For each incoming message this means

- Logging as soon as received
- Time tagging
- Address checking
- Format and, where possible, content error checking
- Receipt acknowledgment after logging and checking
- Diagnosing and acting constructively on errors
- Maintaining statistics on errors
- Stopping transmissions from terminals and lines that send an excessive numbers of errors.

For outgoing messages it is important to

- Log at time of transmission
- Require acknowledgment and act constructively if acknowledgment is not received
- Provide a priority scheme to ensure that outgoing overloads are worked off in a rational manner
- Test the integrity of lines and terminals and maintain statistics on results
- Provide a rational means of disposing of messages that cannot be sent because of line or terminal errors.

The final step in ensuring message accountability combines inputs and outputs. The relationships that tie inputs and outputs together (that is, "input message type A yields exactly one output message type B") must be put to work by coupling them to the input and output logs. Such logs provide a continuous statistical accounting of work in process, detect any failures in obeying the relations, and provide the proper basis for recovery processes invoked after failures.

The next step is to conduct a *failure mode analysis.* In this process the software designer examines the consequences of each possible failure in the system for message integrity. These failures must be presumed to occur successively at each stage of program execution. Possible countermeasures are evaluated and selected for implementation by the software. This is an easy process to describe and a difficult and exhausting one to perform, but it is absolutely necessary if the resulting system is to have even minimal initial viability. It is likely that most of the software in the system will be devoted to dealing with the exceptions occasioned by communication, hardware, and people failures rather than to accomplish the "mainline" system functions. The failure mode analysis, therefore, easily can turn out to be the major component in the design task.

Finally, all the principles of good noncommunication program design also apply. The chief of these are

- **Modularity:** Break the functional job into small, "neat," functional modules and match the program structure to the functional structure.

- **Hierarchy:** Recognize the hierarchy of the functional modules and mirror it in the module calling relationships.

- **Generality:** Look for the truly "primitive" functions, generalize their definitions, and clearly identify their basic parameters so that modules can be defined to perform groups of similar functions, rather than proliferating specialized modules.

SOFTWARE TESTING PRECEPTS

The same factors that make design of data communication software uniquely difficult also tend to make testing of that software more complex than in a batch environment. Because the timing and sequencing of "real-world" inputs are not always predictable, it is difficult to build confidence that any testing procedure has exercised the time relationships sufficiently to reveal all time-dependent pathological behavior in the program under test. Similarly, in testing the response of the program to environmental factors, such as communication errors and hardware failures, the system implementer finds it difficult to create a sufficient variety of these events to ensure that the program is tested thoroughly.

What can be done to ease these problems? There are three areas on which the systems implementer should focus.

- Test planning
- Test execution
- Test documentation

First, *test planning* is a function that often is forgotten until it is too late. The software test plan should be developed as part of the *software functional specification*. This approach helps ensure that

- The test focuses on proving the performance of the software, rather than proving that the programmer's concept of the design matches the design.

- Test support facilities, such as computer time, special test data, communications, and special test generation or data reduction programs, are identified early enough to plan their acquisition intelligently and efficiently.

- "Testability" of the software will be a design criterion. This also will improve the "diagnosability" of errors and the overall maintainability of the software.

- All parties in the system development effort know the criteria by which the suitability of the software will be judged.

Second, *test execution* should be handled, if possible, by different personnel from those who developed the software. The objectivity introduced by an independent software test group pays off in improved performance and software integrity. Users must participate in the conduct of the testing if it is to be done effectively. Their involvement should increase progressively as the "bugs" are removed from the software and the detailed functional characteristics become more apparent. During this time, two things tend to happen. First, the user can offer immediate, first-hand pragmatic judgments about discrepancies between specifications and actual performance. Often such discrepancies can be removed in simple ways, that is, by using the application knowledge of the user. Second, the user is building knowledge of, and confidence in, the software.

Finally, *test documentation* often is an overlooked activity. "Coming events cast their shadows before them" is an adage that predates the software business by about 150 years,[1] but it has real importance in this current context. Almost all the failures in a software system, whether they are found during testing or weeks, months, or years after the software has been declared operational, "cast their shadows" during implementation and testing. Thorough, careful documentation of test planning, preparation, execution, and post-test analysis provides the best possible groundwork from which to analyze the failures and prevent their recurrence. It is important to recognize that frequently more money is spent on "maintaining" real-time programs (that is, fixing failures not found during testing and upgrading functional capabilities) than was spent on developing the programs in the first place. Good documentation is the foundation of any effort to keep software maintenance costs in line.

PROTOCOL/SOFTWARE/ARCHITECTURE

Before discussing the details of protocols, software, and architecture, we want to define their differences. It is a somewhat blurred distinction because they all

[1] Thomas Campbell (1777–1844), "Lochiel's Warning."

work together. In fact, a full data communication network cannot operate without their interaction.

Protocol This is a strict set of rules or procedures that are required to initiate and maintain communications. Protocols exist at many different layers in the network, such as link-by-link, end-to-end, subscriber-to-switch/packet switching networks. These layers will become evident in the section on the OSI Seven-Layer Model given later in this chapter. Certainly the most famous protocols are the international packet switching protocol X.25 and IBM's protocols, Synchronous Data Link Control (SDLC) and Binary Synchronous Communications (BSC).

In day-to-day life we experience person-to-person social protocols. These *person-to-person protocols* are followed to determine when it is permissible to interrupt someone, acknowledge the presence of a speaker, offer the social graces such as thank-yous, and the like. By contrast, we have *person-to-network protocols* in the world of data communications. For voice telephone calls, these person-to-network protocols are simple. They basically involve lifting up a telephone receiver, listening for a dial tone, dialing a telephone number (noticing that the dial tone stops immediately after you dial the first number), waiting for the telephone to ring or produce a busy signal, and taking whatever action is appropriate after this point. These protocols allow you to interconnect with a network in order to pass on information signals between you and the other network users.

Another type of data communication protocol is a *machine-to-machine protocol.* The data link control protocols in layer 2 of the OSI model (explained later in this chapter) are machine-to-machine protocols and are the ones discussed most commonly in the classroom. These protocols make possible the exchange of information after a call (information channel) has been established. A *data link control protocol* specifies

- Which terminal transmits at a given time
- How a message signals its beginning and end
- How a receiver knows when a character or block of data begins and ends
- How a receiver recognizes a transmission error and asks for a message retransmission
- How a receiver sequences messages or characters within the message itself
- How a receiver distinguishes between instruction bits (such as those that indicate end of message) and bits that represent the message itself
- How a transmitter or receiver recovers when an unexpected catastrophe occurs, such as a break in the flow of data

The trend is toward standard protocols developed by standards bodies such as the

Consultative Committee on International Telephone and Telegraph. (Telecommunication standards are described in the last section of this chapter.)

Software These programs are located at various points in the network. For example, the host computer might have such software packages as the operating system, the teleprocessing monitor, database management systems, security packages, and application programs. At the front end communication processor there probably are telecommunication access programs for network control. At a remote switch/statistical TDM/concentrator there can be switching software, store and forward software, and control software packages that perform a subset of the telecommunication access program functions that are located at the front end communication processor. Farther out in the network there might be software packages located at or within intelligent terminals, and of course the whole range of software that might be available in a remote microcomputer.

Microprocessor chips have blurred the definition of software, which originally referred to computer programs. The term *firmware* is used increasingly to refer to that halfway point between hardware and software. Firmware is a microcircuit chip that contains the program functions to be performed. When the program functions are placed in an electronic chip, they can operate much faster and are more secure from unauthorized change or modification.

You also can find protocols within various software, such as within the telecommunication access programs. This further blurs the distinction between protocols and software.

Architecture System network architectures attempt to facilitate the operation, maintenance, and growth of the communication and processing environment by isolating the user and the application programs from the details of the network. Network architectures use both protocols and software in their operation. The architectures package the software and protocols together into a usable network architecture system.

Many people ask, "What is the reason behind network architecture when you have protocols and other software packages that actually perform the networking functions?" Quite simply, network architecture is the most cost-effective way to develop and implement a coordinated set of products that can be interconnected. The architecture is the "plan" that connects protocols and other software programs. This is beneficial to both the network users and suppliers of hardware and software. The nine basic characteristics of a network architecture are

- *Separation of functions.* Because user networks and vendor products evolve over time, there must be a way to provide enhanced functions to accommodate the latest technology. With a network architecture, the system is designed with a high degree of modularity so that changes can occur incrementally and with a minimum of disruption.

- *Wider connectivity.* The goal of most networks is to provide optimum connection between any number of nodes, taking into account whatever levels of security might be required.

- *Resource sharing.* Network architectures allows shared resources, such as printers and databases. This, in turn, makes a network both more efficient and more economical to operate.

- *Network management.* The architecture must allow users to define, operate, change, secure, and maintain the network.

- *Ease of use.* With a network architecture, the designers can focus their attention on the primary interfaces in the network and therefore make it user friendly.

- *Standardization.* A network architecture encourages software developers and vendors to use standard hardware and software. The greater the standardization, the greater is the connectivity and the lower the cost.

- *Data management.* Network architectures take into account the management of data and the necessity to interconnect with various database management systems.

- *Interfaces.* The architectures also define interfaces, such as person-to-network, person-to-person, and program-to-program interfaces. In this way, the architecture blends together the appropriate protocols (which are written as computer programs) and other appropriate software packages to produce a functioning network.

- *Applications.* Network architectures separate the functions required to operate a network from the organization's business applications. It is more efficient when business programmers do not need to be concerned with network operations. This will become clear when you read about LU (logical unit) 6.2 in the section Systems Network Architecture (SNA).

Some of the better-known network architectures, such as IBM's BSC and SNA, are discussed in this chapter, and a section briefly describes eleven other network architectures.

TELECOMMUNICATION ACCESS METHOD

Figure 5-1 listed the various types of software. Now we can discuss two of these packages that are related directly to data communications. These are the telecommunication access programs and teleprocessing monitors.

The *telecommunication access programs*, which used to reside in the host computer, now reside primarily in the front end communication processors, the switching nodes (SN) of a packet network, the remote intelligent controllers, or microcomputers. This software package provides for some of the following capabilities.

- Polling and selecting of terminals in a central control network
- Automatic dial-up and answering of calls
- Code conversion
- Message switching and store and forward (although this might be in the host computer's teleprocessing monitor)
- Circuit switching and port contention
- Logging of all inbound and outbound messages
- Error detection and the ordering of retransmission when an error is detected

The list of functions for the front end communication processor (enumerated in Chapter 3) can provide other ideas as to the type of software functions that are performed in many of today's front end communication processors.

The most familiar communication software packages are the telecommunication access programs that started with the IBM 360/370 series, some of which have been in use since the late 1960s. They cover a range of capabilities. There are five such packages, and they are presented here as examples of the range of products with which the designer can work.

- Basic telecommunication access method (BTAM)
- Queued telecommunication access method (QTAM)
- Telecommunication access method (TCAM)
- Virtual telecommunication access method (VTAM)
- Network control programs (NCP)

The *basic telecommunication access method* (BTAM) provides the basic functions needed for controlling data communication circuits in IBM 360/370 systems. It supports asynchronous terminals, binary synchronous communications, and audio response units. BTAM is a set of basic modules that are used to construct communication programs. It is recommended for use when there are ten or fewer circuits to support or when a specialized communication control program is required. BTAM requires knowledge of the terminal's operation, link discipline, and a basic knowledge of programming. A BTAM user must write routines for the scheduling and allocation of facilities. The basic flow control and data administration routines are also the responsibility of the BTAM user. It is the least sophisticated of the five data communication software programs listed above, but it does contribute the lowest system overhead. BTAM provides facilities for polling terminals, transmitting and receiving messages, detecting errors, automatically retransmitting erroneous messages, translating code, dialing and answering calls, logging transmission errors, allocating blocks of buffer storage (OS360/370 only), and performing online diagnostics to facilitate the testing of terminal equipment. BTAM resides in the central computer and is the interface between the front end communication processor and the user-written application programs.

The *queued telecommunication access method* (QTAM) is an extension of BTAM and includes all the BTAM facilities except that it does not support binary synchronous communications; it supports only asynchronous terminals. QTAM provides a high level and flexible macrolanguage for the control and processing of communication data, including message editing, queuing, routing, and logging. It can schedule and allocate facilities, poll terminals, perform error checking routines, reroute messages, cancel messages, and the like. QTAM is not used much any more and has been replaced primarily by TCAM.

The *telecommunication access method* (TCAM) replaces and extends the older QTAM. The most significant features of TCAM are those for network control and system recovery. An operator control facility also is provided for network supervision and modification. It supports asynchronous terminals, binary synchronous communications, and audio response units. TCAM performs all the functions of BTAM and QTAM and handles the data communications in a system that utilizes a high degree of multiprogramming. Unlike the prior basic data communication software, TCAM has its own control program that takes charge and schedules the traffic-handling operations. In some cases it can handle an incoming message by itself without passing it to an application program—for example, routing a message to another terminal in a message switching system. TCAM also provides status reporting on terminals, lines, and queues. It has significant recovery and serviceability features to increase the security and availability of the data communication system. The checkpoint and restart facilities are much more capable than those of QTAM. TCAM has prewritten routines for checkpointing, logging, date and time stamping, sequence numbering and checking, message interception and rerouting, and error message transmission, and it supports a separate master terminal for the data communication system operator. TCAM can manage a network structured on Systems Network Architecture (SNA). Advanced Communication Function/TCAM (ACF/TCAM) also accommodates SNA/SDLC terminals.

The *virtual telecommunication access method* (VTAM) is the data communication software package that complements IBM's advanced hardware and software. VTAM manages a network structured on SNA principles. It directs the transmission of data between the application programs in the central computer and the components of the data communication network. It operates with front end communication processors. The basic services performed by VTAM include establishing, controlling, and terminating access between the application programs and the terminals. It moves data between application programs and terminals and permits application programs to share communication circuits, communication controllers, and terminals. VTAM controls the configuration of the telecommunication network, has the capability of creating virtual connections, and permits the network to be monitored and altered in addition to performing all the basic functions of the other three data communication software packages.

When VTAM establishes sessions, one end of the session is understood to be the host and the other the terminal. VTAM makes the mainframe the primary end of the session and the other the secondary end. In technical terms, the host program is said to be the *primary logical unit* (PLU), and the terminal or microcomputer is considered to be the *secondary logical unit* (SLU). Only the primary logical unit can start a session, end the session, and perform key aspects of error recovery. When personal computers are linked to SNA hosts, they are considered to be secondary logical units just like terminals.

Even though VTAM has kept pace with such supports as the LU 6.2 peer-to-peer protocols, one problem remains. That is its inability to connect two personal computers. Many companies build their data communication networks around the company mainframe and use VTAM. A link to the mainframe often is the only common denominator for the diverse computers installed within a firm. Perhaps there are many PCs and a System/36, all of which are connected in some way to the VTAM network so they all can have sessions with the host. The problem is that the PCs cannot have sessions with each other. A company that has 100 personal computers connected to the same VTAM host cannot put those PCs into sessions with each other through VTAM.

Why does VTAM impose such a restriction? We mentioned above that VTAM requires a session partner that is not a host to be a secondary logical unit. This has not changed over time. Even though VTAM has been updated to include LU 6.2 and peer-to-peer communications, VTAM still enforces the rule that every nonhost be secondary and that every session must have a primary.

Network control programs (NCPs) route data and control its flow between the front end communication processor and any other network resources. These other network resources can be the host mainframe computer or an intelligent control unit located either locally or at the remote end of the communication link. IBM's primary network control program is the Advanced Communications Function/Network Control Program (ACF/NCP). Network control programs reside in the front end communication processors, primarily IBM's 3704, 3705, and 3725. NCP is *not* a replacement for TCAM or VTAM; it interfaces with TCAM and VTAM.

In summary, BTAM is the least sophisticated access method and, to be effective, requires the most effort on the part of a user. QTAM has been replaced with

TCAM which provides the facilities for a complete communication system. It supports a wide range of terminals, provides network control and significant recovery facilities, and now supports SNA. VTAM is the most advanced data communication software package and is intended to support the virtual computer systems using SNA. NCP is the only one of these five programs that actually resides in an IBM front end communication processor.

TELEPROCESSING MONITOR

As has been stated, some functions overlap between the front end telecommunication access programs and the host computer teleprocessing monitor. IBM is a good example of this overlap. The IBM software package Telecommunication Access Method—TCAM (this is a telecommunication access program) and the IBM package Customer Information Control System—CICS (this is a teleprocessing monitor) both reside in the host computer. IBM's front end has its own set of programs called Network Control Programs—NCP (these are telecommunication access programs). TCAM and NCP overlap; therefore, functions such as polling/selecting can be performed from either the host computer or the front end communication processor.

Teleprocessing monitors are software programs that directly relieve the operating system of many tasks involved in handling message traffic between the host and the front end or the host and other internal central processing unit (CPU) software packages (such as the host database management system). Generally speaking, teleprocessing monitors can perform functions such as line handling, access methods, task scheduling, and system recovery.

The telecommunication access programs need to access and move data into and out of the host computer. The teleprocessing monitor must interface with the telecommunication access programs on one side and with all of the host computer's software on the other side. Teleprocessing monitors must interface with various network architectures, operating systems, computer architectures, database management systems, security software packages, and application programs.

Typical teleprocessing monitors offer features such as the ability to interface with X.25 and other protocols, the ability to interface with various operating systems and hardware, and interfaces to database management systems and telecommunication access programs.

Teleprocessing monitors should offer some type of security. For example, a security package should be designed to accommodate a unique password and identification for each terminal operator and allow access only to the specific functions assigned to that operator password and/or terminal identifier. It can assign highly sensitive functions to a specific terminal or a group of terminals. Some security features might be security sign-on fields, darkened password fields, and a complete log of terminal sign-ons, including any security violations.

Other tasks that can be performed by teleprocessing monitors are logging of all messages (both input and output), accounting procedures for cost control, restart and recovery procedures in case of failure, utility features in order to perform special maintenance tasks, and queue management of both inbound and outbound message queues, as well as the ability to place priorities on messages and/or queues. A teleprocessing monitor should have the ability to interface with multiple front end processors, terminals, microcomputers, and various code structures, as well as various data communication transmission speeds. The monitor provides input/output job task queue management, various methods of instituting priorities for certain transactions or jobs, file/database management, application program management, task and resource control, restart and recovery procedures in case of failure, special utilities that perform tasks done often enough to warrant setting them up as a utility feature (OSI model layer 6) or in a program library. It keeps track of various accounting features and operating statistics and isolates various programs or parts of the system from other programs or parts of the system. In other words, a teleprocessing monitor can be considered a "mini" operating system with data communication interfaces.

The world's most widely used teleprocessing monitor is the customer information control system (CICS). It was developed by IBM in the late 1960s to meet the emerging demand for high volume transaction processing systems. In today's world it may not be perfect, but even IBM's direct competitors implicitly acknowledge that there are few alternatives to CICS. Whatever comes along to supplant it will be an evolving CICS rather than a totally new replacement. CICS is a table-driven teleprocessing program that offers 64 layers or systems that it can service. Like any other teleprocessing monitor, CICS is a program that runs in conjunction with the operating system to handle communications with remote terminals. CICS takes over the communication-related tasks that previously were handled by the operating system, thus allowing the operating system to concentrate on other control tasks.

The teleprocessing monitor will increase in importance and almost equal the operating system for performing job tasks as we move toward more online data communication-oriented networks, and especially as we move into the world of distributed systems and databases.

THE OSI SEVEN-LAYER MODEL

One of the most important standards-making bodies is the International Organization for Standardization (ISO), which makes technical recommendations about data communication interfaces. (The abbreviation ISO comes from its French name, which sometimes causes confusion in English-speaking countries.) During the late 1970s, ISO created the Open Systems Interconnection (OSI) subcommittee whose task was to develop a framework of standards for computer-to-computer communications. The resulting effort produced the *Open Systems Interconnection Reference Model*, which is referred to as the OSI Reference Model. It serves as a framework around which a series of standard protocols are defined.

The membership of the ISO is comprised of the national standards organizations of each ISO member country. In turn, ISO is a member of the Consultative Committee on International Telephone and Telegraph (CCITT), whose task is to make technical recommendations about telephone, telegraph, and data communication interfaces on a worldwide basis. On issues of telecommunication standards, ISO and CCITT usually cooperate, but they are mutually independent standards-making bodies and they are not required to agree on the same standards.

The seven-layer OSI model was conceived in 1979. It should be noted here that the CCITT also is moving toward the independent development of a separate seven-layer architecture. There are some functional differences between these two seven-layer architectures, but the ISO standard is now so deeply entrenched that it probably will be the long-standing international standard used as a framework to develop data communication software/protocols.

The widely implemented OSI model facilitates control, analysis, upgradability, replacement, and management of the resources that constitute the communication system. It also makes it much easier to develop software and hardware that link incompatible networks because protocols can be dealt with one layer at a time. Whichever standard is followed, the use of layers in designing network software and applications is strongly recommended.

The seven layers of the OSI model are shown in Figure 5-2. A layered approach to computer communications offers the following benefits.

- Network software and hardware engineers can allocate tasks among network resources more easily and effectively.
- Network managers can assign responsibilities within their departments more effectively.
- It is easier and less costly to replace any network layer with its equivalent product from another vendor.
- It is easier to upgrade a network by replacing an individual layer rather than all of the software.
- Networks can be converted to international and industry standards on a progressive layer-by-layer basis as the standards become available.

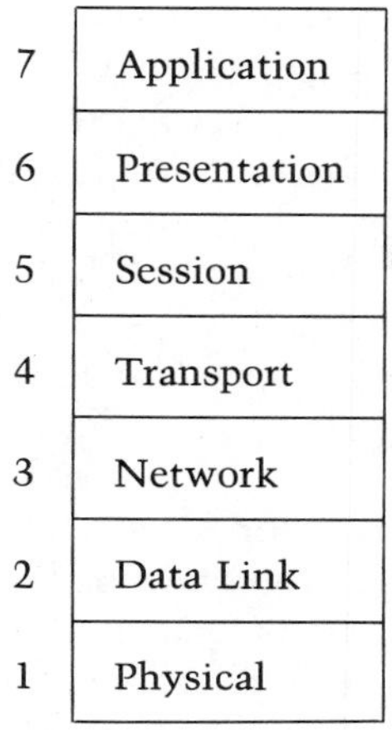

7	Application
6	Presentation
5	Session
4	Transport
3	Network
2	Data Link
1	Physical

FIGURE 5-2 Seven-layer OSI model.

- Many network functions can be offloaded from the host mainframe computer into a front end communication processor or other remote network control devices.

Before describing the details of the seven-layer OSI model, let us summarize the functions of each of the seven layers. Follow along by looking at Figure 5-2 where we will start at the top, which is the application layer.

The *application layer* (layer 7) serves the end user directly by providing the distributed information service to support the applications and manage the communications. The utilities being defined at this layer include assistance for applications programs, virtual terminals, file transfers, job transfers, user directories, and password authentication.

The *presentation layer* (layer 6) allows the application to interpret the meaning of the exchanged information. This layer performs format conversions that allow otherwise incompatible devices to communicate. Encryption, decryption, character set translations, and videotex syntax translations are examples of presentation layer services.

The *session layer* (layer 5) manages the dialogue between the two cooperating applications by providing the services needed to establish the communication, synchronize the data flow, and terminate the connection in an orderly fashion. Examples might be windows on a microcomputer or those services associated with applications in a host mainframe.

The *transport layer* (layer 4) provides end-to-end control and information interchange with the level of reliability requested by the user. This transport layer chooses a protocol that considers both user requirements and knowledge of the underlying network service being provided. This layer ensures reliable communications between the transmitting and receiving devices, possibly over multiple links and nodes, by providing functions that are more complex than, but similar to, those provided by the data link layer on a single link. This layer insulates the upper three layers (layers 5 to 7) from the telecommunication details resident in the bottom four layers (layers 1 to 4), thereby allowing communication facilities to change without requiring modifications in the upper-layer procedures or application programs.

The *network layer* (layer 3) provides the means to establish, maintain, and terminate the switched connections between the end users' systems. Addressing routines, routing functions, and the like are included in this layer. The network layer is responsible for the packetizing and efficient routing of information through a network. Typically, the network layer in each node uses a predetermined routing algorithm to select the best data link over which to send information.

The *data link layer* (layer 2) ensures the reliability of the transmission medium by providing error checking, retransmission, carrying out of flow control, and sequencing capabilities for use by the network layer. It also provides synchronization and error control for the information that is transmitted over the physical link of layer 1. One example is IBM's Synchronous Data Link Control (SDLC) protocol.

The *physical layer* (layer 1) provides the electrical, mechanical, functional, and procedural characteristics required to move data bits between each end of the communication link. At this point you should note that *all* physical movement of data bits takes place at this physical level and includes parameters such as voltage levels, number of pins required on a connector plug, and circuit activation/deactivation.

To expand further on these concepts, look at Figure 5-3, which shows the seven layers of software that are located at the host end of a network and the seven layers that are located at the remote terminal end of a network. Notice that there are physical communications between the two machines (host and terminal) *only* at layer 1, the physical layer. In other words, the actual message data bits must move down from layer 7 to layer 1, across the communication circuits that interconnect layer 1 at the host end to layer 1 at the terminal end, and then back up to layer 7 at the other end of the communication link.

All other connections between layers 2 through 7 are known as *virtual links* because they are only theoretical; physical data bits do *not* move between them. For example, even though a software program at layer 5 at one end may "think" that it is sending data directly to layer 5 at the other end of the link, it is not. This is only a virtual link. Another way of looking at virtual links is to view them as

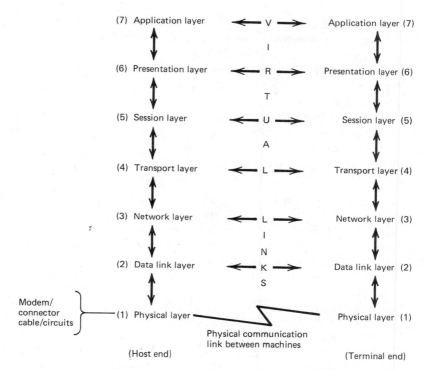

FIGURE **5-3** Communications using the seven-layer OSI model.

logical links or logical connection paths. Logically, data may flow between any of the layers at either end of a communication link, but physically it must flow down to layer 1, across the communication media, and back up to layer 7 (or the layer to which it is addressed) at the other end of the communication circuit.

The purpose of these seven layers is to define the various functions that must be carried out when two machines want to communicate. For example, suppose you want to start a conversation with your friend. You probably would *not* carry out the dialogue by saying, "Hi. Do you want to talk with me?", then wait for your friend's answer, and if he or she said yes, start the conversation. Machines must do things like this first!

In a real-life, human-to-human situation, the protocols used to get someone's attention or obtain permission to speak are looking at the person, using body language, or just beginning to speak. These types of informal human "protocols" cannot be used in a data communication network situation because the terminals are not human and they are not in face-to-face contact. As you learned when you read about establishing a dial-up telephone call, you must set up the communication path (communication circuit) first. Next, you establish your session by whatever it takes to identify the person or machine with which you want to communicate. Then you carry out the session by sending and receiving the business information (this is the purpose of your communication). Finally, you terminate the session (end your conversation with the other person or machine) and end your link (disconnect the communication circuit).

With the foregoing as your introduction to the OSI seven-layer model, remember that this is a *model* used by designers and programmers for developing communication software and protocols; it is not a program. The following sections provide a more in-depth perspective of each of the seven layers.

Layer 1: Physical Layer This layer is concerned primarily with transmitting data bits (0s or 1s) over a communication circuit. The main purpose of this layer is to define the rules by which one side (host) sends a 1 bit so that the other side (terminal) defines it as a 1 bit when it is received. This is the physical communication circuit layer. At this layer we are concerned with very basic things like voltages of electricity, timing factors such as 1200 bits per second being equal to 833 microseconds per bit (1 sec./1200 bps = 833 μsec.), full duplex or half duplex transmission, rules for establishing the initial connection, how to disconnect when the transmission is complete, and connector cable standards such as RS232C, RS449, or X.21.

At this layer we are concerned with how the physical, electrical, and functional interchange takes place that establishes, maintains, and disconnects the physical link between DTEs (data terminal equipment) and DCEs (data circuit terminating equipment).

Remember that layer 1 is the basic link over which all data must pass. Communication between layers 2 to 7 at a host and layers 2 to 7 at a terminal are only virtual (appearing to exist) communications. In reality, the messages must be passed down to layer 1 (physical layer) for the actual movement of the message between the host computer and a remote terminal.

Layer 2: Data Link Layer This layer manages the basic transmission circuit that was established in layer 1 and transforms it into a circuit or link that is free of transmission errors. This error-free transmission link interfaces closely with layer 3, the network layer. The data link layer accomplishes its tasks by breaking the input data into data frames, transmitting these frames, and processing acknowledgment frames that are sent back to acknowledge the received data.

Because layer 1 accepts and transmits only a serial stream of bits without any regard to meaning or structure, it is up to the data link layer to create and recognize frame boundaries and check for errors during transmission. As you can see in Figure 5-3, the layer 1 protocol is handled in conjunction with the modem, connector cable standards, and communication circuits.

Layer 2 requires intelligence (software) and therefore is located in some type of programmable device, such as a front end communication processor or possibly the host mainframe computer itself. At the remote end, layer 2 is located in an intelligent terminal controller or a microcomputer. The data link layer establishes and controls the physical path of communications before sending your message down to the physical layer below it. The data link layer takes the data, which has been divided into packets by layer 3 above it, and physically assembles the packet for transmission. This assembly involves adding error detection, message type, and other control characters. If the protocol does not use packets, this layer assembles the message into a frame. Basically, a frame is just another name for the block that is transmitted. Both packets and frames contain the message, along with other control/error detection characters.

It is up to layer 2 to solve the problems caused by damaged, lost, or duplicate message frames so the layer above it (layer 3) can work with error-free messages. This includes error detection, correction, and retransmission, definition of the beginning and end of the message itself, resolving competing requests for the same communication link, and flow control. *Flow control* involves keeping a fast transmitting device from "drowning" a slow receiving device. Some mechanism or procedure must be employed to let the transmitting terminal know that the available buffer space at the receiver is filling to a critical level. This procedure and the error handling are integrated, although the problem of one terminal overrunning another also is handled in some of the layers above this layer.

Finally, some of the other functions that were discussed in regard to the physical layer also have to be acted upon in the data link layer, such as whether transmission is in full duplex or half duplex. Some of the typical data link level protocols are X.25, High-level Data Link Control (HDLC), Synchronous Data Link Control (SDLC), and, of course, Binary Synchronous Communications (BSC).

Layer 3: Network Layer This layer provides for the functions of internal network operations such as addressing and routing. In other words, it provides services that transport data through the network to its destination node/terminal. The network layer actually controls the operation of the combined layers 1, 2, and 3. This sometimes is called the *subnetwork* or the *packet switching network* function.

This layer provides control from one node (terminal) to another across the network. Basically, this layer of software accepts messages from the host computer, converts them to packets, and ensures that the packets get directed to their proper destination. A key issue is how the route of the packet might be determined. It can be based, for example, on dynamic tables containing the various circuit routes that are updated frequently to show possible down circuits or circuits that currently are overloaded with data transmissions. It is at this layer that your message, which starts out in the host mainframe computer (let us say at layer 7), is cut into packets. Then it is passed down to the data link layer which frames it and passes it down to the physical layer, which in turn pumps the data bits over the communication circuit to wherever the packets are addressed.

The network layer also is concerned with enabling simultaneous use of multiple links to increase information transmission performance. Among the issues addressed here are routing, flow control, end-to-end acknowledgments on the network for multilink paths, and host mainframe-to-network interfaces.

One major issue is concerned with front end communication processors, packet switching nodes, and host computers. Which of these three physical hardware devices should ensure that all packets are received correctly at their destination and in the proper order? Even though the software takes care of this function, the problem lies in the fact that the software can be located physically in any one of these three devices.

Because this is the packet switching layer, the software accepts messages from a host computer, converts them to packets, and then sees to it that the packets get addressed and directed toward their destination. Routing of the packets is another task. At this layer there may be a database of routing tables used to keep track of the various routes a packet can take and to determine how many different circuits there are between any two individual packet switching nodes. Packet routing also involves load-leveling the volume of transmission on any given circuit, as well as knowing whether a circuit has failed.

There even can be an accounting function built into layer 3. An example is one that keeps track of how many messages are transmitted for each organization so that each group can be billed correctly; however, accounting functions usually are built into one of the higher layers of the seven-layer protocol scheme.

It should be noted that X.25 only defines the requirements of a connection between data terminal equipment (DTE) and a public data network; it is not a standard connection between sets of data communication equipment from different vendors. Each equipment manufacturer must implement X.25 on its own hardware by using appropriate software.

Layer 4: Transport Layer This layer often is called the *host-to-host layer*. It establishes, maintains, and terminates "logical" connections for the transfer of data between end users. It is responsible for generating the address of the end user, ensuring that all the packets of data have been received, eliminating duplicate packets, and ensuring that packets have not been lost during transmission. This layer provides the facilities that allow end users to pass messages between themselves across several intervening circuit links, stations, or nodes. The layers below this layer (layers 1 to 3) are transparent (invisible) to the end user. Layer 4 includes facilities to do all user addressing, data assurance (control), and flow control of messages from source to destination across either simple or complex networks. It sometimes is called the end-to-end or host-to-host layer. Even though the layer below this one (the network layer) actually is responsible for routing the packets, the transport layer initially establishes the distant terminal address that is used for packet routing.

The transport layer deals with end-to-end issues, such as network addressing, establishment of virtual circuits, and procedures of use for entering and departing from the network. Only when we get above this layer do we start to discuss issues that are visible directly to the end user.

At this layer we have moved out of the message protocols and into other software programs, peer-to-peer protocols between layers, and the vendor's network architectures. Probably the best known is IBM's Systems Network Architecture (SNA).

This layer might include the specifications for broadcast messages, datagram-type services, accounting information collection, message priorities, security, response times, and a recovery strategy in case of failure.

The transport layer is a source-to-destination or end-to-end layer because a program at the source machine can carry on a virtual conversation with a similar program on a destination machine using message headers and control messages. The physical path still goes down to layer 1 and across to the destination machine, however.

At the lower layers (layers 1 to 3), the protocols are carried out by each machine and its immediate neighbors, not by the ultimate source and destination machines. These source/destination machines always are separated by many other pieces of hardware such as front end communication processors, concentrators, multiplexers, message switches, and modems. Layers 1 to 3 are chained together in a sequential fashion, whereas layers 4 through 7 are end-to-end or computer-to-remote terminal software interfaces. The message actually is passed down from layer 7 to layer 1 and then transmitted.

The transport layer also can multiplex several streams of messages onto one physical circuit by creating multiple connections that enter and leave each host computer. The *transport header* delineates which message belongs to which connection. Also at this level there is a mechanism that regulates the flow of information so that a very fast host cannot overrun a slower terminal or especially an overburdened host. The flow control at the transport layer is a little different from the flow control at the lower layers. The transport layer prevents one host from overrunning another by controlling the movement of messages, whereas in the lower layers the physical flow (speed) of packets or frames is controlled. In fact, a lower layer can hold back data sent out by layer 4. Today layer 4 is physically in the host computer, although it may be drifting slowly out toward the front end. Layer 4 functions usually are carried out by software, but they are rapidly becoming part of the firmware as we get standards that are defined more clearly.

Layer 5: Session Layer The session layer is responsible for initiating, maintaining, and terminating each logical session between end users. In order to understand the session layer, think of your telephone. When you lift the receiver, listen for a dial tone, and dial a number, you create a physical connection (layer 1). When you start speaking with the person at the other end of the telephone circuit, you are engaged in a session. In other words, the session is the dialogue that the two of you carry out.

In addition, this layer is responsible for managing and structuring all session-requested data transport actions. Session initiation must arrange for all the desired and required services between session participants. Required services can be logging on to circuit equipment, transferring files between equipment, use of various terminal types or features, security authenticators, the software tasks for half duplex or full duplex, and the like.

Sometimes the session layer is referred to as the *data flow control layer* because it is responsible for establishing the connection between two applications or processes, reestablishing the connection if it fails, enforcing the rules for carrying on the session, and maintaining data flow control. The session layer also is concerned with establishing communications between given pairs of users and starting, stopping, and controlling those communications. For example, if the host mainframe is sending data to a printer that has a limited-size buffer, the established rules might be to send only one buffer-size block of data to the printer at any one time and then wait for the printer to signal that its buffer is empty before sending the next block of data. This layer is responsible for controlling this flow in order to avoid a buffer overflow (loss of data) at the printer.

This layer provides for session termination, which is an orderly way to terminate the session. It also provides the facilities to abort a session prematurely through items such as a break key. The session layer also might keep track of various accounting functions so the correct party receives the bill later. It may have some redundancy built in to recover from a broken transport (layer 4) connection in case of failure. This layer ensures that, when a database management system is used, a transaction against the database is never aborted. This precaution is necessary because an abort leaves the database in an inconsistent state; concurrence and deadlock may preclude further use of the database unless that specific transaction is carried to its completion. If there is an abort, software at the session layer has to restore the data to its original condition (before the database transaction).

The session layer is very close to the transport layer, although it has more application-oriented functions than does the transport layer. Because the session layer usually is handled by the host computer operating system supervisors, it would be easy to merge the session and transport layers into a single layer.

Layer 6: Presentation Layer The presentation layer performs a selectable set of message transformations and formatting in order to present data to the end users. This layer has items such as video screen formatting, peripheral device coding, other formatting, encryption, and compaction.

This layer defines the end user's port into the network in terms of the code being utilized, format, and any other attributes. Its job is to accommodate the totally different interfaces seen by a terminal in one node and what is expected by the application program at the host mainframe computer. For example, the layer 6 presentation services in IBM's Systems Network Architecture perform data compression, additions (such as column headings), translation (for example, program commands such as clear screen) from its machine language into local terminal commands, and so on. The teleprocessing monitor customer information control system (CICS) is a layer 6 service located in a host mainframe computer, although a product like CICS has many other functions beyond the presentation layer. At the presentation layer we are concerned with the display, formatting, and editing of inputs and outputs for a user.

Basically, any function (except those in layers 1 to 5) that is requested sufficiently often to warrant finding a general solution for it is placed in the presentation layer. Even though some of these functions can be performed by library routines, placing them in the library routines might overwork the host computer operating system and slow throughput. More generally, different computers have different file formats, so a file conversion option is useful as well as protocol conversion between incompatible hardware. Other items such as number of printed lines per screen, characters per line, and cursor addressing can be handled by this layer.

Layer 7: Application Layer The application layer is the end user's access to the network and therefore is developed by the individual user organization.

At this layer we are concerned with what the user is trying to do, namely, perform a business function. This task may be the generation of a form, creation of a policy, a complete financial report, or any other number of products, such as spreadsheets. The application layer is intended to provide a set of utilities for application programs. Each user program determines the set of messages and any action it might take upon receipt of a message. Other considerations at the application layer include network management statistics, remote system initiation and termination, network monitoring, application diagnostics, making the network transparent to users, simple processor sharing between host computers, use of distributed databases, and industry-specific protocols such as you might have in banking.

In summary, the OSI model defines a framework for the development of protocols and communication system architectures. The four lower layers are guides for moving and receiving information between two systems; the three upper layers (5, 6, and 7) are concerned with setting up and performing applications.

ARQ (AUTOMATIC REPEAT REQUEST)

A system that detects an error in data and has it retransmitted automatically is called an ARQ (automatic repeat request) system. ARQ systems are of two types: stop and wait ARQ and continuous ARQ.

With *stop and wait ARQ*, after sending a block, the transmitting terminal waits for a positive acknowledgment (ACK) or a negative acknowledgment (NAK). If it is an ACK, the terminal sends the next block; if it is a NAK, the terminal resends the previous block. Another possible response is a WAK, which means positive acknowledgment but do not transmit any more at this time.

With *continuous ARQ*, the transmitting terminal does not wait for an acknowledgment after sending a block; it immediately sends the next block. While the blocks are being transmitted, the stream of returning acknowledgments is examined by the transmitting terminal. If a NAK is received, the transmitting terminal usually (depending on the specific vendor's protocol programs) retransmits all the blocks from the one that was in error to the end of the stream of blocks sent. The terminal also may be able to retransmit only the block that was in error, although this method requires more logic and buffering at the terminals. Continuous ARQ requires a full duplex circuit or a reverse channel modem.

Older protocols like BSC (Binary Synchronous Communications from IBM) usually employ the stop and wait ARQ. Newer protocols like X.25 and IBM's Synchronous Data Link Control (SDLC) use continuous ARQ for error recovery procedures. SDLC is the name of the protocol in IBM's Systems Network Architecture (SNA). To make this usage clear and to refresh your memory, we suggest that you go back to the section on satellites in Chapter 2 and reread the part on satellite delay and satellite delay compensators.

When using various protocols with microcomputers, you may hear the term *sliding window protocol*. This is another type of continuous ARQ protocol.

X.25 PACKET PROTOCOL

The most popular bit-oriented protocol is the international standard X.25. It defines the structure, contents, and sequencing procedures for the transmission of data among DTEs, DCEs, and a public data network. It also defines the techniques used for error detection and recovery.

Flag 01111110	Address	Control	Message	Frame check sequence	Flag 01111110

FIGURE 5-4 X.25 frame.

Look back to the section that described the seven layers of the OSI model and note that the X.25 protocol is involved only in layers 1 to 3. Layers 4 to 6 are more concerned with other software and network architecture, while layer 7 involves user application programs.

Figure 5-4 shows a typical frame from the X.25 protocol. Each frame begins and ends with a special bit pattern (01111110). This is known as the *beginning* and *ending flag*. The beginning flag references the position of the *address* and *control* frame elements and initiates error checking procedures. The ending flag terminates the error checking procedures. When you have contiguous frames, it also may be the beginning flag for the next frame.

The *address field* is used to identify one of the terminals. For point-to-point circuits it sometimes is used to distinguish commands from responses or to address a specific terminal device on multifunction terminals. This eight-bit field might contain a station address, a group address for several terminals, or a broadcast address to all terminals.

The *control field* identifies the kind of frame that is being transmitted, such as information, supervisory, or unnumbered. The *information* frame is used for the transfer and reception of messages, frame numbering of contiguous frames, and the like. The *supervisory* frame is used to transmit acknowledgments, such as to indicate the next expected frame, indicate that a transmission error has been detected, acknowledge that all frames received are correct, stop sending, and call for the retransmission of specified frames. The *unnumbered* frame is used for other purposes, such as to provide a command "disconnect" that allows a terminal to announce that it is going down, or to indicate that a frame with a correct checksum has impossible semantics. Since frames used for control purposes may be lost or damaged, just like information frames, they also must be acknowledged. A special control acknowledgment frame is provided for this purpose; it is called an *unnumbered acknowledgment*.

The *message field* is of variable length and is the user's message or request (data packet). This field may include a general format identifier, logical channel group numbers, logical channel numbers, packet-type identifiers, internal message DTE addresses (calling DTE and called DTE), and, of course, the message that is being transmitted.

Notice that we have *two* types of sequence numbering here. In the previously mentioned control field we sequence number the individual contiguous frames.

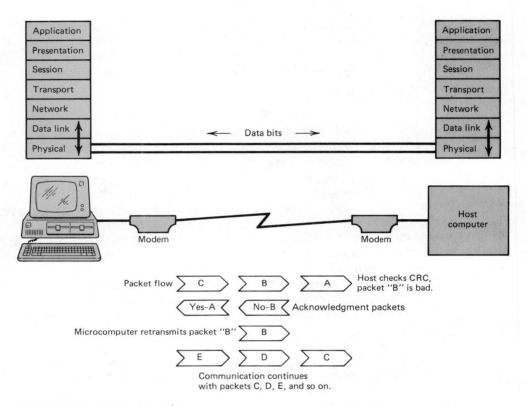

FIGURE **5-5**   X.25 protocol.

In the message field we sequence number the individual packets when the system breaks the message into multiple packets.

The *frame check sequence field* provides a 16-bit cyclic redundancy checking (CRC) calculation that is placed in the field by the transmitting station. Upon receipt, the receiving station recalculates the CRC values and matches them to determine whether any errors occurred during data transmission. If the CRC values do not match, a request for retransmission is sent to the sending terminal station/node.

In summary, X.25 handles call setup and termination, identifies the logical channel number chosen by the call initiator, addresses the called party, counts packets for charging methods, starts the transmission over a virtual circuit (remember that the transmission goes down to layer 1, the physical circuit), breaks the message into various packets/frames, prevents network overload by controlling the flow of packets, determines the path over which the packet is transferred, and provides various security features. The security features for packet switching are discussed in Chapter 9, Security and Control.

IBM's Systems Network Architecture with its protocol (Synchronous Data Link Control—SDLC) is compatible with the international standard X.25. The frames of SDLC are virtually equal to the frames of X.25.

Figure 5-5 illustrates the flow of data on an X.25 communication circuit and shows the process of error detection and retransmission. This error detection and correction service is provided in layer 2 (the data link layer). In this case, a microcomputer is accessing a remote host computer via modems. As the microcomputer user types requests to the host computer, the protocol puts the data into packets and sends packets A, B, and C down the communication circuit. The protocol on the host side recalculates the cyclical redundancy check (CRC) on the packets and finds that packets A and C are complete, while packet B was corrupted during transmission. Packet B is rejected and a negative acknowledgment is sent back to the microcomputer. At that point, the microcomputer retransmits packet B and then resumes communication with packets C, D, and E. Note that packet C had to be transmitted again even though it was correct the first time. Sometimes this is called "go back *n*" error correction because transmission resumes from the corrected packet. The above explanation assumed that the entire message was in one packet. If a message was in two or more packets, the entire message would be rejected as compared with rejecting a single packet.

BINARY SYNCHRONOUS COMMUNICATIONS (BSC)

The binary synchronous protocol is still in wide use because many organizations have not had the time or money to upgrade to the newer bit-oriented protocols such as X.25. BSC is a byte-oriented protocol because it takes an entire 8-bit byte to send a command signal to the receiving station. With a bit-oriented protocol such as X.25, the changing of a single bit within a frame's control byte sends a command to the receiving station. BSC transmission takes place in a two-way, half duplex transmission mode with 8-bit byte commands.

Figure 5-6 shows a BSC message format. The (SOH) start of heading (first character of message) is a fixed 8-bit character. The SOH is 10000001 in ASCII coding. It is followed by other header control characters, which are used by the system for terminal addressing or other control purposes. The start of text (STX— 01000001) is another special 8-bit character, as is the end of text (EXT— 11000000). These two characters sandwich the message text characters so that they can be identified easily. Finally, there is a block check character (BCC), which is a checksum at the end similar to the cyclical redundancy check of X.25. The purpose of this block check character is to detect any errors that occurred

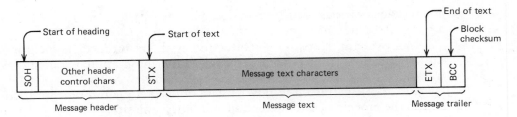

FIGURE **5-6** BSC message format.

during data transmission. The list of special control characters used in the binary synchronous protocol can be seen in Figure 2-7, ASCII code structure.

Notice that the BSC protocol does not place messages into tightly defined frames, nor does it handle any packet switching. It is an inefficient protocol because it is designed for half duplex circuits. BSC does not support loop circuits or mixing of terminal types. Different code structures cause serious problems. For example, the BSC protocol was designed to be used with Extended Binary Coded Decimal Interchange Code (EBCDIC), American Standard Code for Information Interchange (ASCII), or a 6-bit, 64-character code. BSC does not support hub go-ahead polling or some of the fast select polling scenarios.

Under BSC, the initiation of an error recovery procedure at one terminal station can cause line unavailability to other terminal stations until the recovery procedure is complete.

BSC half duplex operation works on the basis of sending one message block at a time to a distant terminal. The terminal must acknowledge the successful or unsuccessful receipt of each block before another one can be transmitted.

BSC does not provide complete error checking for all control and information transfer messages as does X.25. It does not provide any system-assigned block sequence numbering except for odd/even acknowledgments. BSC cannot be used on satellite circuits because the propagation delay time affects the stop and wait ARQ. Messages are delayed too long in satellite circuit transmission when you have to send a message block and receive an acknowledgment before you can send the second message block. This is because there is an approximate 0.5- to 0.6-second round-trip propagation time delay on a satellite circuit.

SYSTEMS NETWORK ARCHITECTURE (SNA)

SNA describes an integrated structure that provides for all modes of data communications and upon which new data communication networks can be planned and implemented. SNA is built around four basic principles. *First*, SNA encompasses distributed functions in which many network responsibilities can be

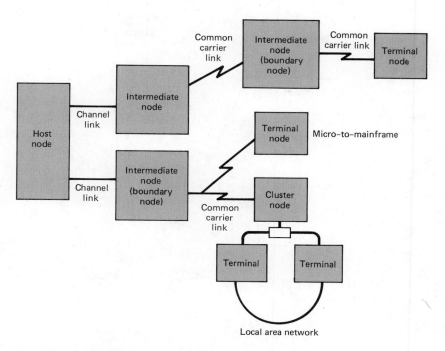

FIGURE 5-7 SNA network.

moved from the central computer to other network components, such as remote concentrators. *Second*, SNA describes paths between the end users (programs, devices, or operators) of the data communication network separately from the users themselves, thus allowing network configuration modifications or extensions without affecting the end user. *Third*, SNA uses the principle of device independence, which permits an application program to communicate with an input/output device without regard to any unique device requirements. This also allows application programs and communication equipment to be added or changed without affecting other elements of the communication network. *Fourth*, SNA uses both logical and physical standardized functions and protocols for the communication of information between any two points. This means that there can be *one* architecture for general purpose and industry terminals of many varieties, and *one* network protocol.

The appropriate place to begin understanding the concept of SNA is to look at it from the viewpoint of the end user (see Figure 5-7).

The end user (terminal operator) talks to the network through a logical unit (LU). These logical units are implemented as program code or microcode

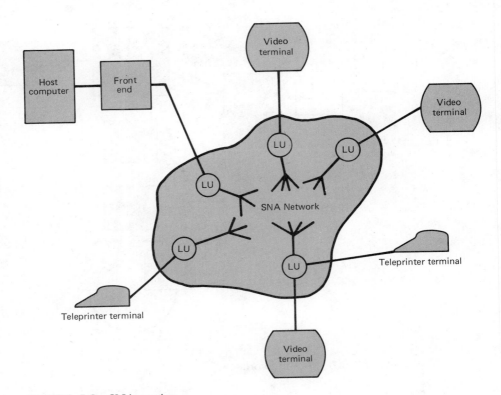

FIGURE 5-8   SNA session.

(firmware), and they provide the end user with a point of access to the network. The program code or microcode can be built into the terminal or implemented into an intelligent terminal controller/concentrator/remote front end.

Before one end user of an SNA network can communicate with any other end user, each of their respective logical units must be connected in a mutual relationship called a *session.* Because a session joins two logical units, it is called a *LU-LU session.* Figure 5-8 depicts the interconnection of logical units when one end user wants to talk to another.

The exchange of data by end users is subject to a number of procedural rules (protocols) that the logical units specify before beginning the exchange of information. These procedural rules specify how the session is to be conducted, the format of data, the amount of data to be sent by one end user before the other end user replies, actions to be taken if errors occur, and other types of protocols that usually are handled by body language when two human beings converse.

Each logical unit (LU) in a network is assigned a network name. Before a session begins, the SNA network determines the network address that corresponds to each LU network name. This scheme allows one end user (for example, a terminal operator) to establish communication with another end user (for example, an application program) without having to specify where that end user is located in the network. These network names and addresses are used for addressing messages.

The flow of data between users actually moves between two logical units in a session. This flow of data moves as a bit sequence (frame) and generally is referred to as a *message unit*. The message unit also contains the network addresses of the logical unit that originated the message and the logical unit that is to receive the message. These are the basic protocols at work. Refer back to Figure 5-4, which depicted the X.25 frame. (This is identical to the IBM Synchronous Data Link Control frame.)

A session between a pair of logical units is initiated when one of them (the end user) issues a REQUEST TO SEND message. Once a session has been activated between a pair of logical units, they can begin to exchange data. This is where the basic IBM protocol Synchronous Data Link Control (SDLC) handles the movement of data to have an orderly data flow. SDLC controls such items as the rate at which data flows, whether the sending LU expects to receive a response after every message unit (frame) it sends, sequencing, and the like.

A session between a pair of logical units is deactivated when one of them sends a deactivation request or when some other outside event interrupts the session. This outside event can be intervention by a network operator or failure of some part of the network.

The logical organization of an SNA network, regardless of its physical configuration, is divided into two broad categories of components: *network addressable units* and *path control network*.

Network addressable units are sets of SNA components that provide services which enable end users to send data through the network and help network operators perform network control and management functions. Physically, network addressable units are hardware and programming components within terminals, intelligent controllers, and front end communication processors. Network addressable units communicate with one another through the path control network.

The network addressable units in SNA contain three kinds of addressable units. The first has already been introduced, and it is called a *logical unit* (LU). The second kind is the *physical unit* (PU). This is not truly a physical device; it is a set of SNA components that provide services to control communication links, terminals, intelligent controllers, front end communication processors, and host computers. Each terminal, intelligent controller, front end processor, and the like contains a physical unit that represents the terminal, intelligent controller, and the like to the SNA network. The third kind of network addressable unit is the *System Services Control Point* (SSCP). This also is a set of SNA components, but

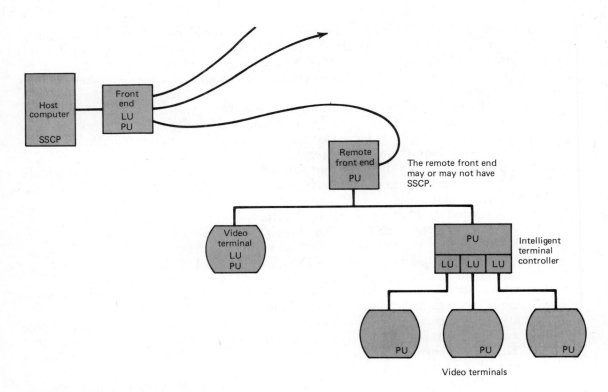

FIGURE **5-9**   SNA SSCPs/LUs/PUs.

its duties are broader than those of the physical units and logical units. Physical units and logical units represent machine resources and end users, whereas the SSCP manages the entire SNA network or a significant part of it called a *domain*.

Just as sessions exist between logical units, sessions can exist between other kinds of network addressable units such as SSCP-LU. Even an SSCP-PU session is possible. Figure 5-9 shows the location of LUs, PUs, and SSCPs.

Systems Network Architecture defines a *node* as a point within an SNA network that contains SNA components. For example, each terminal, intelligent controller, and front end communication processor that is designed into the SNA specifications can be a node.

An expanded definition of a *node* is any microcomputer, minicomputer, mainframe computer, or database that constitutes a point on the network at which data might be stored, forwarded, input into the network, or removed from the network as output. Depending on which vendor's literature you are reading, they might refer to a node as a *station*, an *intelligent microprocessor-based device*, a *terminal*, or a *workstation*.

Each SNA node contains a physical unit that represents that node and its resources to the System Services Control Point. When the SSCP activates a session with a physical unit (SSCP-PU session), it makes the node (terminal, intelligent controller, or front end communication processor) that contains that physical unit an active part of the SNA network. It is convenient to think of an SNA node as being a terminal, intelligent controller, or front end communication processor within the network. Certain more powerful nodes also can have an SSCP.

The *path control network* provides for routing and flow control; the major service provided by the data link control layer within the path control network is the transmission of data over individual links. Logical units must establish a path before a LU-LU session can begin. Each SSCP, PU, and LU has a different network address, which identifies it to other network addressable units as well as to the path control network. Path control provides for the following.

- Virtual routing so all sessions send their messages by different routes
- Transmission priorities
- Multiple links to maximize throughput
- Message pacing to keep a fast transmitter from drowning a slow receiver
- Ability to detect and recover from errors as they occur
- Facilities to handle disruption because of a circuit failure
- Facilities to inform network operators when there is a disruption in the network

The *path control network* has two layers: the *path control layer* and the *data link control layer*. Routing and flow control are the major services provided by the path control layer, whereas transmitting data over individual links is the major service provided by the data link control layer. It is in this area that the Synchronous Data Link Control (SCLC) is used by links for serial bit-by-bit transmission of frames. SDLC is a discipline for the management of information transfer over data communication links. The SDLC function includes the following activities (layer 2 of the OSI model).

- Synchronizing or getting the transmitter in step with the receiver
- Detecting and recovering from transmission errors
- Controlling the sending and receiving stations
- Reporting improper data link control procedures

The SDLC procedures take each message and sandwich it into a frame for transmission. In the SDLC concept, the frame is the vehicle for every command

and response and for all information that is transmitted using SNA. Figure 5-4 depicted the X.25 frame, which is identical to the SDLC frame. All messages are put into this frame format for transmission from one node to another node (PU-PU). The error checking for each message is incorporated in the frame check sequence portion of the SDLC frame.

SNA describes an integrated structure that provides for all modes of data communication and on which new data communication networks can be planned and implemented. It is a network architecture that incorporates protocols and various software packages. SNA encompasses distributed functions in which many network responsibilities can be moved from the host computer to other network components such as the front end communication processor, remote intelligent control units, or terminals. SNA describes paths between the end users (LU-LU). This allows multiple network configurations. SNA uses the principle of device independence which permits an application program to communicate with an input/output device, regardless of unique device requirements. This also allows application programs and communication equipment to be added or changed without affecting other elements of the communication network. SNA uses standardized functions and protocols (now compatible with the international standard X.25) between logical units and physical units for the communication of information between any two points. This means that there can be one architecture for general purpose network systems which can use terminals of many varieties interconnected through one network protocol.

A data communication network built on SNA concepts consists of the following (see Figure 5-10).

- A host computer
- A front end communication processor (intermediate node)
- Remote intelligent controller (intermediate node or boundary node)
- A variety of general purpose and industry-oriented terminals (terminal node or cluster node)
- Possibly local area networks or micro-to-mainframe links

A newer part of IBM's Systems Network Architecture is Advanced Program-to-Program Communications (APPC). APPC often is referred to by its two components, *logical unit 6.2* (LU 6.2) and *physical unit 2.1* (PU 2.1). It has been defined by various users as an architecture, a high level program interface, a network operating system, and a protocol. IBM describes APPC as a program interface and an operating system. As a high level program interface, LU 6.2 provides a set of conversion verbs that a programmer can embed into a program when that program needs to communicate with another application, such as to access a database.

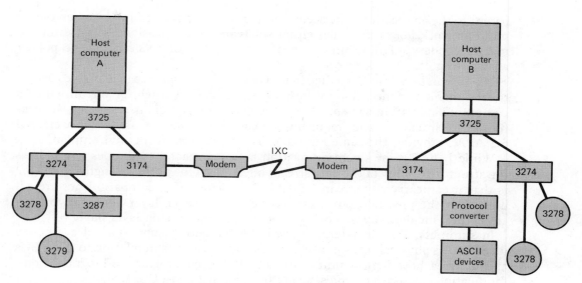

FIGURE **5-10** SNA network devices. The numbers refer to model numbers of IBM devices: 3725 front end processor; 3174/3274 controller; 3278/3279 terminals; 3287 printer.

To accomplish this conversion, the programmer issues the LU 6.2 verb called "allocate" which in effect says, "I would like to converse with this other program." The programmer then issues a "send" command, which starts the program at the other end of the communication link. That program carries out its task (let us assume in this case that it is a database hookup) and returns the data to the original transmitting device. LU 6.2 translates these commands and destination data retrieved from a directory into messages that can flow through the network. The advantage of this is that the user of a high level programming interface such as LU 6.2 does not have to be sensitive to the location of the destination or to networking, in terms of efficiency, encoding, or flow control. That is all taken care of by LU 6.2.

Until the advent of LU 6.2, or APPC, data communications meant two or more computers exchanging data in a predefined format. In the hands of skilled software developers and system designers, APPC/LU 6.2 allows programs on different computers to manipulate data interactively. This sometimes is referred to as providing users with protocols for peer-to-peer communications.

The term *peer to peer* can have two meanings when referring to communications. The first meaning is communication between two or more processes or programs by which both ends of the conversation exchange data with equal privilege and in which any physical differences between the computers is transparent to the application. The second meaning is communication between two or more network nodes in which either side can initiate sessions because no primary-secondary relationship exists, and in which either side is able to poll or answer to polls.

LU 6.2 can be viewed as a distributed operating system because it gives access to, and manages contention for, resources. An analogy might be one in which a computer's operating system allows the sharing of different resources, such as files and printers, without requiring the user to be aware of any other users. LU 6.2/APPC provides the same kind of capability in a network environment.

While LU 6.2 allows communication between machines, it does not address file structures or any of the data stream questions. One of the strengths of LU 6.2 is that it can carry any stream of data. Data structure incompatibilities are addressed at the presentation layer (layer 6) of the seven-layer SNA network architecture. Incidentally, LU 6.2 is contained in the session layer (layer 5).

In summary, IBM's Systems Network Architecture supports nodes that are identified as physical units. It also defines logical entities in relation to these PUs and dictates how logical units interact. Advanced Program-to-Program Communications assumes the presence of both LU 6.2 and PU 2.1.

The physical unit protocols are those that govern the manner in which a node on a SNA network can operate. (Remember that a node is any end point in the network.) The type of physical unit one can support dictates which node in the communication does polling, the status of the node (primary or secondary), whether multiple links can join adjacent nodes, and so forth.

Every node contains one physical unit that manages the telecommunication links connecting it to adjacent nodes. The component of PU 2.1 nodes that activates sessions with other PUs and LUs is called a *peripheral node control point* (PNCP). The PNCP becomes active when the node is operational.

The basic physical units include the following.

- PU 1.0 is an obsolete terminal controller protocol.
- PU 2.0 is an end node associated with minicomputers, 3274 display controllers, remote job entry (RJE) stations (such as 3770), and printers (such as the 3820) that attach only to host computers.
- PU 2.1 is the same as PU 2.0, but it allows connections to both hosts and other PU 2.1 nodes.
- PU 4.0 is a communication controller or front end processor.
- PU 5.0 is a host processor.

PU 2.1 enables peer devices to communicate without host intervention. Each individual node with PU 2.1 capability can bring up a connection with other PU 2.1 intermediate nodes. LU 6.2 is a common set of programming verbs allowing communication among devices, but complete office networking cannot be achieved without the additional capabilities of PU 2.1. Before the advent of PU 2.1, the mainframe would direct traffic through an intermediate node if a direct route from Point A to Point B was unavailable. Thus, a terminal wishing to converse with a non-IBM mainframe had to pass through an IBM host. PU 2.1 is the mechanism that allows you to use LU 6.2.

It is easier to understand PU 2.1 if we compare two APPC implementations: IBM's APPC/PC and Digital Equipment's DECnet SNA/APPC. Both products support LU 6.2, but APPC/PC also supports PU 2.1, while DECnet APPC does not.

A personal computer using APPC/PC can have sessions with a mainframe host (PU 2.0 to PU 5.0) or any other node that can support PU 2.1, a list that includes the System/38, System/36, the Series 1, another PC, a Displaywriter, and many other devices.

In contrast, a DECnet APPC node can have sessions only with a mainframe host, connecting PU 2.0 to PU 5.0. The host acts as the primary station and is responsible for polling, most error recovery, and so on. Beyond that, DECnet APPC nodes cannot have sessions with any other SNA devices, not even themselves. They cannot communicate directly with any other minicomputer or even a personal computer through LU 6.2 sessions.

Digital Equipment Corporation's lack of support for PU 2.1 severely limits the use of their VAX minicomputer in APPC applications. DEC could enhance its APPC product by adding PU 2.1 support. With that support, it could have LU 6.2 sessions with almost any other computer. As you can see, PU 2.1 is important. It is the type of support you do not appreciate until it is not there when you need it.

The support of lower SNA layers is determined primarily by the node type, by whether it has an intermediate routing capacity, and by the types of addressing and formatting support it provides.

The LU type specifies the higher level protocols. If the seven layers of the SNA are split into two groups, a lower level and a higher level, the lower transport level is determined by the PUs and the higher level is the LU.

In summary, an SNA network is a collection of logical units, or LUs, linked by sessions. An LU is a "port" through which a user gains access to the services of a network. It can be a terminal, a personal computer emulating a terminal, a printer, or an application program. It works at the application level and performs two tasks: it activates a session and it uses the session to communicate. A session is a logical connection between LUs. It is independent of the physical nature of the connection. Think of it as a conversation between applications. The protocol that is used to perform the conversational duties is called an *LU session type*. There are a number of LU session types, as shown below.

- LU 0 is a product-specific unit type that is used for devices such as specialized store controllers. Each product-specific LU 0 is unique.
- LU 1 is used for printers and remote job entry stations.
- LU 2 is used for the 3270 line of display stations.
- LU 3 is for printers that use the 3270 data stream.
- LU 4 was a precursor to LU 6.2, designed for low entry cost for terminal-to-terminal communications and terminal-to-host communications.
- LU 6.0 and LU 6.1 are called intersystems communications and have been implemented in IBM's most popular teleprocessing monitor (CICS) and the IBM Information Management System (IMS) database.
- LU 6.2 represents a merger of the requirements of LU 0, LU 4, LU 6.0, and LU 6.1. APPC is LU session type 6, release 2, or LU 6.2. The purpose of LU 6.2 is to eliminate product-specific protocols in order to provide a common set of protocols that can be used for program-to-program (peer-to-peer) communications. At the low end, LU 6.2 is for terminal-to-terminal type functions. At the high end, it is for IMS-to-CICS type communications. From the low end to the high end, it is for a personal computer talking to a mainframe.
- LU 7.0 is for 5250-type workstations.

APPC/LU 6.2 allows any two computers to communicate over any type of link, such as a local area network or telephone circuits. LU 6.2 eliminates the master-slave (primary-secondary) relationships in micro-to-mainframe communications, thus allowing microcomputers to speak as equals to the host mainframe computer. LU 6.2 may be an opportunity to break the bonds of micro-to-mainframe emulation that are confining the microcomputer to virtually dumb terminal status in its communications with host mainframes.

Readers who would like to obtain a more detailed explanation of Systems Network Architecture should order IBM Manual GC30-3072, titled "Systems Network Architecture: Concepts and Products."

SYSTEMS APPLICATION ARCHITECTURE (SAA)

IBM has a broad architectural outline called *Systems Application Architecture* (SAA). It enables programmers to develop applications that can run on IBM personal computers, System/3X, and System/370 mainframes. SAA is IBM's attempt to address the shortcoming of software incompatibility across its major hardware lines. As such, it represents the company's most important strategic direction since Systems Network Architecture (SNA) made its debut over a decade ago. SAA consists of four related components, one of which, common communication support, should be a major improvement in the interaction between microcomputers, minicomputers, and host mainframes. The four components are:

- Common User Access. IBM has defined the basic elements of the user interface and how programs use them. The PC and PS/2 window-based interfaces will use a mouse and be very icon-oriented.
- Common Programming Interface. IBM has selected several programming languages—initially Cobol, Fortran, and C—and services for use in writing consistent applications. Several programming services available on some IBM systems will be made available for use on other systems.
- Common Communications Support. IBM designated existing communication methods in the architecture, including the 3270 data stream, Document Content and Document Interchange Architectures, SNA Distribution Services, LU 6.2, SDLC, and the Token Ring LAN. The Enhanced Connectivity Facility and the PC Network LAN are not on the list.
- Common Applications. IBM plans to use the first three elements to develop office applications that will be common across hardware lines. IBM plans to write a single product that will work on diverse machines to perform document creation, document management, electronic mail, and decision support functions.

SAA will gather all of IBM's incompatible operating systems, communication links, applications, and processors into one container. As an analogy, look at SAA as though it were an automobile dealership that has vehicles of all makes, sizes, engines, colors, and so on. SAA will provide a common methodology for garaging these vehicles (storing data), as well as instructions for driving each vehicle from Point A to Point B (data transmission).

SAA's primary goal is to make IBM's incompatible processors, operating systems, and applications compatible for use. Its purpose is to give users the option of sitting down at any PC, Personal System/2, or terminal and apply the same approach to access any data in the system. Figure 5-11 summarizes the features of Systems Application Architecture.

OTHER NETWORK ARCHITECTURES

System network architectures attempt to facilitate operation, maintenance, and growth of the communication and processing environment by isolating the user and application programs from the details of the network. An ideal general pur-

Common User Access
Consistent application access across
systems
- Icons
- Color
- High resolution graphics
- Mouse support

Common Programming Interface
Languages and services to be used
for multisystem programs
- Cobol, Fortran, and C programs
- Application generator like Cross
  System Product
- Procedure language like REXX
- Database interface like SQL
- Query interface like QMF
- Presentation interface like GDDM
- Dialogue interface like EZ-VU

Common Communications Support
Methods of communicating between
systems
- 3270 Data Stream
- Document Content Architecture
- Document Interchange Architec-
  ture
- Intelligent Printer Data Stream
- SNA Distribution Services
- LU 6.2
- Synchronous Data Link Control
- IBM Token-Ring Network

Common Applications
Single product for all systems incor-
porating
- Document creation
- Document library
- Personal services mail
- Decision support

FIGURE **5-11**   Features of Systems Application Architecture (SAA).

pose data communication network architecture should have the following characteristics.

- It should be adaptable for multivendor network use.
- It should not make any restrictions on the network configurations (topol-
  ogy), and it should allow for easy changes in configuration.
- The end user should not be required to know how network functions are
  implemented.
- It should be independent with regard to both program and device location.
- It should allow for interprocess or task-to-task communications.
- It should detect all errors and recover from them when possible.
- It should be able to recover from hardware or software failures.
- Users should not have to perform any complicated control procedures.
- It should be developed in a systematic layered manner; strict interfaces and
  protocols should define the interaction of any given layer with its adjacent
  corresponding layer (seven-layer model).

The selected architectures that will be discussed below are manufacturing automation protocol (MAP), digital network architecture (DNA), distributed systems environment (DSE), distributed communications architecture (DCA), Burroughs network architecture (BNA), distributed network architecture (DNA), distributed systems (DS), transmission control protocol/internet protocol (TCP/IP), open network architecture (ONA), Xerox network standard (XNS), and UNIX.

Manufacturing Automation Protocol (MAP) The manufacturing automation protocol satisfies the special needs of factories, such as the networking of dissimilar devices, very fast access times, immunity from electrical or other interference on the communication circuits, reliability, ruggedness, and ease of maintenance and reconfiguration. This protocol is an overall organizational network that also can be used in an office setting for multiple tasks, such as word processing and other business functions. Its special use is in factory settings where, for example, it can enable scheduling from one of the terminal nodes, while simultaneously measuring the paint thickness on a piece of metal with automatic sensing devices.

MAP is envisioned to be an ideal network that provides high level communications among computers of various sizes and brands, along with a variety of programmable factory floor devices, such as robots, programmable controllers, vision systems, and microcomputers. MAP describes the network system that links everything together over a broadband backbone cable. Sometimes MAP is referred to as MAP/CIM. CIM, or *computer integrated manufacturing*, means that computers are used to integrate the manufacturing, design, and business functions of an organization. Since it really describes concepts, it is a strategy or a road map rather than a system. Another name for the manufacturing automation protocol is MAP/TOP (for *technical and office protocol*).

The final version of "agreed upon" specifications probably will not be available until 1989. To date, the MAP model has six layers rather than the standard seven as in the OSI model. The layer that is missing is the presentation layer (layer 6). The basic backbone network endorsed by MAP is a token passing broadband bus local area network that transmits at 1, 5, or 10 million bits per second.

In summary, MAP is a type of OSI standard network that uses local area networking, and it optimizes factory applications. It endorses the IEEE 802.4 token passing broadband bus network type using coaxial cable, broadband modulation, and token passing access. MAP intends to provide a multivendor device connection by using amplifiers, cable taps, splitters, terminators, and cable that is off the shelf rather than custom made.

Digital Network Architecture (DNA) This is Digital Equipment Corporation's distributed network architecture. It is called DECnet and has five layers. The physical layer, data link control layer, transport layer, and network services layer correspond almost exactly to the lowest four layers of the OSI model. The fifth layer, the application layer, is a mixture of the OSI model (see Figure 5-2) presentation and application layers. DECnet does not contain a separate session layer.

DECnet, like IBM's SNA, defines a general framework for both data communication networking and distributed data processing. The object of DECnet is to permit generalized interconnection of different host computers and point-to-point, multipoint, or switched networks in such a way that users can share programs, data files, and remote terminal devices.

DECnet supports the X.25 international protocol standard and has packet switching capabilities. An emulator is offered that permits Digital Equipment Corporation systems to be interconnected with IBM mainframes running in an SNA environment. The digital data communication message protocol (DDCMP) is DECnet's byte-oriented protocol, which is similar in structure to IBM's Binary Synchronous Communications (BSC) protocol.

Distributed Systems Environment (DSE)   The DSE is Honeywell's concept in communication processing/architecture. It is not a rigid framework for network implementation. Any two remotely connected Honeywell processors can be considered a DSE system. This network includes a host, a front end processor, one or more remote host processors, and various terminals. High-level Data Link Control (HDLC) is the standard protocol for interprocessor links. HDLC is a bit-oriented, synchronous protocol and is implemented in accordance with the ISO standard. Because this is a distributed processing architecture, some of the application processing is performed at the user site, thus relieving the host computer of some of its processing burden. In addition, this architecture supports distribution of databases and database management functions.

Distributed Communications Architecture (DCA)   DCA is Sperry Univac's network architecture, and it embraces a concept of distributed networking similar to Honeywell's DSE philosophy. DCA attempts to relocate communication control logic away from the host computer. One of its distinguishing features is its ability to support an environment consisting of non-Univac terminals, processors, and networks. DCA is compatible with IBM terminal handlers and has protocol compatibility with X.25 packet switching services. Univac's standard protocol is the Universal Data Link Control (UDLC), which is a bit-oriented, synchronous link protocol based on the ISO standard. Now that Sperry Univac has merged with Burroughs to become UNISYS, the two companies' protocols may be merged into one.

Burroughs Network Architecture (BNA)   BNA consists of two major layers. The host services layer provides logical access to the communication network.

The network services layer provides physical access to the communication system.

The *host services layer* provides the user with a transparent interface to the network system by making use of an enhanced version of the master control program (MCP). The master control program is the teleprocessing monitor and control interface to the host computer operating system. The *network services layer* has three distinct levels: port level, router level, and station level. The *port level* contains several system ports with a port level manager to allocate and deallocate the ports as needed. The ports provide individual access to the network services layer and the underlying network through the host services layer. This level is responsible for breaking a message into a series of packets, coordinating message sequencing, and chaining a series of messages. The *router level* is responsible for routing as well as storage and forwarding of various packets. The router has a database of active hosts used to determine required routing information. The *station level* manages the physical interface between adjacent hosts. The station module checks packets for errors and requests retransmission of a packet if there is an error. Connection to an X.25 public packet network is supported. The new UNISYS company (Burroughs and Sperry Univac) may merge the protocols of these two organizations.

Distributed Network Architecture (DNA)   This is NCR Corporation's network architecture. It operates through two software modules: the telecommunication access method and the data transporting network.

The *telecommunication access method* software provides all queuing and addressing of information functions.

The *data transporting network* has three separate logical layers: (1) communication system services, which manage all messages that pass between systems by breaking them into packets and passing them in a controlled manner to the route manager; (2) the route manager, which dynamically routes each packet through the system; and (3) the data link control layer, which then builds transmission packets according to an NCR protocol (Synchronous Data Link Control protocol) or X.25 protocols.

This network architecture supports point-to-point or multipoint connections.

It should be noted that the NCR Corporation has acquired Comten, an advanced company that produces IBM-compatible front end communication processors. The DNA philosophy has been altered since then and now is referred to generally as the Communication Network Architecture (CNA). Many further enhancements can be expected from NCR because Comten is a very advanced fourth generation front end manufacturer.

Distributed Systems (DS)   DS is Hewlett-Packard's distributed network system. DS allows the flexibility of adding an optional front end communication processor to the communication nodes as a means of removing much of the communication processing load from the host computer. The hardware interfaces are all microcoded (firmware). This microcoded software builds the transmission protocol units and forms the lowest level (the communication access method) in the DS network architecture. This layer is accessed by the communication management layer, which provides all local system management routing and queuing functions.

There are two higher level layers which effectively create a master-to-slave relationship between any message generating and receiving node. In a *master-to-slave* relationship, when a slave terminal wants to talk, it must first ask permission from a master terminal which initiates the session. In a *master-to-master* relationship, any terminal can initiate a session. As IBM moves away from the master-to-slave relationship in its SNA, Hewlett-Packard also will because the X.25 protocol uses a master-to-master rather than a master-to-slave relationship.

Transmission Control Protocol/Internet Protocol (TCP/IP) The transmission control protocol/internet protocol probably is the oldest working standard. It was developed for the U.S. Department of Defense's Advanced Research Project Agency network (ARPANET). It allows reasonably efficient and error-free file transmission between different systems, and it has been in place for about ten years. Because it is a file transfer protocol, it can send large files of information across sometimes unreliable networks with great assurance that the data will arrive in an uncorrupted form.

Open Network Architecture (ONA) Open network architecture is under development by a new organization calling itself the Open Network Architecture Forum. This organization is run by Bellcore, a research group owned jointly by the seven Bell Operating Companies. ONA is not fully defined as yet, but it has two key parts. First, the *protocol model* describes a detailed methodology by which information and instructions are provided to, received from, or communicated within the network. Second, the *architecture model* is a conceptualization of the basic network functions required to make it more "open" or "closed." By openness ONA's developers mean the ability of any hardware or software to connect to and transmit over the network. The ONA model is described as a layered protocol for use between different information systems at their interfaces. The ONA model will be different conceptually from the internationally recognized seven-layer OSI model.

XEROX Network Systems (XNS) Xerox network systems allows transparent connections between machines and the sharing of applications on a network. Certain parts of the standard, such as the Ethernet specification, are extremely well developed, well known, and widely used. Except for its Ethernet portion, however, XNS is not generally understood outside the Xerox Corporation.

UNIX   This general purpose, interactive, timesharing *operating system* was developed by Bell Laboratories. The UNIX operating system has become more popular as the power of minicomputers and microcomputers has increased and their cost decreased.

The UNIX operating system is a candidate to control local network applications of the future. With regard to office automation, local networks allow different functions in the automated office to communicate with one another and to perform in an integrated manner. The UNIX operating system is well suited to support office automation because it can be transported easily from host computer to minicomputer and on to microcomputers. In addition, UNIX can control many kinds of hardware, including terminals, copying machines, facsimile devices, and typesetting machines.

Office automation has four major aspects: data processing, database administration, automation of office administrative functions, and data communications. In an integrated office system these four areas must be combined to allow efficient performance.

With UNIX, the end user can control multiple operations on different hardware/terminals. It allows communication between many different brands of computer hardware ranging in size from microcomputers to host mainframes.

The UNIX system contains the networking and telecommunication software that is necessary to tie many brands and types of computers together. It allows the prospective end user to deal with a common language, regardless of the size of the computer.

Also of interest to the data communication user is the ability to use the output of an executing program as the input to another program. This information is transferred through software channels called PIPES. A special kind of PIPE, called a TEE, can be used to allow output data to be transferred simultaneously to two other programs. A FILTER, which is an extension of PIPEs and TEEs, is a feature that can extract selective information from one file, modify it, and send the result to another file or into a PIPE. A typical use of a FILTER is to help a data communication user monitor data from several nodes in a local network.

UNIX users need to become familiar with only six basic groups of commands: User Access Control, File Manipulation, Manipulation of Directories and File Names, Running Programs, Status Inquiries, and Communication. We will discuss only Communication here.

All of the commands in the Communication group are of importance to the prospective data communication-oriented UNIX user. The three most commonly used commands are: MAIL, WRITE, and MESG. The MAIL command is used to mail a message to one or more users, and it also can be used to read and dispose of incoming mail. At log-in, the user is advised that there is mail. The WRITE command allows the user to establish direct terminal communication with another user. The MESG command inhibits receipt of messages from the WRITE command when invoked by the user.

The most powerful commands in the Communication group are the CU and UUCP commands. The CU command is used to communicate with another timesharing system even though the remote computer system may not be running under UNIX; the user interface is transparent. This command allows the user to transmit ASCII files to the remote machine at standard bit rates (the default is 300 bits per second) or to engage in interactive conversation. In addition, the CU command allows the user to take remote input from a file on the local system or to place remote output into a file on the local system. The UUCP command is used for file transfer between two UNIX systems. It provides automatic queuing until a circuit is available and the remote machine is ready to receive data. The syntax of the command is quite complex, and the available options may vary depending on which UNIX version is being used and how much it has been customized.

TELECOMMUNICATION STANDARDS

The world of standards rarely represents current realities. To set a standard, industry leaders must agree on the specific way something will be performed. A telecommunication standard may take one of many forms. It may exist as the definition of an architecture, the specification of a physical design, a set of protocols for interaction between two devices, or a set of conventions for behavior. Regardless of the form, the goal of the standard usually is to define interfaces between different stations or nodes. The driving force for defining standards is economics.

Standards are necessary in almost every business and public service entity. For example, before 1904 fire hose couplings in the United States were not standard, which meant that a fire department in one community could not help in another community. In addition, the transmission of electric current was not standardized until the end of the nineteenth century. As a result, customers had to choose between Thomas Edison's direct current (dc) electricity or George Westinghouse's alternating current (ac) electricity. Before 1927, traffic signs and signals in the United States varied according to the plan adopted by each state. Drivers from one state often could not recognize signs when driving in other states. As an example, people who were color-blind might drive through red lights because red was the top of the three signal lights in their state, whereas green was at the top of the three signal lights in the state they were visiting.

Standardization is achieved in three stages: specification, identification of choices, and acceptance. The *specification* stage consists of developing a nomenclature and identifying the problems to be addressed. *Identification of choices* involves choosing from among the various solutions and identifying the optimum solution from among the alternatives. *Acceptance*, which is the most difficult stage of the three, consists of defining the solution and getting the recognized leaders of the industry to agree upon a single, uniform solution.

To understand why data communication functions are arranged in architectural layers (remember the seven-layer OSI reference model presented earlier in this chapter), let us review the functions that communication networks support to provide effective communications among end users (nodes). These functions generally include

- Establishing a transmission path
- Coding information into analog or digital form for transmission
- Ensuring that successive messages (groups of frames or packets) arrive successfully at the receiving node
- Establishing how a retransmission must be performed if a frame or packet is in error
- Sending messages to the correct network node
- Incorporating optimization devices into the network, such as multiplexers, concentrators, port sharing devices, front ends, and switching devices
- Bypassing failed circuits or nodes
- Providing screen and printer formats, code conversion, programming languages, and protocol conversions for end users

Telecommunication standards, therefore, are protocols (rules) that have been agreed upon by pertinent industry and government standards-making bodies. As we discuss the most relevant standards, it is helpful to name and describe the most important organizations that define standards. The following organizations specify, identify, and agree upon telecommunication standards.

International Organization for Standardization (ISO) As noted earlier, the International Organization for Standardization, which is based in Geneva, Switzerland, developed the OSI seven-layer reference model. The American National Standards Institute (ANSI) is the United States' voting participant in ISO. The seven-layer reference model was discussed at length earlier in this chapter.

American National Standards Institute (ANSI) The American National Standards Institute is the coordinating organization for the United States' national system of standards and is comprised of about 900 companies. ANSI is a standardization organization, not a standards-making body. Its role is to coordinate the development of voluntary national standards and interact with ISO to develop national standards that comply with ISO's international recommendations. For example, the X.3 (PAD) Packet Assembler/Dissembler Committee recommends national standards for computers and office equipment suppliers.

ANSI also is developing a Fiber Distributed Data Interface (FDDI) for large backbone LANs. An FDDI network essentially is a counter-rotating token ring with a throughput of 100 million bits per second. This speed is at least ten times faster than the three IEEE 802 LAN standards; it is the fiber's ability to accommodate the speed that mandated the FDDI standard. Even though IEEE 802.3 (Ethernet), IEEE 802.4 (token bus and MAP), and IEEE 802.5 (token ring) all can be implemented with fiber, they do not take advantage of fiber's capacity. Ethernet and MAP specify a throughput of up to 10 million bits per second, and token ring specifies 4 million bits per second.

The FDDI standard contains two specifications: one for the physical layer and one for medium access control. The physical layer details the characteristics of the optical fiber transmission medium. These characteristics are determined primarily by the choice of signaling rate and fiber dimension. The specified data rate is 100 million bits per second using an encoding scheme called 4B/5B. The dimensions of optical fiber cable are specified in terms of the diameter of the core of the fiber and the outer diameter of the cladding layer.

Consultative Committee on International Telephone and Telegraph (CCITT)
The Consultative Committee on International Telephone and Telegraph is the technical standards-setting organization of the United Nations International Telecommunications Union. It too is based in Geneva, Switzerland, and is comprised of representatives from over 150 Postal Telephone and Telegraphs (PTTs), private telecommunication agencies, as well as industrial and scientific organizations. ISO is a member of CCITT. The PTTs are telephone companies outside of the United States. For the United States, we can refer to them as the Bell Operating Companies (BOCs) or American Telephone & Telegraph (AT&T) or common carriers. CCITT establishes recommendations for use by PTTs, other common carriers, and hardware and software vendors.

Institute of Electrical and Electronics Engineers (IEEE)
The Institute of Electrical and Electronics Engineers is a professional society in the United States whose standards committees focus on local area network standards. The IEEE 802.3 standard often is referred to as the Ethernet standard. It was the first and most entrenched standard of the three local area network standards. The 802.3 uses a bus topology, carrier sense multiple access/collision detection (CSMA/CD) access methodology, and it has been implemented using baseband media or fiber optics for broadband circuit links. The IEEE 802.5 token ring standard uses a token passing access mechanism with baseband signaling over twisted pair wire media. The IEEE 802.4 is a token bus scheme used primarily on broadband cable.

This is the standard transmission methodology being pushed for the manufacturing automation protocol (MAP) network mentioned earlier.

Another standard that the IEEE is working on is the 1003.1 trial use standard for portable operating systems in computer environments. It is based on the UNIX operating system and may be the first standardization of an operating system environment. Other countries have their own national standards-making bodies, too. For example, the British counterpart of IEEE is the Institution of Electrical Engineers (IEE).

Electronic Industries Association (EIA) The Electronic Industries Association is an ANSI-accredited standards organization that develops a variety of standards, including equipment standards. Possibly its most prominent standard is the RS232C connector cable/plug. It also endorses other standards, such as the RS449 connector plug. Membership is drawn from manufacturers of telecommunication equipment and other electronics components.

National Bureau of Standards (NBS) The National Bureau of Standards in Washington, D.C., an agency within the U.S. Department of Commerce, develops federal information processing standards for the federal government. Among its many test facilities is the Network Protocol Testing and Evaluation Facility which contains eight laboratories for research in the design, implementation, and testing of computer network protocols. This facility is used to develop prototype implementations of protocols and then test them in a variety of communication environments.

Exchange Carrier Standards Association The Exchange Carrier Standards Association's T-1 Committee (accredited by ANSI) is comprised of representatives from domestic telecommunication manufacturers, carriers, users, and other interested parties. The T-1 Technical Subcommittee develops North American telecommunication standards. This committee, ANSI, and CCITT are developing standards for ISDN circuits. The I430 basic user network interface (also known as 2B+D, which was discussed in Chapter 4) details the physical and electrical specifications for an interface that supports two 64,000 bits per second digital channels and one 16,000 bits per second signaling channel. The I440/I441 details the message formats of the Q.921 and Q.920 message.

Corporation for Open Systems (COS) The Corporation for Open Systems is a nonprofit corporation established in 1986. The organization has members from computer and communication equipment vendors and users. It was formed under the auspices of the Computer and Communication Industry Association. While it is not a standards-setting body, it was established to accelerate the introduction of products based on international standards, principally those based on the seven-layer OSI model.

The COS charter states that its purpose is to monitor OSI and Integrated Services Digital Network (ISDN) standards. It will develop standards for implementing layer 7 (the application layer) of the OSI model and the X.400 electronic message-handling features (OSI layer 6). X.400 is an electronic mail standard.

The X.400 recommendation from the CCITT often is referred to as the *electronic envelope*. It specifies the format in which messages can be exchanged between various systems. The X.400 standard describes an electronic mail system using message transfer agents (MTAs) and user agents (UAs). The UAs are nothing more than mailboxes, including the software that helps with composing and addressing a message. The MTA gets the message from a mailbox to its various destinations.

The part of X.400 that has attracted the most attention is the P1 protocol, which specifies the rules for communication between one MTA and another MTA. The aim of the P1 protocol is to allow X.400 electronic mail systems to interconnect easily so that the users of each system have mail access to the users of all other systems. Among other things, it defines basic message delivery, date and time stamping, delivery notification, and type of information in the message. Most, if not all, of the hardware vendors are endorsing the X.400 standard. X.400 also is being endorsed in both the United States and Europe by the various telephone companies in each country. In the world of digital transmission, ISDN can be regarded as a uniform voice and data transport scheme that incorporates a worldwide X.400 standard for transmission of electronic mail.

COS is a very powerful group because its members are important leaders in the world of communications. The COS goal is to stimulate the development of interoperable communication products. In addition to promoting standards, COS operates a test center which focuses on such tasks as testing compatible ISO hardware/software and the ISDN network.

COS members contribute funds to the COS Strategy Forum, the primary decision-making body at the technical level. The Forum's numerous subcommittees deal with specific technical issues. These subcommittees are responsible for submitting recommendations to the Strategy Forum for its approval. Whereas the standards developed by the international standards-making bodies generally concentrate on a particular layer of the OSI model, COS users assemble a number of standards to produce a profile that supports a particular application or hardware. For example, COS is deeply involved in interconnecting and implementing various standards to achieve a successful MAP network.

Legally Enforceable Standards Legally enforceable standards are defined and enacted into the law by the governments of various countries. In the United States, we have a law that provides criminal punishment for three types of computer criminals: those who gain access to federal computers, those who gain unauthorized access to computers at financial institutions that are covered by federal laws, and those who gain access to computers that hold national security data.

The Federal Wiretap Statute, which was enacted in 1968, protected only voice communications from interception. In 1986 an electronic privacy law was passed relating directly to data communications. The Electronic Communications Privacy Act of 1986 makes it a federal crime to intercept electronic communications, such as data communications or electronic mail, or to tamper with the computers in a data network. The 1986 law prohibits the interception of data and video communications on private networks and the unauthorized access of network computers if stored messages are obtained or altered. Under this law, individuals are subject to penalties of up to $100,000 and ten years in prison if the crime is committed for commercial gain or malicious reasons. Fines for organizations can be up to $250,000. The privacy of network users also is protected under this law. Law enforcement officials need a court order to obtain electronic messages, and electronic mail services cannot disclose the content of messages transmitted over their service without the sender's authorization.

Congress also passed the Computer Fraud and Abuse Act in 1986. This law expands federal jurisdiction of computer crimes. It refines and builds on the 1985 statute that covers only federal computers; the new law covers computer crimes involving private sector computers located in two or more states. It also hits the so-called pirate bulletin board systems that exchange computer passwords. Specifically, the law makes it a federal offense to access a computer in a fraudulent scheme to steal, and it makes it a felony to alter or destroy data, hardware, or software without authorization. The law also makes it a federal misdemeanor for an individual to traffic in computer passwords belonging to others, if there is a clear intent to defraud. The Computer Fraud and Abuse law imposes fines of up to $100,000 and ten years in prison for people who intentionally gain unauthorized access to computer systems to damage records or to steal records or cash.

CCITT X.nn and V.nn Standards The CCITT's X.nn and V.nn standards are recommendations for how particular electrical connections should be carried out. As you read the following recommendations, you should remember that terminals or microcomputers are referred to as data terminal equipment (DTE) and modems as data circuit terminating equipment (DCE).

The X.nn series concerns the connection of digital equipment to a public data network that employs digital signaling.

- X.1 User classes of service for public data networks.
- X.2 User facilities in a public data network.

- X.3   The packet assembly/disassembly (PAD) facility for asynchronous transmission in a public data network.

- X.4   General structure of the signals of the international alphabet number 5 code.

- X.20   Interface between DTE and DCE for asynchronous operation on public data networks. This interface cable connects modems to cables.

- X.20bis   Used on public data networks for DTE that is designated for interfacing to asynchronous V series modems. This interface cable connects modems to cables.

- X.21   Interface between DTE and DCE for synchronous operation on public data networks. This interface cable connects modems to terminals.

- X.21bis   Used on public data networks for DTE which is designated for interfacing to V series modems. This interface cable connects modems to terminals.

- X.24   List of definitions for interchange circuits between DTE and DCE on public data networks.

- X.25   Interface between DTE and DCE for terminals operating in the "packet mode" on public data networks.

- X.26   Electrical characteristics for unbalanced polar or bipolar circuits for general use with data communications (identical to V.10—see V.*nn* series below). In Europe, polar or bipolar circuits may be called double current interchange circuits.

- X.27   Electrical characteristics for balanced polar or bipolar circuits for general use with data communications (identical to V.11). In Europe, polar or bipolar circuits may be called double current interchange circuits.

- X.28   DTE/DCE interface for start-stop data terminal equipment interconnections with the packet assembly/disassembly (PAD) facility in public data networks. It defines the interface between a nonintelligent terminal and PAD.

- X.29   Procedures for the exchange of control information and user data between a packet mode DTE and a PAD facility. It defines the procedure for governing the exchange of data between a PAD and a packet mode terminal.

- X.32   Defines a dial-up connection between the DTE and the DCE using an X.25 link (packets).

- X.75   Terminal and transient control procedures for data transfer systems on international circuits between packet switched networks using the X.25 protocol. It governs the interface between public packet switched networks (X.25). It is the *gateway* that is at each of the two networks.

- X.121   International numbering plan for public data networks.

The V.*nn* series concerns the connection of digital equipment to a public telephone system that employs analog signaling.

- V.3   International alphabet number 5 (of which ASCII is one case).

- V.10   Same as X.26. In this case, when they talk about electrical characteristics, they are referring to communication between modems, which also may be called data circuit terminating equipment (DCE).

- V.11   Same as X.27. In this case, when they talk about electrical characteristics, they are referring to communication between modems, which also may be called DCE.

- V.21   Modem electrical characteristics for transmissions up to 300 bits per second. These modem electrical characteristic standards must be defined because the standard tones used in the United States seem to "discombobulate" the European telephone billing systems. As a result, many European countries have outlawed the use of standard American modems.

- V.22   Electrical characteristics for modems that operate up to 1200 bits per second in an asynchronous format on two-wire circuits.

- V.22bis   Electrical characteristics for modems that operate up to 2400 bits per second using asynchronous or synchronous transmission on two-wire circuits.

- V.24   List of definitions for interexchange circuits between DTE and DCE.

- V.25, V.25bis, V.26, V.27, V.28, and V.29.   As you move up the V-numbered series for electrical characteristics between modems, the speed moves up through 2400, 4800, and on to 9600 bits per second. In addition, various standards are set regarding half or full duplex, dial-up or lease lines, automatic dial capabilities, automatic call receipt capabilities, and so on. Of course, one of the primary purposes is to ensure that the modem does not interfere with the public dial-up telephone system of the country in which it is operating or to which it is transmitting data.

- V.32   Electrical characteristics for a modem that operates at full or half duplex over two-wire direct dial networks and transmits at a speed of 9600 bits per second. It is capable of forward error correction. The modem signals at 2400 baud and encodes 5 bits per baud; 4 bits are for data, and 1 bit is for forward error correction. Therefore, even though the total transmission rate is 12,000 bits per second, only 9600 bits per second actually contain data. Full duplex transmission over two-wire circuits is not a new development because the V.22 and V.22bis modems had this capability. The secret behind the V.32 modem's ability to transmit at 9600 bits per second full duplex over two-wire circuits is a technique called echo cancellation. *Echo cancellation* isolates the received signal from the remote modem by eliminating interference caused by the transmitting signal, as well as reflections of the signal echoed back from the near-end and remote-end central offices. Echoes can be canceled because there is a small time difference between when they are received and when the main signal is received. This time difference is caused by propagation delays. Delay times are established when the modems independently clock (or train) between the local office and the remote location.

- V.33 Electrical characteristics using a leased line at 14,400 bits per second.
- V.35 Electrical characteristics for a modem using a high speed transmission rate of 48,000 or 56,000 bits per second.
- V.36 Electrical characteristics for modems using synchronous digital data transmission at 56,000 bits per second.
- V.54 Specifications for loop test devices for modems.

SELECTED REFERENCES

1. Christian, Kaare. *Unix Operating System.* New York: John Wiley & Sons, 1983.

2. Conrad, James W. *Standards & Protocols for Communications Networks.* Madison, N.J.: Carnegie Press, 1982.

3. Deasington, R. J. *X.25 Explained: Protocols for Packet Switching Networks.* New York: Halstead Press, 1985.

4. Folts, Harold C., ed. *McGraw-Hill's Compilation of Data Communications Standards,* 3rd ed. New York: McGraw-Hill Book Co., 1986.

5. Lim, Pacifico Amarga. *CICS/VS Command Level with ANS COBOL Examples,* 2nd ed. New York: Van Nostrand Reinhold Co., 1985.

6. Schwartz, Mischa. *Telecommunications Networks: Protocols, Modeling, and Analysis.* Reading, Mass.: Addison-Wesley Publishing Co., 1987.

7. *Systems Network Architecture: Concepts and Products.* White Plains, N.Y.: IBM Corporation, January 1981 (IBM Publication GC30-3072-0).

8. *Systems Network Architecture: Technical Overview.* White Plains, N.Y.: IBM Corporation, March 1982 (IBM Publication GC30-3073-0).

9. Ungaro, Colin B., ed. *Network Software.* Brooklyn, N.Y.: Data Communications, 1987.

10. Wood, Patrick H., and Steven G. Kochan. *Unix System Security.* Hasbrouck Heights, N.J.: Hayden Book Co., 1985.

QUESTIONS/PROBLEMS

1. Using the following as a model, invent a line protocol for transmission between a central computer site and a remote terminal site.

Steps	Central Computer Site		Remote Terminal Site
1.	Poll Terminal A	⟶	
2.		⟵	Negative response code
3.	Poll Terminal B	⟶	
4.		⟵	Positive response code (changes terminal mode from idle to text mode)
		⟵	Text (first block of text)
		⟵	End of block code
		⟵	Parity check
5.	Parity received correctly	⟶	
6.		⟵	Text (second block of text)
		⟵	End of block code
		⟵	Parity check
7.	Parity received correctly	⟶	
8.		⟵	End of message
9.	Poll Terminal C	⟶	
10.		⟵	Negative response code

2. What does the following describe? The central computer sends a message saying, "Terminal A, do you have anything to transmit? If so, send it." If Terminal A has nothing to send, it replies negatively and the central computer goes on to Terminal B.

3. In the section titled Software Design Precepts, "acting constructively on errors" is cited as important for handling incoming messages. Assume a banking situation in which a withdrawal message is sent from a terminal to the central computer. Describe a "constructive action" for each of the following unusual conditions detected at the computer.

   - Account number is garbled.
   - Requested withdrawal exceeds account balance by a small amount, such as $100.
   - Requested withdrawal exceeds account balance by a very large amount, such as $1,000.

- The incoming transaction is unreadable.

4. Discuss the four basic principles of SNA, concentrating on the advantages they provide.

5. In the Software Design Precepts section, it is suggested that both incoming and outgoing messages be logged. Explain how these logs can be used to recover from a computer failure in a store and forward message switching system.

6. Propose a set of procedures for a store and forward message switching system that meets the requirements given in the Software Design Precepts section. Provide a rational means of disposing of messages. Assume there are three priorities of messages: delivery as soon as possible, delivery in 2 hours, and delivery in 12 hours. Assume terminals can be repaired in 6 hours and circuits in 1 hour if they fail. Assume that the messages are in English and that their average length is 20 words.

7. What are the three basic factors that make data communication programs different from batch programs?

8. What is the first step to ensure that the software provides proper integrity of the messages being transmitted?

9. Can you define the difference between a protocol, software, and network architectures?

10. What is the biggest drawback of the Binary Synchronous Communications (BSC) protocol?

11. Which protocols use a continuous automatic repeat request (ARQ)?

12. When Systems Network Architecture (SNA) is used, can a remote front end communication processor have all three addressable units, such as LU, PU, and SSCP?

13. In UNIX, there are specific communication commands. Can you name them?

14. Summarize the basic functions of the X.25 protocol.

15. What is the difference between ISO and OSI?

16. Which standards-making body developed the RS232 standard?

17. What is X.400?

18. Name some "legally enforceable standards."

19. What are X.*nn* and V.*nn* standards?

6

MICROCOMPUTERS AND COMMUNICATIONS

This chapter introduces microcomputers and describes their use in networks. It covers microcomputer hardware, software, protocols, modems, electrical protection, the independent microcomputer in a network, and micro-to-mainframe links.

MICROCOMPUTERS

After the invention of the transistor by American Telephone & Telegraph's Bell Laboratories, the world of data communications and telecommunications began to move forward and dominate business operations at an unprecedented rate. Although the transistor is the heart of a data communication network, it was not until Intel's introduction of the microprocessor circuit chip that business data communication functions changed significantly. Data communications is now changing the very way a business enterprise operates.

Today, a business uses many microcomputers in areas such as the automated office, micro-to-mainframe links, local area networks, standalone microcomputers, disk/file server database systems, local area network server systems, and microcomputer control systems that monitor and control almost anything, but especially data communications. In addition, hardware such as front end communication processors, modems, multiplexers, protocol converters, switches, and automated digital switchboards depend on microprocessor chips and the ability to interconnect with microcomputers. Today's powerful microcomputer has a 32-bit microprocessor chip in it. Popular microprocessor chips of this type are the Intel 80386 and the Motorola 68030. (IBM uses the Intel chip, and Apple Computer uses the Motorola chip.)

Because this textbook focuses on data communications, we will present only a short overview on microcomputers and move directly to the interrelationships between microcomputers and communications. (Local area networks are covered in the next chapter.) For the business microcomputer, there are many different configurations. A configuration oriented toward business use (see Figure 6-1) might be a microcomputer that has approximately 640,000 bytes (characters) of internal memory, disk operating system software—DOS (the software that runs the microcomputer), a parallel input/output port for connection to a printer, a serial input/output port for connection to a modem for data communications, and disk storage. The disk storage may be an internal high speed hard disk able to store 10 to 100 million bytes of information or one or two diskette drives able to store 360,000 to 1.4 million bytes of information on each diskette. It is possible to connect either the parallel input/output port or the serial port to a data communication network, although generally it is the serial port that is connected to the modem which connects in turn to the network required for the business application.

The widespread use of these microcomputer systems is accelerating the need for adequate communications throughout the entire business organization, as well as between the business organization and outside entities, such as public networks, specialized databases, other companies, and government agencies. The individual standalone microcomputer can be made to be at least ten times more effective when it is networked with other microcomputers and, of course, other business entities.

The two most important items with regard to networking microcomputers are the modem and communication software. For example, the first piece of software you encounter is the microcomputer operating system. For this discussion, we will use the IBM Personal Computer (PC/AT) or PC-compatible microcomputer. In today's world the terms *microcomputer* and *PC* are used interchangeably.

Before moving on to the microcomputer communication software, modems, and micro-to-mainframe links, let us discuss the internal operation of a typical microcomputer. The microcomputer depicted in Figure 6-2 has a central processing unit (CPU) with both random access memory (RAM) and read only memory (ROM). *Random access memory* is the read and write memory in the mi-

FIGURE **6-1** Microcomputer.

crocomputer that stores programs and business applications. You can enter commands to write new programs or read data into this memory. This is your microcomputer's working memory. When you turn your microcomputer off, all the contents of the RAM are lost. *Read only memory* is a memory location in your microcomputer where data is stored permanently; it cannot be altered by the microcomputer operator. If you want to change the ROM memory, the circuit chip has to be removed from the microcomputer and erased electronically or with ultraviolet light. The ROM contains the basic startup (boot) instructions. The information contents in a ROM can be copyrighted and even patented, making one microcomputer unique among others.

There are two 5¼-inch diskette drives (A: and B:) and one 30-megabyte hard disk

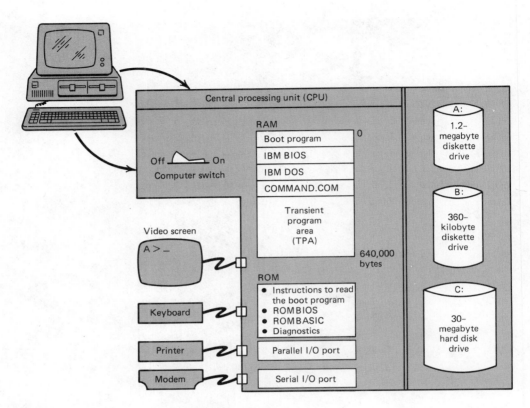

FIGURE **6-2** Microcomputer configuration.

(C:) in our example (Figure 6-2). You also have a video screen, both a serial and parallel input/output (I/O) port, and a connector plug for a keyboard. If you have ever wondered what happens when you turn on your microcomputer, let us go through the disk operating system (DOS) software boot scenario.

Loading the DOS software and starting it is called *booting* the system. When you boot the microcomputer to use it, the following operations take place (follow along by looking at Figure 6-2).

- The power switch is turned on. This gives control to the ROM chip in the central processing unit.
- Fixed instructions in the ROM chip initiate diagnostic equipment checks to ensure that all components of the microcomputer are operating.

- The ROM instructions check the diskette in Drive A to determine whether the gate is open or closed. The *gate* is the lever you turn after inserting the 5¼-inch diskette.

- If the gate is closed, the boot program is read from the Drive A diskette into the RAM and control is given to this program.

- If the gate is open and you have a hard disk, the boot program is read from the hard disk (Drive C) into the RAM and control is given to this program.

- If the gate is open and there is no hard disk, an IBM PC enters the IBM ROM BASIC programming language. An IBM PC-compatible (clone) probably displays a message from the ROM on the video screen telling you to insert the DOS boot disk into Drive A.

- The boot program also reads the ROMBIOS (Basic Input/Output Systems) into the RAM. ROMBIOS handles the input and output commands for your microcomputer.

- The boot program (now in RAM) initiates a "bootstrap load" of the remainder of the DOS programs from either drive A: or C: into the RAM storage. Control then is given to DOS in the RAM.

- DOS looks for its primary program COMMAND.COM and displays the time and date prompt so that you can enter the current time and date, unless your PC has a battery to keep this information current. DOS also looks for two other key files: the CONFIG.SYS that configures your system, and the AUTOEXEC.BAT that executes any special programs you would like to have operational every time you start up your PC. Notice that DOS also starts to use what is called the A> prompt if you booted from Drive A: or the C> prompt if you have a hard disk. The A> or C> prompts tell you that you are talking with DOS, the Disk Operating System software that runs your PC (microcomputer). DOS has three important files.

  COMMAND.COM   A file that contains the program or instructions required to carry out a DOS command. It also is called DOS' command processor because this program interprets and carries out any command you enter into the microcomputer.

  CONFIG.SYS A file that is used to set up your system. It controls various configuration features, particularly some aspects of the way memory is partitioned, loads special files called device drivers, sets up buffers, and performs various other functions.

  AUTOEXEC.BAT   A batch file that allows you to automate routine tasks that need to be done before you begin working. It is used to load automatically programs used for customizing your hardware and software when starting (booting) your system. This enables special programs to be available for use without entering them each time you start (boot) your microcomputer.

You now have control and can enter DOS commands to run your system. DOS controls every part of your computer system and is the link between you and the computer. As you already know, any application-level programs, such as spreadsheet programs, accounting packages, and word processing packages, must be able to interface with DOS because DOS controls everything—the hardware and the software.

Some of the DOS commands are known as internal commands and some as external commands. An *internal DOS command* resides in RAM (memory resident) and is available whenever DOS has been loaded. The COMMAND.COM file contains the internal DOS commands. *External DOS commands* are available, but they are contained on the disk (external to RAM) and must be called in from the disk every time you need to use them. When COMMAND.COM cannot find the command that you entered via the keyboard in RAM, it calls it from the external commands on the disk. Typical internal commands include DIR, COPY, RENAME, ERASE, TYPE, CLS, TIME, and DATE; typical external commands include DISKCOPY, FORMAT, and CHKDSK. These are described in your DOS documentation.

One other point should be made for clarification. There are two boot programs. The first and most basic one has been programmed permanently into the ROM. This boot program can only go out to one of the disk drives (whichever one contains the DOS software) from where it calls the much larger DOS boot program into RAM. It is this second boot program that has a broader capability and can call in all of the rest of the DOS software programs.

Now that your computer is running, you have use of a video screen connected by a cable to the central processing unit, a keyboard for data entry, a serial port connected to a modem (although in a later section you will see that modems also can be installed inside the computer box itself), and a parallel input/output port that can be connected to a printer. Again, review Figures 6-1 and 6-2.

The internal workings of a microcomputer are configured in Figure 6-2 so that you can understand the flow of information through the microcomputer. Microcomputers communicate with networks (the rest of the world) through the serial port, although it is quite possible to communicate with a network through the parallel port. The programs you utilize with the microcomputer are located in the *transient program area* (TPA) shown in Figure 6-2. Notice that our microcomputer has a total memory space of 640,000 bytes. In RAM, the boot program, IBM BIOS, IBM DOS internal commands, and COMMAND.COM occupy some of the 640,000 bytes of memory space. The remaining unused memory space, up to 640,000 bytes, is the transient program area. Your application programs or data reside in the TPA when you are using them. Examples of application programs are word processors, spreadsheets, BASIC programs, and communication software.

Microcomputers are assigned a certification number by the Federal Communications Commission (FCC). Microcomputers sold in the United States must have their electronic circuits certified to ensure that they do not emanate electrical signals that interfere with through-the-air broadcasts, such as radio, television, short wave transmissions, cellular radio, and police/fire transmitters. The FCC has a special computer bulletin board (called Public Access Link—PAL) that provides equipment authorization application status (the certification number) and related technical information on microcomputers. If you want to verify a microcomputer's FCC certification number, call (301) 725-1072. Users can access this public bulletin board for up to five minutes per session. The HELP section of the main menu provides assistance in using the system. In addition to microcomputers, the system also contains the certification status of other devices, such as printers, graphics cards, and modems.

Before proceeding to the specific aspects of communications and microcomputers, let us briefly examine the new versions of IBM microcomputers. These are referred to as Personal Systems (PS) or, more accurately, Personal System/2™ (PS/2). These PS/2 microcomputers replace the current PC, PC/XT, and PC/AT distributed by IBM. There are four models, as summarized below.

	Model 30	Model 50	Model 60	Model 80
Microprocessor	8086	80286	80286	80386
Standard memory	640KB	1MB	1MB	Up to 2MB
Expandable to		7MB	15MB	16MB
Diskette size and capacity	3.5 inch 720KB	3.5 inch 1.44MB	3.5 inch 1.44MB	3.5 inch 1.44MB
Fixed disk	20MB	20MB	44, 70MB	44, 70, 115MB
Additional options			44, 70, 115MB	44, 70, 115MB
Maximum configuration	20MB	20MB	185MB	230MB
Expansion slots	3	3	7	7
Operating system(s)	PC DOS 3.3	PC DOS 3.3 and Operating System/2	PC DOS 3.3 and Operating System/2	PC DOS 3.3 and Operating System/2

All models include integrated display support, 256-color graphics capability, clock/calendar, and ports for serial, parallel, and pointing devices. All systems use a common IBM enhanced keyboard and accept any IBM Personal System/2 monochrome or color display. All models accept the 200MB IBM 3363 Optical Disk Drive option.

Notice that these microcomputers run with either DOS 3.3 or a totally new operating system called Operating System/2 (OS/2).

IBM Disk Operating System (DOS) Version 3.30 is the latest single-tasking DOS for IBM Personal Computers and supports the new IBM Personal System/2 computers. The following are highlights of DOS 3.3.

- DOS 3.30 is compatible with DOS 3.20.
- Three new commands (FASTOPEN, CALL, APPEND) have been added.
- BACKUP/RESTORE, DATE/TIME, ATTRIB, and SYS commands are enhanced to improve performance and usability.
- Other enhancements have been made to increase the number of open files, to speed up disk drive input/output, and to provide a faster and more secure method of writing to a disk file in multiuser environments like networks.
- The number of supported asynchronous ports has been increased to four, 1.44-megabyte diskettes are supported, and fixed disks greater than 32 megabytes are supported in partitioned mode.
- DOS 3.30 supports 11 national languages.

The IBM Operating System/2 supports large memory, multiple applications, graphics and windowing capability, and IBM Disk Operations System Version 3.30 (DOS 3.3) compatibility. IBM Operating System/2 Standard Edition (Version 1.0) will not support graphics and windowing, which is provided in Version 1.1. IBM Operating System/2 supports the following IBM Personal Computer Systems: IBM Personal System/2™ Models 50, 60, and 80; IBM Personal Computer AT® Models 099, 239, 319, and 339; and IBM Personal Computer XT™ Model 286. The following are highlights of the IBM Operating System/2 Standard Edition, Version 1.0.

- 16-megabyte addressable random access memory (RAM) support
- Concurrent processing of multiple applications
- High level programming interface
- Presentation manager
- Enhanced ease-of-use facilities
- Compatibility with IBM PC DOS Version 3.30
- Systems Application Architecture (SAA) compatibility

IBM Operating System/2, Extended Edition (Version 1.1), extends the capabilities of the IBM Operating System/2, Standard Edition, supports the IBM Personal System/2 Models 50, 60, and 80, the IBM Personal Computer AT®, and the IBM Personal Computer XT™ model 286. The Communications Manager provides a wide range of concurrent connectivities and protocols, concurrent emulation of multiple terminal types, file transfer under terminal emulation, multiple programming interfaces, and communications and systems management support. The Database Manager is consistent with the IBM family of relational database products, DB2, SQL/DS, and QMF. IBM Operating System/2, Extended Edition, (Version 1.1) is a participant in the IBM Systems Application Architecture. (SAA was described in Chapter 5.)

The following are highlights of IBM Operating System/2, Extended Edition, Version 1.1.

- The system includes all of the IBM Operating System/2, Standard Edition functions plus:

- Communications Manager:

  Concurrent communications over links selected from a wide range of connectivities (SDLC, DFT, IBM Token-Ring Network, IBM PC Network, X.25, and asynchronous links) using LU 6.2, IBM 3270 Data Stream (LU 2), and asynchronous protocols

  Concurrent emulation of multiple terminal types (IBM 3270, IBMB 3101, or DEC VT100™) and the 5250 workstation feature.

  Communications and Systems Management (C&SM) support

  Programming Interfaces (PIs), including Advanced Program-to-Program Communications (APPC), Server-Requester Programming Interface (SRPI) for Enhanced Connectivity Facilities, a 3270 program interface, Asynchronous Communications Device Interface (ACDI), IBM NET-BIOS, and IEEE 802.2

  SNA local area network gateway support

- Database Manager:

  Relational model of data consistent with IBM Database 2 (DB2) and IBM Structured Query Language/Data System (SQL/DS) on IBM host systems.

  IBM Structured Query Language (SQL) based on implementation of SQL in DB2 and SQL/DS

  Queries and reports that are extensive subsets of those in the IBM Query Management Facility (QMF) on IBM host systems

  Local area network support for multiuser LAN database access with no additional user programming.

Now let us examine the communication software for microcomputers, microcomputer protocols, modems for micros, electric protection for microcomputers and LANs, and micro-to-mainframe connections.

COMMUNICATION SOFTWARE FOR MICROCOMPUTERS

This section discusses why you might use your microcomputer for communications, what communication facilities are built into DOS, and other communication software packages. Microcomputers are used for many data communication tasks. The primary uses are for accessing timesharing networks, transferring programs and/or data, communicating with other microcomputers such as in an office automation environment, using them on a local area network, and communicating with minicomputers and mainframe computers.

Microcomputer users may have databases of information available within their own organizations, but they also may want to use some of the many commercial databases that are available—for example, The Source, Dialog Retrieval Service, Dow-Jones News Retrieval Service, and CompuServe Information Service. By using these commercial databases, you can obtain services such as airline schedules, electronic mail, medical advice, business financial information, educational information, news, weather, games/entertainment, home shopping/banking, and bibliographic information on specialized subjects. To connect your microcomputer to any of the commercial databases or other microcomputers, you need both a modem (modems for microcomputers are covered in a later section) and a communication software package.

The data or information that you transmit or receive moves through the serial I/O port of your microcomputer. You can use DOS to control this information flow, but it is more likely that, along with DOS, you will have a specialized communication software package because it is easier to use.

DOS/NETBIOS   First, let us discuss the capabilities of DOS, and then we will move on to specialized communication software packages. With DOS (Version 3.1 or higher), there is a communication command called MODE that controls data transfer. The transfer of data is done in a serial fashion, and it is controlled by several characteristics or parameters that define how fast and in what form the data is transmitted. The communication parameters of your serial port must be equal to the serial ports of other microcomputers or commercial database services with which you might want to communicate. Before you can use your communication port, you must set these parameters with the DOS MODE command. The communication parameters that must be set include

- **Baud:**   How many bits per second are sent or received.
- **Parity:**   The kind of error checking, such as odd, even, or no parity.
- **Databits:**   The number of bits per character. (DOS assumes seven bits per character.)
- **Stopbits:**   The number of bits at the end of a character (usually one, but there can be two). Notice that you use asynchronous transmission.

When you use DOS to *initialize* (start) a serial communication port, the MODE command is entered in the following form:

MODE<port><baud>,<parity>,<databits>,<stopbits>

The < > symbols indicate the parameters you must enter. Port is the name of the communication port; usually it is described as COM1: or COM2:. The remaining parameters are separated by commas, and they might look like this:

300 for 300 bits per second

E for even parity

7 for seven databits

1 for one stopbit.

The entry of the actual command when you are at the A> prompt appears as follows:

A>MODE COM1: 300,E,7,1

As you can see, DOS allows you to enter what is needed for data communications, although this is from the viewpoint of DOS Version 3.1. Earlier versions did not offer complete communication facilities. DOS Version 3.1 or later also includes three interfaces that are significant for networking applications: a file redirector, a file server, and NETBIOS.

The *file redirector* intercepts file requests and determines whether the file is local (resident on the PC from which the request was issued) or remote (located on a PC somewhere else in the network). If the file is remote, it routes the request to the *file server*, which can transmit, retrieve, and write to files anywhere in the network. Thus, the file server is a routing and retrieval program.

It is important to differentiate the DOS 3.1 standard from the NETBIOS standard. DOS 3.1 is the PC's operating system interface for applications. NETBIOS, on the other hand, is a pure communication interface that is independent of the PC's operating system. Most multiuser applications written for networks use the DOS 3.1 interface and do not use NETBIOS at all. DOS 3.1 is sufficient for applications that merely need to share centralized data. It provides such file service functions as file opens, reads, writes, and file locks.

The *NETBIOS* is a circuit chip that usually resides on the network communication circuit card. This circuit card occupies one of the extra slots in your microcomputer. These slots allow you to add capabilities such as networking to your microcomputer. NETBIOS implements several layers of the International Organization for Standardization's Open System Interconnect Reference Model for Data Communications (see Chapter 5). For example, in IBM's PC Network, NETBIOS resides in the ROM on the network communication circuit card.

The acronym NETBIOS stands for *Net*work *B*asic *I*nput/*O*utput *S*ystem. It works at the session layer of the ISO model, the layer responsible for establishing, managing, and terminating connections for individual applications.

NETBIOS is a peer-to-peer communication protocol. In this case the term *peer to peer* refers to a system designed to support communications among intelligent machines. It allows microcomputer-based workstations (PCs) on a network to establish a connection between themselves and to communicate directly without having to go through a central host computer, file server, or other hardware device. It allows applications to talk directly to the network, instead of talking to DOS, which in turn talks to the network operating system. That is why NETBIOS is a separate communication circuit card. It operates faster because it bypasses the DOS operating system. IBM uses the NETBIOS protocol (developed by Sytek for IBM) for communications on both the PC Network and the Token-Ring Network. (These are local area networks.)

Like the ROMBIOS, NETBIOS provides an interface for applications and operating systems that need to address directly some other piece of system hardware. Applications programmers call this interface *Interrupt 5Ch*. An application that issues an Interrupt 5Ch is telling the operating system that it needs to access the network interface card directly and use NETBIOS to communicate with another microcomputer via the network. Applications needing this type of interface are primarily communication programs like SNA gateways-to-mainframes and electronic mail systems (X.400). NETBIOS compatibility in an application is important only if that application has some need for direct and immediate network communications. All other application programs can use the DOS 3.1 interface to the network. The NETBIOS communication cards have one major limitation: their inability to bridge incompatible networks. NETBIOS must be augmented with additional protocols to allow transmission across two incompatible networks. Remember that in this area you might need a protocol converter.

DOS 3.1 was the first version of DOS to include these multiuser function calls. A *function call* (also called a *primitive*) is a hook that DOS provides for application commands. For example, when a word processing application wants to open a file, it says "Open file" to DOS and DOS does it by using a function call. The function call is the part of DOS that understands the command from the application. DOS 3.1 has expanded the range of commands it understands to include multiuser commands. In other words, it has multiuser function calls for communications and networks. These function calls, particularly the extended open and the lock/unlock calls, are defined specifically for networks.

To a programmer, the most important part of this interface is known as *Interrupt 21h*. An *interrupt* is a call that an application program issues to the operating system for a specific set of functions associated with that interrupt. When an application issues an Interrupt 21h call, it lets the operating system know it needs access to the functions that allow it to open and lock/unlock a file.

The two most important functions available via this Interrupt 21h are the extended open and the lock/unlock (also known as a physical lock). The *extended open function* allows an application program to open a file and specify how it will be used. For example, open the file and use it in a read only status. Or you can open the file and use it with a read/write status. The *lock/unlock function* goes to the disk and locks a physical range of bytes with the restriction specified via Interrupt 21h. It also is possible to lock portions of a file. This allows an application to open some part of a file, manipulate it freely, and leave the rest of the file open for others. This function is very important for database applications with multiple user micro-to-mainframe connections or local area networks.

Now that we are getting some de facto network standards for servers, such as NETBIOS, DOS 3.1 or higher, and the ISO standards, simple networks, micro-to-mainframe connections, and local area networks will be proliferating in even greater numbers. Many people, preferring not to work at the DOS or NETBIOS level, purchase independent communication software packages that take care of all of these functions (sending commands to DOS and NETBIOS) and offer easy use through menu-driven user friendly screens.

Other Communication Software   The independent software packages that are used for communications do not require the use of the DOS MODE command. These software packages offer provisions for logging on to a timesharing network, transmitting data files, logging on to other microcomputers or local area networks, and automatic dial and automatic answer facilities if someone else dials your microcomputer. More than 30 of these packages are available. Typical brand names of currently available communication software packages are ASCOM IV, CROSSTALK XVI, DATALINK, PC-DIAL, PCTALK III, KERMIT, LOGON, MICROLINK II, SMARTCOM II, MOVE-IT, and WATSON. The easiest way for you to understand what a communication software package does is to follow along as we describe how to use one straightforward software package by the name of CROSSTALK.

CROSSTALK XVI is a command-driven communication program. Its main status (menu) screen is divided in two, the top half exhibiting the protocol associated with the presently active service, and the bottom half containing helpful user aids, listings of available services, online help, and some of the more complex protocol information required. You can enter commands either as whole words or as two-letter mnemonics. You can change some of your parameters at any time. For example, a user who wants to change the modem speed can type SPEED, and the program responds with a listing of the available speeds from which to choose. You also can build and store automatic log-on "scripts." This is the entire log-on sequence that contains all the necessary parameters for logging on to a specific database service. The purpose is to make the log-on as fast and painless as possible.

Because CROSSTALK is one of the fastest communication packages available, it can work with the new generation of ultrafast modems that communicate in the range of 20,000 bits per second. It also supports transmissions as fast as 115,000 bits per second (for direct wire connections to other computers), although at speeds above 57,600 bits per second it may require a faster computer than a standard PC (such as a PC AT or a COMPAQ 386).

The current version of CROSSTALK (Version 3.61) supports three kinds of error checking file transfer protocols: XMODEM (the checksum version only), KERMIT, and a proprietary protocol developed by Microstuf (the developers of CROSSTALK). With both XMODEM and KERMIT available, the needs of almost any remote end modem are met. CROSSTALK also supports an answer-back feature with which you can establish a string of characters that are sent automatically when CROSSTALK receives an ENQ (control E) character. In communications, this is used to confirm that you are the person the caller wanted to reach. For example, if Friend B calls your modem and sends the ENQ character, your modem sends back the string of characters that Friend B is expecting. This is similar to confirming who you are talking with at the other end of a voice communication circuit. In voice communications, the confirmation comes because the person verbally identifies himself or herself as the person you asked for, and, of course, there is a secondary confirmation if you can recognize someone's voice.

CROSSTALK allows you to dial into a host computer system and act as though your computer is a terminal in that system; this is the *terminal program* part of CROSSTALK. CROSSTALK also allows you to call another CROSSTALK compatible system and exchange files with that system; this is the *file transfer program*. The terminal program allows your microcomputer to emulate many popular terminals. By emulating these terminals, your microcomputer can capture incoming data from the host computer or another microcomputer and send files from your disk back to the host or microcomputer system (called *uploading* a file). The file transfer program allows you to transmit files to another system using CROSSTALK, as well as to request files from another system using the CROSSTALK software package.

To use CROSSTALK, give the following commands (other packages are similar).

- Load the CROSSTALK program by entering XTALK.
- Load the command file by entering 1.
- Enter the telephone number you want to call.

- Select the modem speed by entering 1200, 2400, and so on.
- Begin communication by entering GO LOCAL or by pressing ENTER at the command prompt.
- To exchange operator-to-operator messages, enter your message and the microcomputer at the other end enters its messages.

Asynchronous communication (see Figure 2-4 in Chapter 2) between microcomputers requires a pair of modems or a null modem cable (described in a later section) plus a software package for the microcomputer. Most asynchronous packages claim to support file transfer. Let us expand on this last point.

Most of the available microcomputer communication software handles the task of capturing data as it comes across the screen through the use of a memory buffer. When the memory buffer is full, the programs write the data to disk in a format specified by *your* microcomputer. This is not a true file transfer because the data is now in the operating format of *your* microcomputer rather than in the format of the sending microcomputer. This "capture buffer" system of downloading from a host computer allows you to store information from bulletin boards and other electronic databases, even though this information may not be in the exact file structure format that is needed for use in one of your programs, such as a spreadsheet or statistical package. These asynchronous communication software packages usually allow you to upload information (to another microcomputer) by sending a text file over the telephone lines. Again, this is not actually a file transfer because the receiving machine presumably is using the capture buffer method of saving the data.

True file transfer, either uploading or downloading, requires that *both* microcomputers use the same protocol. That is easy if the machines are identical. The problem arises when the machines at each end of the communication circuit are different or when different communication packages are used.

This consideration is mentioned here because one microcomputer user may develop something like a spreadsheet and want to transmit it for use by another person. Unless they have the same machines using the same protocol (asynchronous or synchronous communication software), the receiver may not be able to use the spreadsheet data without reformatting the information for his or her own spreadsheet. Use of the term *file transfer* implies that the data is transferred and received in its original format. *Capture buffer* means the data is transmitted, but it is reformatted as it moves from the memory of the computer to a disk. While this may not affect you now, it might if you want to transmit various applications, such as a spreadsheet, among different microcomputer users.

Whereas a software package like CROSSTALK is used for transmission of data with another manufacturer's modem, there also are combination packages (voice or data). One such package is called WATSON, and another is PCDIAL LOG. As an example, WATSON's capabilities can be divided into the areas of telephone management, the modem and its software, and voice messaging. Even though this system uses the PCTALK III software and a 300/1200 Hayes compatible modem, it also runs other packages that can use a Hayes compatible modem.

WATSON is a microcomputer compatible, single full-slot circuit board and software that combines voice mail/telephone management with the standard data communication facilities. You get telephone management, data communications, and voice mail. Applications include a 500-entry electronic telephone book and database, a built-in 300/1200 baud modem, automatic telephone dialing and redialing, dictation machine with true voice editing, and electronic calendar with alarms. WATSON VIS (for Voice Information System) upgrades the voice mail capabilities of the system to provide information to callers. The best way you can see how this package operates is to dial 1-800-6WATSON. (In Massachusetts, the number is [617] 651-2198.) Do call by using a touchtone telephone so that you can control this demonstration, and record and play back your own voice. (WATSON digitizes your voice to record it onto the disk.)

MICROCOMPUTER PROTOCOLS

Regardless of which popular microcomputer communication package you use, it probably will utilize one of the following serial transmission protocols.

X-ON/X-OFF   When two microcomputers communicate, some form of flow control is required. *Flow control* provides a mechanism for regulating data transfer between two communicating systems. For example, if one microcomputer is sending a long file to a printer, there is a good chance that the printer cannot print as fast as the data can be sent. To avoid losing data, the printer sends an X-OFF character as its communication buffer fills. This instructs the transmitting microcomputer to stop sending the file. After the printer has caught up and its communication buffer is empty, the receiving printer sends an X-ON character to resume the file transfer where it left off.

Under this protocol, when the receiving device is no longer willing to receive a string of characters, it sends an X-OFF character to the sending device. (The X-OFF character usually is the ASCII character DC3 or control S.) To restart transmission, the receiving device sends the X-ON character to the sending device (which usually is the ASCII character DC1 or control Q). Because the X-ON and X-OFF characters can be destroyed or accidentally generated by communication line errors, this simple scheme can lead to confusion. As a result, more sophisticated protocols have been developed that use various cyclical redundancy error checks to turn on or off the flow of data between two modems.

XMODEM This protocol takes the data being transmitted and divides it into blocks. Each block has a start of header character (SOH), a 1-byte block number, 128 bytes of data, and a 1-byte checksum for error checking. To start the communications, the sending device must receive a negative acknowledgment (an ASCII NAK character) telling it to start sending the data blocks. After receiving a block of data, the receiving device must wait for the circuit to clear before sending either a NAK character (this indicates an error was found through use of the checksum) or an ACK character if the block was received correctly. Blocks that are acknowledged with a NAK character are retransmitted. At the completion of the session, the transmitter sends the ASCII character EOT and waits for an ACK character before terminating the session.

Even though this protocol was developed for micro-to-micro communications, it often is used for micro-to-mainframe communications where the host mainframe can support the XMODEM protocol. XMODEM is the type of protocol that uses a stop and wait ARQ (automatic repeat request). The 1-byte checksum for error checking is obtained by adding the ASCII values of each character in the 128-character block, dividing the sum by 255, and retaining the remainder as the checksum that is transmitted to the other end of the communication circuit. As you already know, the receiving microcomputer calculates its own checksum and compares it with the transmitted checksum. If the two values are equal, the receiving computer sends a positive acknowledgment (ACK) that informs the transmitter to send the next sequential message (block of data). If the two values are not equal, the receiver sends the transmitter a negative acknowledgment (NAK) to request a retransmission of the last message (block of data).

Because the XMODEM protocol is in the public domain, it is readily available for software designers to incorporate into communication programs. Some of the advantages of the XMODEM protocol are that it is easy to implement in a high level language, it only requires a 256-byte buffer, it can transmit 8-bit characters, and it has a very effective error detection scheme. PCTALK III is one of the communication packages that has incorporated this protocol for uploading and downloading of files. The primary disadvantage of XMODEM is that it is slow compared to some of the higher speed protocols. It is slow because it uses start-stop asynchronous transmission with the stop and wait ARQ (ACK/NAK).

KERMIT KERMIT probably is the single most popular protocol used with microcomputers. The KERMIT software is distributed free and unlicensed by Columbia University's Center for Computing Activities in New York City, although there may be a minimal charge ($5 or so) for the documentation and disk.

Various versions of KERMIT also can be located on public bulletin board systems that can be downloaded to your microcomputer. The KERMIT protocol was developed and is still copyrighted by Columbia University, which put it into the public domain as a free software communication package.

KERMIT is a RS232-based communication protocol designed to accommodate similar types of computers, operating systems, and file systems. It is suited especially to micro-to-mainframe connections for both IBM and non-IBM systems, but it works equally well with microcomputer-to-microcomputer or mainframe-to-mainframe connections. KERMIT communication-based programs provide error checked transfer of text and, in most cases, binary files using both 7- and 8-bit codes. The only requirements are an asynchronous serial connection and KERMIT software running on each of the two communicating computers.

Currently, there are over 200 versions of KERMIT, so there probably is at least one version that can meet each individual's needs. If you need a protocol to connect two dissimilar machines, contact Columbia University. It probably has one that works with your hardware. Generally, Columbia sends a copy for your own use. Various other versions have been developed by KERMIT users. Columbia University does not support KERMIT or offer telephone support.

KERMIT operates much like XMODEM. In sending a file, both KERMIT and XMODEM break the data into packets, and both protocols send the packets one at a time, along with error checking information. XMODEM does require an 8-bit data character, which may cause a problem because the front ends to some mainframe host computers and minicomputers support only a 7-bit data character. Even some of the systems that can handle 8 bits will not accept something they see that matches a control code pattern. KERMIT solves this problem by using a 7-bit character length. When the character requires an 8-bit character length, KERMIT still codes the data into 7-bit lengths and then sends the eighth bit separately. The KERMIT protocol on the receiving end decodes the packet.

Many corporations use KERMIT because it is a file transfer protocol. In addition to allowing file transfers between micros, minis, and mainframes, this protocol is handy for storage purposes. You can upload 8-bit characters with KERMIT and use the mainframe later for storage when you download to another microcomputer.

Some of KERMIT's other features include terminal emulation, speeds from 300 to 38,400 bits per second, a command that moves you into DOS without breaking the telephone connection, and wild card (DOS wild cards are * and ?) support for file transfers. In addition, the KERMIT protocol may act as a server, depending on which interpretation of KERMIT you have. Under normal circumstances you have to give two commands to transfer a file: one is to the host telling it to send and the file name, and the second one is to your own system telling it to receive and the file name. With KERMIT as a server, you give the command only once and your system's KERMIT controls the server.

Remember, KERMIT is only a protocol, and it does not offer all the features of a complete communication package. There are no color support and no directory to store frequently dialed numbers, nor are there auto dial and redial features.

One thing all KERMIT versions are well suited for is the exchange of files between IBM mainframes and personal computers. This exchange can be difficult with some protocols because IBM mainframes are built to talk to dumb terminals that have no storage and no need to hold files. An additional problem is that IBM mainframes use the 256 EBCDIC characters to represent the bytes, while microcomputers use the 128 standard ASCII characters. In KERMIT, the ASCII-EBCDIC conversion and all error checking are performed exclusively on the mainframe side. The KERMIT version running on the IBM mainframe takes EBCDIC characters and turns them into ASCII characters to add or remove its error checking bits.

X.PC This protocol, developed by Tymnet, can construct packets on a microcomputer and send them to a network node that converts them to the synchronous X.25 standard format. Tymnet developed the X.PC protocol specifically for connecting asynchronous devices to packet switching networks. X.PC also can be used as a file transfer protocol for direct communication between microcomputers. X.PC provides error checking with automatic retransmission; it allows several different sessions to take place concurrently using only one modem and one telephone circuit; and several different host mainframes can be accessed from a single microcomputer.

Some people call X.PC an asynchronous version of X.25, even though the X.PC packet is not packaged in the typical X.25 fashion. Thus, special "translation software" is required for each dial-up transmission between an X.25 network and the Tymnet network using X.PC protocol. The X.PC protocol uses a more comprehensive error detection scheme (cyclical redundancy check polynomial) than XMODEM. It also includes recovery procedures based on packet sequence numbers to ensure data integrity. Up to 15 logical channels can be accessed simultaneously allowing a microcomputer to be logged into 15 simultaneous applications, and it implements flow control through a rotating "window" algorithm, which allows up to 15 packets per window and variable size packets up to 256 bytes long.

BLAST This protocol is a *BL*ocked *A*synchronous *T*ransmission program that incorporates the power and efficiency of High-Level Data Link Control (HDLC) or Synchronous Data Link Control (SDLC) synchronous transmission protocols, but for use with asynchronous transmission. In this case, the blocks can be acknowledged negatively (NAK) in an out-of-sequence fashion. In other words, it uses continuous ARQ, thereby increasing data transmission efficiency. This protocol uses a cyclical redundancy error detection scheme (the ANSI CRC-16 polynomial), which is more effective than the one used by XMODEM. Both BLAST and X.PC result in a higher throughput than that of XMODEM. Because both BLAST and X.PC use sliding "windows," they allow propagation delays of up to approximately two seconds at 1200 bits per second transmission speed. For this reason, these protocols work on satellite links which have a longer propagation delay as messages move from earth to the satellite and back to earth. BLAST also works in full duplex mode. BLAST currently does not support any modem by name, but modem control parameters can be configured to match most available modems.

The latest version of BLAST is called PC-BLAST II, and it works between microcomputers, minicomputers, and mainframes. It automatically converts data between different computers, as from MS-DOS to VAX/VMS to UNIX or even to IBM mainframes. BLAST can be operated in full duplex (simultaneous transmission in both directions), and it allows *scripts*. When a software package allows scripts, it means you can set up a script of repetitive functions or automated routines and they are carried out when you call upon the script. The word script is derived from movies and television; a movie or a television show has a written script which is followed by the actors. In the context of protocols, a BLAST script allows repetitive functions to be set up so that they can be carried out in an automated and unattended fashion. BLAST is available to MS-DOS, UNIX/XENIX, MacIntosh, IBM mainframes, DEC VAX/VMS, and other popular computers.

As you now must realize, a microcomputer is a very powerful source of communications. Through its modem it can be connected to another microcomputer, a local area network, a micro-to-mainframe connection, or any of the large commercially available public networks or databases of information. Now that you understand the communication sofware that is required, let us discuss the microcomputer modem and its features.

MODEMS FOR MICROCOMPUTERS

From reading the earlier section on modems in Chapter 2, you should have a complete understanding of what a modem is and how it operates. We now want to present some specifics on microcomputer modems, although they perform the same basic tasks as the more powerful modems discussed in Chapter 2.

As your modem passes data between itself and the distant modem, the protocol must carry out some type of flow control. To avoid loss of data, the average rate at which the receiving modem is able to accept data must be equal to or greater than the average rate at which the sender is sending the data. Therefore, if one modem is set to transmit at 1200 bits per second, then the receiving modem also must be able to transmit and receive at 1200 bits per second or greater. In addition, a mechanism must exist whereby the receiving modem can let the transmitting modem know that it should stop transmitting, or at least that it should slow down. This mechanism is known as flow control, and it is handled by the communication software package and is referred to as the protocol. These were explained in the previous section (X-ON/X-OFF, XMODEM, X.PC, BLAST, and so on). The modem for your microcomputer can be either an internal modem or an external modem.

Internal Modems   Internal modems are built into a circuit board that fits into one of the slots inside your microcomputer box. This circuit card contains the typical RJ-11 telephone jack plug, such as you have in your home telephone. By using the standard RJ-11 plug, you can connect the circuit card containing the modem directly to the telephone line wall socket. Internal modems do not offer external indicator lights depicting the status of the modem or the circuit.

External Modems   As the name implies, external modems are external to your microcomputer. They come in a small metal box and are connected to the serial I/O port by using a 25-pin RS232C or a 9-pin PC/AT connector plug and cable. External and internal modems are the same, the only difference being their physical location (inside your microcomputer or as a separate external box). Once the external modem has been connected to the microcomputer, you will find a RJ-11 telephone jack on the external modem. Use this jack to connect the external modem to the telephone company plug on the wall at your home or place of business.

Typical modem brand names are Hayes Smartmodem, Novation Auto-cat, Signalman, Racal Vadic, and Codex. Today's microcomputer modems usually operate at a speed of 300, 1200, and 2400 bits per second. Because 300 bits per second is very slow, it is recommended that you buy at least a 1200 bits per second modem, possibly with the option of being able to transmit at 2400 bits per second or even higher. As more of the public database systems offer 2400 bits per second transmission rates, the microcomputer user is advised to transmit at this higher speed to save on telephone line usage costs and on the per hour charge rates of public database information systems.

Micro/Modem Connectors   The serial port on your microcomputer may have a RS232 connector plug or, if it is a PC/AT or PS/2 microcomputer, a 9-pin plug. All you must do is ensure that whatever plug you need comes on the modem you purchase. Several types of plugs are available for connecting cables to your microcomputer serial port. You already saw the RS232 connector in Figure 2-16 (in Chapter 2) and a 9-pin connector in Figure 2-17. In the world of microcomputers, these also may be referred to as D-type connectors (DB-25 and DB-9) because they are shaped somewhat like the letter D. Figure 6-3 shows the pin configurations for both the 25- and 9-pin connectors for IBM microcomputers.

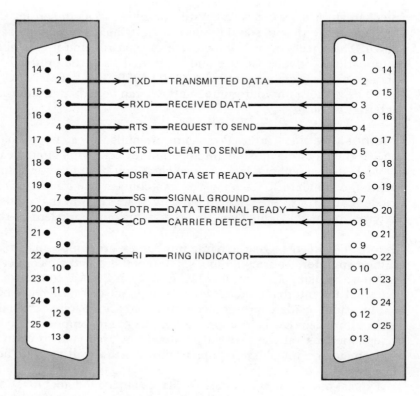

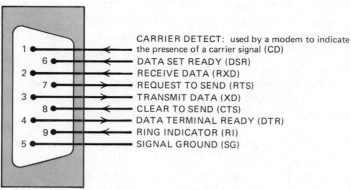

FIGURE 6-3 Typical IBM 25-pin (RS232) and 9-pin (PC/AT) modem connector.

If you are plagued by extraneous radio or television transmission signals on your telephone circuit, you can ask the local telephone company to install a suppressor. This device helps suppress extraneous radio signals picked up by the telephone line that might interfere with data transmissions or voice calls. Internal modems usually cost less because they do not have the external boxes and the RS232 plug/cable.

As an example of one of the many multifunction modems available for microcomputers, let us describe the ZOOM/MODEM PC 2400. This modem transmits at 300, 1200, or 2400 bits per second. It is Hayes compatible, and it dials the telephone for you. It can be programmed to dial one number repeatedly and then another until it gets through. It even has the ability to dial a number by seizing the line, waiting for a dial tone, dialing a local access number (such as US Sprint or MCI), waiting for the US Sprint or MCI return tone, and then dialing the rest of the number of the service bureau or computer center to which you wish to make a connection.

This modem has additional features that allow it to answer an incoming telephone call. If your computer is not set up to respond, the modem takes a message because it comes with either an 8,000- or a 32,000-byte buffer in which to store incoming messages. You can retrieve your messages from this buffer at a later time. Using its own internal clock, the modem stamps the messages with the time and date. The modem has three security levels and it can be programmed to require a password before passing control over to the communication software program in the microcomputer. This modem even has a provision for audio-out, which means that your computer can send short messages to callers through this modem. Another provision for audio-in allows the recording of a message. A clever person could build an inexpensive digital messaging system around this modem. Sophisticated microcomputer-oriented modems like this one are priced in the range of $300, as compared with $100 for barebones (no optional features) microcomputer modems.

Null Modem Cables Null modem cables allow transmission between two microcomputers that are next to one another without using a modem. If you discover that the diskette from your microcomputer will not fit into another one, that transmitting over telephone lines is not possible, or that you cannot transmit data easily from one microcomputer to another for some reason, then it is time to get a null modem cable.

First, bring the two microcomputers close together. Next, obtain a null modem cable (more on the pin connections in a moment). The cable runs from the RS232 communication port on the first microcomputer to the RS232 communication port on the second one. The cable is called a "null" modem cable because it eliminates the need for a modem. You can either build a null modem cable or buy one from any microcomputer store. Null modem connector blocks are available to connect between two cables that you might own already.

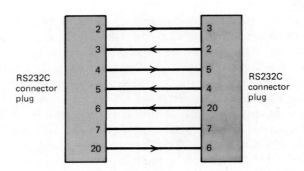

FIGURE 6-4 Null modem cable configuration
(synchronous).

To transfer data between two microcomputers, just hook the null modem cable between them and call up one of the computers by using the communication software you normally use. To do so, put one microcomputer in answer mode and get the other one to call it. After the receiving computer has answered that it is ready, the data can be sent, just as you would on a normal dial-up connection.

With a null modem cable, a higher bits per second rate is accomplished easily; transmission can be 9600 bits per second, for example. This may be a great advantage for high volume data transfers from microcomputer to microcomputer because your modem might limit transmission to 1200 bits per second or less. Basically, a null modem cable switches pins 2 and 3, the TRANSMIT and RECEIVE pins of the RS232 connector plug. A null modem wiring configuration is shown in Figure 6-4.

ELECTRICAL PROTECTION FOR MICROS AND LANS

If you believe your microcomputer or local area network communications are immune to electrical power fluctuations, you may find that you have serious problems, and possibly even physical damage to your hardware or data files. Electric power utilities only protect building wiring systems from surges of over several thousand volts. Telephone companies only protect buildings from surges of over several hundred volts on a communication circuit. Surges can be catastrophic to your microcomputer, and even more so to files on a hard disk.

Communication Circuits Communication circuits may have an electrical surge that can destroy data on a disk. The normal voltage on a telephone line is 48 to 96 volts direct current, and it sometimes reaches a few hundred volts for ringing the telephone bell. The newer 32-bit microprocessor chips that operate at 16- and 32-megahertz speeds may be affected by the shorter and weaker spikes (surges of electrical power) that did not affect earlier versions of these chips. Spikes are read as fault signals, thereby causing malfunctions, loss of a data communication circuit, or damage to data on a disk. What you may need is some method of protecting the communication system from surges, sags, and dirty power that *both* the electric power utility and telephone circuits let slip through to your building premise. You need this protection between the microcomputer and the incoming electrical power, as well as between the microcomputer and the incoming telephone circuits. You should be aware that the telephone company's protection devices may not respond until the voltage on an incoming telephone line (*not* the electricity from the public utility) approaches several hundred volts. Keep this fact in mind when you think about protection devices for your telecommunication equipment.

Several commercial devices are available that can perform the surge protection operation on telephone circuits/modems. One device has two modular telephone plugs (RJ-11 type) housed in a plastic body with a three-pronged 120-volt electrical plug sticking out from the back. The user plugs the electrical plug into the 120-volt ac power outlet, then plugs the modem into one side, and connects the other side to the telephone line wall socket.

Because some of these devices cost only $20, it may be cost effective to purchase surge and sag protection devices for the modem on your individual microcomputer, as well as for a larger microcomputer installation or network that uses communications such as a local area network or a cluster of microcomputers that have a micro-to-mainframe connection. Small devices are available that plug into the telephone jack for protection of modems from spikes, surges, and sags. While this type of equipment usually is purchased for the larger digital PBX switchboards, it can be used to protect your individual modem, software, and microcomputer from an incoming surge or sag that could burn out your microcomputer chips or destroy data on a disk.

To give your microcomputer, modem, and disk units complete protection from fluctuations in electricity, you may need *four* different types of protection, although they can be built into the same box.

Surge/Sag Device   First consider a device that protects against surges and sags in electrical power. A *surge* is an increase in power, and a *sag* is a decrease (lowering) in power. This type of device may cost anywhere from $20 to $100. It protects equipment from *transients*, which are short-duration high amplitude surges, and spikes of voltage. These are caused when a utility switches lines, nearby power lines are hit by lightning, there is intermittent operation of nearby electrical motors, a large group of lights is turned on, and there is other electrical noise (which is electrical interference superimposed upon the normal voltage sine wave).

Electrical noise occurs when power lines pick up high frequency waves and "hash" from automobile ignition systems, transformers, fluorescent lights, static electricity, and lightning storms. Voltage fluctuations (sags and surges) are variations on the magnitude of incoming voltage. These sags are fluctuations caused by voltage loss during power transmission, uneven regulation of the voltage, and intentional lowering of voltage by utilities during peak operating periods (commonly called a brownout). A surge is a sudden intense increase in the voltage. A surge and sag protection device provides instantaneous regulation of voltage input, but you should not expect too much from such inexpensive devices. They are designed primarily to protect against surges; little protection is provided against sags because it takes some type of battery or capacitor to do this effectively. All such devices have a minimum of filtering to clean up electrical noise.

For these devices to work effectively, the ground should not have a resistance greater than 5 ohms. (An *ohm* is a unit that determines the amount of difficulty an electric current encounters when trying to move. An insulator such as glass has a high ohmic value, while a conductor such as copper has a low ohmic value.) A surge device connects the active electrical wires and the ground. If the surge protector detects a surge, it very quickly becomes a short circuit (very low resistance path) between the conductor on which the surge appears and the ground conductor, thus draining away the surge before it can get to a sensitive piece of electronic equipment. The device should clamp the line (shunt the surge) at a value of voltage 10 to 15 percent over the value expected on the line. Some surge protectors that are built to shunt a surge of electricity from the 120-volt ac power also have RJ-11 plugs so they can shunt a surge coming in on the telephone lines supplied by the telephone company.

Power Line Conditioner (PLC)   You also should consider power line conditioners that protect against "dirty" power and undervoltages or sags in power levels. They also protect against surges. You can see what dirty power does when a power drill or blender causes interference on your television set. A full-featured PLC is designed to isolate electrical noise while also regulating the voltage to the microcomputer. In other words, these devices control both the quality and quantity of power to microcomputers. Typically, PLCs use an isolation transformer to filter out any electrical noise. Voltage regulators also maintain power quantity by keeping power within what is considered an acceptable range.

Microcomputers are quite susceptible to sags, which are undervoltages lasting 8.3 milliseconds or longer. Undervoltages can cause electronic systems to lose memory if the voltage falls below approximately 87 percent of its rated voltage. A typical power line conditioner converts ac input of 95 to 130 volts into a constant 120-volt level.

Some power line conditioners can handle very short voltage variations as great as $+15$ or -15 percent, regulating them down to a $+3$ or -3 percent level of electrical variation. What you get is protection against sags, surges, and very short temporary brownouts, which in reality are long-duration sags.

Uninterruptible Power Supply (UPS)   A third device you should consider for protecting microcomputers, communications, and local area networks is an uninterruptible power supply that allows your system to continue functioning, even if there is a total loss of power. These devices also protect against sags and surges and may include built-in power line conditioners. When the power goes to zero, the system automatically switches over from the normal utility-furnished electrical power to the unit's built-in batteries. Because this function involves a switching time, the faster the UPS system can switch over to batteries, the less the likelihood that you will lose valuable data or cause hardware damage. Switchover might take from 2 to 8 milliseconds.

On the other hand, more expensive UPS systems have no switchover time because they operate on batteries and generate true sine wave power at all times. You want true sine wave power output instead of a square wave because square wave power may damage electronic circuits.

UPS backup systems usually are rated in watts; you might see them advertised at 90 watts, 200 watts, 400 watts, 600 watts, and so forth. Some of the systems are rated in volt-amps (VA), which are approximately the same as watts. If you want to convert volt-amps into watts accurately, the phase angle (0) between the voltage and current waveform must be known. The cosine of this angle or "power factor" times volts times amps equals real power in watts:

$$\text{Watts} = \text{Volts} \times \text{Amps} \times \text{Power Factor}$$

For instance, computer installations usually have an average power factor of 0.89 lagging. The current waveform lags behind voltage. In other words, the typical computer loads are inductive rather than capacitive because most loads are inductive or resistive. Capacitive loads are rare and usually are canceled out by the larger inductive component. Therefore, if 400 watts of power are needed, the UPS should be rated at 450 volt-amps. A quick conversion involves dividing the watts by 0.89 to get the volt-amps:

$$\frac{400 \text{ watts}}{0.89} = 449.44 \text{ volt-amps}$$

The needed UPS size is determined by adding the number of watts used by your microcomputer, modem, video monitor, and printer. (Many users do not include the printer because it requires so much electrical power.) If this total is 400 watts, then you should buy a 400-watt UPS system to serve as backup.

If you intend to run the system on backup batteries for a long period after the loss of power, let us say more than 20 minutes, then you should buy a UPS backup that has a rating approximately 25 percent higher than the total wattage of the system. Using the previous example, you would want a 500-watt UPS even though your equipment uses only 400 watts. This extra wattage is needed to avoid overheating the UPS system electrical components and possible burnout of the unit. UPS systems are far more expensive than sag and surge devices or power line conditioners. Backup power supply systems may cost anywhere from $250 for a 200-watt system to $700 for a 450-watt system. Prices vary depending on whether there is a high quality built-in power line conditioner and whether the UPS switches over to batteries or uses its own batteries all the time (no switch-over time).

The importance of surge/sag devices was highlighted by a recent Bell Laboratories statistic which found that the number one electrical problem faced by microcomputers is voltage sags. They account for 87 percent of all electrical problems monitored.

Static Electricity   One other form of electricity that may "zap" your system is static electricity. Now that we have discussed how to protect ourselves from incoming electrical problems, whether they originate from the telephone circuits or from electrical power, we need to consider our own self-generated static electricity. As we walk across a floor, we can generate a charge of static electricity that can be as high as 15,000 volts. Touching a circuit board or a disk unit while carrying that much static electricity can cause problems that result in permanent damage, erasure of data from a disk, or just an ordinary misread by a disk unit. To avoid such problems, always make sure that the equipment is grounded and that you have discharged any static electricity from your body (by touching a ground) before handling new circuit boards or removing them from your microcomputer. The best practice is to touch anything that is metal and grounded before you start working inside the computer.

If you have such problems as unexplained data losses, unexplained lockup of your microcomputer keyboard, scrambled information, and intermittent unexplainable bugs, the microcomputer operators may have too much static electricity in their bodies. To solve this problem, make sure their chairs are placed on an antistatic grounding mat. This is a floor mat that is grounded in such a way that stepping on it or wheeling a chair across it causes the person or chair to be "drained" of any static electricity. Also check to make sure the microcomputer electric plugs are grounded properly.

MICRO-TO-MAINFRAME CONNECTIONS

Organizations that rely heavily on microcomputers often find that rapid growth requires substantial new investments in hardware and software to allow these microcomputers to talk among themselves and especially to the mainframe host computer with its massive corporate databases. Obviously, the local area network is one solution, and this will be covered in the next chapter. At this point it is more appropriate to discuss the idea of a single PC micro-to-mainframe connection, or the clustering of microcomputers together in such a way that the clustered group of microcomputers possesses a micro-to-mainframe link.

Various options can be used to connect microcomputers and mainframes. The simplest and least expensive way is to use a terminal emulation capability without software. The hardware circuit boards used to facilitate terminal emulation are inexpensive, but if there are several different types of microcomputers or mainframes you may need several different terminal emulation circuit boards.

A more sophisticated way of making a micro-to-mainframe link is to employ a system that uses both a hardware board and some software. This type of terminal emulation is a link that permits file transfers, while also allowing users to capture and manipulate the mainframe-extracted data. Such links allow the microcomputer to emulate the video data terminal (VDT), the printer, and a control unit over the remote link.

Probably the most used data transmission method is the download environment for batch processing. *Downloading* involves the transfer of a program or a data file from the host mainframe computer to a remote microcomputer. On the other hand, *uploading* involves the transfer of a program or data file from the remote microcomputer to the host mainframe.

In today's data processing environment, we have moved from batch processing, through real-time individual transaction processing, and now we are in a file transfer environment. The massive volumes of data transmitted when file transfers instead of individual transaction-type transfers of data take place has exerted a great amount of pressure on our communication links. Software resides at both the host and the microcomputer to handle downloading and uploading. For example, these communication links allow a microcomputer to emulate an IBM 3270 terminal and capture information for storage in another format. These products usually are based on IBM 3270 protocols or Systems Network Architecture (SNA).

Before connecting to an IBM 3270-based network, the PC must be able to emulate an IBM 3270 terminal. One popular method is to use an emulation board like an IRMA circuit board from Digital Communication Associates. (Other vendors also sell 3270 emulation boards.) Remember, however, that all hardware/software products *MUST* be tested prior to acquisition to determine whether the product actually can perform all the protocol requirements. This testing can be accomplished by contacting the network control center to request that the technicians monitor the connection via a data line monitor and watch for protocol violations or other problems. The best way, however, is for the mainframe owner to provide a list of tested and authorized products that can be used for micro-to-mainframe connections. Establishing this micro-to-mainframe connection is a three-step procedure.

First, the systems programming/communication groups at the host mainframe site must generate (sysgen) a new set of system software that describes the microcomputer as a participating 3270 terminal. The term *sysgen* is used when system software is generated for a mainframe computer.

Second, the hardware emulation board must be installed and the circuit (usually coaxial cable) installed between the microcomputer and the IBM 3174 or 3274 control unit. The emulation board is installed into an expansion slot in the microcomputer, and the coaxial cable connects to the emulation board. The emulation board communicates with the 3174/3274 control unit, which in turn communicates with the host mainframe. This communication is independent of whatever programs may be running in the microcomputer.

Third, you must run the emulation program software to make the microcomputer act like a 3270 terminal. This software can be customized to emulate any kind of 327X terminal and/or keyboard. Because the emulation board has its own microprocessor chip and memory, it functions separately from its own host microcomputer DOS environment. With this emulator, your microcomputer now can operate in two modes: regular microcomputer mode (using DOS) or terminal emulation mode (using the emulation board) when you use the micro-to-mainframe communication link. You can switch between these two modes without losing your screen of data because the emulation board's memory is independent of the microcomputer's memory. It is analogous to having in one box two microcomputers that share the keyboard and video screen.

To download data from the host mainframe computer, the micro-to-mainframe link must be able to support the mainframe access methodology (protocol), which is what the emulation board does. For example, if the micro-to-mainframe link can communicate with a certain DBMS that resides in the host mainframe, it may access files located in that database. Another task that must be performed is to reformat the files from their form as stored in the database at the host mainframe to a form that is manageable at the microcomputer. This process usually is performed at the mainframe. After the data is reformatted, it is transferred to the microcomputer. In other words, the mainframe converts the data files into data interchange format (DIF), document content architecture (DCA), ASCII, or worksheet (WKS) file formats and then downloads them to the microcomputer or one of the emulator boards residing in the microcomputer.

When choosing a micro-to-mainframe link, you should consider a number of factors.

- Which mainframes are supported by the software?
- Which mainframe operating systems are supported?

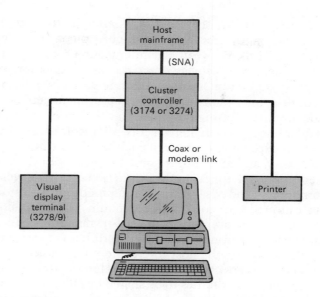

FIGURE **6-5**   Single PC (micro-to-mainframe connection).

- Which microcomputers are supported at the remote end?
- Which communication protocols are supported?
- Which microcomputer operating systems are supported?
- Which mainframe database file structures can be accessed?

In addition, determine

- The amount of memory required in the emulation board
- Whether the data is reformatted automatically before being transferred from the host mainframe to the microcomputer
- The total price for the software and any circuit boards required at the microcomputer end of the link

At the microcomputer end of the micro-to-mainframe link, there may be a single microcomputer or a clustered set of microcomputers. Figure 6-5 shows a single personal computer on a micro-to-mainframe connection. This connection depicts an IBM PC connected to an IBM 3174/3274 cluster controller by means of a coaxial cable or a modem link. The cluster controller is then attached to the host mainframe computer. Integrated hardware and software packages are available for the IBM PC; these packages allow it to emulate a 327X visual display terminal and plug directly into the cluster controller (either a 3174 or a 3274). Usually, the link between the microcomputer and the cluster controller is a coaxial cable, but it also can be a data communication modem attachment. This is the most popular micro-to-mainframe connection because data is transferred to and from the mainframe system by capturing or generating screen images on the microcomputer. Terminal emulation expansion boards are not required.

In order to set up a clustered set of microcomputers, the user need only plug an emulation expansion board into a powerful host microcomputer (probably a PC/AT or 80386) and attach to it some inexpensive PCs or dumb terminals. Such a cluster promotes the complete functionality of several fully configured microcomputers. Today's advanced equipment allows clusters of up to 31 users to be built around one host microcomputer. The expansion board in the host microcomputer provides a microprocessor, random access memory, and two serial ports. Because each new expansion card has its own processor and memory, it puts little strain on the host microcomputer. Multiple user software for clustering runs on the host microcomputer, along with the host copy of DOS, thereby allowing two disk volumes to be shared by all users for data storage. The software allows shared use of printers attached to the host microcomputer and permits ordinary DOS requests to be executed on the host.

A more sophisticated clustered set of microcomputers is one in which a specialized piece of hardware called a network server is the host at the microcomputer end of the micro-to-mainframe link. The *network server* is dedicated to providing any required data or file conversions, shared printer or shared disk storage facilities, and the tasks required for communication between the host mainframe and the clustered set of microcomputers on the micro-to-mainframe link. As circuit cards get more powerful, it is difficult to tell the difference between a dedicated network server for clustering microcomputers on a micro-to-mainframe link, and a more powerful host microcomputer installed with the proper network server circuit cards and software. These clustered microcomputer groups are referred to as a *multiuser system micro-to-mainframe link.*

Figure 6-6 depicts several microcomputers interfaced with a single host mainframe. The host or master microcomputer acts as a cluster controller and controls the other microcomputers. Each of the other microcomputers (often known as slave microcomputers) is attached to the host microcomputer by means of a high speed data communication link. With the proper software and hardware emulation boards in the host microcomputer, data can be transferred to and from the host mainframe computer, and the data can be modified while it is in the possession of any of the microcomputers on this link. A major advantage of this method is that it requires only a single remote modem to connect all of the locally multidropped microcomputers to the host mainframe computer.

Notice that the host microcomputer in Figure 6-6 has a hard disk. When there

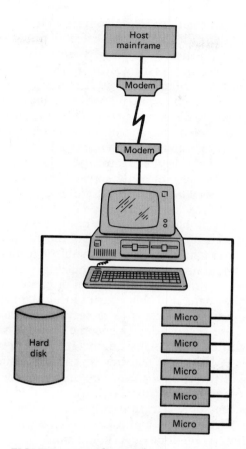

FIGURE **6-6** Clustered microcomputers (micro-
to-mainframe connection).

is a clustered set of microcomputers, one of them (usually the host) can serve as
either a disk server or a file server. Any of the other microcomputers attached to
the host microcomputer therefore can store data on this disk and retrieve data
from it, although there is a difference between disk servers and file servers.

A *disk server* is a hardware/software combination that treats the hard disk of
the clustered group of microcomputers as though it were one large disk usable by
all the microcomputers in the cluster. Each of the local microcomputers treats
the disk as though it were a local peripheral device. Disk servers allow file
sharing only by designating certain areas of the disk as public (each microcomputer can use a different public area), and these public areas are able to be read
by any microcomputer in the cluster. Disk servers are somewhat dangerous in

that anyone can easily write on someone else's files, destroy their files accidentally, or read their files. Control is effected primarily by being a "good neighbor."

A *file server*, on the other hand, has the further software intelligence to decide what to do with requests for writing to the disk storage and for reading from the disk storage. It can manage the storage more efficiently, and it can handle problems of security, such as putting various "locks" on records or files for security. File servers handle file serving on a file-by-file basis, meaning a user can set up his or her own files with read and/or write protection enabled or denied. File servers are far more elegant and convenient than disk servers. In a business environment, it is recommended that the organization always purchase software that enables it to use the philosophy of a file server. This should be done in order to protect valuable business data from purposeful theft, accidental erasure or modification, and accidental breaches of privacy.

A significant problem in the use of micro-to-mainframe links is in the area of control and security. When a portion of the organization's data is downloaded to a microcomputer, this data becomes subject to manipulation outside of the host mainframe programs that were intended to control and validate it. It frequently happens that the movement of data or programs to a remote microcomputer, the editing of that data, and the return of that data to the host mainframe database bypass many host edit controls or security features. This means the host mainframe database may gradually become polluted with corrupted data. This problem is magnified when many microcomputers are on a micro-to-mainframe link.

Some of IBM's newer features link microcomputers and mainframes in new ways. One of these, *application program interface* (API), allows applications on personal computers and mainframes to speak directly to each other. API should help in the area of application program development because it allows PCs into the connectivity at the application software level. Many vendors support their own proprietary versions of API, and some have added IBM API compatibility.

Corporate users are examining how LU 6.2, an IBM communication protocol, will affect their micro-to-mainframe links. LU 6.2, as further defined by IBM's APPC/PC (Advanced Program-to-Program Communications/PC), helps determine the requirements for establishing a program-to-program link between microcomputers and mainframes. When you are at the program-to-program level, you are doing much more than transmitting screens (remember emulation) between the host mainframe and the microcomputer. APPC/LU 6.2 allows cooperative applications to exchange data more easily when they are running under different operating systems and when the systems are written in different languages. With APPC/PC, microcomputers should be able to communicate with mainframe systems while retaining full standalone processing capabilities. This protocol is IBM's input to the industrywide effort to establish a program-to-program communication link (see Chapter 5, Systems Network Architecture). As you might imagine, this would be a great improvement over terminal emulation, which just passes the screens back and forth.

IBM's Operating System System/2, the first product to support the company's Systems Application Architecture, will let its new PS/2s be integrated into IBM's SNA environment. OS/2 consolidates user access to a group of application program interfaces, including Advanced Program-to-Program Communications and NETBIOS. The new operating system includes connections for the IBM Token-Ring Network and the PC Network.

Another ongoing development in the area of micro-to-mainframe links is the speed at which data is transmitted. The 1200 or 2400 bits per second speed that we use in microcomputer communications usually is inadequate for going between a mainframe and a remote microcomputer or a clustered set of microcomputers. In practice, 9600 bits per second is barely adequate for a clustered set of microcomputers; 19,200 is better. We are moving very rapidly toward the fully digital 56,000 bits per second and the 64,000 bits per second Integrated Systems Digital Network (ISDN) speed of transmission for micro-to-mainframe connections. For those micro-to mainframe links that go over a coaxial cable (with its approximate 2,000-foot limitation), it is quite easy to transmit at 19,200 bits per second. The point is that as we progress technologically with our micro-to-mainframe links, your microcomputer will have the ability to talk directly from a program within itself to a program in the host mainframe computer, and this will be accomplished at faster and faster data transfer rates. Terminal emulation is slowly but surely being phased out as the method for micro-to-mainframe connections.

Even though the micro-to-mainframe connection is a powerful communication link in the business environment, the next chapter covers an even more powerful one: local area networks.

SELECTED REFERENCES

1. Andriole, Stephen J., ed. *Microcomputer Decision Support Systems: Design, Implementation, and Evaluation.* Wellesley, Mass.: QED Information Sciences, 1986.

2. Bugg, Phillip W. *Microcomputers in the Corporate Environment.* Englewood Cliffs, N.J.: Prentice-Hall, 1986.

3. *Byte: The Small Systems Journal.* Published monthly by McGraw-Hill, 1 Phoenix Hill Lane, Peterborough, N.H. 03458, 1975– .

4. *Datapro PC Communications.* Published monthly by Datapro Research Corp., 1805 Underwood Boulevard, Delran, N.J. 08075, 1986– .

5. Durr, Michael, and Dwayne Walker. *Micro to Mainframe: Creating an Integrated Environment.* Reading, Mass.: Addison-Wesley Publishing Co., 1985.

6. Gofton, Peter W. *Mastering Serial Communications.* Berkeley, Calif.: SYBEX, 1986.

7. *Info World: The PC News Weekly.* Published weekly by Popular Computing, 1060 Marsh Road, Suite C-200, Menlo Park, Calif. 94025, 1979– .

8. Kamin, Jonathan. *MS-DOS: Power User's Guide.* Berkeley, Calif.: SYBEX, 1986.

9. Kopeck, Ronald F. *Micro-to-Mainframe Links.* Berkeley, Calif.: Osborne/McGraw-Hill, 1986.

10. *PC Tech Journal for the IBM Systems Professional.* Published monthly by Ziff-Davis Publishing Co., 1 Park Avenue, New York, N.Y. 10016, 1983– .

11. *PC Week: The National Newspaper of IBM Standard Microcomputing.* Published weekly by Ziff-Davis Publishing Co., 1 Park Avenue, New York, N.Y. 10016, 1983– .

12. *Personal Computing.* Published monthly by Hayden Publishing Co., 10 Mulholland Drive, Hasbrouck Heights, N.J. 07604, 1976– .

13. Schwaderer, W. David. *Modems and Communication on IBM PCs.* New York: John Wiley & Sons, 1986.

14. Wolverton, Van. *Running MS DOS: The Microsoft Guide to Getting the Most Out of the Standard Operating System for the IBM PC and 50 Other Personal Computers,* 2nd ed. Bellevue, Wash.: Microsoft Press, 1985.

15. Wolverton, Van. *Supercharging MS DOS: The Microsoft Guide to High Performance Computing for the Experienced PC User.* Redmond, Wash.: Microsoft Press, 1986.

QUESTIONS/PROBLEMS

1. How many bits wide is the transfer path in the Intel 80386 and Motorola 68030 microprocessor chips?

2. If you have a microcomputer, compare the specifications of your system with those enumerated in the third paragraph of this chapter.

3. What is the difference between RAM and ROM?

4. Define booting (use the Glossary, too).

5. Define DOS.

6. What is DOS's command processor called?

7. What is a TPA?

8. What is NETBIOS?

9. What is the difference between a "file transfer" and a "capture buffer" transfer of data?

10. What does flow control do?

11. Which is the simplest (least sophisticated) protocol described in this chapter?

12. If a RJ-11 telephone jack plug is used for a two-wire circuit (this is the same line as the one on your telephone at home), what plug is used on a four-wire circuit? *HINT:* See Glossary.

13. Name the pins (wires) used in the null modem cable in Figure 6-4.

14. Figure 6-4 shows a null modem cable for synchronous communication. Can you guess which three pins can be used for an asynchronous null modem cable?

15. Can you identify four different types of electrical protection needed for your microcomputer?

16. In micro-to-mainframe, what is the basic function of the emulation board during a download from the mainframe?

17. Figure 6-6 uses one modem pair to connect six microcomputers to the host. How many modem pairs are required if the philosophy of Figure 6-5 is used?

18. Which is more powerful, a disk server or a file server?

7

LOCAL AREA NETWORKS (LANs)

This chapter introduces the concept of local area networks and defines their use. It describes topology, protocols, baseband/broadband, software, bridges and gateways, cabling, costs, implementation, management, and selection of LANs.

LOCAL AREA NETWORK

Today's "hot button" is linking personal computers for a system that provides the standard business office functions of word processing, electronic mail, file sharing, file transfer, shared printers, and processing capabilities. A large organization may have numerous local area networks connected together, and, of course, they tend to be connected to the organization's host mainframe computer. A local area network/host mainframe connection is just like the micro-to-mainframe connection, except that it has the additional capability of being able to connect many microcomputers to form an interdepartmental or intradepartmental network that is independent of the host mainframe.

INTRODUCTION TO LANS

A *local area network* is a group of microcomputers or other workstation devices located in the same general area and connected by a common cable. The network is located within a small or confined area. Although it is not confined like a computer input/output bus or a micro-to-mainframe link, it definitely is more restricted than the larger corporate or government backbone networks that go between cities and even countries. A LAN is a major portion of an overall communication network that is designed to interconnect microcomputers, word processors, minicomputers, facsimile machines, voice/data PBXs, executive workstations, and other hardware, for the purpose of communicating among themselves and ultimately with the host mainframe computer or public networks.

A LAN covers a clearly defined local area: a single building, a group of buildings within a business firm's property, a campus, or a confined area within a city if the conditions are correct. It is not limited to communication within that defined local area because LANs provide easy access to the outside world through a bridge or gateway to other networks. Aside from their limited spread of a few miles, local area networks have a capacity for very high speed transmission. Since lower speed transmission currently is available on standard telephone circuits, a LAN has added functionality because it can provide speeds from 50,000 bits per second up to a planned-for capacity of 100 million bits per second. Today's local area networks commonly operate at 2 to 10 million bits per second. Bandwidths are available to support very high speed transmission for fully animated motion of graphics, full-color video, digital voice telephone conversations, or any other high data rate analog or digital signals which an organization may want to place onto a LAN.

Local area networks operate outside of the government's "regulated" environment; therefore, an organization that wants to install a LAN does not have to be licensed or obtain approvals from any federal or state communication regulatory bodies.

In a utopian environment, you would be able to attach an infinite number of users without major modifications to the LAN system. Access would be as simple as picking up and dialing your telephone; the network would be very user friendly. It would have protocol conversion to allow any device to talk to any other device, and, of course, it would be available 100 percent of the time for sending or receiving messages that were completely free of errors. We are not in Utopia, however, and find that we must make tradeoffs to optimize the network so it meets specific user requirements. These tradeoffs are discussed later in this chapter.

Figure 7-1 shows a local area network. This LAN is built to interconnect a

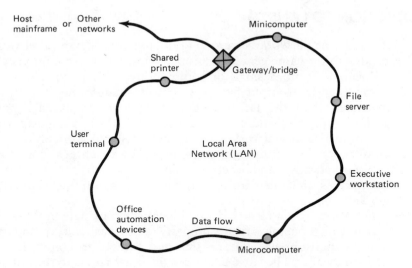

FIGURE 7-1   Local area network (ring topology).

variety of electronic office equipment, microcomputers, minicomputers, mainframe host computers, databases, and other equipment within an organization. Local area networks are owned and operated by the individual organization. There are at least 50 vendors of local area networks.

Several methodologies compete with LANs. One of these is the micro-to-mainframe connection, especially with clustered microcomputers (see Chapter 6). Another competing methodology is the all-digital private branch exchange (PBX) switchboard, which was defined in Chapter 4. Multiuser systems are yet another alternative to LANs.

The local area network designer might view the LAN as far more flexible and more easily expandable than the micro-to-mainframe link. The LAN also is much easier to install when it is small (five or ten nodes) and located in an isolated area of a building. Obviously, this type of LAN does not require a connection to the organization's host mainframe computer. The same designer, when comparing a LAN to a micro-to-mainframe connection, might argue that the local area network offers a far greater networking ability among the various workstations attached to the LAN. Figure 7-1 illustrates some of the devices that might be connected to a LAN. Obviously, both micro-to-mainframe connections and local area networks have their place, but a local area network does offer far greater future growth capabilities.

Local area network designers might look at the PBX as a single component connected to the LAN because designers provide a gateway or bridge between the local area network and the PBX. PBX manufacturers might hold the opposite view, contending that the PBX is the hub of all communications within the organization and, accordingly, that the local area network should be configured within the PBX by using software. (Review Figure 4-22 in Chapter 4.)

Which view will prevail? Probably both. At this point, the advanced digital PBXs are moving toward a distributed switching architecture that is very similar to the architecture used by most local area networks. Perhaps, then, the LAN and PBX technologies ultimately will combine into a single configuration. Now let us proceed with our in-depth discussions of local area network topologies, baseband/broadband, protocols, bridges/gateways, software, cabling, costs, and implementing, managing, and selecting LANs.

TOPOLOGY

The *topology* of a system refers to the way the pieces of the network are connected together. In other words, a topology is the shape or geometric arrangement of network stations, which in turn determines the flow of information across the network. Local area networks are constructed using three basic topologies: the ring, the bus, and the star. Each of these three topologies can be used alone, or they can be combined when configuring a local area network.

Ring A *ring topology* connects all workstations in a closed loop, and messages are passed to each node or station in turn. Figure 7-1 is a ring topology. The ring approach connects devices much as a bus topology does, but it connects the cable in a ring or circular pattern rather than in a straight line. With this approach data is transmitted around the ring in one direction only. The time required for the data to travel around the ring between the interconnected devices is called the *walk time*. Each workstation in the ring has a unique address.

Bus A *bus topology* connects all stations to a single cable running the length of the network. The bus approach physically connects devices by means of cables that run between the devices, but cables do not pass through a centralized controller mechanism. In this situation, data *may* pass directly from one device to another, or it may be routed through a head end control point. The head end controller (see Figure 7-2) turns the message transmission around and sends it back down the cable in the opposite direction. With some bus networks, the message always must go to the head end and then back down the cable to whichever node or workstation it is addressed. Other bus networks allow the message to go directly to whichever node or workstation it is addressed without going to the head end first.

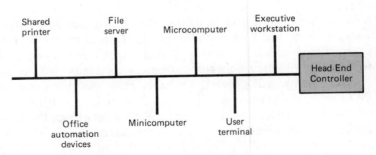

FIGURE 7-2   Local area network (bus topology).

Star   A *star topology* (sometimes called a *hub topology*) connects all workstations to one central station that routes traffic to the appropriate place. The star configuration is a traditional approach to interconnecting devices in which each device is linked by a separate circuit through a central connection point or controller. This configuration is typified by many existing data terminal systems and most PBX switchboards in which all transmitted data must pass through a central control point to be transferred (switched) from the sender to the receiver. Figure 7-3 depicts a star topology as the primary configuration. In reality, it shows a combination topology because the star network has a ring network at the end of one of its links.

The physical topology is slightly different from the electrical topology. The *electrical bus topology* is designed so that every station receives every signal generated on the network media. In the *electrical ring topology*, the signal is regenerated and repeated as it is passed from station to station. The ring and the bus electrical topologies, however, can have one or both of two very different physical topologies. In each electrical topology, you can choose between running cables from station to station around the ring. Or, at the same time, you can bring one cable from each station back to a central point called a *wiring hub*. The station-to-station and star physical topologies predominate in microcomputerbased LANs.

The station-to-station arrangement uses less wire and may be easier to install, but if the cable fails because of a bad connector or damage at any point along the way, the entire network goes down. The advantage of the star arrangement is that one bad cable or connector normally does not disrupt the network; if it does, you can restore the system quickly by unplugging the bad cable from the hub.

Configurations/Standards   Local area networks may have a variety of *configurations*. For example, networks with star topologies have a central controller, but a ring topology network connects each station to those on its left or right. Networks with the bus topology sometimes are called *backbone networks* because

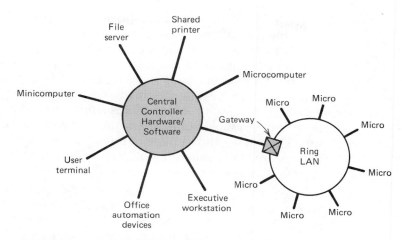

FIGURE **7-3** Local area network (star topology).

they connect each station to a central cable called the backbone. Sometimes large business networks are called backbone networks because they use a bus topology to connect everything in the organization to a central network. You can view these larger backbone networks as being one long cable to which the business connects each of its mainframe computers, a number of local area networks (without regard to the topology of each LAN), individual microcomputers, office automation workstations, outside public networks the corporation may want to access, and so forth.

With each topology (ring, bus, or star), every message contains a destination address and each station on the network "listens" for its address in each message. If the address is present, the station picks up the message; if the address is not present, the station ignores the message. Several different local area network methodologies/standards are evolving.

- The methodology or standard that has been available the longest is *Ethernet*, which is a 10 million bits per second carrier sense multiple access/collision detection (CSMA/CD) baseband technology. (At this point, do not be concerned about the detailed definitions of CSMA/CD and baseband because they are defined later in this chapter under Protocols.) Ethernet is a bus-oriented wiring scheme. It offers the highest bandwidth, or data rate, using wires, and it transmits direct electrical voltages over the wire. It takes fiber optics to have a wider bandwidth or greater data rate.

- Another methodology or standard is the *IBM PC Network*. This standard is appropriate for a smaller LAN and is based on a 2-megabit transmission speed and CSMA/CD broadband technology. IBM's PC Network uses two channels. One channel is for data sent to the head end, which is a device that serves as a message translator (see Figure 7-2); the other is for information sent from the head end. Because many channels are left over, you can put more information on the network than IBM supports, but you have to use a third party vendor to supply hardware to use this added capacity. The wiring scheme for IBM's PC Network is similar to cable television; it uses coaxial cable in a branching tree structure. A branching tree structure is just one form of network similar to the bus topology. Both Ethernet and the IBM PC Network are applicable to high speed scientific applications as well as the business data environment.

- A more recent local area network standard is General Motors' *manufacturing automation protocol* (MAP), a broadband technology using a token bus philosophy. The MAP protocol specifies either a 5 or 10 million bits per second data rate. (MAP was discussed in Chapter 5.)

- A fourth LAN methodology/standard is American Telephone and Telegraph's *STAR local area network*. This standard is based on a 1 million bits per second CSMA/CD scheme that uses twisted pair telephone wires. One advantage of the STARLAN is that these twisted pair telephone wires already are in most office buildings. The IBM wiring scheme can take the same approach, although when IBM uses the same type of topology you are required to pull new data grade twisted pair wires through the building in place of the original twisted pair telephone wires.

- IBM has a second LAN methodology called the *IBM Token-Ring Network*. This network is for a larger environment than the previously mentioned IBM PC Network. The topology of this network is a single baseband ring. As many as 72 microcomputers can be connected over twisted telephone wires or up to 260 microcomputers if data grade cabling is used.

    Bridges are available to connect different rings so that you can build an almost unlimited set of interconnected Token-Ring Networks. The IBM Token-Ring Network can be connected to a host mainframe through the IBM communication controller. Microcomputers usually are spaced no more than 660 feet apart, but this can be extended to 2,400 feet by using a repeater/amplifier. This network also can be designed to use optical fiber cables.

    IBM has a program to manage this network by logging network error and status information on a fixed disk by date, time, and station address. This network manager program monitors the Token-Ring Network on a continuous basis. The Token-Ring Network microcomputers are attached to one another with network adapters and access units that are connected in a ring configuration. A high speed electronic token travels around the ring continuously, going from device to device. When you want to access the network, your microcomputer intercepts the token as it goes by. The token then carries your message or information around the ring until it reaches the station or microcomputer you have specified. After dropping off the message, the token returns to the sending microcomputer and then is freed for use by other microcomputers on the network.

BASEBAND VERSUS BROADBAND

As you may have noticed in the previous section, current vendor offerings for local area networks may be divided into two categories—baseband and broadband.

Baseband is a single channel, whereas *broadband* provides multiple channels. Baseband can carry only a single transmission at any one moment, whereas broadband splits the bandwidth into many channels. As a result, with broadband more than one transmission can occupy the LAN cable at the same time. Another view is that broadband is placed on a multiplexed signal (review Figure 3-13 in Chapter 3). Baseband transmits by using basic digital electronic voltages and, therefore, occupies only one channel for which it uses the entire bandwidth.

Baseband is less complex than broadband, and usually it is less expensive because it does not need modulation devices (modems) for each microcomputer connected to the LAN. Baseband, however, is limited to a single channel (one message transmission at a time). This means that we cannot intermix all signals (voice, data, and video) on the same network if it is baseband.

Baseband systems pose a few other problems, some of which are caused by the laws of physics and others by vendors. In the laws of physics category, we find a lot of natural and humanly created noise in this band, as well as signal losses (attenuation) that limit the cable length more than with broadband. These restrictions affect layout and flexibility of use. Baseband cable lengths range from a few hundred feet to a mile or two, without using repeaters/amplifiers. A *repeater* or *amplifier* is a device that picks up the weak (attenuated) signal and increases its signal strength before passing it down the next cable segment or link. For example, some repeater/amplifier devices are able to connect cable segments to create a 5- to 10-mile-long network. In local area networks, baseband may be best for organizations that constantly rearrange terminals, computers, and other devices. Reconfiguration is easier because there are no modemlike devices.

The broadband world presents somewhat more variation because of its ability to provide multiple channels. Vendors therefore can offer a combination of access methods, such as frequency division multiplexing (FDM), which is television-type channel selection, time division multiplexing (TDM), or any other approach on a given channel of the LAN. Whereas baseband involves electrically pulsing the cable with a voltage or current switched between two different voltage levels (review Figure 2-24), broadband involves dividing the signaling into allocation slots, such as we see on television channels (review Figure 3-13). Broadband also may involve frequency modulation, amplitude modulation, phase modulation, or pulse modulation (electrical or optical) in its modemlike devices.

Baseband uses only a small portion of the coaxial cable capacity and achieves a highly efficient data throughput with a straightforward circuitry. This technique, however, precludes the use of the cable for other services such as video. By contrast, broadband uses relatively complex modemlike modulation devices to subdivide the coaxial cable into many channels, each of which can have the capacity of a baseband network. As an analogy, think of these modulation devices as similar to multiplexers that allow a cable television system to carry 25 different channels of television over a single coaxial cable into your home. If they used baseband, the coaxial cable would be able to carry only one channel at a time; therefore, you would need 25 cables to receive 25 television channels.

The primary drawback of broadband is the expense of the modulation devices (modems). Broadband allows unlimited distance because these modulation devices can be connected easily with amplifiers to amplify the signal, just as is done with the dial-up telephone networks used worldwide. This is no longer a very great advantage because repeaters/amplifiers now are available for baseband LANs, although they are somewhat more limited in distance. Because of these amplifiers, cable layout is very flexible, and cable distances are no longer as important as they are with a baseband signal. For example, if another segment of cable is needed to install a new user workstation, we simply put an amplifier on the cable to bring the signal to that workstation up to a usable level.

In summary then, the network can be baseband, which may be somewhat less expensive but limited to a single channel and usually more limited in distance, or it can be broadband, which gives multichannel capability and unlimited distance but is more expensive and more complex.

PROTOCOLS

As noted in earlier chapters, a *protocol* is a formal set of conventions governing the format and control of inputs and outputs between two communicating devices. Put another way, protocols are the rules used by two machines to talk to each other. Some LAN functions require specific network access methods (protocols), while others use any method, from one end of the spectrum (dedicated channels) to the other (random access). The user must consider carefully the advantages and disadvantages of each method.

Dedicated channels may be frequency channels (such as those found in television broadcasting) or dedicated time slots (for example, the channel is yours between five after the hour and ten after the hour, every hour). The concept of dedicated channels is very simple, but it can be inefficient. Assume the network is yours for five minutes every hour. This means that it is yours even during the lunch hour when it is not in use.

In some situations dedicated channels are necessary. For example, if you are holding a video teleconference, a standard 66 megahertz channel might be used to provide a constant update of your full-motion, full-color picture. Dedicated channels, then, have their special place in the network of the future. They should not be chosen as a general approach to network access, however, because they are not efficient enough for typical office or distributed data processing functions.

Random access schemes, on the other hand, appear to provide a simple, yet flexible, method that can accommodate changing user needs dynamically. Pure random access, which was used in the ALOHA system in Hawaii, commonly is called the transmit and pray approach. That is, if you want the channel, then take it. If no one else is on it, there is no problem; if someone else already is using it, keep trying—you will get through eventually. This seemingly chaotic approach works well for a small number of users who have short messages sent on an intermittent basis.

CSMA/CD This protocol might be termed "somewhat more ordered chaos." It encompasses one of the access methods that has been supported by the IEEE 802 Local Area Network Committee. The *carrier sense multiple access* (CSMA) protocol is very simple in concept. Do not get on the network and transmit at random—listen first! If anyone is there, refrain from transmitting. When it gets quiet, take the network and send your message. This eliminates a certain number of collisions (messages interfering with each other), but a small problem remains. Because of network propagation delay, two users who are located some distance from one another can both listen to the channel, find it empty, and begin to transmit simultaneously. They follow the carrier sense rule, but their messages still collide. Therefore, let us add another piece to this concept.

Instead of just listening before we talk, we should listen while we talk. If we detect a collision during transmission, we wait for the other message to end and then retransmit. This is *collision detection* (CD). Two users still can attempt to retransmit at the same time, so to rectify this problem we add an algorithm into the network. Instead of each user retransmitting immediately after the end of the message, it is held back until some random time interval after the colliding message disappears. This does not eliminate collisions completely, but it reduces them to manageable proportions. This approach allows rather high utilization of the network by many users while still providing the flexibility necessary to accommodate intermittent traffic. This access method is called *CSMA/CD*.

As an analogy of the CSMA/CD protocol, let us suppose that you are involved in a conversation with a small group of friends (four or five people) sitting at a restaurant table. As the discussion progresses, each person tries to interrupt or "get the floor" when the previous speaker completes whatever it is he or she is saying. For example, you might be listening to one of your friends, while trying to judge when the person is going to stop talking so that you can immediately get into the conversation to say whatever it is you want to say. To determine when your friend is about ready to stop talking, you take into account the subject matter being discussed, whether the idea is reaching a logical conclusion, your friend's facial expressions, body language, voice intonations, and the overall attention level of the other members of this small group. Usually, the other members of the group "give the floor" to the first person who interrupts at the precise moment that the previous speaker finishes talking, with an adequate amount of volume (not argumentative), and with a proper answer or a complementary comment related to the "jist" of the conversation.

Notice how a small group of people tend to use a CSMA/CD protocol when talking. The point is that the carrier sense protocol of a LAN is patterned on a real-life situation. The way your conversation works is identical to the way a collision detect system works. A terminal listens, interrupts, and sends its message. As long as no other terminal interrupts during the short time period that the message is traveling on the local area network, everything is all right.

For the sake of comparison, do you remember how a central control protocol works? It is used by large host mainframe computer networks with front end communication processors for polling the terminals on the network. With central polling, the front end or host mainframe computer polls each individual terminal or terminal controller and gives it permission to transmit. In other words, central control polling is analogous to a classroom situation in which the professor calls on the students who raise their hands. In a classroom situation, the professor is the front end central controller. To gain access to the class, the students raise their hands or make some other appropriate motion and the professor recognizes them so they can ask a question or make a comment. When they have finished, the professor again takes charge, or possibly recognizes someone else and allows that person to make a comment or ask a question. As you can see, even the central control protocol (polling) is patterned after a real-life type of protocol.

Look at Figure 7-4 and follow along while we examine how the transmission speed (10 million bits per second) helps to get your messages through the local area network. If you have a message length that is 1000 bits long and a collision detect protocol with a transmission rate of 10 million bits per second, you theoretically can transmit 1,000 messages per second and have very few collisions.

The upper half of Figure 7-4 shows that if you take a 10 million bits per second transmission speed and divide it by 1000 bits per message, you have a theoretical 10,000 messages per second capacity on the network. The lower half of the figure shows that if you take this theoretical 10,000 message per second capacity and

$$\frac{10,000,000 \text{ bits/second}}{1,000 \text{ bits/message}} = 10,000 \text{ messages/second}$$

$$\begin{array}{r} 10,000 \text{ messages/second} \\ \times\ \underline{0.10} \text{ utilization percentage} \\ 1,000 \text{ messages/second} \end{array}$$

FIGURE **7-4** Local network capacity.

utilize only 10 percent of that capacity, you have the ability to send 1,000 messages per second. Now, assume that the network designer has specified only a 10 percent network utilization. In this case, 90 percent of the time the network is empty (that is, no messages are being transmitted). In this situation you could transmit 1,000 messages every second, and the network still would be empty 90 percent of the time. In reality, it would not be totally empty because some of that time would be used to make up for collisions, retransmission of data that was received with an error, and the like.

The point is that if you have 100 terminals on a local area network, then each terminal can send 10 messages each second. This amounts to using only 10 percent of the network's 10 million bits per second capacity. As a result, collisions are reduced significantly.

Local area networks transmit data in packets. Certain bits are added to the user's data to enable correct transmission. *Packet size* depends on the protocol that encapsulates the user data to provide addresses, check fields, message type (whether it is a message for control purposes or a message containing business data, etc.), and other needed information.

Let us use as an example the IEEE 802.3 standard (the Ethernet protocol). If you want to transmit ten characters (80 bits) of user data in a packet, by the time it is transmitted the packet contains approximately eight times that number of bits. First, a 64-bit preamble is used to "wake up" the modem in a broadband network. This is followed by a 48-bit destination address, a 48-bit source address, a 16-bit message-type field, a 32-bit cyclic redundancy error check, and, finally, the 80 bits of user data. Then come a 288-bit pad field (the IEEE 802.3 standard requires a minimum user data field size of 368 bits or enough padding to make up the difference) and a 96-bit guard. The guard, which lasts for 96 bit times (a *bit time* being the time it takes to transmit one bit on a network) at 10 million bits per second, ensures a 9.6-microsecond interval between packets to avoid packet overrun (collisions). This example shows that only 80 out of the total of 672 bit times associated with this packet actually represent user data. In other words, we used a 1,000-bit message length in our Figure 7-4 example because a 50-character-long message is approximately 1,000 bits in length.

The above example demonstrates why networks should be built so no more than 25 percent of their capacity is dedicated to network traffic when CSMA/CD is used. The other 75 percent of its capacity should be reserved for such items as peaks in the message traffic, collisions between packets, or retransmission of packets that are in error. If the number of microcomputers attached to a baseband network grows drastically, it may become necessary to build a second baseband network and connect the two with a bridge or convert to a broadband network that has multiple channels.

Token Access Method In some areas, however, guaranteed access to the network is required. An example might be in the area of process control. Suppose there is a chemical process in which one chemical must be added to a mixture to produce a special plastic or rubber. If the automated chemical plant process cannot gain access to the network to add this ingredient at the exact time it is necessary, the network is worthless. Because of the need for guaranteed access (no collision delay), the *token access method* has been developed. As a result, token access networks have become very popular in the process control area. These networks can accommodate a number of users, provide a certain amount of flexibility, and give a guaranteed "known" maximum access time to access the network. The token access method is the second major standard supported by the IEEE 802 committee.

This method can be likened to a relay race in which the track belongs to you as long as you have the baton. When your run is finished, you hand the baton to the next runner and the track then belongs to that runner. In a token passing network, the baton is called a *token*, which is a unique short message that is generated when the network is turned on. During the network's daily operation, some intelligent terminal device keeps checking to be sure the token has not been destroyed by electrical noise or some other type of error-producing phenomena.

Token networks can be operated on either a ring topology or a bus topology. For example, when a ring topology is used, the microcomputers are attached to each other around the ring (see Figure 7-1 as an example). A very high speed electronic signal (token) travels continuously around the ring, going from device to device. As noted earlier in this chapter, the time required to travel completely around the ring is the *walk time*. When you want access to the network, your microcomputer intercepts the token as it goes by. The token then carries your message around the ring until it reaches the microcomputer address you have specified. After dropping off the message, the token returns to your microcomputer and is freed for use by someone else. This return of the token can be used to carry an acknowledgment back to the message sender as to whether the other microcomputer received the message properly (error checking).

Because the network is selected on the basis of overall network performance characteristics, the protocol is only one criterion provided by the vendor. In general, we select a network vendor and then specify that any equipment purchased between now and the foreseeable future must match that vendor's protocol. We really do not want to discard all of the equipment that has been gathered over the years, however. The network vendor, therefore, must provide protocol conversion to allow the use of most, if not all, of the existing equipment. Compatibility is one of the biggest problems in local area network selection. The development of protocol conversion hardware and software is a large business area in itself.

BRIDGES AND GATEWAYS

Microcomputer users on a local area network gain access to shared resources and host mainframe computers through the use of bridges and gateways. Basically, a bridge links two similar networks, whereas a gateway links two dissimilar networks. With bridges, the protocol that is passed between networks is transmitted in its original format; therefore, no protocol translation is required.

Bridges are used to increase the number of addressable nodes on a network or to link two geographically distant but similar networks that use the same protocol. For example, users on one LAN can address another LAN as though it were another node in their own network. This is done without signing on to a communication service, dialing through a modem, or learning any new software commands or menus.

Bridges are a combination of both hardware and software. A typical bridge consists of a black box that sits between the two network servers and has its own processor, memory, and software; its operations are transparent to the network user. A simple bridge can be used to connect two Ethernet local area networks located in the same building. If a LAN cable gets too long (that is, beyond its operational length limits), you can divide one LAN into two LANs and connect them with a bridge. Figure 7-1 shows a bridge going to another network or host mainframe computer.

Gateways are more complex than bridges because they interface two dissimilar networks. Gateways translate one network protocol into another, thus interfacing both hardware and software incompatibilities. Gateways are used in micro-to-mainframe and micro-to-mini communications. Like bridges, gateways are combinations of hardware and software products. Gateways are designed for transparent operation, and users do not need to know the protocol of the network from which they are requesting or sending data.

More complex gateways even take care of such tasks as code conversion (ASCII/EBCDIC). An example is a Systems Network Architecture (SNA) gateway that allows microcomputer network users to access mainframe applications, data, and peripherals. Without this SNA gateway on their local area network, each microcomputer would have to have its own 3270 hardware emulation card, coaxial cable, and mainframe controller port. The SNA gateway eliminates the need for additional hardware for the microcomputer, and it requires only one connection to the host computer because all data is sent through the local area network central gateway.

IBM has three types of gateways: SDLC (Synchronous Data Link Control, which is the protocol used in SNA), DFT (Distributed Function Terminal), and the Token-Ring's ability to link directly to a front end processor.

In an *SDLC gateway*, one microcomputer on the local area network emulates an IBM 3174 cluster controller. This usually is implemented by putting an SDLC card into the microcomputer and executing some software. This software does link-level (layer 2 of the OSI model) communication with the host front end processor, and then acts as a relay point between data from other microcomputers on the LAN and the mainframe. The gateway microcomputer talks to the other microcomputers through an interface such as NETBIOS. These links to the host run at speeds between 4800 and 19,200 bits per second.

With a *DFT gateway*, one microcomputer on the LAN is designated as the gateway. It contains a board that makes it look like a multisession workstation (such as a 3290 or 3194). In addition to being connected to the LAN, the gateway microcomputer is connected by a coaxial cable to a 3274 cluster controller, which then is connected to a mainframe (parallel) channel. The speed of the link usually is between 50,000 and 150,000 bits per second.

When using an IBM *Token-Ring* network, the IBM 3725 and 3720 front end processors can act as nodes on the Token-Ring LAN and can be accessed through the IBM 3270 Emulation Program Version 3. To make the connection between the LAN and the communication processor, you install a Token-Ring Interface Coupler (TIC) on a 3725 or a Token-Ring adapter on a 3720. Once that has been done, any microcomputer on the LAN can talk directly to the mainframe front end at 2 million bits per second, without any additional hardware or cable other than the LAN hardware.

IBM's advanced program-to-program communications (LU 6.2) may have major implications for all gateway vendors. LU 6.2 (described in Chapter 5) is a type of gateway because it allows messages to be transmitted between networks. Many people may be disappointed by LU 6.2 because their expectations are for broad network connections, but the protocol currently is limited to direct connections. In spite of this limitation, LU 6.2 gateways will become popular during the next few years because most computer vendors probably will offer LU 6.2 compatibility. As this happens, applications can become more protocol independent. Product development program teams spend a great deal of time solving protocol incompatibilities.

Because gateways can be used to link two different LANs, they eventually may ease the incompatibility problems between local area networks. Gateways are indispensable if an organization wants to develop a hybrid network. *Hybrid networks* combine a variety of network environments and protocols into a single network, utilizing each network type's strong points to the fullest advantage. Gateways connect user nodes to hybrid networks.

Gateways may be a standalone box or a subsystem within a mainframe computer. Two systems might not be able to communicate because of restrictions on connectivity caused by addressing constraints or differences in input/output protocols such as the connection of an asynchronous system to a synchronous system. In both cases, some type of gateway is necessary to allow the flow of data between them. Three examples of gateways are presented: network-to-network gateways, system-to-network gateways, and system-to-system gateways.

In the *network-to-network gateway*, two fairly compatible X.25 networks need a gateway because of internetwork domain problems. An X.25 network typically is designed to route data and calls only within the boundaries of its own network definitions. As a result, each X.25 network is distinct and is controlled separately. If two such networks seek to communicate with one another, it is necessary for the calls to go through a special gateway called an *X.75 gateway* node. The primary purpose of this X.75 gateway is to provide any necessary translation (particularly terminal address translation) for a call originating in one network that is to go to a destination in another network. Such translation is mandatory because the X.25 standard does not specify how an address is mapped into a user port. When a user in one X.25 network specifies an address in another X.25 network, a large portion of that address may not be comprehensible to the receiving network. For this reason the message is routed to a gateway for call processing. X.75 gateways are an integral part of the hardware furnished by packet switch manufacturers.

The primary function of the *system-to-network gateway* is to make the X.25 network available to a minicomputer system, as well as to some secondary functions related to the handling of asynchronous PAD functions. In either of these instances, the source of any data is assumed to be compatible with what the minicomputer is expecting. When a minicomputer system wants to communicate with the network, however, it may encounter problems because of incompatibilities. For example, the gateway might connect directly into the minicomputer system bus on one side and into the synchronous line of the X.25 network on the other. The primary function of the system-to-network gateway is to take a protocol that is not understandable to the minicomputer and convert it to a form that the minicomputer can understand. In this minicomputer example, a gateway handles the X.25 protocol and provides the interface to the operating system of whatever minicomputer is utilized.

By using a *system-to-system gateway*, you can connect one vendor's computer system to another vendor's computer system. When you are trying to interface two different computer systems, the gateway must provide both the basic system interconnection and the necessary emulation in both directions. The gateway probably would be made to look like the terminals or terminal controllers of one of the two computer systems, allowing the terminals of the second computer to attach to the first computer's controller.

The major difference between a system-to-system gateway and a system-to-network gateway is that the system-to-network gateway assumes that the source of any data is compatible with whatever the computer system is expecting. By contrast, the system-to-system gateway assumes a basic difference in the systems themselves, such as character-oriented (asynchronous) versus block-oriented (synchronous), or other differences between computer manufacturers.

LAN SOFTWARE

Software represents the major limiting factor in network availability or capability; LAN network hardware is available today. The software provides the applications and, in addition, provides the conversion from protocol to protocol or equipment to equipment. As a result, it is the software vendors who ultimately drive the LAN market. Manufacturers provide hardware, but, just as happened in the computer industry, many software houses have emerged that specialize in application software, protocol conversion software, and operating system software. This specialized software in turn enhances network capabilities and makes integrated networks a reality.

Just hooking some microcomputers together with a cable does not make a network. Software programs must be developed to

- Move data back and forth between the server and the microcomputers
- Offer some type of data security at the server
- Direct traffic through the transmission medium (the cabling)
- Allow the microcomputers to connect and disconnect from the network
- Manage and compile network statistics for management and control of the local area network

At this point you might want to review the section titled Communication Software for Microcomputers in Chapter 6.

Currently, MS-DOS Version 3.1 or higher can be used to set up a network, assuming you have the proper hardware and software. These networks may be inadequate because DOS cannot handle multiuser/multitasking applications. (Only one user at a time can access the disk in a sequential fashion.) To install the LAN software using Microsoft's MS-DOS requires the following steps.

- Create a file on the Microsoft network server disk that contains the network name for each microcomputer in the LAN.
- Copy the appropriate Microsoft network program and data files onto each server and each microcomputer or workstation's system disk.
- Add a few lines of information to the DOS CONFIG.SYS file on each server and each workstation's system disk.

After completing these steps, you have created a network system disk for each server and each microcomputer workstation.

As another example, Novell NetWare™ is a microcomputer network operating system that runs on 24 different makes of LAN hardware. When a user turns on his or her microcomputer, in one NetWare™ application, the AUTOEXEC.BAT file signs them on to the network. The LOGIN command then maps the file server directories to the network disk drive (like virtual diskettes), diverting LPT1 to a shared/spooled printer and LPT2 to a laser printer, if one is available. LPT1 and LPT2 are the names the disk operating system (DOS) uses to refer to the printer ports. The NetWare™ then determines whether any electronic mail is waiting for that user; if mail is present, it pauses to allow the user to check the mail. As you can see, LAN software is a layer of software that interfaces the network to the microcomputer's disk operating system and to the user. In fact, LAN software has taken over many of the functions that previously were performed by the microcomputer's operating system (DOS). This takeover is similar to the takeover by telecommunication monitor software, in a larger centrally controlled backbone network, of some of the functions of the host mainframe operating system.

One of the most important functions of LAN software is to offer various types of access security among the many users of the network. Many network operating systems simply split the server's hard disk into sections called volumes, which then are assigned for user access (called a *disk server*). Volumes usually are assigned public status (meaning everyone has read only access to these files), private access (a single user has both read and write access), or shared access (all users have both read and write access). More sophisticated LAN network software allows a greater number of access levels (called a *file server*). For example, a user might have the power to delete or create files, have no access to certain volumes or files, or just read access or write access. Access is a key network issue when many users want to share data.

Under an access level known as *volume locking*, a user who opens a volume can bar all other users from accessing any part of that volume by utilizing a secret password. Some networks allow several users to view and modify different files within the same volume. This technique, which is called *file locking*, determines who can use each file within the volume. Certain file locking schemes permit several users to read but not write on the same file. Under the most precise level of network data access, known as *record locking*, only parts of files are locked. Users can read the same file in a volume simultaneously but are prevented from writing over the same records. True record locking is rare. All locking schemes depend on both the network operating system and the application software that runs under it. Some network operating system software supports record locking, but application programs must be written to take advantage of this feature. Some LAN software allows the LAN network manager to keep an audit trail of which network volumes a network user is working on and which network stations are exchanging information.

For example, a multilevel secure LAN software package may have a security center that is a dedicated workstation offering centralized management of the network's security system. The types of security implemented might be

- Encryption of each packet before it is transmitted
- Physical or electronic keys that must be inserted into a network security device to gain access to the network
- Security windows, which define what type of data can be received and transmitted by a given user or a given microcomputer
- Volume, file, and record locking
- Password protection

Over the next three to five years, the LAN software market will become as important as the market for telecommunication monitor software and telecommunication access software for large-scale host mainframe-based networks.

CABLING

The selection of a local area network topology can be influenced greatly by the type of cable that already exists in the building in which the LAN is to be installed. Just as superhighways carry all kinds of traffic, from the tiniest economy cars to the largest trucks, the perfect cabling system also should be able to carry all kinds of electronic transmissions to all corners of the building in question. Ideally, there should be only one kind of cable and one kind of connector. The problem is, until such a system arrives, users must be content with existing choices.

Most LANs are formed with a blend of twisted pair wires (ordinary telephone wire), fiber optic cable, and coaxial cable. You even can obtain a wireless local area network. The Federal Communications Commission (FCC) has granted approval for a radio LAN called Freenet. Freenet runs at a top speed of 19,200 bits per second. It is being offered by Sterling Networks and is priced at approximately $400 per node. An advantage might be that it runs on radio frequencies, thus eliminating the purchase of wires, cables, or optical fibers. Obviously, a disadvantage is the lack of security. To protect the privacy of messages, users have to encrypt them before transmission. There also are several other petitions before the FCC for wireless networks. One of them is a packet switching digital network for microcomputers operating in the 902- to 928-megahertz frequency range.

Of all cabling types, fiber optic and twisted pair appear to be emerging as the more important types, although coaxial cable should not be counted out. Coaxial cable is still viable because baseband coaxial Ethernet cable is firmly entrenched in many local area network installations. Furthermore, broadband coaxial cable has an established position on the factory floor where it is used for robotics and other factory controls. Fiber optics ultimately may win out on the factory floor because it is not affected by electrical noise and dirty power, both of which corrupt data as it is passed through the cable or wire.

IBM Cabling   IBM has introduced the IBM Cabling System, a wiring scheme based primarily on twisted pair telephone wire, data grade twisted pair, and fiber optic cable. Seven combinations of cable types are included, each of which handles different voice and data communication requirements. Although several cable types consist of a combination of data and voice grade twisted pair wires, voice and data always are transmitted on separate pairs of wires.

The IBM Cabling System wall outlet has two connections, one for data and the other for voice transmissions. The data connector allows workstations and host processors to attach to data grade twisted pair wires, while the telephone connector provides telephone support on three voice grade twisted pair wires. The IBM connector is a specialized, IBM-developed connector plug that incorporates pin connections for RS232, coax connectors, or RJ-11 telephone jack plugs. The IBM cabling system is designed to be wired in a star topology, with a central *wiring closet*. The wiring closet is the originating source of all the cables. This star-based cabling scheme can be used to interconnect with Token-Ring networks, PC Networks, micro-to-mainframe connections, and PBXs (switchboards).

AT&T Cabling   AT&T's Premises Distribution System (PDS) is a uniform cabling system that uses combinations of existing or new twisted pair wires and optical fibers. PDS is intended for installations where the equipment of various vendors is used. It can accommodate from 1 to 4,200 twisted pair wires and up to 144 optical fiber strands. The twisted pair wires currently may be used for concurrent, but not simultaneous, data and voice communications.

PDS is designed to incorporate STARLAN, which is AT&T's twisted pair local area network. This cabling system supports not only AT&T's own LANs (Information System Network—ISN and STARLAN) but also their System 75 and 85

PBXs, as well as Wang workstations, IBM 3720s, RS232 connector plugs, RJ-11 telephone jacks, and coaxial cables. The most interesting aspect of PDS is that AT&T uses regular unshielded twisted pair telephone wires. PDS does call for a fiber optic backbone system. Like the IBM cabling system, AT&T has devised a universal connector which is a standard modular wall jack except for being able to accommodate its RJ-48 jack plugs used with the Integrated Services Digital Network (ISDN). The AT&T cabling system appears to be oriented more toward telecommunications (combinations of voice, data, and video) than toward straight data communications. It also is built more around PBX switchboards than around LANs because it uses a fiber optic backbone, and, like IBM's wiring plans, PDS is configured as a star with centralized wiring closets.

DEC Cabling Digital Equipment Corporation has its DECconnect, which includes twisted pair and fiber optic cable, but also offers baseband coaxial Ethernet cable. Coaxial cable is used as a backbone from which either new thin-wire Ethernet cable or twisted wire pairs run to workstations and terminals. A special wall outlet offers four network connections: telephone, intelligent workstation/microcomputer/video equipment, asynchronous terminals, and LAN.

DECconnect has a different architecture than either IBM or AT&T. The backbone of the DEC plan is Ethernet, running on a coaxial or fiber optic cable, and it is a bus-based topology. Actually, it is a hybrid network because each floor of a building can be wired in a star topology, but a bus topology goes vertically up and down through the building to connect each star. The DECconnect wall plate has four separate jacks. These jacks support a 10 million bits per second Ethernet connection, a 19,200 bits per second asynchronous connection for terminals, a voice telephone connection, and intelligent workstations or other equipment.

All three of the above wiring systems—the IBM, AT&T, and DEC systems—have proprietary wall outlets that may limit the choices for installing equipment on systems in spite of vendor claims of openness. Although IBM, AT&T, and DEC, along with other vendors, have released building wiring plans meant to simplify support of both telephones and data devices, considerable user confusion remains. Basic differences in the vendors' approach, LAN topology, and cable media may cause the price to wire a building to vary from $400 to $2,500 per office.

LAN Cabling Media Three transmission media dominate today's network installations: twisted pair wires (which is current telephone technology), fiber optics, and coaxial cable. (All types of media were defined in Chapter 2, under Circuits/Channels.)

Today most office buildings already have twisted pair wires in place. While initially it appeared that twisted pairs would not be able to meet long-term capacity and distance requirements, today this medium is one of the leading LAN cabling methodologies. Its extremely low cost (it often is in place already), the availability of data grade twisted pair wiring that can handle higher transmission speeds, and the high data rates attainable on twisted wire pairs through the use of digital transmission methodologies make the cabling very useful. In addition, low cost repeaters/amplifiers allow the signal to be sent over greater distances, which overcomes the attenuation.

Twisted pairs are connected easily to a coaxial cable through the use of a BALUN. *BALUN* is an acronym for *BAL*anced *UN*balanced and refers to an impedance-matching device used to connect balanced twisted pair cabling with unbalanced coaxial cable. A BALUN is a small device 1 inch in diameter and 3 inches long. One end has a standard RJ-11 telephone jack plug, and the other has a standard screw-in type of coaxial connector lead. One use for a BALUN may be made when two microcomputer workstations are located some distance away from one another. At one workstation, you place a BALUN to direct the signal from the twisted pair wires onto a coaxial cable. At the other workstation, you use a second BALUN to take the signal off the coaxial cable and onto a twisted pair for entry into the microcomputer. The signal does not attenuate (lose power) as rapidly when it is going over a coaxial cable as it does when twisted pair wires are used; therefore, a slightly greater distance between the microcomputers can be attained by using coaxial cable.

Twisted pairs also are replacing coaxial cable in a number of applications in larger LANs. While twisted pairs never will supplant coaxial or fiber optic cables in a large broadband network, many cabling designers use them as feeder or drop cable to microcomputer workstations. Because of BALUNS, existing coaxial cable that is bulky, inflexible, and therefore troublesome to pull (install) can be replaced with more manageable twisted wire pairs. This is accomplished by cutting the coaxial cable (coax) and installing a tee connector with a BALUN that allows a twisted pair of wires to be routed to a microcomputer workstation.

Twisted pairs can be unshielded as in regular telephone wiring, or they can be shielded with braiding or foil as shown in Figure 7-5. This figure also depicts an enlarged RJ-11 telephone jack plug that is connected to the telephone wire. The twists of the twisted pair act as a shield against *radio interference* (called RFI) and *electromagnetic interference* (called EMI). They do this by aligning the two magnetic fields of the two wires in such a way as to diminish interference. The number of twists in a pair is important. In other words, the more twists, the less interference. *Data grade twisted pair* wiring has a higher number of twists per foot than the regular twisted pair wiring used in voice telephone applications.

The other type of noise that twisting controls is cross-talk. *Cross-talk* is the interference a wire exerts over other nearby wires. RFI and EMI interference also can be controlled by shielding twisted pairs with some type of noise-reducing material such as foil or braiding. Cross-talk can be controlled in a multipair cable bundle by shielding each individual pair from the other pairs.

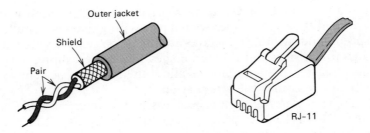

FIGURE **7-5** Twisted pair (shielded) and a RJ-11 telephone jack plug.

The second transmission medium, *fiber optics*, uses a glass fiber that is drawn into a long cylinder to act as a "waveguide" for the light. Once the light is put into one end of the fiber, it becomes trapped in that fiber, bounces back and forth as though it were hitting a series of mirrors, and eventually comes out the other end. The light injectors generally are lasers or light-emitting diode (LED) transmitters. Lasers provide higher power than LEDs, but the LED is a less expensive method. The advantages of fiber optics are straightforward. It has tremendous bandwidth capability, it is impervious to electrical noise, and it is far more secure against tapping. It has more bandwidth available than we could fill with everything currently being transmitted, as well as anything we can think about in the future. The fibers are small, they do not pick up electrical noise, and they operate in extremely hostile environments such as open flames.

To make fiber optic cable readily usable for a multiple-user environment, it must be spliced. Cutting into fiber cable is relatively difficult. A few years ago, this task was performed by a technician with a microscope. Today, it is done by a piece of precision hardware that cuts the fiber, aligns it, and fuses it end to end. While this equipment is a major improvement, the disadvantage is that it is quite costly and it still requires significantly more work than splicing copper wire pairs or coaxial cables.

Fiber optic cable also is noisier than wire cable, so we need a higher signal power to get the same signal-to-noise ratio. This is not really a practical limitation today, but it means that the higher powered, and therefore more expensive, laser injector (rather than LEDs) may have to be used to obtain the higher signal power of light.

Fiber optic cable is thinner than single twisted pairs (even unshielded ones) and therefore takes far less space when cabled throughout a building. It also is much lighter, weighing less than 10 pounds per 1,000 feet. Fiber optics can handle data rates of 90 million bits per second, and some vendors are boasting that 200 million bits per second speeds will be achieved soon. The American National Standards Institute (ANSI) standard, Fiber Distributed Data Interface (FDDI), may be passed soon, thereby setting a speed of 125 million bits per second. A fiber

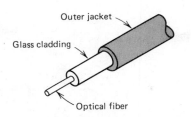

FIGURE 7-6   Fiber optic cable.

optic cable can handle over 1,300 two-way conversations. By contrast, two twisted pairs are limited to about 24, although 96 is possible as you may remember from your reading about T-1 circuits in Chapter 2.

Because of the high bandwidth, fiber optic cabling is perfect for broadband LANs. With regard to distance between amplifiers, laboratory tests have achieved distances of over 600 miles, but most of the time the distance is still only about 2,000 feet to 30 miles in the absence of an amplifier. From a data loss standpoint, fiber optics is not affected by dirty power, static electricity, or lightning strikes. Figure 7-6 shows the hair-thin glass fiber running down the middle of its glass cladding.

Fiber optic transmission is classified by how the light travels down the fiber. In *multimode* fibers, the light bounces off the walls of the glass in the pathway. In *monomode* fibers (also called single mode), a single concentrated light beam travels straight down the smaller diameter fiber. Monomode has a higher capacity and can handle greater distances than multimode, but it requires a more powerful laser light source, which is more expensive than the light-emitting diodes used in multimode fibers. Figure 2-46 in Chapter 2 shows both multimode and monomode (single) fibers.

Coaxial cable, the third type of transmission medium, comes in all sizes, shapes, and characteristics. Each has some unique properties that can help or hinder design efforts. Cables are available in 50 ohms, 75 ohms, 93 ohms, and a number of impedances in between. (See the Surge/Sag Device sections in Chapter 6 for a definition of ohms.) Coaxial cable usually supports frequencies up to 300 or 400 megahertz. If the correct cable is selected, it even can support frequencies into the gigahertz range. Coaxial cable sets the requirements for almost unlimited bandwidth support for LANs and is available in the marketplace at a reasonable price. As you might suspect, twisted pairs are the least expensive and fiber optic cables the most expensive, leaving coax somewhere in between.

Coax is physically the largest of the three cable types. Depending on which coax you use, it may weigh anywhere from 20 pounds per 1,000 feet up to an astounding 90 pounds per 1,000 feet, which can have a detrimental effect on a dropped overhead ceiling, especially if it collapses because of the cable's weight!

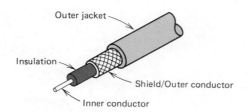

FIGURE 7-7   Coaxial cable.

Coax is not very flexible because it cannot be bent around sharp corners easily. Unlike twisted wire, coaxial cable allows you to send information a greater distance before an amplifier is required. Coaxial cable is, by definition, shielded because its outer conductor also is a shield (see Figure 7-7). Some coaxial cables are doubly shielded. An example is Ethernet whose coaxial cables generally have both a foil shield and the second braided conductor shield. With regard to this outer conductor, by now you should know that all electrical wiring requires two conductors to complete a circuit. For example, look at the wiring in your home or the twisted pair of a telephone wire. Coaxial cable uses its inner conductor as one of the two conductors and the braided shield (outer conductor) as the second conductor to complete the circuit.

Installing Cables   You need to consider a number of items when installing cables or when performing cable maintenance.

- Before moving any cable, always inspect the ceiling above or the floor beneath, as well as the surrounding area, to see how the cables enter, exit, and are attached. In other words, check the "blind side."

- Before reconnecting any cables back into a network, test for continuity, polarity, or any shorts.

- Conceal all cable as much as possible to protect it from damage and for security reasons.

- When disconnecting a cable, always disconnect it at the central computer (CPU) end first, and always remove and tape one lead at a time. *Always* work with just one wire at a time.

- When connecting a cable, make all required connections on the floor end (remote) of the cable before effecting the final connection into the central computer (CPU) end.

- Properly number and mark both ends of all cable installations as you install them.

- Always update your cable documentation records immediately upon installing a new cable.

To establish a management responsibility focused on cabling, you should adopt the following policies and procedures.

- Assess where the local area network fits into the total cabling distribution of the building or within the entire organization.
- Obtain documentation relating to the existing wiring.
- Perform a physical inventory of any existing cabling systems and document those findings.
- Create diagrams of existing cable distributions.
- Establish an ongoing management software package to keep track of any current and all newly installed cable.
- Establish a long-term plan for the evolution of the current cabling system to whatever cabling system will be in place in the future.

Single/Dual Cable   Vendors now cause us to pose the next question. Will we use a *single* broadband cable, which dictates a frequency division and thus limits the number of channels? Or should we use a *dual cable* approach, which allows full use of the frequency spectrum but may cost more because of the dual cable and installation requirements?

Early networks selected the dual cable approach (Figure 7-8) because bidirectional amplifiers were difficult to find. In the dual cable approach, single direc-

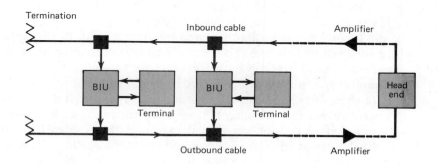

BIU: Bus Interface Unit

FIGURE **7-8**   Two-cable (dual) configuration.

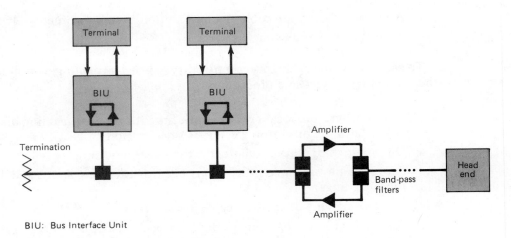

BIU: Bus Interface Unit

FIGURE 7-9 Single cable configuration.

tional amplifiers bring the incoming traffic up to some head-end point which also amplifies the signal, turns it around, and places it on the other cable. Outgoing traffic uses unidirectional amplifiers heading out.

A dual cable configuration with standard cable television (CATV) components allows about 50 usable channels on the network, which should cover all data or analog signals. The 50-channel calculation is derived very simply by taking the normal 300 megahertz band and dividing it by 6 megahertz, assuming the use of standard television channels. Each 6 megahertz channel should be good for a data rate of up to 5 megabits, so we can have 50 channels—each of which is equivalent to a 5-megabit baseband channel on the dual cable network.

If a single cable approach (Figure 7-9) is used, we have to replace the signal path of the second cable with an isolated frequency band. In this approach, outgoing (transmitted) signals use the bottom half of the frequency band, and incoming (received) signals the top half. For example, outgoing signals are from 0 to 150 megahertz and incoming signals are from 150 to 300 megahertz. This means that one of the cables has been eliminated, but 25 potential data channels have been given away. Because our channels are not limited, however, we can afford to lose this number of channels and still maintain the needed level of service. This allows the use of smaller conduits and simpler installations, and it takes advantage of the fact that the television industry is gravitating toward a two-way single cable approach. Thus, we can buy standard components, which should have the lowest cost.

The single cable configuration does not look all that different from a dual cable configuration. The exception is that the BIU (Bus Interface Unit) must have

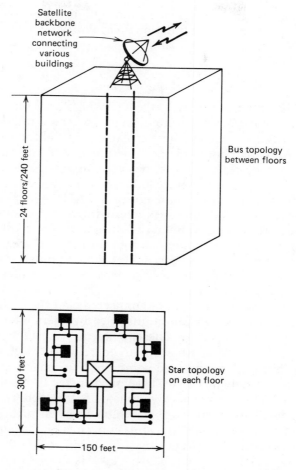

FIGURE **7-10** Sample of a cable layout in a building.

bidirectional components and the cable amplifiers must be bidirectional with bandpass filters. This adds some loss and design complexity to avoid degradation in signal-to-noise ratio.

What, then, will we use—single cable or dual cable? The answer lies in what the vendors supply. Because this characteristic is not critical, it generally does not affect our network selection criteria. The building layout illustrated in Figure 7-10 shows that there is not very much cable in a network design for a high rise facility. The cost of the cable for either configuration is low compared with the cost of the equipment that is placed on the cable, so this presently is not an issue. Complexity also is becoming a nonissue because the CATV manufacturers have solved the single cable problem and equipment is available off the shelf. The key parameter, then, is the functional capability of vendor offerings.

LAN COSTS

Over the last few years, low cost LANs based on communications using the microcomputer's serial I/O port have increased in both number and sophistication. Thus, they have become an extremely viable option for business. These low cost LANs offer most of the functionality of the higher priced networks, but typically they feature a cost per node (microcomputer workstation) of $100 to $250. This price does not include wiring a building. This is about one-quarter to one-fifth the cost of the more powerful LANs. According to manufacturers of low cost RS232 LANs (sometimes referred to as *sub-LANs* or *zero slot LANs*), these inexpensive networks provide a solution to almost all of the business challenges that induce microcomputer users to turn to local area networks in the first place. For example, these low cost LANs provide functions such as sharing printers, plotters, data, and disk storage devices, as well as micro-to-micro file transfer and print spooling.

Some of these LANs can be connected over the building's existing twisted pair telephone wiring and can be installed by users who do not have sophisticated tools or wiring expertise. These inexpensive systems also have capabilities such as a total network length of approximately 1,000 feet, operating at approximately 50,000 bits per second, and some offer automatic error correction (XMODEM protocol) on transmission. These inexpensive LANs use a standard RS232C connector cable with inexpensive twisted pair telephone wires.

The more expensive LANs, which have far greater capabilities, such as 10 million bits per second transmission rates and several miles of cable distance before requiring repeaters, cost from $250 to $1,000 per node (microcomputer workstation). The more expensive LAN is the best option if you want to install it throughout a building, but if you want to install a small two- to ten-node LAN within a single department (one or two rooms), you should seriously consider one of the inexpensive serial port LANs that use twisted pair wires, RS232 connector plugs, and the COM1: or COM2: port of the microcomputers.

When users are asked how much it costs to install a LAN, they often answer $1,000 per connection. While this figure may not be accurate for the small inexpensive serial port LANs, it probably is for larger LANs and the cost may be even higher. This is because there are some hidden costs. For example, the cabling in a large multifloor building can cost from $400 to $2,500 per microcomputer workstation. In addition, when you get a local area network, you begin needing the services of either a part-time or a full-time microcomputer local area network "guru." This is the person who makes sure the network keeps running, teaches people how to use the network, and does all the necessary chores such as updating software, performing various programming tasks, fixing broken cables, and moving microcomputer workstations.

If the LAN is very large, a management system is needed to keep track of the network, error rates on the network, the number of collisions among messages, and all the management tasks that must be performed on large-scale backbone networks. The costs jump even higher when you have a bridge/gateway or protocol converter to another network, such as to public databases or your organization's mainframe centrally controlled backbone network. The cost is higher because users transfer data files (rather than single transactions) between databases, and people must learn new work habits. User training is another hidden cost because, for every hour of user training, there is one hour that an employee is unable to perform his or her normal daily business tasks. Finally, if the LAN has a file server, someone has to oversee security, optimize the disk storage (reorganize disk files), and ensure proper backup and recovery routines at the file server disk drive.

LAN buying decisions can be complicated; but, the following considerations may be helpful.

- Do all users need their own printer or can they share one printer?
- Is a disk server adequate, even though it lacks security, or is a file server needed, which means a greater LAN hardware/software cost?
- What type of workstations will be connected to this network? They may be microcomputers, dumb terminals, executive workstations, minicomputers, production typewriters, and so on.
- Will a local departmental database be in place, with its accompanying cost and extra management requirements?
- Will several operators need to use the same application program?
- What type of cabling will be needed? Twisted pair wire, coaxial, or fiber optic?
- How extensive will the LAN be? Are we designing a LAN that might have only ten nodes, or is there a probability that it could grow to hundreds and be bridged to another LAN?

IMPLEMENTING A LAN

Buying a local area network usually is easier than getting it to work. Before you buy, recognize that the LAN implementation process encompasses cabling the building, installing hardware and software, testing the LAN, training the users, establishing network security, managing the system, and supervising the ongoing maintenance of the LAN.

We already have discussed cabling, but during the LAN implementation you must either install the cabling yourself or contract for someone to pull the wires through your building and extend them to each area where a workstation is to be located. Do not be surprised if the installation of the cabling costs far more than the purchase of the cable itself.

Hardware installation means that you install each microcomputer or other workstation, along with any centralized items such as a shared printer or shared file server. The hardware is not necessarily a cost of the LAN, although it probably is the largest cost of the entire network.

Software installation means that you install the LAN software and the microcomputer operating system onto the centralized disk/file/network server (if you have one) and onto the microcomputers. You must ensure that the proper printer commands can be passed over to the shared printer, along with the proper printer spooling software so that the file server can queue up numerous print requests from its many users.

By the end of software installation, you should have the cabling in place, the workstations connected, and the software up and running. At that point, testing of the system can begin. Testing is the process that verifies all the LAN features are working as they should.

Training begins simultaneously with the testing of the system. Users must be trained in how to use the new local area network software commands so they can perform such functions as send their files to a shared printer and transfer or modify files in a shared file server environment. It is during this training that the various security restrictions on each user are established. Hopefully, the LAN software package has volume-locking, file-locking, and record-locking capabilities.

The next two items pertain to the "guru" who is responsible for managing and maintaining the network. If it is a substantial LAN, then some type of software package or hardware monitor probably should be purchased for network management. This package helps inventory all of the items in the network, as well as the operating aspects of the network such as how many collisions occur between messages and the transmission error rates.

The amount of energy and money devoted to maintenance depends on the complexity of the LAN you have chosen. A part-time network supervisor probably can manage a small network used primarily for word processing and offering shared printers and disk servers. On the other hand, a large network with, let us say, 100 users dispersed throughout a multistory building requires the full-time effort of a LAN manager with top-notch technical and programming skills. Large sophisticated LANs require special monitoring hardware that may cost as much as a network workstation. Large corporations may even have a small group of

LAN management and maintenance personnel who install, manage, interconnect, and maintain all of the individual LANs throughout the corporation.

MANAGING A LAN

Local area networks are being used in many organizations as a "backbone" for communications among many communicating devices, such as terminals, microcomputers, shared printers, file servers, and office automation devices. As a result, LANs quickly are becoming critical networks within the organization. For this reason network management tools are important for the orderly growth and control of these local area networks. Network management encompasses a wide range of features necessary to plan, operate, control access to, and maintain a local area network. (Chapter 8 covers network management in greater detail.) Obviously, a small LAN with three or four microcomputers does not require extensive network management. Larger LANs with 50 to 200 workstation nodes, however, require a full-time network manager with the appropriate network management hardware and testing equipment.

A manager is required for straightforward activities, such as adding new users to a network, removing user workstation devices from a network, reassigning security levels, and installing new software. This manager should be responsible for incorporating new technology into the network and "fixing" the network should a catastrophic problem occur. Again, a small four- or five-node LAN does not require much management, but it seems that today's small LANs have a propensity for growing quickly into very large networks. These rapidly enlarged networks often become critical to the organization because of the organization's dependence on communication among the network's users and/or departments.

For network security, constant monitoring is necessary; it must be possible to keep track of which nodes currently are using the network and when they last logged on to the LAN. This can entail issuing individual passwords. Further, the LAN software must operate properly with regard to volume locking, file locking, and record locking.

Hardware/Software Tools *Network management* involves use of the appropriate tools for monitoring network activity, network loading, network errors, network changes, and network security. The network manager probably has a dedicated microcomputer, appropriate software, and possibly specialized hardware used for the LAN management and analysis system. This system shows on the microcomputer screen the various facts and figures necessary to manage and control the overall LAN.

Network activity displays the number of packets on a network for the current second, minute, hour, or day. It also might show the elapsed time since a token loss was detected on a token ring or, if there is a collision detect system, the number of collisions within a certain period of time. Cumulative network downtime and network demand levels can be monitored.

Network loading displays the average load on a network and provides information on the busiest times of the day or hour. This information on loading can help reconfigure the network as it grows.

The *network error* display provides statistics necessary to identify a deteriorating network operation. It identifies and displays bad format packets, missing control characters, number of messages retransmitted because of error, and so forth.

Changes to the network show network additions or deletions and keep track of the overall network configuration with regard to its users and the pieces of hardware attached to the network.

Network security includes restricting access to the LAN, backup/recovery, physical/data security, and volume, file, and record locking. Chapter 9 is devoted to the security and control of networks.

LAN Analyzer Software Organizations that use local area networks find that managing them can be quite a challenge. Network managers need to know the answers to such questions as

- What are the network loads at any particular time?
- What channels are the most active?
- What types of errors are occurring?
- What will happen to the network when more terminals are added?

Several LAN management software packages are available. We wish to present here one example of a local area network management and control package. The LANalyzer™ EX5000E software package can monitor or capture up to 1,000 packets per second, and it is independent of the protocol being utilized. It graphically displays statistics in real time, generates data traffic for testing and maintenance, searches for and captures packets according to user-defined criteria, time-stamps data, and the like.

This software is a second generation network analyzer system that enables network users to monitor network activity, troubleshoot problems, debug protocol and application software, and fine-tune the performance of their local area network. Figure 7-11 shows the LANalyzer circuit board, software, and a COMPAQ portable 286 microcomputer. This network management package runs on a microcomputer and works on any local area network that adheres to the IEEE 802.3 Ethernet standard. Figure 7-12 is a LANalyzer monitor screen. The

FIGURE **7-11** LANalyzer™ network analyzer.

bulleted items list some of the real-time statistics that are available with LANalyzer.

Instructors who use this text may obtain a free copy of the DEMO DISK for the LANalyzer from John Wiley and Sons, or you can call EXCELAN (San Jose, California) at (408) 434-2300. Appendix 4 describes how to use the LANalyzer™ demonstration diskette.

SELECTING A LAN

Selecting a LAN for your organization is no simple task. Today's LAN market-place is crowded with at least 50 different vendors, all claiming their product is the one you need. Further confusion abounds because there are few fixed standards, de facto or otherwise. To complicate the situation, you may be sure that the simple small LAN you start with today will turn into a large multifloor, multibuilding, or multifacility network at some time in the future! To choose the proper LAN, you should answer some basic questions, such as how many users there will be, how much data will need to be stored, how easy it will be to add workstations, what cabling is needed, whose software should be selected, and how much security is needed. Over 20 *key issues* for selecting local area networks are examined in the following discussion.

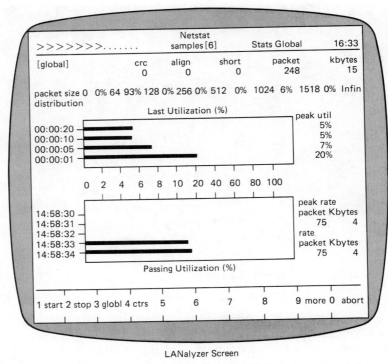

LANalyzer Screen

- Network and channel utilization
- Packet counts
- Error counts
- Packet rates
- Packet size distribution
- Total packets captured
- Counts/distribution of packets transmitted with and without collissions

FIGURE **7-12** Real-time statistics for network analysis.

One of the first issues that must be addressed is the *configuration* of the network. Is it best to have a star, bus, or ring network? This issue of topology is influenced greatly by the decisions as to whose cable scheme is chosen and which topology is supported by which vendor. While a star network may fit in quite well with sophisticated file servers and digital PBXs (switchboards), a bus topology may work equally well when wiring a large building. Ring topologies also work well, especially when it is a one-floor installation, although rings can be wired up and down throughout a large building. Remember, DECconnect uses a bus topology vertically (between floors) and a star topology on each floor.

In the *cabling* decision, you need to determine whether the existing telephone wiring (twisted wire pair), coaxial cable (93-ohm for IBM and 75-ohm for CATV television cable), or fiber optics will be used. The cost of pulling the cables through the building may be greater than the cost of the cable itself.

The maximum *number of nodes* (microcomputer workstations) probably will be an issue in the future, and it is one you must plan for now. Today you may be developing a local area network with only five or ten nodes, but will the LAN you are selecting now be able to grow to the size that may be required five years in the future?

Along with the number of nodes, the practical *number of users* must be considered. In other words, how many people might use this local area network? Remember, you may have multiple users at a single node. Even though the number of nodes may not increase, as the number of users increases there may be an overload at the file server. This overload might mean that the LAN software and operating system are unable to handle user requirements and specific business needs such as database retrievals, electronic mail, and file transfer. Remember that multiple users at a single node can overload a LAN.

The *protocol* is an important issue because it is here that you must decide whether you are going to use the token passing approach or the carrier sense multiple access approach. Associated with the protocol decision is the decision on whether the system will be *baseband* (one message at a time) or *broadband* (multiple data transmission channels). These two considerations are critical when related to future growth because when you choose a protocol and baseband/broadband, you determine the LAN's future growth as to whether voice, data, and video transmissions can be added to it.

Another critical decision is the *server*. You must decide whether you are going to use a disk server, a file server, or a network server.

- *Disk servers* divide the disk storage capability into fixed-size disk volumes. Each user is given one or more private volumes on which to store data. Because size is fixed and many volumes are required, it becomes a difficult administrative problem to add new users, manage existing users, and allocate disk space efficiently. In addition, there is a grave security risk because these files are only designated as public and private. Public means other users may be able to read or write on someone's volume. As a result, users may lose data inadvertently, experience breaches of privacy, or have data modified improperly.

- With *file servers*, a single file volume can be shared by many users and can be any size, limited only by the size of the disk storage itself. Each user can have one or more files. There is no administrative overhead for the movement of users and files. Moreover, security is addressed because the LAN software controls unique security locks such as volume locks, file locks, and record locks.

- A *network server* is a separate computer (microprocessor-based) that provides special network application services for all the attached microcomputers. These services can include a shared file system, shared printers, and an electronic mail system. Network servers extend the facilities offered by file servers, local microcomputer applications, and operating systems. For example, a network server can provide a large-capacity file system extension to MS-DOS or PC-DOS that allows microcomputer users to share files more easily and have larger file storage areas available for their use. Network servers also can contain the necessary software to keep track of network resources, handle all the details of communications, and have network management software available.

Another consideration when selecting a LAN is the range or *distance between nodes.* On this topic you will have to check with each vendor, but you will find figures such as 100 to 1,000 feet per segment between each node, 3,000 feet from the network/file server to the microcomputer node, or 7,000 feet end-to-end total network length. Remember, in today's world this is not an absolute limitation because you always can add amplifiers to reamplify the signal and extend the medium for several thousand more feet.

The *workstation hardware* is a consideration. You already may have many of the workstations that must be connected to the local area network. In this case, you must determine whether each piece of hardware that your organization already owns and/or intends to connect to the local area network will function properly. Remember, sometimes one vendor's LAN software or hardware (network server) may not handle another vendor's hardware or software protocols.

Another consideration with the hardware is *whether the required circuit boards can plug into the equipment you already own.* Are there enough slots? Do you need full- or half-length slots? The serial port probably will not cause any problem because you always can use a RS232C connector, but remember that PC/AT or PS/2 microcomputers may have a 9-pin plug instead of the RS232. (See Figure 6-3 for a 9-pin connector plug.) If you intend to use the parallel port for connecting to the LAN, however, make sure that the proper connector plugs and cables are available.

The *software* is another critical consideration. You must be sure that the microcomputer operating system, the network operating system in the network server (if you use a network server), and the LAN's vendor-supplied software package are all compatible. They all must be able to work together and interface with the various nodes (microcomputer workstations or whatever hardware is connected to the LAN), as well as interface with your organization's application programs and host mainframe.

Do not overlook the area of *bridges* and *gateways.* You must determine whether the LAN is going to interface with another corporate network, the organization's host mainframe, and/or some of the public dial-up networks outside of the organization. The question to ask is, What gateways are available to connect our network with the corporate host mainframe or other LANs within the organization?

A number of miscellaneous features also need to be addressed.

- *Security features* must be installed to protect each user's data. Remember, each user may have data of differing sensitivity or risk values. Another security consideration is whether users should have microcomputers without any disk storage devices attached to them.

- The various *print features* must be available at the shared printer. Will some users need to buy special purpose printers?

- The proper *backup and recovery* procedures must be taken into account. Certainly the server disk must be backed up, and recovery features for each user's disk files must be provided.

- Both *electronic mail* and *voice mail* are big features on a local area network, although you may opt not to put voice transmission on the LAN.

- A decision must be made as to whether some users will be allowed to have their own *local disk files,* as well as access to the disk files on the network server, file server, or host mainframe.

- Ensure that there is proper written *documentation* covering all of the hardware, software, nodes, users, and so forth.

- Determine whether *backup electrical power* or *surge/sag protection* will be required on both the utility power lines and any connecting telephone lines. What is being done to protect against *static electricity*?

- *Training* must be considered. You must examine both the cost of the training and the time required by each user to attend the various training sessions. Another consideration is who will perform the training.

- You must take into account the overall *costs* of hardware, software, cabling (including installation), possibly a full-time network manager, and training.

By using the above checklist, you may rest assured that you will not overlook any key issue when planning for a local area network or when making modifications to your current one.

To convince management that this advanced communication system should be added to the repertoire of equipment and services, we have to prove that it is cost effective. This is necessary because corporations are in business to make a profit; unprofitable firms are in business for only a short time.

New communication networks, then, either have to save money directly or have to provide a means for making more money faster. In the case of a LAN, it can do either or both, depending on current corporate communication capabilities.

One way a LAN helps in corporate economics is in the centralization of communications management. Right now we may have a data network and a voice network, with each one administered by a separate department. Generally, there is little if any contact between departments regarding networks because the networks do not tend to interact. With an integrated corporate network viewpoint made possible by a LAN, we can centralize the management of these resources, operate them more efficiently, and do it with fewer people.

A second economic benefit is gained in sharing resources. Again, many corporate networks are separated functionally with terminals tied to specific jobs. Many organizations also have standalone word processors with built-in memory and microprocessor controls. With an integrated local area network, a host computer can be accessed by any terminal or workstation. Its computing capability can provide text editing, calculations, management information, or data storage and retrieval for everyone in the organization, thereby combining all the independent functions into a single integrated activity. This simplifies control and data access and provides a single resource to meet many needs.

A third economic benefit is that our voice network (telephones) can be incorporated into our local area network.

SELECTED REFERENCES

1. Bridges, Stephen P. M. *Low Cost Local Area Networks*. New York: John Wiley & Sons, 1986.

2. *Glossary of LAN Terms*. New York: LAN Magazine, 1986.

3. Hammond, Joseph L., and Peter O'Reilly. *Performance Analysis of Local Computer Networks*. Reading, Mass.: Addison-Wesley Publishing Co., 1986.

4. Hopper, Andrew, Steven Temple, and Robin Williamson. *Local Area Network Design*. Reading, Mass.: Addison-Wesley Publishing Co., 1986.

5. *LAN: The Local Area Network Magazine*. Published monthly by Flat Iron Press, 12 West 21st Street, New York, N.Y. 10010, 1986– .

6. Lehrman, Stevanne Ruth. *Local Area Networking with Microcomputers: A Guide for the Business Decision-Maker*. West Nyack, N.Y.: Brady Computer Books, 1986.

7. Mayne, Alan J. *Linked Local Area Networks*, 2nd ed. New York: John Wiley & Sons, 1986.

8. McNamara, John E. *Local Area Networks: An Introduction to LAN Technology.* New York: Telecom Library, 1985.

9. *Smart LAN Performance Test.* A software package available for $50 from Innovative Software, Attn: LAN Test, 9875 Widmer Road, Lenexa, Kans. 66215, (800) 331-1763.

QUESTIONS/PROBLEMS

1. Look at Figure 7-4. Determine how many messages per second the local network can handle if it is designed for 100 percent utilization. (Technically, this is impossible because there would be too many collisions.)

2. What other two methodologies compete with LANs?

3. Define topology.

4. What might be an efficient combination of topologies for a LAN in a multistory building?

5. What is Ethernet?

6. _____ uses direct electrical voltages, and _____ uses a modulated signal.

7. How many transmission channels are on a broadband LAN?

8. Name two popular LAN protocols.

9. How many messages per second can your LAN handle if it has a speed of 2 million bits per second, 70-character messages using USASCII code, no start-stop bits, and utilization is 25 percent?

10. How large is the "bit pad field" for an IEEE 802.3 standard packet if the message is 46 characters using USASCII?

11. What is walk time?

12. What is the basic difference between a bridge and a gateway?

13. What is the international standard gateway that links X.25 networks?

14. What are the most popular cable types for LANs?

15. What is a BALUN? Also see Balanced and Unbalanced in the Glossary.

16. Why are wire pairs twisted as they run through the building's walls?

17. What is data grade twisted pair wiring?

18. Run the EXCELAN software demonstration package.

19. Discuss which of the more than 20 key issues for selecting a LAN would be the most important at your college/university or company.

PART TWO

NETWORK MANAGEMENT AND SECURITY

Part two of this book is devoted to the human and organizational aspects of managing networks as well as security requirements and the controls that are necessary. There are chapters on . . .

- **NETWORK MANAGEMENT**
- **SECURITY AND CONTROL**

8

NETWORK MANAGEMENT

This chapter discusses the basic management skills required to be a successful network manager. It also describes departmental functions, how to manage the department, required reports, error testing, and test equipment. Network control and troubleshooting, combining voice and data communications, and the Chief Information Officer also are discussed.

THE DATA COMMUNICATION FUNCTION

In both government and private business, one of the major growth sectors has been the internal "service function." The service functions have grown rapidly, and their management techniques need to be strengthened. These service functions include staff assistance, research departments, planning groups, coordinators, data processing departments, data comunication networks, and the like. Such functions are organized to support manufacturing, sales, or the specific product for which the organization was conceived.

The data communication function has the primary responsibility of moving and conveying data/information. This transfer of information takes place within or between departments of an organization. Remember that *data* are nothing more than meaningless characters, whereas *information* takes these meaningless characters and assembles them into a fact or idea that can be used for decision making by managers. Information presupposes adequate communication because information is useless if it is not available when needed.

The manager of a data communication function should always remember that data or information transmitted over any network must CATER to the overall needs of its users. CATER is an acronym that stands for *c*onsistent, *a*ccurate, *t*imely, *e*conomically feasible, and *r*elevant. While the manager of the data communication function may not have direct responsibility for *consistency* or *relevancy* (those are the responsibility of the data or information owner/gatherer/developer), the data communication manager is responsible for ensuring *accuracy, timeliness,* and *economic feasibility.*

Information activities present a special organizational problem because they have to be both centralized and decentralized. The original developers or gatherers of data may be either centralized or decentralized, but the total organization is served best if the data communication function is centralized. This may seem to be a dichotomy in today's world because we are moving so quickly toward distributed data processing and distributed databases. Nevertheless, the data communication function should be centralized. It can be likened to the nervous system of the human body because it controls the paths over which all control messages and data/information flow. When viewed in this manner, it is obvious why centralized control is necessary to interconnect all the various terminals, CPUs, databases, and so on.

The individual manager who is responsible for the data communication function must be adept at performing the five key management tasks of *planning, organizing, directing, controlling,* and *staffing.* These functions require the following expertise.

- Planning activities require . . .
  Forecasting
  Establishing objectives
  Scheduling
  Budgeting
  Allocating resources
  Developing policies
- Organizing activities require . . .

  Developing organizational structure
  Delegating
  Establishing relationships
  Establishing procedures
- Directing activities require . . .
  Initiating activities
  Decision-making
  Communicating
  Motivating
- Controlling activities require . . .
  Establishing performance standards
  Measuring performance
  Evaluating performance
  Correcting performance
- Staffing activities require . . .
  Interviewing people
  Selecting people
  Developing people

Because the information-based society is dominated by computers and communications, a manager's value to the organization is increased not only by knowledge but also by the *speed* at which that knowledge moves. Today's information economy is vitally concerned with the movement of voice conversations, data/information movement, and image (video/graphics) transmissions. For this reason, today's manager is concerned with the overall telecommunication function, even though it may be referred to as data communications. The point is that effective corporate or government communications managers must be aware of voice transmissions, data transmissions, and image transmissions; the information systems manager can no longer be concerned solely with data transmissions.

NETWORK ORGANIZATION

Network organization, as discussed here, focuses upon the management and organization of the *people* running the network rather than upon the physical organization of the network communication circuits. Management must define a central control philosophy with regard to the overall network functions. This means that there is a single control source for all emergency problems, testing, and future planning. Details on how to implement this policy are discussed in the next section, Network Management.

The data communication network organization should have a written charter that defines its mandate, operational philosophy, and long-range goals. These goals must conform both to the parent organization's information processing goals and to its organizational goals. Along with its long-term policies, the organization must develop individual procedures with which to implement the policies. These policies and procedures provide the structure that guides the day-to-day job tasks of people working in the data communication function.

The ultimate objective of the data communication function is to move data from one location to another in a timely fashion, and to provide and make available the resources that allow this transfer of data. All too often this major objective is sacrificed to the immediacy of problems generated by factors thought to be outside the control of management. Such factors might be problems caused by unexpected circuit failures, pressure from end users to meet critical schedules, unavailability of certain equipment/circuits, or insufficient information (on a day-to-day basis) to ensure that the network provides adequate service to all users. In reality, network managers must develop their own decision-making information in order to perform such essential tasks as measuring network performance, identifying problem areas, isolating the exact nature of problems, restoring the network (how to do this is discussed in the next section), and predicting future problems.

Too many managers spend too much time on the management function of *control* because they must contend daily with a series of breakdowns and immediate problems. These managers do not spend enough time on the management functions of *planning* and *organizing* which are needed to develop a proper information base so they can foresee problems and reduce the need to drop everything to fix a breakdown (sometimes called firefighting).

Combining Voice and Data A major organizational challenge is the prospect of combining the older voice communication function with the somewhat newer data and image communication functions. Traditionally, voice communications were handled by a manager who oversaw the telephone switchboard systems and also coordinated the installation and maintenance of the organization's voice telephone networks. By contrast, data communications have been handled by the data processing function (information systems department) because the staff installed their own communication circuits as the need arose, rather than contacting and coordinating with the voice communications management staff.

While this separation of voice and data has worked extremely well over the years, changing communication technologies are causing enormous pressures to combine these two functions. These pressures are magnified by the high cost of maintaining separate facilities, the lower efficiency/productivity of the organization's employees because there are two separate network functions, and the potential political problems within an organization when neither manager wants to relinquish his or her functional duties or job position. A key factor in voice/data integration might turn out to be the elimination of one key management position and the merging of two staffs into one.

We cannot present a perfect solution to this problem because it must be handled in a unique way within each organization. Depending on the business environment and specific communication needs, some organizations may want to combine these two functions and others may find it better to keep them separate. We can state unequivocally that an organization that avoids studying this situation might be promoting inefficient communication systems, lower employee productivity (especially in an automated office environment), and increased operating costs for its voice and/or data networks.

The preceding statement may seem exaggerated, but it is important because we predict that by 1990 the total cost of both voice and data communications will equal or exceed the total cost of the data processing function. Sometimes this cost factor is overlooked, ignored, or underestimated. Typically, voice communications can require 10 times the budget needed for data communications. For example, an organization with a $1 million annual budget in data communication costs might find itself spending $10 million per year in voice communication equipment and transmission, and this may not count lost work time of employees that is caused by inefficient use of a voice telephone system. In communications we are moving from an era in which the computer is the dominant information systems function to one in which communication networks are the dominant information systems function.

The integration of voice and data combines factors in voice technology, the office equipment market, the workstation market, and the network market. As a result, the manager who controls voice and data operations must be knowledgeable about workstations, office equipment, and network devices so they can be interconnected efficiently. In addition, this person needs to have an in-depth understanding of voice communications and especially digital PBX switchboards.

If you were to become the manager of a combined voice/data communication organization, some of your responsibilities and tasks would be to

- Plan, organize, direct, control, and staff the entire voice/data network operation
- Acquire knowledge of public data networks and how to connect them to the organization's networks
- Learn about the workstation and office equipment markets

- Develop and control the organization's backbone networks, including satellite, microwave, and other bypass (DTS) technologies
- Be responsible for micro-to-mainframe network connections
- Manage the organization's local area networks, as well as connect them to the organization's hybrid backbone network
- Be responsible for PBXs (switchboards)
- Plan, understand, and keep abreast of the latest technological developments in telecommunications
- Manage the day-to-day operations of all network functions
- Acquire communication-oriented hardware and software
- Manage the communication budget, with emphasis on controlling costs
- Develop a strategic (long-term) communication plan to meet the organization's policies and goals

You may ask, How does an organization decide whether to integrate its differing types of communications? One way is to implement the following nine-step plan to help in making this decision.

1. *Survey the existing flow of information throughout the organization.* Create an *inventory* of this current information flow. This inventory should include both data movement and voice movement, along with all the associated hardware and software. Do not overlook the people who are involved and their functional responsibilities.

2. *Compile a complete picture of the communication costs for both voice and data communications.* Do not forget to include both capital and expense items. This means putting a dollar cost on the inventory compiled in item 1, including the value of the information, people costs, and any hardware or software costs.

3. *Develop an information movement five-year plan.* This plan is the *needs analysis* for both voice and data communications. It begins with today and projects at least five years into the future. Major items that should be addressed are the organizational structure of the voice and data groups, whether they should be combined, how hardware and software will be affected by technology over the next five years, and what to do with the current people in the voice and data organizations.

4. *Design the required detail plan that is needed to move from today to five years in the future.* Make this functional plan consistent with your organization's needs, long-term goals, the direction it is taking or may take over the next five years, and the individual objectives of the people who work in

both voice and data communications. This really is the organizational structure and network design required for the organization to achieve its five-year plan goals.

5. *Create a time and cost plan to implement the detail plan outlined in item 4.* Make sure that you account for both the overall calendar time (five years) and the chargeable time (person hours required to implement the plan). This should be related to the hardware, software, and people costs that are required to manage and operate the voice/data network. These time frames and costs must be consistent with the organization's needs and realistic regarding costs the organization can support.

6. *Install a central control and administrative system for information movement.* This involves two separate tasks. First, develop the management capability to oversee and ensure that the five-year plan is implemented. Second, develop a network control center that is responsible for monitoring, switching, and managing both voice and data communications during implementation of the plan and through the life of the new organization-wide information system.

7. *Establish a business plan to achieve your goals.* This plan should be a detailed written description of why, how, and when the organization will achieve a stated level of profitability, accountability, and cost effectiveness. With this plan you are helping future management understand and concur with the evaluation criteria that should be used to judge whether the five-year plan implementation was successful.

8. *Implement the plan.* The decision now has to be made on whether to maintain separate voice and data functions or have one combined function. After that, the task of implementing the five-year plan involves building any new systems, testing them, installing them, and managing them as permanent arrangements.

9. *Document the effort as changes occur.* Train the staff to understand current functions and future directions, as well as how to document "where we came from" and "where we are today." Manage the new system as a business rather than as a day-to-day activity. Although there always will be some unforeseen problems, both the manager and staff should plan ahead to foresee and act on problems before they become critical. The need to spend inordinate amounts of time "firefighting" can be avoided by forethought and planning.

Merging voice and data promises many potential benefits, with some of the more important ones being lower costs, a competitive edge in meeting the organization's business needs, better strategic planning to anticipate the long- and short-term telecommunication needs of the organization's end users, improved use of new communication technology, and better cross-training between voice and data professionals.

Cross-training is very important in today's deregulated communication market. In the past, telephone companies took care of most of the organization's circuit, equipment, and planning needs. Today, deregulation has forced both voice and data communication users to deal with a variety of vendors. As a result, sometimes it is almost impossible to pinpoint clearly where a problem lies or who is responsible for it. Accordingly, a modern telecommunication function must have a staff that can cope with the wide range of available equipment and at the same time understand the vast cost implications.

As communication costs increase between now and 1990, more consideration should be given to the idea of having a communication department that is totally separate, not only from data processing but from all other departments as well. As microcomputers proliferate throughout the organization, the major centralized thread of continuity will be combined voice and data communication networks. It is conceivable that in the next few years 10 to 25 percent of an organization's employees might work at home several days a week instead of at the office. In such a situation, the organization relinquishes some control over its microcomputers, but it *must* maintain control over the communication functions.

The Chief Information Officer   In our previous discussion on the possibility of combining voice and data communications, we alluded to the problem of where the communication organization should be located within the business or government entity. This subject raises two other issues. Will the surviving organization be data communications or voice communications? To which higher organization will the communication manager report? This person could report to information systems (data processing), an administrative vice president, or to some other function.

While any of these reporting relationships would work, we believe that the information control function will be placed higher in the hierarchy of both private companies and government agencies as these oganizations recognize that one of their strategic resources is the movement and control of information. In today's business environment we have accounting departments and senior vice presidents of finance to control the strategic resource, "capital." It will be only a matter of time until we have a senior vice president and a single functioning department to control the organization's newest strategic resource, "information."

Because information is such an important strategic resource to organizations, a new executive function is beginning to surface in the business world. This function, the Chief Information Officer, probably will be a "czar" who oversees all information within the organization. Figure 8-1 depicts a tentative organizational

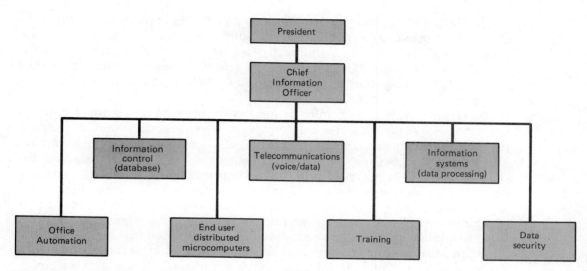

FIGURE **8-1**   Chief information officer organization chart.

structure in which we have combined the voice and data functions. Notice that we have included the new vice-presidential-level position called Chief Information Officer. Also notice that, within this organizational structure, voice and data communications do *not* have to be combined. The Chief Information Officer could have both a Manager of Voice Communications and a separate Manager of Data Communications in the box marked Telecommunications. Also observe in the figure that the traditional data processing department has been broken into three major categories.

- *Telecommunications* (voice and data) is responsible for ensuring that voice, data, and image transmission networks are available and optimized for all organizational users.
- *Information Systems* is responsible for running production jobs and operating the organization's centralized computers, as well as system design and programming.
- *Information Control* (database) is responsible for identifying and controlling who can have use of which data, as well as being the custodian over all databases (data/information) within the organization.

The other four organizational boxes in Figure 8-1 depict the complete scope of management for the Chief Information Officer.

Chief Information Officers would perform three basic functions. *First*, they would oversee all the company's communication technology, including the traditional data processing functions, office automation systems, and telecommunications. *Second,* they would report directly to a high ranking executive such as the Chief Executive Officer, Chairman, or President of the organization. *Third,* they would concentrate on long-term strategy and strategic planning and leave day-to-day operations to subordinates.

The creation of the Chief Information Officer position is the direct result of new technology and of the rise of information as a strategic resource. In the past, the traditional data processing department could handle the entire organization's needs for data processing, but this is no longer true. In many organizations the strategic resource *information* now is more important than the strategic resource *capital.* This shift dictates a realignment of responsibilities, the single most important one being in the area of voice and data communications. Having data or information that cannot be communicated or that is communicated too late is equal to not having the information at all. Most of the time managers are drowning in information; they have too much. The problem then becomes one of how to organize the information (database) and communicate it (voice/data communications) more effectively. Today we are an information-based society, and that factor alone will cause us to restructure totally the communication management functions.

Another major topic that concerns the Chief Information Officer is security of the network and, even more so, security of data/information. An outside organization (data security in Figure 8-1) should develop security standards and ensure that they are implemented and followed adequately. This organization should be separate from voice communications, data communications, and information systems. The internal audit function should review the security standards to ensure compliance. The next chapter covers the topic of security and control in greater detail.

NETWORK MANAGEMENT

In this section we address the day-to-day management of the organization's networks, regardless of whether there is a Chief Information Officer. Today's network managers are faced with a growing number of users, more microcomputer-based workstations, an increased use of applications, and tremendously heavy traffic on all the communication circuits.

Many network managers use the "old" rule of thumb that designates one person in the network department for every 50 to 100 terminals in the network. This yardstick probably is still adequate for today's networks because management would be assigning five to ten people to the day-to-day network management tasks needed for a 500-terminal network. The rule fails, however, with very large networks that have grown to, let us say, 5,000 terminals. In this case, the network manager cannot afford a corresponding increase in the number of staff to 50 to 100 people. This problem is not a serious one because both software and hardware network management systems are available.

A large company's nationwide network typically includes a host-based management system that monitors the mainframe computer and stations operating under its direct control, a modem management system designed to collect operational statistics from the network devices, and various local area network management systems that monitor the devices and traffic on LANs at different sites around the nation.

All of these systems, and any others that exist on the network, operate under their own proprietary protocols and often involve one-of-a-kind procedures to run them. Aside from putting a strain on network operators, this diversity impedes a company's attempts to monitor and control its entire network in a cohesive manner from a central command site.

The parameters monitored by network management systems fall into two distinct categories: physical network statistics and logical network information. The communication device suppliers typically monitor the *physical* network. That is, they monitor the operation of the modems and multiplexers and the lines linking them. Most of these vendors perform such monitoring via a "secondary" communication channel that transmits statistical information from the intelligent network devices to a central site. This monitoring uses a different frequency from the one carrying the actual network traffic.

Logical parameters traditionally have been addressed both by the computer vendors who have management schemes related to their proprietary network architectures, and by independent vendors who sell management systems that overlie on networks containing a mixture of different vendor products. Independent suppliers sell such products, which monitor the actual traffic flowing on the network rather than the diagnostic signals sent from modems and multiplexers over a secondary channel. In reality these are performance measurement systems, and they keep track of user response times, the destination of data routed across the network, and other indicators of the network's level of service.

Because the physical and logical types of management systems measure different parameters, they can coexist within the same network and perform their distinct functions alongside one another. Such an approach usually requires that network operators buy two sets of equipment and learn two different diagnostic procedures.

Not all of the available software for network managers can be covered in this section. Instead, selected descriptions are offered to provide some idea of the types of network packages available.

IBM hopes that its network management program *NetView* will become the industry standard for managing multivendor voice and data networks. This program combines and enhances the functions of five other IBM programs into a single software product that automates many network management tasks. It is designed for managing host network management services for Systems Network Architecture (SNA) networks. Prior network management programs that have been combined to form NetView include Network Communication Control Facility (NCCF), Network Logical Data Manager (NLDM), Network Problem Determination Application (NPDA), and some of the functions from Virtual Telecommunication Access Method Node Control Application (VNCA) and Network Management Productivity Facility (NMPF).

We mention these five NetView programs to demonstrate that numerous network management programs are on the market. Among the functions that NetView helps to automate are

- The ability of a centrally located person to test and monitor the status of analog communication circuits
- The ability to view a set of interactive displays and to execute commands against the displayed circuit/hardware in order to change something
- The online help facility that provides current dynamic network information status
- The help desk facility that isolates failed network components, provides suggestions on ways to fix the failure, and records incidents
- The ability to examine any network log and check the definitions to ensure that parameters are set properly
- The ability to monitor the network for out-of-service conditions and automatically reactivate all devices downstream from the failed resource

Another automated network management system is offered by one of the Bell Operating Companies (Pacific Bell) and is called *PACTEL Spectrum Services.* This communication service is one that a user might subcontract to use. It uses proprietary hardware and software to diagnose system problems remotely, coordinate service restoration, detect degrading communication circuits, and maintain all the relevant description information about a communication system, including system configurations. This service eliminates the user's need to track down the cause of a problem, deal with multiple communication service providers, determine responsibility for a problem, and correct the problem. Test engineers at the Pacific Bell response center test the user's system by means of remote hardware devices located at the user site. PACTEL Spectrum Services gathers complete information about each user's communication system, stores this information in a database, monitors the user's system, and handles the basic failure control, testing, and problem management functions for the user's network.

There are even specialized software/hardware management programs that manage communication switches. Switches are pieces of hardware that switch circuits or messages among different terminals. They might be used by a corporate network, packet networks, cellular telephone services, and your normal dial-up telephone service. For example, the AT&T 5ESS switch can be used for circuit switching. To manage these switches, the user must check a separate operations system that collects diagnostic messages from the switch to determine its health. This other system monitors telecommunication traffic moving through the switch to ascertain how well telephone calls are moving over the network.

To overcome the problems of coordinating some of these user tasks, AT&T has developed a *Multi-Function Operations System* (MFOS) package for the automated management, operation, administration, and maintenance of switches. This switch management control system uses small microcomputers to collect diagnostic messages for monitoring, and the MFOS software has a set of menus with English-like commands.

This MFOS management system has six modules. *Switch management* monitors the vital signs of switches to be sure they are operating properly. *Alarm processing* alerts the user's network control center to a switching malfunction. *Traffic data collection and analysis* tells users how much traffic is on the network and alerts them to abnormal conditions. *Trouble ticketing* tells the user what went wrong, where, why, who fixed it, and when. *Switch database administration* allows the user to change subscriber information stored in a switch's database, including telephone numbers and enhanced features such as call forwarding. *Network management* enables the user to control network operations, including routing traffic for more efficient use when a network is overloaded.

Another network management tool, called NETMAN, was developed by California Software. This product has the ability to provide invoice verification for acquired network products, configuration data on the network (which is helpful in conducting network inventories), billing for network usage, and change control for managing changes on the network. Appendix 4 presents another software product for local area network management.

To put the day-to-work network management functions in perspective, Figure 8-2 depicts a typical organization chart for a data communication function that is organized within the information systems or data processing departments. In this organization chart (contrary to the Chief Information Officer organization chart shown in Figure 8-1), the data communication network manager reports directly to the director of information systems (the highest ranking person in the Information Systems Department). Remember that this function may report ouside of the Information Systems Department if there is a Chief Information Officer, or it can include the tasks of managing the voice communications if they are a combined voice/data function reporting to the director of information systems. We will assume here that the network manager is part of the Information Systems Department and that voice and data are not combined. This assumption is made so

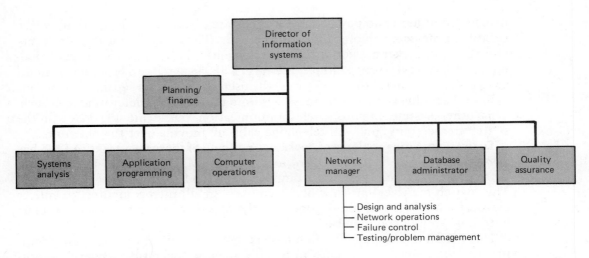

FIGURE **8-2** Organization chart for information systems.

that we can identify and explain the four basic job functions of a data communication network manager: design and analysis, network operations, failure control, and testing/problem management.

Each of the functions involves a set of specific tasks that require the utmost in management expertise and control of personnel, as well as in-depth technical knowledge. Let us examine these four functions.

Design and Analysis   This function involves planning overall network design and continual analysis of the network. Management requires, for example, ongoing statistics with regard to network performance and feedback with regard to user satisfaction. Obviously, these statistics require close interaction with the other three job functions.

This function should be responsible for developing operations procedures and standards for the network personnel. The network designers use measurement tools such as network models, simulators, statistical measurements, daily data collection routines, and other tools to manage the ongoing network. Network design should not be viewed as something that is done when the network is created and then forgotten. Network design and analysis is a continuing redesign of the hardware, circuits, and software of the data communication network. This redesign continues for the life of the network.

As the people in this function analyze and redesign their network circuits, switches, other hardware, and software products, they may use either manual or automated design techniques. Chapter 11 presents a thorough 13-step systems approach to designing networks. When redesigning a network, the network designers may need to use only a few of these steps.

Chapter 12 presents an automated approach to network design by using a network optimizer. A *network optimizer* is a system that allows its users to input various parameters such as circuits and nodes. By varying the parameters, the programs redesign a network and produce an optimal solution that is based on meeting service requirements at the most advantageous cost. The network optimizer presented in Chapter 12 can be run on a microcomputer.

Establishing equipment selection criteria is one of the responsibilities of the design and analysis function. When determining the vendor of specific data communication products, a data communication manager must have a list of criteria to be met. He or she must determine which criteria are important for the specific network. Some criteria that should be considered are

- Technical decisions
  Type of usage
    Short haul versus long haul
    Multipoint versus point to point
  Protocol requirements
    Analog versus digital
    Synchronous versus asynchronous
    Electrical protocol (RS232, V.35, and so forth)
    Compatibility with existing equipment
    Transmission speeds
- Vendor concerns
  Problem diagnosis (Mean Time To Diagnose—MTTD)
  Repair record: Mean Time to Repair (MTTR) = Mean Time to Respond (MTTR) + Mean Time to Fix (MTTF)
  Equipment reliability (Mean Time Between Failures—MTBF)
  Repair personnel
    Qualifications
    Location of nearest vendor technician (wait time)
  Documentation manuals
    Availability
    Usability
- In-house staff
  Qualifications
  Training requirements
  Problem diagnosis ability (Mean Time To Diagnose—MTTD)

- Cost decisions

  Lease versus purchase

  Industrial forecast (yesterday's technology at low cost versus tomorrow's technology for long term)

  Personnel costs to troubleshoot equipment

  Cost negotiation

    Volume purchases

    Maintenance

    Free training versus fee training

  Direct versus indirect costs

  Cost-benefit analysis

- Equipment diagnostics

  Diagnostic capabilities (front panel diagnostics or smart equipment)

  Optioning of equipment (manual optioning via strapping or DIP switches versus automatic options via electronic circuitry with downline loading of options)

Network Operations   This function is responsible for the network's day-to-day operations. This person or group of people maintains the ongoing communication services for the organization. They turn on the networks at the start of the business day and turn them off when the business day ends. One major function is the master network monitoring and daily gathering of statistics that are used by the network design and analysis people and, of course, the network manager. This group sometimes may interface with irate users, as well as with the various hardware, software, and circuit vendors. Sometimes this function or group of people is combined with the next one, failure control.

Failure Control   This is the central control group that receives telephone calls when problems occur and records the incidence of problems. This task may be performed by the network operations personnel. Basically, it is a help desk that is called when anything goes wrong in the system. This group has appropriate customer service representatives to record problems, report them to the testing/problem management people, follow up, and generally ensure that the network is back in operation as soon as possible. This group also might be responsible for change scheduling, coordination, and follow-up on any changes, whether they involve hardware, software, or circuits. In other words, this is the user's interface when there is a problem of any kind.

Failure control involves developing a central control philosophy for problem reporting and other user interfaces. This group should maintain a central telephone number to call when any problem occurs in the network. As a central troubleshooting function, only this group or its designee should have the authority or responsibility to call hardware or software vendors or common carriers.

Numerous software packages are available for recording the information received from an incoming telephone call. The reports are known as *trouble tickets*. The software packages assist the help desk personnel so they can type the trouble report immediately into a computerized failure analysis program. This type of package prints out a trouble report for follow-up and correction of the problem. It also automatically records and assembles various statistical reports to keep track of how many failures there have been for each piece of hardware, circuit, or software package.

Trouble tickets must be kept if a manager wants to do any type of problem tracking. Automated trouble tickets are better than paper because they allow management to gather problem and vendor statistics. There are four main reasons for trouble tickets.

- Problem tracking
- Problem statistics
- Problem-solving methodology
- Management reports

Problem tracking allows the network manager to determine problem ownership or, in other words, who has responsibility for fixing any outstanding problems. Why is this important? Problems often are forgotten in the rush of a very hectic day of network problems. In addition, the client may request the status of a problem. The network manager can determine whether the problem-solving machinery is meeting predetermined schedules. Finally, the manager can be assured that all problems are being addressed. Problem tracking also can assist in problem resolution. Are problems being resolved in a timely manner? Are overdue problems being flagged? Are all resources and information available for problem solving?

Problem statistics are important because they are a control device for the network operators as well as vendors. With this information a manager can retrieve and compute end-user availability. The manager can determine whether problem solving by the network operators is excessive. These statistics also can be used to determine whether vendors are meeting their contractual maintenance commitments. Finally, they can be used to determine whether problem-solving objectives are being met.

Problem-solving methodology helps you determine whether your problem priority system is working. You would not want a network operator to work on a terminal problem if an entire multidrop line consisting of dozens of terminals is waiting for a free technician. Moreover, a manager must know whether problem resolution objectives are being met. For example, how long is it taking to resolve problems?

Management reports are required to determine end-user availability, product and vendor reliability (Mean Time Between Fix), and vendor repair responsiveness (Mean Time to Repair). Without these reports a manager has nothing more than a "seat of the pants" estimate as to the effectiveness of the technicians, network, and vendors.

Regardless of whether this information is typed immediately into an automated package or recorded manually in a bound notebook trouble log, the general purposes are the same. If the organization does not have a computerized package, then the notebook format is quite appropriate. The bound notebook, with two carbon copies for each original trouble report, should have prenumbered pages to avoid missing any of these reports. One page always should be kept at the "trouble log desk." Two carbon copies are useful because it may be desirable to give one copy to a vendor who is called in to fix the problem and one copy to the internal testing/problem management personnel.

When a problem incident is reported, the trouble log desk staff should record the following to the best of their ability.

- Who reported the incident
- The problem reporter's telephone number
- The time and date of the problem (not the time of the call) as closely as it can be identified
- Location of the problem
- The nature of the problem
- When the problem was identified
- Why the problem happened (probably unable to identify this in most cases)
- How the problem occurred (probably unable to identify this in most cases)

Once this information is recorded, the failure control personnel should use an electric time-and-date stamp machine to indicate when they received the incident report.

The purpose of this procedure is to mandate central control of all problems and totally eliminate unnecessary service requests to vendors. Remember that there may be many hardware or software vendors for a data communication network. In fact, a typical network might have different vendors for the following equipment: terminals, cable connectors, modems, multiplexers, circuits (sometimes the local loops and the IXC circuits have different vendors), front end communication processors, host computer, and probably two or three different vendors for the various software packages.

The purpose of the bound trouble log volume is to record problems on paper so people will correct them and follow up, as well as to keep track of statistics with regard to problem incidents. For example, after a period of time utilizing a centralized failure control group, the organization might learn that there were 37 calls for software problems (3 for one package, 4 for another package, and 30 for a third software package), 26 calls for modems evenly distributed among the two vendors, 49 calls for terminals, and 85 calls to the common carrier that provides the network circuits. Data of this type is valuable when the design and analysis group begins redesigning the network to meet future requirements. Also, hard statistics like this enable you to put more pressure on the vendor who supplied the software package with the high number of problems.

Testing/Problem Management   This group establishes test and validity criteria and coordinates the various testing functions. It maintains the complex testing equipment that is needed to diagnose problems quickly and sometimes fixes them in house. Probably its single most important function is to interface with the failure control group. This is because the failure control group, when it becomes aware of a problem, immediately passes it to the testing group so they can diagnose the problem and identify what needs to be corrected. Depending on the severity of the problem, the operations group may be notified as well. The complete problem-handling procedure may involve either fixing the problem in house or notifying the appropriate vendor so that corrections can be made and the system can be operating again.

As soon as a problem is reported, the failure control group should immediately send a copy of the trouble log incident report to the testing/problem management group so they can diagnose the problem and possibly fix it themselves. The testing/problem management group should report back to the failure control group as soon as they have diagnosed the problem so that the time required to diagnose the problem can be recorded. In other words, the organization should keep track of *Mean Time To Diagnose* (MTTD), which is an indicator of the efficiency of testing/problem management personnel. This is the first of three different *times* that should be kept for future statistics.

Assume that a vendor is contacted for correction of a problem. Either testing or failure control personnel should keep track of the time the vendor takes to respond. In other words, the *Mean Time To Respond* (MTTR) is identified. This is a valuable statistic because it indicates how well vendors respond to emergencies. A collection of these figures over a period of time can lead to a change of vendors or, at the minimum, can put severe pressure on vendors who do not respond to problems promptly.

$$\text{Network availability} = \frac{\text{Uptime}}{\text{Uptime} + \text{Downtime}}$$

$$\text{Network availability} = \sum_{J=1}^{N} \frac{\text{MTBF}_J}{\text{MTBF}_J + (\text{MTTD}_J + \text{MTTR}_J + \text{MTTF}_J)}$$

N = Total number of network elements.

FIGURE 8-3 Network availability calculation.

Finally, after the vendor arrives on the premises, the last statistic to record is the *Mean Time To Fix* (MTTF). This figure tells how quickly the vendor is able to correct the problem. A very long time to fix in comparison with the time of other vendors may be indicative of faulty equipment design, inadequately trained customer service electronic technicians, or even the fact that inexperienced personnel are repeatedly sent to fix problems.

Some organizations combine the Mean Time To Respond (MTTR) and the Mean Time To Fix (MTTF); this is called *Mean Time to Repair* (MTTR).

One other statistic should be gathered. It usually is developed by the equipment vendor, and it is called *Mean Time Between Failures* (MTBF). The Mean Time Between Failures of vendor-supplied network interface equipment should be very high. Usually, the figure is greater than 30,000 hours. When you ask for the Mean Time Between Failures, always find out whether it is a practical figure or a calculated figure. You want a calculated figure; it is far more accurate and realistic. Sometimes practical figures are developed on a theoretical basis and cannot be depended on fully.

When we use the various mean times, we can work out a formula for calculating network availability (see Figure 8-3). Remember the MTBF can be influenced by the original selection of vendor-supplied equipment. The MTTD is related directly to the ability of in-house personnel to isolate and diagnose failure of hardware, software, or circuits. This means that test personnel need adequate training. The MTTR can be influenced by showing the vendor how good or bad their response time has been in the past. The MTTF can be influenced by the use of redundant interface equipment, alternate circuit paths, adequate recovery/ fallback procedures to earlier versions of software, and the technical expertise of internal staff. Since all four of these mean time statistics are used to calculate network availability, their collection is vital if network performance is to be measured accurately and if performance is to be improved.

Another set of statistics that should be gathered are those collected on a daily basis by the network operations group. These statistics record the normal opera-

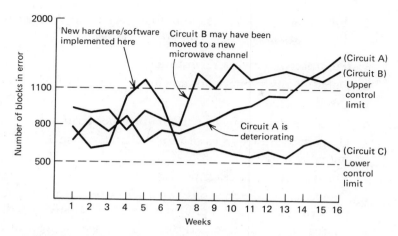

FIGURE **8-4**   Quality control for circuits.

tion of the system, such as the number of errors (retransmissions) per communication circuit, per terminal, or whatever is appropriate. Statistics also should be collected on the daily volume of transmissions (characters per hour) for each communication link or circuit, each terminal, or whatever is appropriate for the network. This data can identify terminal stations/nodes or communication circuits that have higher-than-average error rates. It also can be used for predicting future growth patterns and failures.

Such predictions can be accomplished by setting up simple quality control charts similar to the those used in manufacturing processes. Such programs use an upper control limit and a lower control limit with regard to the number of blocks in error per day or per week. Notice how Figure 8-4 identifies when the common carrier moved a circuit from one microwave channel to another (circuit B), or how a deteriorating circuit can be located and fixed before it goes through the upper control limit (circuit A) and causes problems for the users.

NETWORK REPORTING

Poor network management leads to an organization that is overburdened with today's problems (firefighting) and does not have time to address future needs. Management requires adequate reports if it is to address future needs. Information for these reports can be gathered from host computers, front end communication processors, network monitors, the network management group, vendors, test equipment, and the like.

Technical reports that are helpful to management should contain some or all of the following.

- Cumulative network downtime
- Detail of any subnetwork downtime
- Circuit utilization
- Response time analysis per circuit or per terminal
- Usage by various types of terminal stations such as interactive versus remote job entry
- Voice versus data usage per circuit
- Queue-length descriptions, whether in the host computer, front end communication processor, or at remote terminal sites
- Histograms of daily/weekly/monthly usage, number of errors, or whatever is appropriate to the network
- Failure rates for circuits, hardware, or software
- Utilization rate of critical hardware such as the host computer or front end processor
- File activity rates for database systems
- Local device activity
- Network gateway failure statistics and daily activity
- Distribution of packet volume (for packet networks), or distribution of character volume per circuit link
- Statistical profile of network traffic
- Distribution of traffic by time of day and location
- Peak volume statistics as well as average volume statistics per circuit
- Correlation of activity between today and a similar previous period
- Correlation of queue-length statistics by time and volume to a similar previous period.

NETWORK DOCUMENTATION

Network documentation is a mandatory requirement for the control of any network. Some software packages document your network by managing the various lists of hardware and narrative descriptions. These same packages also can keep equipment lists and even develop flowcharts of network configurations.

In the area of microcomputers, a full-scale word processing software package might be all that is needed to maintain network documentation. On the other hand, perhaps one of the software packages for drawing flowcharts or developing

Circuit Link	Hardware or Software					
	Circuit Cost	Modem Cost	Multi-plexer Cost*	Software Cost*	Front End Cost*	Total
New York to Chicago						
Chicago to Denver						
Denver to San Francisco						
San Francisco to Los Angeles						
Los Angeles to San Diego						
TOTAL						

*Some costs may have to be prorated among several links.

FIGURE 8-5   Network cost analyzer.

graphic presentations would be of assistance. Finally, the use of a spreadsheet program might be useful to keep track of costs for the various circuits, pieces of hardware, and software.

Figure 8-5 depicts how network costs might be analyzed using a spreadsheet software program on a microcomputer. Network documentation can consist of any or all of the following.

- Network maps
    Worldwide
    Within a single country
    Within a state or province
    Within a city
    Within a specific building or facility

- Circuit layout records
- Vendor maintenance records (MTBF, MTTD, MTTR, MTTF) with hardware/software/circuit cross-references
- Software listings by hardware
- Software listings by network tasks performed
- All user site telephone numbers and individual contacts
- Hardware maintenance history logs located at each user site
- Circuit control telephone contact index and log (when possible establish a national account with the common carrier rather than dealing with individual common carriers in separate states and provinces)
- Serial number inventory (property control) of all network components
- Network switching criteria and redundancy locations
- Vendor contractual agreements
- Legal requirements to comply with local or federal laws, control, or regulatory bodies; also have legal requirements for other countries (international)
- Operations manuals for network operations personnel
- Vendor-supplied hardware operation manuals
- Software documentation manuals
- Escalation levels (where to go when the problem cannot be resolved)
- Preventive maintenance guidelines
- Record of user site tests required by network security monitor
- Disaster plan/recovery techniques
- Diagnostic techniques by hardware component or type of trouble.

NETWORK STATUS

As part of the network organization and control, the present network status should be monitored and assessed continually by the design and analysis group. The data collected from a status review can be used both for future planning and for validating the performance of the network manager. Some questions that can be used to review the network status are as follows.

- Is the voice and data communication system combined?
- Does the network manager report at a high enough level in the management hierarchy?
- Is the network manager within the Information Systems Department?

- How many independent data networks are used in the organization?
- Are the networks application dependent or independent?
- What was the system availability for yesterday, the last five days, the last month, and so on?
- Are any of the reports mentioned in the above two lists available?
- Are the security and control aspects described in Chapter 9 available?
- What is the data communication system's annual budget?
- What is the monthly communication cost or the total cost last year? For voice? For data?
- Is the network critical to the organization's revenue-stream management, expense-stream management, cash-flow management, and so on?
- Is network operation erratic or difficult to monitor?
- Can trouble areas be pinpointed quickly (fault diagnosis)?
- Are 95 percent of all response times less than or equal to 3 seconds for online real-time traffic?
- Are management reports timely, and do they contain the latest up-to-date statistics?
- How many hours per day or days per week are utilized for network operation versus network management functions?
- What is the inventory of the current network configuration for all of the pieces of hardware?
- What is the network configuration for all circuits?
- What is the inventory of software and where is it located in the network?
- Does a formal network management organization exist with mandated goals, policies, procedures, and the like?

TEST EQUIPMENT

Just a decade ago many data communication facility managers did not have test equipment. It was quite proper to depend on the telephone company when a circuit failed and on other communication vendors when hardware or software failed. Today everything is changed because of the deregulation of the telecommunication market. For example, you might be using three or four "telephone companies," five or six hardware vendors, and two or three software vendors. This means that proper test equipment is mandatory. Network management *must* be able to diagnose a problem and determine which telephone company, hardware vendor, or software vendor should be contacted for assistance when the problem cannot be fixed by on-site staff.

Basically, testing can be broken into the areas of analog testing, digital testing, and protocol testing. *Analog testing* involves troubleshooting on the analog side of the modem. Specifically, it means testing the analog communication circuits supplied by the common carriers. A number of components can be measured (tested) on an analog circuit:

- *Loss.* Power loss of a signal transversing a transmission medium in both directions expressed in decibels.
- *Noise.* Background noise on an idle communication channel measured in each direction.
- *C-notch noise.* Measurement of metallic noise while a 1000-hertz tone is present through the active components on a circuit.
- *C-message noise.* Noise on a line measured in relation to a particular frequency.
- *Impulse noise.* The component of a received signal that has been band limited and exceeds a circuit mean square noise level in that band by 12 decibels.
- *Envelope delay distortion.* A derivative of the circuit phase shift with respect to the frequency. This distortion affects the time it takes different frequencies to propagate the length of a communication circuit.
- *Return loss.* A measure of the mismatch between the actual transmission circuit impedance and the normal impedance.
- *Phase jitter.* Short-term instability of a signal's phase which makes it very difficult for receiving modems to sense phase change modulation techniques.
- *Amplitude jitter.* Undesired amplitude modulation on a received signal.
- *Intermodulation distortion.* Compression that causes harmonic and intermodulation distortion in the transmitting signal. This may cause modems to lose signal tracking, thereby losing the circuit.
- *Peak-to-average ratio.* Measurement of a particular test signal to determine distortions taking place over a given transmission path.
- *Hits and dropouts.* Rapid changes in the gain or phase of a received signal or loss of that signal.
- *Single frequency interference.* Spurious, steady tones present on the communication circuit, which are heard in addition to the transmitted tones.

Digital testing is similar to analog testing, except that it is aimed primarily at

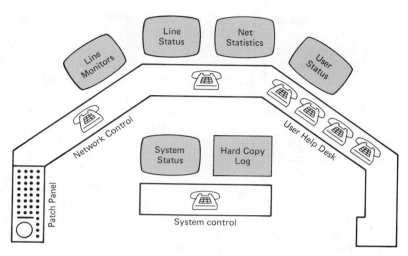

FIGURE **8-6**   Network control center.

testing digital communication circuits. *Protocol testing* is aimed at testing the various sign-on/sign-off procedures (handshaking or line discipline), along with message propagations and other items related to software protocol.

The typical network management tool kit of test equipment can cost $100,000 for a relatively simple network. Before defining specific test equipment, however, let us first present a network control center (see Figure 8-6) and the six basic categories of test equipment.

Monitors and analyzers generally are test sets that allow the operator to simulate specific message streams to test devices, communication circuits, or other workstations. A monitor looks similar to a portable microcomputer.

Analog and digital test sets are found on any network that uses modems in conjunction with telephone company circuits. Most networks require both analog and digital test sets. These devices also look much like a portable microcomputer with a video screen and keyboard for data entry.

Patch panels provide electrical connection to all parts of the network. At the minimum, they provide centralized access to each network communication circuit. They often look like a large panel with a number of plugs or connectors that can be cross-connected between different communication circuits.

Data recorders do not really perform tests. They are used to tap into communication circuits and store on disk pertinent activities about various circuits. Basically, they are a monitor for collecting data, analyzing it, and printing out reports. Data recorders also look like portable microcomputers.

Handheld test sets are the least expensive and simplest type of network equipment. They can be inserted between two network devices to test voltages or to send/receive various test patterns of bits to isolate errors. They also are used to determine whether there is a problem with the RS232 connector cable.

Network management systems may be separate computers (mini or micro) or possibly software running on the host mainframe that services the network. These network management systems are the latest in automated test equipment and may take over some of the functions performed previously by monitors and analyzers. They are used to identify errors, automatically run diagnostic tests, monitor the entire system, keep track of statistics, and prepare real-time management reports for the overall network management. Typical measurements that can be performed by the software in a network management system are

- *Access time.* After requesting service, this is the average waiting time before the network can accept information for transmission. In a dial-up network, this service includes the total time between the user dialing and the answer, polling time, time until the CLEAR TO SEND signal is received from the modem, and so forth.
- *Block transfer time.* This is the time that information is in transit between two end users, including modulation, propagation, and intermediate nodal/storage transfer times if it goes across several network nodes.
- *Bit transfer rate.* This is the total number of bits transferred during a given time interval.
- *Disengagement time.* This is the average waiting time between the user's request for disengagement and the actual disengagement from the network.

Some of today's line monitors and network analyzers are using microprocessor chips to perform sophisticated network tests. In other cases, microcomputers are utilized along with special software purchased to perform network testing and monitoring. Techniques and devices for monitoring network performance can range from comparably simple analog and digital test equipment to highly sophisticated network monitors/analyzers and overall network management systems. Descriptions of various types of test equipment follow.

Breakout Box   The most basic level of data communication monitoring and test equipment is analog test equipment. The breakout box is the next level up. It is a handheld device that can be plugged into a modem's digital side to determine the voltage values for the RS232C connector cable interface (25-pin cable), the V.24 interface, and others.

Bit-Error Rate Tester (BERT)   This piece of digital test equipment is somewhat more sophisticated than a breakout box because it sends a known pseudorandom pattern over the communication circuit. When this pattern is reflected back, the BERT compares the transmitted pattern with the received pattern and calculates the number of bit errors that occurred on the communication circuit. Various test patterns are used, and common pattern lengths are 63, 511, 2047, and 63511 bit patterns. The odd numbers allow simple circuitry in this test equipment.

Bit-Error Rate (BER) measurements can be made with this type of equipment. A BER is the number of bits received in error divided by the total number of bits received. BER measurements are used by service personnel to tune the communication circuit and to make a subjective evaluation as to the quality of a specific circuit or channel. BER cannot be related directly to throughput because error distribution is not taken into account. Assume that 1,000 one-bit errors occur during a time interval of 1,000 seconds. If the errors are distributed evenly (one per second), the effect on throughput will be disastrous; however, if all the errors occur in a single second, the effect will be minimal.

Block-Error Rate Tester (BKER)   This piece of equipment calculates the number of blocks received that contain at least one bit error divided by the total number of blocks received. BKER is more closely related to throughput. Assume a BKER measurement has been made and the BKER value is 10^2 (1/100). This means that out of every 100 blocks received, one contained an error; therefore, you would expect to see one retry for every 100 blocks transmitted (1 percent error rate).

Another error rate parameter (only for digital networks) is Error Free Seconds (EFS). EFS is similar to BKER except that it indicates the probability of success rather than the probability of failure, and the block size is the number of bits transmitted in a one-second time period. For example, for a 4800 bits per second channel, the one-second block would contain 4800 bits.

On digital communication channels, AT&T Communications guarantees that 99.5 percent of all seconds of data transmission will be error-free seconds.

Self-Testing Modems   If a self-testing feature is in a modem, a test pattern is generated inside the modem (after appropriate buttons are pushed) that is as close as possible to the normal digital input (which is, of course, disconnected). The test pattern travels through 90 percent of the modem's circuitry, passes through an artificial telephone circuit, and is returned to its point of origin. The artificial telephone circuit acts as a local analog loop. The returning pattern is compared with the transmitted pattern, and the operator is advised of discrepancies via an indicator lamp.

Some modems also have digital or analog remote loopback testing, whereby the signal actually is sent over the communication circuit and is looped back to the originating modem by the remote modem. Then a comparison of the signal is made, and the operator is advised of any discrepancies.

Finally, some modems have internal circuit diagnostic checks whereby they can diagnose their own failures in case of circuit or chip failure. Self-diagnostics are made possible by the use of firmware and microprocessor chips.

Newer modems contain some of the features of network analyzers. They actually keep track of poll times and other types of network analysis information.

Response Time Analyzer   At this point our discussion of test equipment begins to be less clear-cut because there is so much overlap between the different types of equipment and the names used by various vendors. Basically, a response time analyzer checks the operation of the communication protocols. When the equipment is operating in a polled network, the time that elapses from the initiation of polling until receipt of the response varies. Response time analyzers measure the responses of all hardware in the system. This equipment can determine whether the network equipment is meeting specifications.

A typical piece of this equipment might measure poll-to-poll time, the time from the detection of the poll being sent to the terminal until the time that poll is again sent to the same terminal. This measurement is updated continuously as the polls are generated by the host computer.

Another typical measurement is poll/response time. This measurement starts at the second SYN character of the poll and is terminated by the second SYN character of the response. (You might review the section Data Signaling/Synchronization in Chapter 2).

Other criteria with regard to a response time analyzer are the ability to offer hard copy (print information stored in its capture buffer) or to trap information when certain character sequences appear on the send or receive communication circuits. Some response time analyzers offer performance monitoring operations to help evaluate specific areas of network performance such as response time and link utilization. A totally clear-cut delineation between response time analyzers and data line monitors (the next item below) is impossible because each of these two pieces of test equipment overlaps the other's "territory." Such overlapping is happening throughout the communications industry where many hardware devices are taking over functions of other hardware devices. The primary reason this is happening is to sell more hardware devices, and it is relatively easy because of low cost microprocessor circuit chips. Both response time analyzers and data line monitors tend to look like a portable microcomputer, but with a few extra switches and buttons on the front panel.

Data Line Monitor   As noted above, some data line monitors perform response time analysis in addition to many other tasks; therefore, they are sometimes known as *protocol monitors*. They check the actual data (both control and data characters) on the communication circuit. You can use this device to check the interaction of software and hardware by looking at all the data passing in both directions on a communication circuit. It is mandatory with the new bit-oriented protocols because line control is no longer dependent on an entire byte to transmit the control message, but on a single bit within the 8-bit byte sequence.

Users can capture data in an external tape storage or internal memory, as well as freeze the most current data on the video screen. What makes the line monitor unique is that usually it is the only test equipment that displays all the "control characters" which usually are not seen on the video screen. A line monitor can show when a carriage return or a line feed occurs, as well as when a communication control code is transmitted. The technician can count the number of SYN characters, identify the eight bits of each field/character within a frame, and the like.

There are two basic categories of data line monitors, active and passive. *Active monitors* can generate data, are interactive on the circuit, and can emulate various terminals (they are programmable). *Passive monitors* merely monitor and collect data to be examined later. Active data line monitors contain all the features of a passive data line monitor and more. It should be noted that data line monitors are a security risk because of their ability to generate data, interactively place it on a communication circuit, and do this while emulating another terminal located somewhere else.

A typical data line monitor might cost in the area of $3,200, and it can monitor data speeds up to 64,000 bits per second, trap and count data for gathering communication circuit statistics, offer a video screen and printer, poll various stations, offer BERT capabilities, work with both asynchronous and synchronous systems, analyze various protocols, possess breakout box capabilities, and so forth.

Automated Test Equipment   Automated test equipment consists of hardware (minicomputers or microcomputers) and specialized software packages. All of it has built-in microprocessor chips and programmable testing features. You should note that the programs which perform this testing also can be housed within the host mainframe computer or a remote computer somewhere out in the network. Furthermore, the telephone companies offer centralized automated testing equipment that is located at the telephone company central office and used to monitor your network.

This type of automated testing equipment performs such functions as diagnostic testing, polling, statistics gathering, protocol emulation, measurement of whether the bandwidth (circuit capacity) is being used efficiently, self-diagnosis of its own circuits, both analog and digital circuit testing, testing of centralized and remote switches, and automatic restart and recovery in case of a disastrous situation.

Some real-time network management systems have been mentioned earlier in this section (NetView, Spectrum, MFOS, and so forth). The following is a sample of other network management products.

- Infinet's Series 90 line of centralized network management tools handles hybrid digital and analog networks of up to 1,200 circuits.
- Northern Telecom's Intelligent Matrix 3000 Switch System collects network statistics from a range of vendors' equipment and notifies the manager of network status through graphics presentations and an automatic alarm system.
- Larse Corporation's T-1 Network Diagnostic System offers centralized performance monitoring and diagnostic testing for T-1 networks.
- Keystone Technology's Scoop System offers traffic analysis and performance evaluation, and it runs on an IBM personal computer (XT or AT).
- Gandalf Data's Netscan 2000 Software enables Gandalf-based local and wide area networks to be monitored from a central workstation. It runs on an IBM personal computer (XT).
- Symplex Communications' Maestro Network Management System collects network data from over 500 remote sites. It reports on current network status and user or line-related errors by using audio alarms, thus alerting network management personnel to problems if they are performing other duties.
- EXCELAN's LANalyzer monitors local area networks and displays statistics graphically in real time. It also captures packets and generates controlled amounts of traffic for redesign purposes. It runs on a microcomputer. This automated management tool was described under Managing a LAN in Chapter 7, and a demonstration software package is available from John Wiley and Sons.

SELECTED REFERENCES

1. Abelow, Daniel, and Edwin J. Hilpert. *Communications in the Modern Corporate Environment.* Englewood Cliffs, N.J.: Prentice-Hall, 1986.
2. *Auerbach Data Communications Management.* Published bimonthly by Auerbach Publishers, 6560 N. Park Drive, Pennsauken, N.J. 08109, 1975– .
3. Kasperek, Gabriel. *Troubleshooting the Data Communications Network.* Madison, N.J.: Carnegie Press, 1984.

4. Kaufman, Robert J. *Cost-Effective Telecommunications Management: Turning Telephone Costs into Profits*. Boston: CBI Publishing Co., 1983.

5. Petersohn, Henry H. *Executive's Guide to Data Communications in the Corporate Environment*. Englewood Cliffs, N.J.: Prentice-Hall, 1986.

QUESTIONS/PROBLEMS

1. Data or information transmitted over any network must CATER to the overall needs of the network users. Define the acronym CATER.

2. What are the five key management tasks that must be performed by all managers?

3. If the annual budget for voice telephones at the local university is $106,000, what is a good estimate of their data communication costs?

4. Will technology help or hinder the combining of voice and data communications? Explain.

5. If you were going to initiate a data communication network control department, what would be some of the major job tasks and what organizations would be set up in this department?

6. Identify some of the items that should be reported at the trouble log desk.

7. Define the following and give a description of each: MTBF, MTTD, MTTR, and MTTF.

8. Assume you want to know the availability of a modem. Use Figure 8-3 to calculate the availability if the MTBF = 30,000 hours, MTTD = 4 hours, MTTR = 8 hours, and the MTTF = 2 HOURS.

9. If the modem is located in the village of Old Crow, 210 air miles north of Dawson City in Canada's Yukon Territory, the MTTR might be 96 hours (or possibly longer depending on snowstorms). Now what is the availability of the modem in problem 8?

10. If you were using a bit-error rate tester, would you use a BKER test for asynchronous transmission?

9

SECURITY
AND
CONTROL

This chapter identifies the 18 network control points that must be addressed for security and control. Specific hardware and software/protocol controls are reviewed. Other control areas that are reviewed are management controls, error control, recovery/backup/disaster, and the use of a matrix to identify, document, and evaluate security and control in a data communication network. Hundreds of specific controls that relate to the security and control of networks are discussed in Appendices 1 and 2.

WHY WE NEED SECURITY

Both business and government were concerned with security long before the need for computer-related security was recognized. They always have been interested in the physical protection of assets through such means as locks, barriers, and guards.

The introduction of computer processing, centralized database storage techniques, and communication networks has increased the need for security. Our concerns about security now are focused directly on the computer-related areas of business or government agencies. This emphasis manifests itself in controls to *prevent*, *detect*, and *correct* whatever might happen to the organization through the threats faced by its computer-based systems.

Figure 9-1 shows the typical *component parts* of a computer center such as hardware, software, organization, personnel, data communications, and so forth. Listed below each component of the computer center are a few typical examples of the *threats* faced by that component. For example, notice that typical threats under the component Data Communications are transmission errors, human errors, hardware errors, and taps. The threats shown in Figure 9-1 are not exhaustive; you will find much more complete lists of controls in Appendices 1 and 2, which are described later in this chapter.

Before moving on to network security, we want to present some short summaries of various past cases of computer crimes. The following list will give you some idea of the types of crimes that have been committed. As you read it, try to identify some controls that would have prevented the computer crime if they had been in force at the time the crime was perpetrated.

- A data control clerk in a bank's computer center embezzled $7,200. He did it by diverting and stealing checks that were being processed from a correspondent bank. For each diverted check, he wrote another check for the identical amount against his own checking account. He then deposited the check he wrote into a second checking account, which also was his own. When the check he wrote turned up for processing, he destroyed it and substituted the stolen check. Thus, the check he wrote against his own account was never charged against that account.

- A gunman who was armed with a shotgun and a pistol entered a telephone company central office and took several hostages. He then began blasting away at the central office switch with the shotgun, causing an estimated $10 million in damages. Approximately 15,000 customers lost their telephone service but, to the credit of the telephone company, all service was restored within 22 hours.

- A man was convicted of a welfare fraud that amounted to $73,525. He convinced his former girlfriend, a clerk in a state public aid office, to cooper-

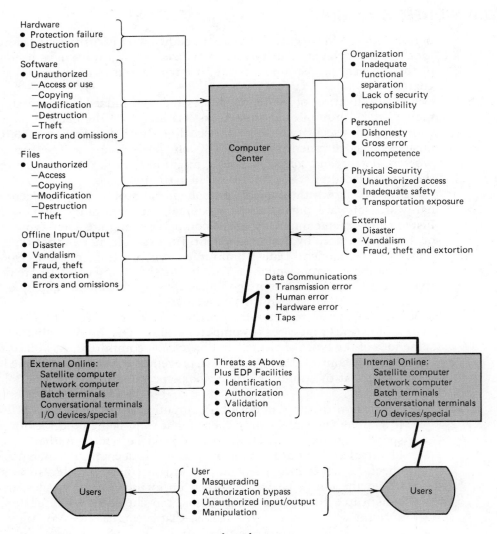

Hardware
● Protection failure
● Destruction

Software
● Unauthorized
 —Access or use
 —Copying
 —Modification
 —Destruction
 —Theft
● Errors and omissions

Files
● Unauthorized
 —Access
 —Copying
 —Modification
 —Destruction
 —Theft

Offline Input/Output
● Disaster
● Vandalism
● Fraud, theft
 and extortion
● Errors and omissions

Computer Center

Organization
● Inadequate
 functional
 separation
● Lack of security
 responsibility

Personnel
● Dishonesty
● Gross error
● Incompetence

Physical Security
● Unauthorized access
● Inadequate safety
● Transportation exposure

External
● Disaster
● Vandalism
● Fraud, theft and extortion

Data Communications
● Transmission error
● Human error
● Hardware error
● Taps

External Online:
 Satellite computer
 Network computer
 Batch terminals
 Conversational terminals
 I/O devices/special

Threats as Above
Plus EDP Facilities
● Identification
● Authorization
● Validation
● Control

Internal Online:
 Satellite computer
 Network computer
 Batch terminals
 Conversational terminals
 I/O devices/special

Users

User
● Masquerading
● Authorization bypass
● Unauthorized input/output
● Manipulation

Users

FIGURE **9-1** Threats to computer-based systems.

ate with him by issuing emergency aid checks to welfare recipients. The emergency aid recipients then shared the gain with the perpetrators. Over a six-month period, 173 unauthorized checks were distributed. The emergency aid disbursement process was performed through online computer terminals and had fewer controls than the normal welfare disbursement process. The crime was discovered by an auditor.

- An illegal bookie in Australia tapped into the communication circuits of the legal bookmaking operations by using his telephone. The purpose was to know race results immediately, enabling him to pay off winners as fast as the legal betting establishments, which are licensed by the government.

- An unknown person gained access to a computer terminal room by asking the janitor to open the door for him. He then picked the locks of the telephones and terminals, thereby gaining unauthorized access and use of timesharing facilities.

- An unknown person poured acid over telephone wires at the point where the wires entered a building that contained data processing equipment.

- By using his employer's computer system, an EDP employee developed his own computer service business in a neighboring city. The perpetrator was discharged.

- An operator at a university computing service center used the university's computer to prepare political campaign literature for a student election. The perpetrator was discharged.

- A computer printer operator was paid to make an extra carbon copy of competitive bidding reports for the purpose of industrial espionage. The printer operator was discharged.

- A programmer changed his firm's dividends payment program and reduced the dividends of eight different stockholders. He then issued a check for $56,000 to a fictitious person. This was the total amount of the reductions from the eight accounts.

- The president of a French software company posed as a professor of computer science on a tour of the United States. During his tour of more than 200 computer centers, he collected many free programs. He then returned to France and sold copies of the programs he had collected. He was caught and fined 5,000 francs for the sale of one program. No action was taken concerning all the other programs he sold.

- A student wrote a program that masqueraded as an operating system. When a user attempted to log on to the system in which he installed this masquerading program, the program obtained the user's account number and then declared the system unavailable. By using the account numbers discovered with this program, the student was able to use the computer without charge.

NETWORK SECURITY

In the previous section, you read some descriptions of computer crimes. Using Figure 9-2 as an example, we want to present a description of a sophisticated network penetration by someone using the public packet networks. In this example, the computer hacker entered communication networks by using stolen MCI, US Sprint, and other voice communication network identification and passwords in the following manner.

The hacker first placed a call to a Philadelphia gateway node. Once that circuit was established, he placed a second call to a US Sprint gateway node in Chicago

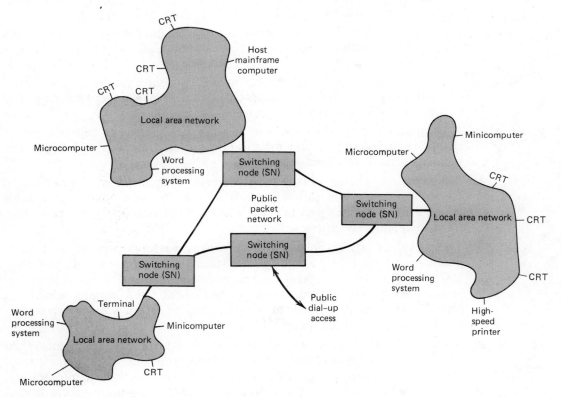

FIGURE **9-2** A typical network.

and used a second stolen access code to enter Sprint. Then he used the US Sprint linkage from Chicago to establish a third circuit with a Telenet gateway node in San Francisco, for which he used still another stolen access code. He then used one of the unprotected host computers maintained by Telenet as a switch before establishing a host-to-host circuit via Tymnet. This hacker then jumped through several other host computers, entered Datapac (a Canadian public packet switching network), and again used several host computers before reentering Telenet to attack the ultimate target computer.

With such a complex communication environment, the chance is very remote that intermediate host computers would ever discover they were being used as switches to go on to another network. In fact, the host computer probably paid unknowingly for this perpetrator's communication charges. The people who are most likely to discover such a computer crime are the individual subscribers, for it is they who receive the dial-up telephone communication bills from the various network vendors.

In recent years organizations have become increasingly dependent on data communication networks for their daily business communications, database information retrieval, and distributed data processing. This commitment to data communications/teleprocessing has changed the potential vulnerability of the organization's assets. This change has come about because the traditional security, control, and audit mechanisms take on a new and different form in data communication-based systems. Increased reliance on data communications, consolidation of many previously manual operations into computerized systems, use of database management systems, and the fact that online real-time systems cut across many lines of responsibility have increased management concern about the adequacy of current control and security mechanisms used in a data communication environment.

There also has been an increased emphasis on computer network security because of numerous legal actions involving officers and directors of organizations, because of pronouncements by government regulatory agencies, and because the losses associated with computerized frauds are many magnitudes larger per incident than those from noncomputerized frauds. These factors have led to an increased vigilance with regard to protecting the organization's information assets from many potential hazards such as fraud, errors, lost data, breaches of privacy, and disastrous events that can occur in a data communication network.

With regard to data communication networks, the organization must be able to implement adequate control and security mechanisms within its facilities, including buildings, terminals, local area networks, local loops, interexchange channel circuits, switching centers, network interface units (gateways), packet networks, hardware (modems, multiplexers, encryption devices, and the like), network protocols, network architecture software, test equipment, and network management control.

For example, Figure 9-2 depicts a typical network that an organization might develop. In such a network all of the areas mentioned above require a positive decision (policies and procedures) as to security and control. With this kind of network the organization is vulnerable to many points of entry from an unwanted intruder. In fact, every terminal in the network is a potential entry point for an unauthorized intruder.

The remainder of this chapter will discuss each of the major portions of a data communication network, such as hardware and software, and describe the various controls that might be used to prevent, detect, or correct threats in that specific area.

Finally, a control matrix methodology for identifying the threats and their associated controls will be presented. The matrix provides a data communication network manager with a good view of the current threats and any controls that are in place to mitigate the occurrence of these threats.

NETWORK CONTROL POINTS

To implement a good security program in a communication network environment, you first need to identify all of the points at which control must be established. These are called *control points*. Once it is determined where these controls should be located, then it is possible to identify what individual controls actually are required in the network.

We have identified 18 control points (that is, areas where control and security mechanisms must be implemented) for a communication network, and they are shown in Figure 9-3. The network manager, quality assurance person, security officer, or the organization's EDP auditor should examine these areas to be sure that the proper controls are implemented and that they are functioning properly. The numbers in the figure correspond to the numbered list below. This list defines the control point and describes the general type of security required at each of these 18 control points.

1. Physical security of the building or buildings that house any of the hardware, software, or communication circuits must be evaluated. Both local and remote physical facilities should be secured adequately and have proper controls.

2. Operator and other personnel security involves implementation of proper access controls so that only authorized personnel can enter closed areas where network equipment is located or access the network itself. Proper security education, background checks, and the implementation of error/fraud controls fall into this area.

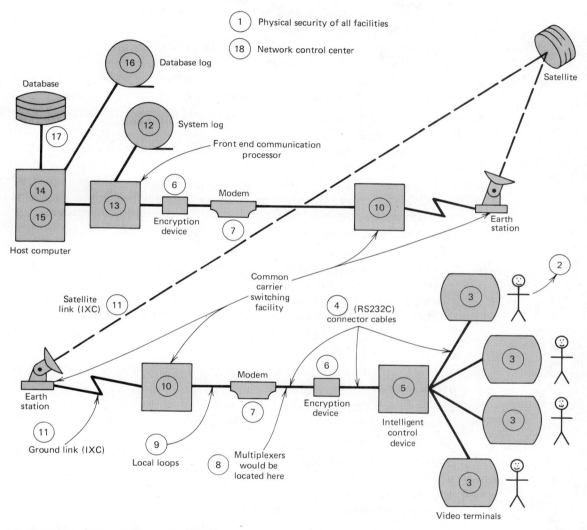

FIGURE **9-3** Network security control points.

417

3. Terminals are a primary area where both physical and logical types of security controls must be enforced.

4. Local connector cables and wire pairs that are installed throughout the organization's facilities must be reviewed for physical security.

5. Local intelligent control devices that control groups of terminals should be reviewed for both physical and logical programmed controls.

6. Hardware encryption is a primary control point, especially as it relates to the security of messages.

7. The modems should be reviewed as they relate to control and security at this point in the network.

8. Multiplexers, if they are used in the network, must be secured to prevent unauthorized entry, and they should contain backup circuits. Physical security and backup electrical power may be required at a remote site.

9. Local loops that go from the organization to the common carrier's switching facility should be reviewed.

10. The physical security and backup of the common carrier switching facility (telephone company central office) should be evaluated. If this facility got destroyed, all the circuits would be lost. This review may include both central offices in a city and earth stations for satellite transmission.

11. The security/control mechanisms in place should be reviewed with regard to the interexchange channel (IXC) circuits.

12. A major control point is the system log that logs all incoming and outgoing messages.

13. The front end communication processor is another major control point to review. There may be a packet switching node (SN) that must be reviewed for security and control.

14. Within the host computer, any controls that are built into the software should be reviewed.

15. Also within the host computer, any controls that are designed into the hardware mechanisms/architecture should be reviewed.

16. Another major control point, but only in database systems, is the database before-image/after-image logging tape. This should be reviewed for any controls that may be in existence. Many other security/control items of data are logged at this point.

17. With regard to database-oriented systems, another control point is the database management system (DBMS) itself. The database management system software may have some controls that help with regard to security of the data communication network and the control of data/information flow.

18. The last control point is the network control center itself. This area has controls that relate to management and operation, test equipment utilized, reports, documentation, and the like (see Figure 8-6 in Chapter 8).

These 18 control points are the specific areas in which control features can be implemented and maintained within a data communication network. To obtain an understanding of the complete impact of a network's complexities, you should compare the following three figures.

- Figure 2-1, the basic technical concepts in Chapter 2
- Figure 5-1, the basic software concepts in Chapter 5
- Figure 9-3, the network security control points in Chapter 9

A comparison of these figures will show you that communications require many technical concepts, numerous software packages, and a multitude of security controls. Each of these three aspects (technical, software, and network security) is located throughout the entire length of the network.

ENCRYPTION

Encryption, a concept introduced in Chapter 3, is the process of disguising information by the use of many possible mathematical rules known as *algorithms*. Actually, *cryption* is the more general and proper term. *En*cryption is the process of disguising information, whereas *de*cryption is the process of restoring it to readable form. Of course, it makes no sense to have one process without the other. When information is in readable form, it is called *clear* or *plaintext*; when in encrypted form, it is called *ciphertext*.

The art of cryptography reaches far into the past and until recently has almost always been used for military and political applications. By today's exacting standards, such ciphers are insecure and therefore obsolete. They usually were alphabetic ciphers (rules for scrambling the *letters* in a message) designed for manual processing. Today's world of binary numbers and the speed of computers have given birth to a new class of cryption algorithms.

The acceleration of new research began during the Second World War and has continued into the present time for four reasons.

- The recognition of the necessity of encrypting communications for military purposes
- The advent of high speed computational electronics (computers)
- A growing interest in cryptography within academic circles
- An interest on the part of private corporations and governments in protecting their proprietary information

Interest in cryptographic protection runs highest in the world of communications. Of all the routes and resting places of information, communicated information is the most vulnerable to disclosure. Data stored on magnetic tapes or disks and in computer memory can be protected to a large extent by physical security, passwords, and other software access control systems.

Modern data communications takes advantage of existing public telephone circuits, microwave transmissions, and satellite relays. As a result, communicated information is highly exposed in a variety of forms. It can be captured at minimum expense and risk to the data thief, and at maximum loss to the organization.

A striking example of this exposure is the daily Electronic Funds Transfer (EFT) of billions of dollars between domestic and foreign banks over public links. The covert alteration of bank account numbers, amount of funds, and the like can have disastrous results.

An encryption system has two parts: the algorithm itself, which is the set of rules for transforming information; and the *key*, which personalizes the use of the algorithm by making the transformation of your data unique. Two pieces of identical information encrypted with the same algorithm but with *different keys* produce completely different ciphertexts. When using most encryption systems, communicating parties must share this key. If the algorithm is adequate and the key is kept secret, acquisition of the ciphertext by unauthorized personnel is of no consequence to the communicating parties.

The key is a relatively small numeric value (in number of bits) that should be easily transportable from one communicating node to another (see item 6 in Figure 9-3). The key is as it sounds: it is something that is small, portable, and with the aid of a good lock, the algorithm, it keeps valuables where they belong.

Good encryption systems do not depend on keeping the algorithm secret. Only the keys need to be kept secret. The algorithm should be able to accept a very large number of keys, each producing different ciphertexts from the same cleartext. This large "key space" protects the ciphertext against those who try to break it by trying every possible key. There should be a large enough number of possible keys that an exhaustive computer search would take an inordinate amount of time or would cost more than the value of the encrypted information.

Almost every modern encryption algorithm transforms digital information. Scrambling systems have been devised for analog voice signals, but it generally is agreed that their algorithms are not as strong as those used for digital signals made up of binary bits. The most recent advances in analog signal protection have not been in newer and better algorithms. Instead, they have been in the technology of high speed conversion of analog signals to digital information bits in preparation for encrypting them with digital algorithms. In any case, the vast majority of today's proprietary information is digital. For this reason we will discuss only digital techniques.

Encryption algorithms may be implemented in software or hardware. The software has some advantages in protecting stored data files and data in the host computer's memory. However, hardware implementations have the advantages of much greater processing speed, independence from communication protocols, ability to be implemented on dumb devices (terminals, telex, facsimile machines, and so on), and greater protection of the key because it is physically locked in the encryption box. Unauthorized tampering with the box causes erasure of the keys and related information. Hardware implementations have been reduced to the chip level because they are simply specialized microprocessors housed in small hardware boxes.

By far the most widely used encryption algorithm is the Data Encryption Standard (DES). It was developed in the mid-1970s by the U.S. government in conjunction with IBM. DES is maintained by the National Bureau of Standards (NBS) and often is referred to as NBSDES or DEA (Data Encryption Algorithm). The U.S. government recommends that DES be used for the encryption of commercial and *un*classified military data. The American Banking Association has endorsed its use for the commercial banking industry.

This combination of credentials makes DES the technique of choice by private institutions. This concept of "choice" is somewhat misleading. DES is the *only* algorithm endorsed by the government. The academic literature is full of alternatives, but practical reasons such as the necessity of obtaining insurance against third party fraud and the lack of mathematical sophistication on the part of encryption system users presently leave little choice.

DES is classified as a *block cipher*. In its simplest form the algorithm encrypts data in independent 64-bit blocks. Encryption is under the control of a 64-bit key. DES expects a full 64-bit key, but it uses only 56 of the bits. (Every eighth bit may be set for parity.) Therefore, the total number of possible keys is 2^{56} or over 72 quadrillion combinations.

To put quadrillions into perspective, let us introduce the metric unit of femtoseconds. A *femtosecond* is one quadrillionth of a second, a unit so small that it is very difficult to associate with it in any meaningful way. Femtoseconds are incredibly brief. To look at it another way, there are as many femtoseconds in one second as there are seconds in 30 million years. If you are looking at 72 quadrillion (72,000,000,000,000,000) combinations, you can see that this is an unbelievably large number of combinations.

If you could test just one of these combinations each second, then one quadrillion combinations might take 30 million years to test! Seventy-two quadrillion combinations equals a time period of 30 million years times 72. Do not be concerned at this point, however, because cryptographers who specialize in breaking these block ciphers only have to test half the combinations to obtain a 50/50 probability of arriving at the secret key. This testing is enhanced further by the use of very specialized mathematical algorithms that are designed to quickly

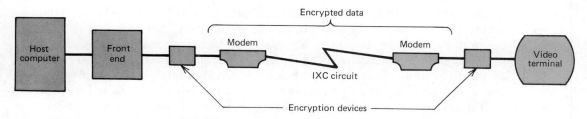

FIGURE **9-4** Encryption devices.

invalidate large groups of these combinations. In addition, specially designed computers may be able to test 10,000 combinations each second. As a result, they never do have to test even one-half of the 72 million combinations to "break" the secret key.

DES ciphertext is composed of blocks containing highly randomized bit sequences. The algorithm is so thorough in its randomizing of any 64-bit block (almost without regard to the cleartext of the key) that ciphertext almost always passes standard tests for randomness. The random quality of ciphertext is a crucial factor in the design of communication networks that convey ciphertext. Communication control characters (for message routing or error detection) cannot be mixed with ciphertext because there is always some probability that DES will generate one of these control characters and thwart the communication system.

As a result, DES hardware usually is employed as shown in Figure 9-4. Communication protocols, parity, and checksums are in place with the message *before* it enters the originating DES hardware device. As is shown, this information may originate from a terminal, a front end, or a variety of communicating devices. The hardware encryption boxes usually are utilized on a link-to-link basis as depicted in Figure 9-4.

Placing the DES device between the modems can present a number of problems. First, most DES boxes are digital devices, and they usually do not accept the analog signals output by modems. Second, in asynchronous communications at least the start bit must be sent in the clear. Encryption can, and usually does, begin with the first data bit and end with the last. Similar problems can occur if synchronous timing signals are encrypted.

The randomized information is transmitted now to a network switch, computer, terminal, or other receiving device. The receiving DES hardware, which must be loaded with the *same* key as the originating DES hardware, then decrypts the information before it enters the receiving terminal device. Any communication protocols are verified *after* the decryption.

In some ways DES provides better error detection than standard parity or checksum techniques. If a single bit of any 64-bit ciphertext block is flipped during transmission, on decryption of that block the result will be 64 bits of random nonsense. This "error propagation" virtually ensures that parity and checksum will fail after decryption.

A more serious problem occurs if a bit is picked up or dropped during communication. The message loses 64-bit block "synchronization" at the point of the dropped or added bit, and the message decrypts into nonsense. The result can be the loss of an entire message.

This magnification of communication errors is not without its price. Since the *minimum* loss of information is usually 64 bits, a retransmission almost always is required if there is a single bit communication error.

DES is a member of a class of algorithms known as *symmetric*. This means that the key used to decrypt a particular bit stream must be the *same* as that used to encrypt it. Using any other key produces cleartext that appears as random as the ciphertext. This can cause some problems in the complex area of key management; keys must be dispersed and stored with great care. Since the DES algorithm is known publicly, the disclosure of a key can mean total compromise of encrypted messages. Therefore, in order for two nodes in a network to establish communication of ciphertext, it is first necessary to define and communicate a common key over a secure channel or send it by personal courier.

The U.S. National Security Agency (NSA) has announced that it will not recertify the Data Encryption Standard at its regular five-year review in 1988. The agency contends that the use of the DES algorithm, which has been a federal standard for ten years and is well established as the American commercial encryption standard, has spread to extremely sensitive applications that make it an attractive target for adversaries of the United States. The National Security Agency currently plans to replace DES with a family of new codes that the agency itself will distribute and regulate.

NSA plans to replace DES with algorithms being developed under the Commercial COMSEC Endorsement Program (CCEP). (COMSEC is military shorthand for communications security.) The CCEP is the result of presidential National Security Decision Directive 145, which was signed by President Ronald Reagan in September 1984. This directive addresses the communication security of government contractors, particularly with respect to national security.

The planned CCEP algorithms represent a new strategy in encryption standards. Rather than certify a single new public algorithm to replace DES, the National Security Agency will develop several proprietary encryption algorithms. The public DES algorithm is being replaced with several secret algorithms to address a problem known as *cross-vulnerability*, which is the compounded risk incurred by all users if someone breaks DES with a general solution. Currently, the only known attack on a DES-encrypted message is to exhaust all possible keys. While the costs of applying massive computer power for such a task are prohibitively high for the majority of potential code breakers, the increasingly widespread use of DES and the declining cost of data processing hardware together are improving the potential cost benefit.

In addition to DES and the algorithms being developed under CCEP, users seeking data transmission protection have at least two major alternatives: proprietary algorithms and public key cryptosystems. These systems, however, also are subject to some concern. A *proprietary algorithm* can be developed solely for the use of a single organization. Assuming that such algorithms are strong enough to withstand cryptanalysis, a few potential problems still remain. First, the cost to develop and implement a proprietary system is considerable and may even be prohibitive. Second, any algorithm developed by an outside party always is open to the question of how good it is at resisting attack. Finally, if the algorithm is to be used outside the United States, the National Security Agency must approve it before it can be exported.

Public key cryptosystems are inherently different from private key systems such as DES. Public key systems are based on one-way functions. Even though you originally know the contents of your message and the public encryption key, once they are put together (encrypted) by the one-way function algorithm, they cannot be taken apart again unless you have the other (private) key. One-way functions, which are relatively easy to calculate in one direction, are computationally impossible to "uncalculate" in the reverse direction. The message sender must look up the recipient's one-way encryption function in a public key directory and use it to encrypt the message. Receivers maintain their own secret keys or mating decryption algorithms (private keys) for this one-way function.

The primary difference between a public key system (asymmetric) and a private key system (symmetric) like DES is that in a public key system no secret keys are exchanged between communicating parties. Public key systems, however, do not eliminate the key distribution problem, but key distribution costs less than one-tenth of what it costs for the symmetric class of algorithms like DES. The keys still must be distributed during system startup. A public directory that contains each participant's one-way encryption function is shared by the community. The strength of these systems lies in maintaining the secrecy of the decrypting function and the ability of the algorithm to withstand cryptanalysis. If someone solves the public key one-way function, the system is compromised. The one-way function is based on the difficulty in factoring very large prime numbers. The solution to this problem has eluded mathematicians for hundreds of years.

In public key encryption systems, the key needed to decrypt a message is different from the one used to encrypt it. The two keys are related distantly in a mathematical sense. The security of asymmetric systems depends on the extreme difficulty (analytic impossibility or computational infeasibility) of deriving one key from the other.

Asymmetric algorithms can reduce the key management problem greatly. Each receiving node has its publicly available key (hence the name "public key") that is used to encrypt messages sent by any network member to that node. These public keys may be listed in a telephone book-style directory. In addition, each user has a *private key* that decrypts only the messages that were encrypted by its *public key*. The net result is that if two parties wish to communicate with one another, there is no need to exchange keys beforehand. Each knows the other's public key from the public directory and can communicate encrypted information immediately. The key management problem may be reduced to each user being concerned only with the on-site protection of its private key.

To visualize how a public key algorithm works, look at Figure 9-5. At the top of this figure is a public directory which contains all the public keys for each organization utilizing public key encryption. Our public directory contains five different banks.

In order to use the public key encryption methodology, a bank also has a secret key known as a private key; therefore, there are two separate keys, the private (secret) key and the public key. In this case, the bank places its public key into the public directory and carefully secures its own copy of the private key.

The middle of Figure 9-5 shows what an encrypted message looks like. When Bank 4 wants to send a message to Bank 1, it encrypts the message with the Bank 1 public key, which is obtained by Bank 4 from the public directory. This represents a straightforward encryption of a message between Bank 4 and Bank 1. Obviously, when the message is received at Bank 1, it decrypts the message using its secret private key.

For more complex encryption, Bank 4 can include its signature so Bank 1 also can verify the signature or, in other words, be sure that the message originated from Bank 4.

To perform a signature verification (see the bottom message of Figure 9-5), Bank 4 first encrypts its ID (signature) plus some of the "key-contents"[1] of the message, using the Bank 4 private key. This is its own private key and is known only to Bank 4.

Next, Bank 4 encrypts both the message contents and the already encrypted bank ID using the Bank 1 public key from the public directory. This means that the Bank 4 ID has been double encrypted, first using the Bank 4 private key and then a second time using the Bank 1 public key. The message is then transmitted to Bank 1.

Upon receipt of the message, Bank 1 uses its private key to decrypt the entire message. At this point, Bank 1 is able to read the contents of the message, except for a block of data that still is encrypted (unidentifiable). Because the message was received from Bank 4, Bank 1 assumes that Bank 4 secretly encrypted its ID

---

[1] The term *key-contents* means unique information from the message such as date, time, or dollar amount. It does not refer to the public/private keys.

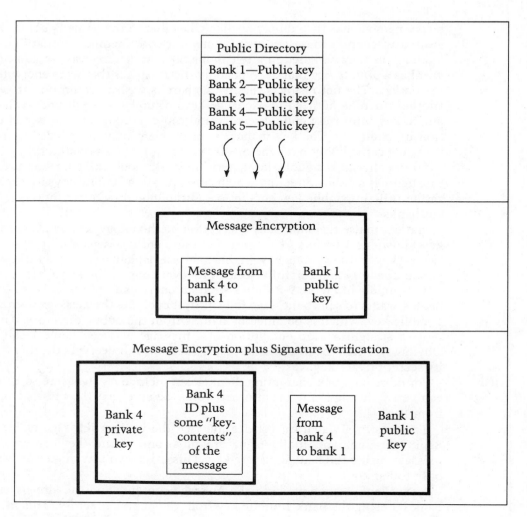

FIGURE 9-5 Public encryption.

plus some "key-contents" of the message for signature verification purposes. At this point, Bank 1 takes the Bank 4 public key (from the public directory) and decrypts the trailing block of data that contains the Bank 4 ID plus some "key-contents" of the message.

In this way, the public key system encrypts messages and also offers electronic signature identification without an exchange of keys among all the thousands of banks around the world. The public directory need only be updated as often as necessary. You can encrypt with a public key and decrypt with a private key, or you can encrypt with a private key and decrypt with a public key.

The algorithms for public key systems are very different from symmetric algorithms like DES, but in practice their ciphertexts are similar when viewed from the data communication standpoint. Each produces ciphertext consisting of randomized bit patterns. There is almost always some degree of inherent error propagation; therefore, the practicalities of handling communication of both types of ciphertexts are for the most part identical.

The world of cryptology is full of controversy and debate, caused in part by tension between governments and independent academic cryptologists. National security is always the issue. The underlying cause of this tension, however, is the fact that cryptology is an art rather than a science. Except for a few noteworthy exceptions, it is impossible to mathematically *prove* whether an encryption/decryption algorithm can be broken. Thus, the only route to breaking a cipher is an artful (and perhaps time-consuming and expensive) trial and error approach. Debates about the security of ciphers often end with mathematical generalizations and seat-of-the-pants type expressions.

This combination of embryonic but exciting mathematics and a dramatic increase of interest in encryption by corporations and governments means that it indeed will be an exciting arena in the years to come.

HARDWARE CONTROLS

Network hardware is discussed here in terms of the controls that relate to them. We will review controls that relate to front end processors, packet switching controllers, modems, multiplexers, remote intelligent controllers, terminals, and voice telephone security.

Front End Processors   The front end processor that controls a centrally controlled data communication network can be one of the single most important areas for security and control. It is only a piece of hardware, but within it are software programs/protocols that control the access methods for data flow.

Some specific controls that might be housed within the front end communication processor are

- Polling of terminals to ensure that only authorized terminals are on the network
- Logging of all inbound and outbound messages (systems log) for historical purposes and for immediate recovery should the system fail
- Error detection and retransmission for messages that arrive in error

- Message switching that reduces the possibility of lost messages (there also can be circuit switching or packet switching)
- Store and forward techniques that help avoid lost messages (although store and forward opens up the possibility of a network programmer's copying messages from the storage disk)
- Serial numbers for all messages between all nodes
- Automatic call-back on dial-up facilities to prevent the host computer from being connected to an unauthorized dial-up terminal
- Systems editing such as rerouting of messages and triggering of remote alarms if certain parameters are exceeded or if there is an abnormal occurrence
- Collection of network traffic statistics for long-term control of the total network
- Also review the list of functions for the front end in Chapter 3.

Packet Switching Controllers A packet switching controller or switching node (SN) is similar to a front end communication processor, but it has some specialized features that pertain to the operation of a packet network. A packet switching controller can perform any of the control functions previously mentioned for front ends. In addition, it performs other specific control functions such as the following.

- Keeps track of messages between different nodes of the network.
- Controls the numbering of each packet to avoid lost packets, messages, or illegal insertions.
- Routes all messages. It may send different packets, containing parts of the same message, on different circuits (unknown circuit path). This may prevent an unauthorized user/perpetrator from receiving all parts of a sensitive message.
- Contains global and/or local databases that contain addresses and other sensitive data pertaining to each node. These databases can be cross-referenced with other written documentation when network nodes are reviewed for security.
- Keeps track of the sender of each message that is delivered on dial-up packet networks for control and billing purposes.
- Restricts the users to dial-up or allows use of leased circuits into the packet network.

Modems The modem may be an interface unit either for broadband (analog) communication circuits or for baseband (digital) communication circuits. It does not matter which because these hardware units can perform any of the controls listed below, depending on the features installed by each manufacturer.

Modems can offer loopback features that allow the network manager to isolate problems and identify where they are occurring in the network. Some modems contain automatic equalization microprocessor circuits to compensate for electronic instabilities on transmission lines, thereby reducing transmission errors. Some modems have built-in diagnostic routines for checking their own circuits. Mean Time Between Failure (MTBF) statistics should be collected for modems because low MTBF indicates that downtime is excessive.

Some dial-up modem controls include changing the modem telephone numbers periodically, keeping telephone numbers confidential at both user sites and the central data center, possibly disallowing automatic call receipt at the data center (using people to intercept), removing telephone numbers from both local and remote dial-up modems, and requiring the use of terminals that have an electronic identification circuit for all dial-up ports. Finally, it may be desirable to utilize a *dial-out-only facility*, whereby the act of dialing into the network and entering a password automatically triggers a disconnect; the front end or host computer then dials the "approved" telephone number that matches the password used during the original dial-in. In other words, dial-in triggers a dial-out.

Multiplexers Because many multiplexer sites are at remote locations, a primary control is to prevent physical access to the multiplexer. Another consideration is whether the multiplexer should have dual circuitry and/or backup electrical power since loss of a large multiplexer site can knock out several hundred terminals. Because time division statistical multiplexers have internal memory space, and some have disk storage, special precautions must be taken. Memories and disk storage make illegal copying of messages easier. Other controls include logging all messages at the remote multiplexer site before transmission to the host computer and manually logging all vendor service call visits.

Remote Intelligent Controllers A remote intelligent controller can be a special form of multiplexer or a remote front end communication processor that is located several hundred miles from the host computer. These devices usually control large groups of terminals. All the controls that were mentioned for multiplexers also apply to remote intelligent controllers.

A review of software controls that can be programmed into this device is suggested. For example, daily downline loading of programs can help ensure that only authorized programs are in this device. Another control is the periodic counting of bits in the software memory space. This identifies a minor program change so that a new one can be downline loaded immediately. Each controller should have its unique address on a memory chip (instead of software) to thwart anyone who wants to change hardware addresses. Remote logging of each inbound/outbound message should be considered seriously. If hardware encryption boxes are located in the same facility as the remote intelligent controllers, then access to these devices should be controlled by implementation of strict physical control procedures and locked doors.

Terminals—Human Error Prevention   Two basic areas must be considered with regard to the control of terminals or microcomputers in a data communication network. The first involves mitigating human errors while running business application systems, and the second is providing terminal security. Controls to prevent, detect, and correct human errors are necessary because, of these two areas, human errors cause the greatest dollar loss. First, we will present a list of human error prevention controls, and then we will move into the second area, which is providing security restriction controls at remote terminals/ microcomputers.

- Adequate operator training should be provided through self-teaching operator manuals and the periodic updating of these manuals.
- Dialogue between the operator and the application system should be kept simple. (Menu selection might be utilized.)
- Terminals should be easy to use and have functional keyboards.
- Preprinted forms for printing terminals and a fill-in-the-blank format (preprinted forms on a video screen) for video terminals should be considered.
- Instructions should be preprogrammed and available for recall when an operator needs help. Secured systems, where assistance should be more difficult to obtain, may be an exception.
- Operators should have restart procedures that can be used for error recovery during a transaction.
- Work area extremes in light, noise, temperature, and so on must be minimized if operators are to reduce errors to a minimum.
- Reasonably fast response times reduce errors because longer response times produce error-causing frustration in operators. Long response times also reduce productivity.
- Intelligent terminals can edit for logical business errors and verify data before transmission.

For further controls, review the section on video terminals in Chapter 3.

Terminals—Security Controls   The remote terminal or remote microcomputer may be the single most important point for controlling security; specific controls *must* be enforced here. There are three general ways of restricting access to computer systems or databases, based on something you know, something you have, or something you are.

First, restricting access by *something you know* involves the use of some sort of secret/restricted identification code or password. These codes/passwords are issued to authorized users of the system, who should keep them confidential. To enter the system, users need enter only their ID/password.

Second, access can be restricted by *something you have,* such as a key. The key might be a physical key like the one used to unlock the front door of your home or to start your automobile, or it can be an electronic key (secret encrypted numbers contained in a small circuit chip). One such product is a domino-sized *ACCESS KEY*™ which adds user authentication to existing network security. With this ACCESS KEY, the user enters his or her ID/password, and if it is valid, the central system generates a flashing pattern on the microcomputer/terminal screen. Thus, an ACCESS KEY will not work without a valid ID/password. The user who wants access holds the domino-sized ACCESS KEY against the flashing pattern, and the ACCESS KEY generates a unique (one-time use) password that must be keyed in and transmitted back to the central system. The central system connects the remote user only if this unique password is correct.

This ACCESS KEY is composed of optical sensors, a custom integrated circuit, a quartz crystal real-time clock, a liquid crystal display, and a five-year lithium battery. Proprietary optical sensing and data scrambling (encryption) techniques are built into the electronic chip to prevent the device from being duplicated. Furthermore, the password generated by the ACCESS KEY is unique for each access, and the real-time clock automatically changes the encryption key contained in the user's ACCESS KEY every three days. Because each ACCESS KEY is unique to a specific user ID/password, there is positive identification as to which user accesses the system. This means that specific users can be given limited access once they gain entry into the network. This system was developed by Gordian Systems in Palo Alto, California. As you can see, "something you have" can be a physical key or an electronic key.

Third, to restrict access by *something you are,* one or more of your own physical characteristics is used as the unique identifier that permits your entry to the system. This might be a fingerprint, a handprint, an eye retina print, a voice print, signature verification, and so forth. Today these techniques may be too expensive, but in the future they will be the norm.

The following are specific security controls that might be implemented at the remote terminal/microcomputer location.

- Make sure that terminals have a unique electronic chip built in that provides positive identification. With chips, the front end or host can identify each terminal electronically.
- Physically lock terminal on/off switches or have locks that disable the screen and keyboard.
- Keep terminals in a physically secure location.

- Lock off all of the communication circuits after hours (positively disable the communication circuits).

- Assign an individual password to each system user.

- Make available to each user a plastic identification card that runs through an identification card reader. Such cards replace the need for individual passwords.

- Use special log-in numbers that can be entered only by a key person in the department.

- Consider using one of the newer types of personal identification such as signature identification, fingerprint identification, voice identification, or hand image identification.

- Transaction-code each terminal. This prevents any transaction that is not related to the work area in which the terminal is located. In other words, the terminal is made transaction specific.

- Develop a security profile of the types of data being entered and the user log-in procedures. If a violation occurs, the terminal that was used can be shut down automatically. In addition, a terminal security report should be delivered the next day to the manager of the user work area.

- Restrict terminals to read only functions.

- Sequence-number, time-stamp, and date all messages.

- Ensure that passwords do not print when they are typed.

- Ensure the proper disposal of hard copy terminal output.

- Allow intelligent terminals to perform editing on transactions before they are transmitted.

- When looking at the control and security of dial-up terminals, review the controls for dial-up modems that were listed previously in the section on modems.

Voice Telephone Security Voice telephone security is a means of restricting users when they use dial-up access to computer systems. We presented call-back restricters in the section on Port/Line Security Device in Chapter 3 and again, with the ACCESS KEY, in the previous section on Terminals—Security Controls. At this point we want to describe a small, inexpensive device that can be used to restrict the potential telephone user from even dialing certain numbers.

This device, called a *telephone sentinel*, is placed between the telephone and the wall jack into which the telephone plugs. It is about as large as a half carton of cigarettes, and it is placed directly over the current telephone wall plug. Using the RJ-11 telephone jack, you plug the sentinel into the wall plug, and the telephone is plugged into the telephone sentinel. This little box is programmed by using DIP switches. It is able to keep telephone users from dialing various numbers such as 0, 411, 555, and 976. As you may know, not being able to dial 0 keeps the user from calling a Bell operator; 411 from calling local directory assistance, for which organizations are now charged; 555 from calling long distance directory assistance; and 976 from calling certain services for which there is a charge, such as a religious message or the joke of the day. In case you did not recognize these numbers, most are the first three digits of a telephone exchange number such as 555-1212, which is long distance directory assistance when preceded by an area code.

This system also gives the caller a tone after a call has lasted for three minutes; this feature is used to keep telephone calls as short as possible in order to save costs. As you might have surmised by now, someone who is intent on violating the system can remove the telephone sentinel from the wall with the appropriate tool—a simple screwdriver. In an unattended area, the telephone sentinel therefore has to have some type of heavy-duty wire case around it which is bolted to the wall. The telephone sentinel is intended for areas in which company employees are present and where numerous telephone users might use the telephone for voice calls.

Because voice telephone calls using cellular radio (discussed in Chapter 2) are transmitted over radio waves, they can be intercepted easily with readily available equipment. To avoid interception, you can buy cellular telephones that have built-in voice scrambling. California recently passed a law that makes it illegal to eavesdrop on a telephone call of this nature. Unfortunately, this is an impossible law to enforce. Who knows when someone is eavesdropping? What about the computer hacker who hears your credit card number or some other vital information?

When you place a cellular telephone call, the telephone sends out a burst of data that gives your telephone number, the number called, and your telephone serial number. Anyone who has a radio scanner can intercept this information because it is broadcast over the radio. A person who wants to spy on telephone calls does not have to sit next to a radio scanner to find out who is calling whom. With a little know-how, a microcomputer can do the job. One simply links up the microcomputer and has it record all of the calls that are made. This is called *vacuum cleaning* because it sucks up the data transmitted during a telephone call. Vacuum cleaning can provide hackers with cellular telephone numbers that can be used easily for another breach of security called spoofing.

Spoofing is the relaying of telephone numbers gained through vacuum cleaning or other means, and reprogramming the telephones to make calls at another person's expense. With spoofing, people can get erroneous charges for calls placed fraudulently on their numbers. Spoofing with cellular telephones has a potential for causing far greater financial loss than the fraudulent use of stolen codes from traditional telephones. Losses can be great because you may know that a call came from a certain telephone number, but the number does not indicate the caller's location, which may change from moment to moment.

In addition to fraud, plain old-fashioned *eavesdropping* can become a serious breach of privacy. People should be careful of what they say when using cellular telephones.

Another potential problem is the ease with which the electronic serial numbers of cellular telephones can be changed. Serial numbers are nothing more than an integrated circuit chip glued into a socket. They can be pried loose easily and replaced with other chips that contain concocted or stolen serial numbers. As you can see, cellular telephone serial numbers do little more than give a false sense of security.

CIRCUIT CONTROLS

Some of the communication circuits that must be reviewed are the wire pairs and cables that are placed throughout the user facility, the local loops that go between the user facility and the common carrier (telephone company), and the interexchange channel (IXC) circuits between cities.

The wire pairs and cables within the user facility should be made as physically secure as possible because this is where anyone wanting to tap the system would enter. It is 100 times easier to tap a local loop than it is to tap an interexchange channel. Ensure that the lines are secured behind walls and above ceilings, and that the telephone equipment and switching rooms (wire closets) are locked and the doors equipped with alarms.

With regard to local loops, little can be done except to visit the common carrier switching facility. The visit can provide you with some idea as to the physical security, fire protection, and disaster prevention controls implemented by the common carrier. If these controls are inadequate, about the only thing you can do is split local loops among your facility and two or three different common carrier switching facilities (telephone company end offices).

For security on interexchange channels, encryption of messages is the only dependable method. If the data/information is so sensitive that a breach of privacy or the insertion/modification of a message cannot be allowed, then encryption must be considered.

With regard to internal cables within your user facility, the use of fiber optics might be considered. Fiber optic cable uses light-emitting diodes or laser light to transmit pulses of light through hair-thin strands of plastic or glass. These devices offer security through their immunity to electrically generated noise, resistance to taps, isolation, and small size. They also have some very special benefits when used in a harsh environment. All of these were discussed in the Chapter 2 section on Fiber Optics.

One more caution is in order with regard to the control and security of connector cables against surreptitious taps. The maximum 50-foot cable length of the RS232C or the 4,000-foot cable length of the RS449 could be prime targets. The RS449 offers extra control features such as special circuits for moving from a primary private line service to a packet switched service when backup is needed or simply to access another database that normally is not used. This eliminates manual patching, switching keys, and so on. The RS449 can invoke tests to isolate problems with either the local or remote data circuit terminating equipment (DCE) or the communication circuit itself.

MICROCOMPUTER CONTROLS[2]

The driving force behind the use of microcomputers, especially micro-to-mainframe connections and local area networks, is the desire of end users to use mainframe data with their own microcomputers which are located in their own departments. Users want to manipulate that data and have the ability to update mainframe databases in real time, as well as to interact with centralized host mainframe application systems. Neither economics nor competition between different organizations is going to stop this development. As a result, microcomputer security is an extremely important issue in today's business, government, and scientific environments.

The security of a microcomputer workstation can be divided into three major areas. *First*, all the physical security features that might be in place at a central host mainframe site also must be considered at the remote microcomputer location. Many of the security features have to be scaled down because of cost or the "smallness" of the microcomputer. For example, a large halon fire suppression system might be located at a host mainframe site, but a remote microcomputer site might have a small handheld fire extinguisher.

Second, data security is the protection of data and programs at the central site, during transmission, and at the microcomputer site. The level of data security required determines whether you allow dial-up connections to the central site, which pathways through the network are allowed, whether the data is encrypted, and what security is imposed on the database at both the central and remote sites. Database controls are discussed in the following section of this chapter.

Network access security is the *third* issue related to the microcomputer connection. It does not matter whether the microcomputer is being used in a micro-

---

[2] The book *Online Auditing Using Microcomputers* describes how to conduct a "hands-on" security review or audit of a microcomputer system by using 28 audit utility programs. These programs run on IBM PC, XT, AT, PS/2 or compatibles. See item 10 in the Selected References section at the end of this chapter for more information.

to-mainframe link, a local area network, or a dial-up link through a public packet switched network. Network access security is critical in any of these uses. Network security involves restricting access to the data/information and programs. To realize this network security in relation to the microcomputer, the general protection objectives should be

- Verification of message origin and destination.
- Verification of the timeliness of a message.
- Detection, by the receiver, of any modification to the message and/or the sequence of messages.
- Detection by the sender of a fraudulent acknowledgment of the receipt or nonreceipt of a message.
- Prevention of unauthorized disclosure of messages to any person not having unique identification, such as passwords or cryptographic keys.
- Positive acceptance of all messages. Once the system gives evidence of acceptance, it should never be lost, distorted, without proper authorization, or duplicated.
- Special network security requirements that are in effect to prevent an illegal entry at network restart time after the system has gone down or if it has been partially out of service.
- Notification to operational personnel, and specialized control reports as follow-up, when errors in message transmission, routing, formatting, or other anomalies occur.
- Control of access to the network and its facilities so that only authorized personnel have access. Complete audit trails should show who accessed what and when the access occurred.
- Rapid and automatic detection and notification of the type and location of any security breach.
- Safe storage of all messages in a front end/host system's log.

Later in this chapter a matrix of controls is discussed. Use of this matrix approach is recommended as a convenient method of identifying the specific threats that relate to microcomputers. These threats then are cross-related to the various component parts of each microcomputer location. Finally, the necessary controls are cross-related to the previously identified threats and components to build a two-dimensional matrix that shows the controls required at each microcomputer location. The matrix concept is augmented with various lists of controls in Appendices 1 and 2.

DATABASE CONTROLS

In the previous section, we discussed three major kinds of security as they relate to microcomputers. The subject is so important that we want to add more material on the security of data. Security experts maintain that there is no foolproof way to protect the data in a microcomputer. Even so, databases must be protected, and so the following are some guidelines for making databases secure.

- Use the central host mainframe management software security features. Along with them, use the volume, file, and record locking features that might be available in the file server or network server installed at the microcomputer. Finally, use all the security features in the "security software package" that is installed on the host mainframe computer.

- Restrict entry to the system based on something the user knows (password), something the user has (key), or something the user is (fingerprints).

- Once a user is in the database, define the user's limits such as the ability to read, write, create, delete, search, modify, and the like.

- Appoint a database administrator who can grant or revoke various security privileges and also interface between different users with regard to the databases. This person is the custodian of the databases and also ensures proper backup and recovery procedures.

- Use dial-back security devices to restrict entry.

- Consider data encryption. In addition to encrypting data for transmission over a network, if the need is for highly secure data, you can encrypt it prior to storage on the diskette and then decrypt it when it is read from the diskette. This ensures that the data is useless if someone steals the diskette.

- Monitor and limit data downloading to microcomputers and the uploading of that data back to the host mainframe. When the data is in the host mainframe database, it may be protected adequately, but once it gets to the remote microcomputer, many of the controls that might prevent errors or other data corruption are lost.

- Do not keep extremely sensitive data online.

- Screen and classify both users and data. Not everyone must see everything. Some organizations, especially in government, assign different security clearance levels to users and also to data, thus permitting users to see only what they "need to know."

- Create a secure physical environment in the area around the database/microcomputer workstation.

PROTOCOL CONTROLS

Protocols are simply the rules by which two machines talk to each other. The word protocol comes from the Greek *prōtokollon,* which was the first sheet glued to a papyrus roll; it was the table of contents.

As discussed in Chapter 5, the International Organization for Standardization (ISO) has developed a seven-layer model (OSI Model) for protocols. A brief description of the tasks performed by each layer, along with the *controls* that should be in each layer, follows. (Chapter 5 contains a detailed description of the *tasks* performed by each layer.) The OSI model is the most popular protocol model that has been approved by international standards-setting organizations.

Layer 1—Physical Link Control This layer is the cable, modem, and physical circuitry. It is concerned with transmitting a serial stream of data bits over the communication circuit. It cannot interpret and does not know that the data may be in packets/frames, nor does it know about error checking, and so forth. This is the physical link over which all data bits move.

Controls are needed to physically protect the connector cable, modem, and communication circuits. The primary goal at this layer is to control physical access by employees or vendors to the connector cables and modems. This includes restricting their access to the wiring closets in which all the communication wires and cables are connected. All local loops leaving the building should be physically secured and out of harm's way to prevent physical damage or an easy telephone tap. Formal procedures should be in effect to help identify breaches of security or illegal entries to the system when made at the physical layer. Finally, physical protection of the telephone circuits (IXC channels) is the responsibility of the common carrier (telephone company). You cannot do much about it, except to audit the telephone company's physical security procedures and possibly encrypt the data before it leaves your building to go out onto the public network.

Layer 2–Data Link Control This layer is responsible for packaging the data into frames or packets and getting it onto the RS232 cable so it can get to layer 1. It also is responsible for calculating the error checking polynomial (CRC) and inserting addresses and so forth into the frame or packet.

Controls at this layer should have features such as sequence counting of frames, error detection and retransmission capabilities, identification of lost frames, and reduction of possible duplicate transmissions to zero. It should solve problems caused by damaged/lost/duplicate frames, prevent a fast transmitter from drowning a slow receiver in data, provide limited restart capabilities in case of abnormal termination situations, ensure that some of the transmitted data are not misinterpreted as line control characters, increase flow control efficiency to ensure that the maximum number of frames can be sent without requiring an acknowledgment, properly terminate a session, and the like.

Layer 3—Network Control   This layer is responsible for translating logical addresses (names) into physical addresses (nodes or devices). It also chooses the best circuit route if more than one is available and packetizes the original message into the designated packet lengths (usually 128 characters).

Controls to be questioned in this layer involve who should ensure that all packets are received correctly and in their proper order at their destinations. This layer of protocol should accept messages from the host, convert them to packets, and ensure that the packets get directed toward their destination. Packet routing should be controlled here. Also, there might be some global or local databases at this layer that should be kept secure. Control of congestion, such as too many packets on one channel, should be controlled by this layer. This layer also can contain billing routines for charging users and should be reviewed for possible problems such as error, theft of time, or improper message charges.

Layer 4—Transport Control   Layer 4 is responsible for establishing and maintaining the connection and initiates the retransmission request when notified by layer 2 that a packet/frame arrived in a corrupted state. It sometimes is called the host-to-host layer.

Controls that should be checked at layer 4 are related to network connections because the transport layer might have to create multiple network connections in order to get the required number of circuit paths. At this layer multiplexing might be invoked, so multiplexing controls should be reviewed. Also at this layer a program on a source machine carries on a conversation with a similar program on the destination machine using headers and control messages; therefore, some of the controls might be in the application programs. At the lower layers (layers 1 to 3) the protocols are carried out by each machine and its immediate neighbors rather than by the ultimate source and destination machines, which may be separated by many hardware devices and circuit links. Another needed control is one that determines whether the software at this level can tell which machine belongs to which connection. Other controls that are performed at this level, even though they may be performed elsewhere as well, are source/destination machine addressing and flow control (here it is flow of messages rather than flow of packets) so one machine cannot overrun another.

Layer 5—Session Control   This layer supports the user's session and turns communication on and off between two workstations. If a temporary electrical disruption on the network breaks the circuit, it often can get the lower layers (1 to 4) to reestablish the connection before the application program or the operator is aware of a problem.

Controls that should be examined at this layer are the typical controls that relate to a terminal (dedicated or dial-up) such as passwords, log-in procedures, terminal addressing procedures, authentication of terminals and/or users, and correct delivery of the bill. Another control occurs when the transport control (layer 4) connections are unreliable; the session layer may be required to attempt to recover from broken transport connections. As another example, in database management systems it is crucial that a complicated transaction against the database never be aborted halfway through the routine because this leaves the database in an inconsistent state. The session layer often provides a facility by which a group of messages can be set aside so none of them are delivered to the remote user until all of them have been completed. This mechanism ensures that a hardware or software failure within the subnetwork never can cause a transaction to be aborted halfway through its processing. The session layer also can provide for sequencing of messages when the transport layer does not.

Layer 6—Presentation Control This layer translates commands from the application layer above it, providing a syntax (language) that is commonly understood by all devices throughout the network.

Controls at this layer might be software encryption, text compression, text compaction, and conversion of incompatible file formats/file conversions so two systems can talk to one another. This layer also can take incompatible terminals and modify line and screen length, end-of-line conventions, scroll versus page mode, character sets, and cursor addressing to make them compatible. Simple errors at the remote terminal might be caused by the software at this layer.

Layer 7—Application Control This layer does not carry out the application itself. It provides such functions as file transfer and electronic mail. It interfaces your application program to the network.

Controls at this layer are related to the business system application programs used in the organization. These application layer controls consist of the tasks that ensure a valid connection between whatever application program is being utilized in the organization and the addressing or transmission of messages required by this business application. Because the purpose of layer 7 is to insulate application programs from the detailed technical task of transmitting messages, controls should ensure that no messages get lost, misdirected, or improperly modified by this layer of software. Many of the controls that are required here are the typical day-to-day logical controls that are built into the business application system itself, although layer 7 also should contain internal controls within the communication software.

NETWORK ARCHITECTURE/SOFTWARE CONTROLS

Controls that relate to network architecture/software typically are associated with layers 4 through 7 of the OSI model. Additional architecture/software controls relate to computer operating systems, teleprocessing monitors, telecommunication access programs, databases, and security software packages.

Teleprocessing monitors relieve the operating system of many of the tasks involved in handling message traffic between the host and remote terminals such as line handling, access methods, task scheduling, and system recovery. System throughput is increased by offloading these data communication functions from the operating system to the teleprocessing monitor.

Some of the controls that should be reviewed for teleprocessing monitors are access controls, for example, who can sign on to a terminal, and who can access program routines (sometimes called *exits*). With regard to these exits, the code of each exit routine should be checked for correctness and security. These exit program modules should be placed in software-controlled libraries.

As an example, let us discuss IBM's teleprocessing monitor CICS (Customer Information Control System). There are too many security features in this package to provide exhaustive coverage here, but two selected items are the LTERM and the sign-on table.

The *LTERM* is the logical terminal address. You can restrict a specific terminal (LTERM) so it can execute only certain types of functions. Obviously, this is a further restriction over and above the password that the user may have had to use to enter the system. In addition, the LTERM restriction is transparent because the user's terminal is restricted without regard to the user entering any password or identification code. This restriction is based on the terminal itself and how the terminal is defined in the LTERM.

The *sign-on table* in CICS is one of 64 levels or systems. It is cross-referenced to the user's identification (password) and therefore can restrict various user identifications/passwords from using CICS. It is essential that you do not assign the security keys in the sign-on table to operators who may have a conflict of interest and that you restrict their transaction capability based on their identification/password. Notice that LTERM and the sign-on table have the ability to restrict what a terminal might access and they also restrict the terminal operator's access.

CICS security packages also are available from vendors other than IBM. These packages offer a unique password and identification of each terminal operator and allow access only to the specific functions assigned to that operator. For instance, it can assign highly sensitive functions to a specific terminal or a group of terminals. Some also offer encrypted security data, security sign-on fields, darkened password fields, and a complete log of terminal sign-ons that includes any security violations. They offer additional features, such as file security to the record/field level, terminal security, transaction security, batch reports on all activity, automatic sign-off of unattended terminals, and immediate online notification of security violations.

The vendor's "system generation" manuals or the teleprocessing monitor should be reviewed to determine whether the vendor built any security controls into the teleprocessing monitor.

Telecommunication access programs are vendor-supplied programs that control the transmission of data to and from the host computer and various data communication devices. The telecommunication access programs most likely reside in a front end communication processor, but they also can reside in the host computer.

Some controls for the telecommunication access program may be documented in the vendor's system generation manuals. The controls that were built in by the software vendor should be evaluated. As with teleprocessing monitors, user program routines (exits) and access methods need to be examined. Review the list of front end functions that were described in Chapter 3.

Another area that interfaces with the teleprocessing monitor is the very sensitive network control database. For example, there might be a *system database* containing global information about addresses and logical names of all peripheral devices, locations of system files, system timing parameters, task locations and priorities, peripheral device control tables, and system supply command lists. Because of its importance to the security of the entire system, data in the system database must be protected from copying or destruction.

Another database is the *network database.* It contains data such as the number of stations, polling/selecting lists, current station identifier, communication control port addresses, logical terminal identifiers, terminal device list, terminal poll/call sequences, dial-up numbers, and message and process information. The network database also must be protected from unauthorized copying.

The network database can be used to cross-check against manual documentation. Because it contains information such as the number of stations and logical terminal identifiers, a copy of the database can be matched against the written network documentation as a means of verifying the currency of the documentation.

The impact of any security software packages that restrict or control access to files, records, or data items should be reviewed. These packages are independent of the data communication software.

Finally, software should be protected in case of a disastrous situation such as a power failure. Restart/recovery routines should be available, and the system should have only one master input terminal for entering sensitive or critical commands. All default options should be identified, and the impact of default options that do not operate properly should be assessed to determine whether adequate software maintenance is available. In addition, all sensitive tables (passwords) should be protected in the memory.

MANAGEMENT CONTROLS

Network management involves setting up a central control philosophy for handling the overall network functions. The network manager should be independent of the other managers in the data processing environment or even of the information systems department itself. Some of the general responsibilities of this job function are design and analysis, network operations, failure control, and testing/problem management (see Chapter 8).

Some network management controls include the following. The network management team should have a national account with the common carrier when possible. There should be a central call number to log all problems related to who, what, where, when, why, the telephone number, date, and time of a problem. The failure control group should compile statistics for their hardware such as Mean Time Between Failures (MTBF). The network hardware vendor often can supply these data. The network management people also should maintain statistics on the time from failure to recovery. In its most detailed form, this is comprised of the Mean Time To Diagnose (MTTD) plus the Mean Time To Respond (MTTR) plus the Mean Time To Fix (MTTF). (Chapter 8 presents detailed usage for MTBF, MTTD, MTTR, and MTTF.)

In addition to central control for problem reporting on the network, the network management group should maintain other statistics such as the cumulative network downtime, subnetwork downtime, circuit utilization reports, response time analysis, queue-length descriptions, histograms of daily usage (such as number of characters transmitted per day per circuit), failure rates of the circuits (such as number of retransmitted messages), local host and file activity statistics, local device error activities, network gateway failures, distribution of character and packet volume, distribution of traffic by time of day and location, peak volumes, and a statistical profile of all time-related network traffic. These reports, or similar ones, should be available for managing the network.

With regard to network documentation (some of this can be cross-checked for currency in the global or local network databases), a good network management team should have the following: circuit layout record, network maps, hardware/software cross-references, all network vendor maintenance records, software listings by network task and component, all user site telephone numbers and names of individuals to contact, an interface component maintenance history log, circuit-controlled telephone contact index, maintenance history by component, inventory by serial number of network components, network redundancy locations and switching criteria, vendor contracts, vendor contacts, and a current list of personnel working in the network center.

A control review of network management should ascertain the existence of appropriate operation manuals and a comprehensive description of how the network operates. There should be adequate recovery procedures, backup procedures and disaster plans.

Not much can be done to control communication test equipment because it is in continuous use. Network management must recognize that misuse can allow breaches of privacy or the insertion of illegal messages.

A network monitor is mandatory when bit-oriented protocols such as X.25 are used. The only control that can be put on such a device is a keylock for its switch. The keylock should be turned off and the key removed when the equipment is not being used. This prevents people from browsing over data as they pass through the data communication network.

Loopback test equipment should be used to diagnose the location and cause of problems.

Microprocessor-based network analyzers permit checks for poll-to-poll or poll-to-response times. Such checks aid network management in assessing polling efficiency.

Other, smaller handheld test devices, such as breakout boxes, allow test personnel to send test patterns of data bits to a modem through the RS232C or RS449 cable.

The primary control for test equipment is to ensure that only qualified people use this equipment and that they use it only when necessary.

RECOVERY/BACKUP/DISASTER CONTROLS

Establishing a disaster recovery capability requires devising recovery procedures and backup for the entire environment (building) and for hardware, software, data, and communication networks. An extremely detailed and broad-scoped disaster recovery plan can bring the organization back to a functioning business should almost any type of disaster strike. This overall plan should address various levels of response to a number of possible disasters and should provide for partial or complete recovery in the areas of

- The building (environment)
- System software and utilities
- Application programs
- Data center hardware
- Remote microcomputers/terminals
- Manual forms
- Data entry support
- On-site and off-site database file storage/retention
- System operating procedures at the data center and at remote user areas
- Staff assignments and responsibilities during the disaster
- Adequate updating and maintenance of the disaster plan
- All communication networks (private leased backbone, local area networks, micro-to-mainframe, public dial-up, and so forth)

A complete disaster recovery plan covering all these areas is beyond the scope of this text, but we intend to address the formulation of a data communication recovery disaster plan.

Recovery and backup controls within the data communication network encompass many areas. The person who reviews these controls may start at either end of the network (remote terminals or central host computer). The object is to check for recovery procedures and backup hardware throughout the network. Perhaps the most important question to ask is whether it is cost effective to back up each piece of hardware encountered between a remote terminal site and the central host computer. A related question is: Are there software procedures for recovery of data files, network databases, network software, and the like? Use of Figure 9-3 during the review of recovery/backup/disaster controls helps ensure that all network control points are considered.

An important consideration is backup of the communication circuits. One option is to lease two separate circuits (that have been alternately routed) to have one for backup. Another option is to utilize dial-up communication circuits as backup to leased circuits. Of course, another alternative is to have manual procedures that can be used if the circuit is down for a very short period of time, perhaps several hours at the maximum.

There should be recovery and restart capabilities in the event of either a hardware crash or a software crash. Backup facilities should include backup power, possibly at both the local and remote sites.

A data communication network disaster plan should include a separate plan for each of five different areas: (1) the data communication network control center, (2) communication circuits, (3) remote switches/concentrators/intelligent terminal controllers, (4) common carrier (telephone company) facilities, and (5) electric power for the data communication facilities and user terminals/lights.

A data communication network disaster plan should spell out the following details.

- The decision-making manager who is in charge of the disaster recovery operation. A second manager should be indicated in case the first manager is unavailable.

- Availability and training of backup personnel with sufficient data communication knowledge and experience.

- Recovery procedures for the data communication facilities. This is information on the location of circuits, who to contact for backup data circuits and documentation, and preestablished priorities as to which data circuits should be reconstructed first.

- How to replace damaged data communication hardware and software that is supplied by vendors. Outline the support that can be expected from vendors, along with the name and telephone number of the person to contact.

- Location of alternate data communication facilities and equipment such as connector cables, local loops, IXCs, common carrier switching facilities, and other public networks.

- Action to be taken in case of partial damage, threats such as a bomb threat, fire, water or electrical damage, sabotage, civil disorders, or vendor failures.

- Procedure for imposing extraordinary controls over the network until the system returns to normal.

- Storage of the disaster recovery procedures in a safe area where they cannot be destroyed by the catastrophe. This area must be accessible, however, to those who need to use the plans.

MATRIX OF CONTROLS

To be sure that the data communication network and microcomputer workstations have all the necessary controls and that these controls offer adequate protection, it is advisable to build a two-dimensional matrix incorporating all the controls that *currently* are in the network.

This matrix is constructed by identifying, first, all threats facing the network and second, all the network's component parts.

- A *threat* to the data communication network is any potential adverse occurrence that can harm the network, interrupt the systems using the network, or cause a monetary loss to the organization. For example, lost messages are a potential threat.

- A *component* is one of the individual pieces that, when assembled together, make up the data communication network. A component can be viewed as the item that is being reviewed or the item over which we are attempting to maintain control. Thus, the components are the hardware, software, circuits, and other pieces of the network.

In Figure 9-6 several *general* threats to a data communication network are defined. Figure 9-7 defines several *general* component parts for a data communication network.

Identifying and documenting the controls in a network requires identifying the *specific* threats and components that relate to whatever network is used by the organization. After the organization's specific threats and components are identified, the individual controls that are in place can be related to these threats and components.

Once the threats and component parts of the network have been identified, the next step is to place a short description of each threat across the top of the matrix.

- Errors and Omissions—The accidental or intentional transmission of data that is in error, including the accidental or intentional omission of data that should have been entered or transmitted on the online system. This type of exposure includes, but is not limited to, inaccurate data, incomplete data, malfunctioning hardware, and the like.

- Message Loss or Change—The loss of messages as they are transmitted throughout the data communication system, or the accidental/intentional changing of messages during transmission.

- Disasters and Disruptions (natural and manmade)—The temporary or long-term disruption of normal data communication capabilities. This exposure renders the organization's normal data communication online system inoperative.

- Breach of Privacy—The accidental or intentional release of data about an individual, assuming that the release of this personal information was improper to the normal conduct of the business at the organization.

- Security/Theft—The security or theft of information that should have been kept confidential because of its proprietary nature. In a way, this is a form of privacy, but the information removed from the organization does not pertain to an individual. The information might be inadvertently (accidentally) released, or it might be the subject of an outright theft. This exposure also includes the theft of assets such as might be experienced in embezzlement, fraud, or defalcation.

- Reliability (Uptime)—The reliability of the data communication network and its "uptime." This includes the organization's ability to keep the data communication network operating and the Mean Time Between Failures (MTBF) as well as the time to repair equipment when it malfunctions. Reliability of hardware, reliability of software, and the maintenance of these two items are chief concerns here.

- Recovery and Restart—The recovery and restart capabilities of the data communication network, should it fail. In other words, How does the software operate in a failure mode? How long does it take to recover from a failure? This recovery and restart concern also includes backup for key portions of the data communication network and the contingency planning for backup, should there be a failure at any point of the data communication network.

- Error Handling—The methodologies and controls for handling errors at a remote distributed site or at the centralized computer site. This also may involve the error handling procedures of a distributed data processing system (at the distributed site). The object here is to ensure that when errors are discovered they are promptly corrected and reentered into the system for processing.

- Data Validation and Checking—The validation of data either at the time of transmission or during transmission. The validation may take place at a remote site (intelligent terminal), at the central site (front end communication processor), or at a distributed intelligence site (concentrator or remote front end communication processor).

FIGURE 9-6 General threats to a data communication network.

- Host Computer—Most prevalent in the form of a central computer to which the data communication network transmits and from which it receives information. In a distributed system, with equal processing at each distributed node, there might not be an identifiable central computer (just some other equal-sized distributed computer).
- Software—The software programs that operate the data communication network. These programs may reside in the central computer, a distributed system computer, the front end communication processor, a remote concentrator or statistical multiplexer, and/or a remote intelligent terminal. This software may include the telecommunication access methods, an overall teleprocessing monitor, programs that reside in the front end processors, and/or programs that reside in intelligent terminals.
- Front End Communication Processor—A hardware device that interconnects all the data communication circuits (lines) to the central computer or distributed computers and performs a subset of the following functions: code and speed conversion, protocol, error detection and correction, format checking, authentication, data validation, statistical data gathering, polling/addressing, insertion/deletion of line control codes, and the like.
- Multiplexer, Concentrator, Switch—Hardware devices that enable the data communication network to operate in the most efficient manner. The *multiplexer* is a device that combines, in one data stream, several simultaneous data signals from independent stations. The *concentrator* performs the same functions as a multiplexer except it is intelligent and therefore can perform some of the functions of a front end communication processor. A *switch* is a device that allows the interconnection between any two circuits (lines) connected to the switch. There might be two distinct types of switch: a switch that performs message switching between stations (terminals) might be located within the data communication network facilities that are owned and operated by the organization; a circuit or line switching switch that interconnects various circuits might be located at (and owned by) the telephone company central office. For example, organizations perform message switching and the telephone company performs circuit switching.
- Communication Circuits (Lines)—The common carrier facilities used as links (a *link* is the interconnection of any two stations/terminals) to interconnect the organization's stations/terminals. These communication circuits include satellite facilities, public switched dial-up facilities, point-to-point private lines, multiplexed lines, multipoint or loop configured private lines, and many others.
- Local Loop—The communication facility between the customer's premises and the telephone company's central office or the central office of any other special common carrier. The local loop usually is assumed to be metallic pairs of wires.
- Modems—A hardware device used for the conversion of data signals from terminals (digital signal) to an electrical form (analog signal) which is acceptable for transmission over the communication circuits that are owned and maintained by the telephone company or other special common carrier.
- People—The individuals responsible for inputting data, operating and maintaining the data communication network equipment, writing software programs for data communications, managing the overall data communication network, and those involved at the remote stations/terminals.
- Terminals/Distributed Intelligence—Any or all of the input or output devices used to interconnect with the online data communication network. This resource specifically includes, without excluding other devices, teleprinter terminals, video terminals, remote job entry terminals, transaction terminals, intelligent terminals, and any other devices used with distributed data communication networks. These may include microcomputers or minicomputers when they are input/output devices or if they are used to control portions of the data communication network.

FIGURE 9-7 General components of a data communication network.

Likewise, a short description of each component is placed down the left vertical axis of the matrix as shown in Figure 9-8.

When the horizontal and vertical axes have been labeled, the next step is to identify all of the specific controls that currently are being used in the data communication network. These "in-place" controls should be described and placed in a numerical list. For example, assume 24 controls have been identified as being in use in the network. Each one is described, and they are numbered consecutively 1 through 24. The numbered list of controls has no ranking attached to it: the first control is number 1 just because it is the first control identified. Figure 9-9 shows what a list of in-place controls looks like.

Next, each of the controls that has been identified is placed in the proper cell of the matrix. This is accomplished by reading the description of each control in the control list and then asking the following two questions.

1. Which threat or threats does this control mitigate or stop?
2. Which component or components does this control safeguard or restrict?

For example, if the description of Control 1 is "ensure that the system can switch messages for a down station/terminal to an alternate station/terminal," then the number 1 should be placed in the very first cell in the upper left corner (see Figure 9-10). It is given this position because a control that ensures that the system can switch messages when a station is down helps control errors, and it also is a control that safeguards or resides in the host computer and/or front end. Figure 9-10 also shows Control 1 in the cell that intersects between Message Loss or Change and Host Computer. Control 1 also appears in several other cells. The point is that by answering these two questions, you can place each control in the proper cells of the matrix.

The finished matrix with controls (Figure 9-10) shows the interrelationship of each "in-place" control to the threat that it is supposed to mitigate and the component that it safeguards or controls.

For a complete list of all of the controls for Figure 9-10 along with this control matrix, see Appendix 1.[3]

The last step in designing a custom matrix of controls for your specific data communication network is to evaluate the adequacy of the controls. This is accomplished by reviewing each subset of controls as it relates to each threat and component area of the matrix. For example, the subset of controls that are listed down a column below a threat are evaluated. The object of this step is to answer

---

[3] This material was taken from Chapter 4 of *Internal Controls for Computerized Systems* by Jerry FitzGerald, published by Jerry FitzGerald & Associates, 506 Barkentine Lane, Redwood City, Calif. 94065.

THREATS

COMPONENTS	Errors and Omissions	Message Loss or Change	Disasters and Disruptions	Breach of Privacy	Security/ Theft	Reliability (Uptime)	Recovery and Restart	Error Handling	Data Validation and Checking
Host Computer or Central System									
Software									
Front End Communication Processor									
Multiplexer, Concentrator, Switch									
Communication Circuits (Lines)									
Local Loop									
Modems									
People									
Terminals/ Distributed Intelligence									

FIGURE 9-8 Blank matrix with border labels.

1. Ensure that the system can switch messages destined for a down station/terminal to an alternate station/terminal.

2. Determine whether the system can perform message switching to transmit messages between stations/terminals.

3. In order to avoid lost messages in a message-switching system, provide a store and forward capability. This is where a message destined for a busy station is stored at the central switch and then forwarded at a later time when the station is no longer busy.

4. Review the message or transaction logging capabilities to reduce lost messages, provide for an audit trail, restrict messages, prohibit illegal messages, and the like. These messages might be logged at the remote station (intelligent terminal), they might be logged at a remote concentrator/remote front end processor, or they might be logged at the central front end communication processor/central computer.

5. Transmit messages promptly to reduce risk of loss.

6. Identify each message by the individual user's password, the terminal, and the individual message sequence number.

7. Acknowledge the successful or unsuccessful receipt of all messages.

24. Consider the following special controls on dial-up modems when the data communication network allows incoming dial-up connections: change the telephone numbers at regular intervals; keep the telephone numbers confidential; remove the telephone numbers from the modems in the computer operations area; require that each "dial-up terminal" have an electronic identification circuit chip to transmit its unique identification to the front end communication processor; do not allow automatic call receipt and connection (always have a person intercept the call and make a verbal identification); have the central site call the various terminals that will be allowed connection to the system; utilize dial-out only where an incoming dialed call triggers an automatic dial-back to the caller (in this way the central system controls those telephone numbers to which it will allow connection).

FIGURE 9-9 Control list.

the specific question, "Do we have the proper controls and are they adequate with regard to each specific threat?" Using Figure 9-11, look down the column under errors and omissions. The matrix clearly defines the specific subset of controls that relate to the threat area Errors and Omissions. They are 1, 2, 3, 4, 7, 12, 18, and 5.

This type of review also can be performed for various other subsets of controls. For example, individual subsets of controls can be evaluated as they relate to threats (columns), components (rows), individual cells, and empty cells. Looking

THREATS

COMPONENTS	Errors and Omissions	Message Loss or Change	Disasters and Disruptions	Breach of Privacy	Security/Theft	Reliability (Uptime)	Recovery and Restart	Error Handling	Data Validation and Checking
Host Computer or Central System	1, 2, 3, 4, 7	1, 2, 3, 4, 5, 7	1, 8, 11, 13, 16	6, 8, 24	6, 8, 24	1, 13, 16			6, 24
Software	1, 2, 3, 4, 7	1, 2, 3, 4, 5, 7	1, 8, 16	6, 8, 24	6, 8, 24	1			6, 24
Front End Communication Processor	1, 2, 3, 4, 7	1, 2, 3, 4, 5, 7	1, 8, 13, 16	6, 8, 24	6, 8, 24	1, 13, 16			6, 24
Multiplexer, Concentrator, Switch	1, 2, 3, 4, 7	1, 2, 3, 4, 5, 7	1, 8, 13, 16	6, 8, 24	6, 8, 24	1, 13, 16			6, 24
Communication Circuits (lines)	12		10, 15, 16, 18			15, 16			
Local Loop	12								
Modems	12, 18	18, 24	8, 9, 10, 11, 13, 14, 15, 16, 18	24	24	9, 10, 11, 13, 14, 15, 16, 17, 18	9, 10, 11, 14, 15	18, 19, 20, 22, 23	
People	5	5, 7		6, 8, 24	6, 8, 24				6
Terminals/ Distributed Intelligence		2		6, 8, 24	6, 8, 24	1			6, 24

FIGURE 9-10 Matrix with controls.

452

THREATS

COMPONENTS	Errors and Omissions				Security/ Theft	Reliability (Uptime)	Recovery and Restart	Error Handling	Data Validation and Checking
Host Computer or Central System	1, 2, 3, 4, 7	1, 2, 3, 4, 5, 7	1, 8, 11, 13, 16	6, 8, 24	6, 8, 24	1, 13, 16			6, 24
Software	1, 2, 3, 4, 7	1, 2, 3, 4, 5, 7	1, 8, 16	6, 8, 24	6, 8, 24	1			6, 24
Front End Communication Processor	1, 2, 3, 4, 7	1, 2, 3, 4, 5, 7	1, 8, 13, 16	6, 8, 24	6, 8, 24	1, 13, 16			
Multiplexer, Concentrator, Switch	1, 2, 3, 4, 7	1, 2, 3, 4, 5, 7	1, 16			13, 16			6, 24
Communication Circuits (lines)	12		10, 15, 16, 18			15, 16			
Local Loop	12								
Modems	12, 18	18, 24	8, 9, 10, 11, 13, 14, 15, 16, 18	24	24	9, 10, 11, 13	10, 11	18, 19, 20, 22, 23	
People	5	5, 7		6, 8, 24	6, 8, 24				6
Terminals/ Distributed Intelligence		2		6, 8, 24	6, 8, 24				6, 24

Annotations (callouts):

- "This shows the subset of controls that mitigate the threat, Errors and Omissions."
- "Empty cells show a lack of control which may be a serious problem."
- "This shows the subset of controls that control the component, Communication Circuits (Lines)."
- "Some individual cells may be more sensitive to your network or your company; therefore, the controls in these cells should be reviewed very closely."

FIGURE 9-11 Evaluation of the control matrix.

at Figure 9-11, we see a pictorial diagram describing the above four areas that should be reviewed. The matrix approach offers a perfect tool for a detailed microanalysis of the controls in a data communication network. The matrix clearly shows the relationship between various subsets of controls and specific threat areas, component parts, individual cells, and empty cells.

The entire procedure of designing and developing a matrix of controls has been automated for use on a microcomputer. Three packages are available. (A free DEMO disk is available; see item 7 in the Selected References section at the end of this chapter.)

- Package 1 teaches how to design and develop a controls matrix.
- Package 2 teaches how to risk-rank the matrix to show regions of high, medium, and low risk.
- Package 3 allows online drawing of a matrix, interactive searching of the database of controls for controls related to any cell (basic artificial intelligence), and the printing of a matrix with its associated matrix control list.

LISTS OF DATA COMMUNICATION CONTROLS

To help you construct a matrix of controls that relate to your organization's data communication network, we have supplied 14 specific lists of controls.[4] Appendix 2 contains 14 control lists for data communication networks. Each list has a definition of that particular area and is followed by its own set of controls. The areas that are addressed are

- Software controls
- Disasters and disruptions
- Modems
- Multiplexer, concentrator, switch
- Communication circuits (lines)
- Error handling
- Local loop (lines)
- Data entry and validation
- Errors and omissions

[4]These lists of controls were taken from *Designing Controls into Computerized Systems* by Jerry FitzGerald, published by Jerry FitzGerald & Associates, 506 Barkentine Lane, Redwood City, Calif. 94065. The 14 lists of controls in Appendix 2 were extracted from the 101 lists of controls in the above book.

- Restart and recovery
- Message loss or change
- People controls
- Front end communication processor
- Reliability/uptime

SELECTED REFERENCES

1. Carroll, John M. *Computer Security*, 2nd ed. Stoneham, Mass.: Butterworths, 1987.

2. *Computer Fraud and Security Bulletin.* Published monthly by Elsevier International Bulletins, Mayfield House, 256 Banbury Road, Oxford OX2 7DH, England, 1978– .

3. *Computer Security Products Report.* Published quarterly by Assets Protection Publishing, P.O. Box 5323, Madison, Wis. 53705, 1986– .

4. Davies, D. W., and W. L. Price. *Security for Computer Networks: An Introduction to Data Security in Teleprocessing and Electronic Funds Transfer.* New York: John Wiley & Sons, 1984.

5. Delberg, Andre L., Andrew H. Van de Ven, and David H. Gustafson. *Group Techniques for Program Planning: A Guide to Nominal Group and Delphi Processes.* Middleton, Wisc.: Green Briar Press, 1986.

6. *EDPACS: The EDP Audit, Control and Security Newsletter.* Published monthly by Automation Training Center, 11250 Roger Bacon Drive, Reston, Va. 22090, 1973– .

7. FitzGerald, Jerry. *Control Matrix Methodology for Microcomputers* (three software packages). Redwood City, Calif.: Jerry FitzGerald & Associates, 1987. Free DEMO disk available.

8. FitzGerald, Jerry. *Designing Controls into Computerized Systems.* Redwood City, Calif.: Jerry FitzGerald & Associates, 1981.

9. FitzGerald, Jerry. *Internal Controls for Computerized Systems.* Redwood City, Calif.: Jerry FitzGerald & Associates, 1978.

10. FitzGerald, Jerry. *Online Auditing Using Microcomputers.* Redwood City, Calif.: Jerry FitzGerald & Associates, 1987. (Includes a diskette with 28 audit utility programs to be used for a security or audit review of a microcomputer system.)

11. Moulton, Rolf T. *Computer Security Handbook: Strategies and Techniques*

for Preventing Data Loss or Theft. Englewood Cliffs, N.J.: Prentice-Hall, 1986.

12. Parker, Donn B. *Fighting Computer Crime.* New York: Charles Scribner's Sons, 1983.

13. Ruthberg, Zella G., ed. *Audit and Evaluation of Computer Security.* Proceedings of the NBS Invitational Workshop held at Miami Beach, Florida, March 22–24, 1977. Washington, D.C.: National Bureau of Standards, October 1977 (NBS Special Publication 500-19).

14. Ruthberg, Zella G., ed. *Audit and Evaluation of Computer Security II: System Vulnerabilities and Control.* Proceedings of the NBS Invitational Workshop held at Miami Beach, Florida, November 28–30, 1978. Washington, D.C.: National Bureau of Standards, April 1980 (NBS Special Publication 500-57).

15. Van Duyn, Julia. *Human Factor in Computer Crime.* Princeton, N.J.: Petrocelli Books, 1985.

16. Walsh, T. J., and R. J. Healy. *Protection of Assets Manual.* Santa Monica, Calif.: Merritt Co., 1982.

QUESTIONS/PROBLEMS

1. Look at Figure 9-3. To protect your message as it moves from the earth station to the satellite, which of the 18 control points is the primary one that you should consider?

2. Is it possible to first encrypt with a public key and then decrypt with a private (secret) key, as well as to first encrypt with the private key and then decrypt that message with the public key?

3. Do packet switching controllers have job tasks beyond that of a typical front end communication processor?

4. What do you think are the three most important security controls that can be placed on a terminal?

5. Which layer of the OSI model performs controls such as error detection and retransmission or sequence counting of contiguous frames for transmission?

6. As the manager of a major data communication network, how many types of disaster plans might you consider?

7. Define a threat and a component.

8. What if you encrypted a file on your microcomputer and lost the "key." How long would it take you to break the key if

- Your microcomputer could test one key each second.
- There were only 25 million possible keys.
- The correct key showed up after testing 50 percent of the possible keys.

9. Review the section Matrix of Controls and complete the following case study. You should perform the following tasks.

- Read the case and identify five or six general threats. List these threats in the Threats column of Figure 9-12.
- Read the case and identify five or six general components. List these components in the Components column of Figure 9-12.
- Place the names of the threats across the top of the blank matrix that is provided in Figure 9-13.
- Place the names of the components down the left vertical axis of the blank matrix that is provided in Figure 9-13.
- Read the case again and, using some interpretation, underline or identify all of the controls that appear to be in this case.
- Number all identified controls.
- Place the controls (by number) in the appropriate cells of the matrix (Figure 9-13). This was explained in the section Matrix of Controls, in this chapter.
- Analyze the control structure. Determine which controls are adequate and which new controls should be added. Use your own knowledge and the Appendix 2 control lists when identifying new controls that might be recommended.

This is a situation case, one that depicts a specific situation within an organization, rather than a full-length case study of an organization. It is performed best when small teams of two or three students work together. There is no single answer to this case. You must seek a reasonable level of control for each cell. Some cells may have too many controls, while others may have none at all. Use your discretion to determine the "reasonable level."

MULTISYSTEM COMMUNICATION NETWORK

This case depicts the online data communication network that is used by a major bank for several of its business functions. This network is used for online inquiry of the passbook savings system and the demand deposit accounting system (checking).

THREATS	COMPONENTS

FIGURE **9-12** Identify threats and components.

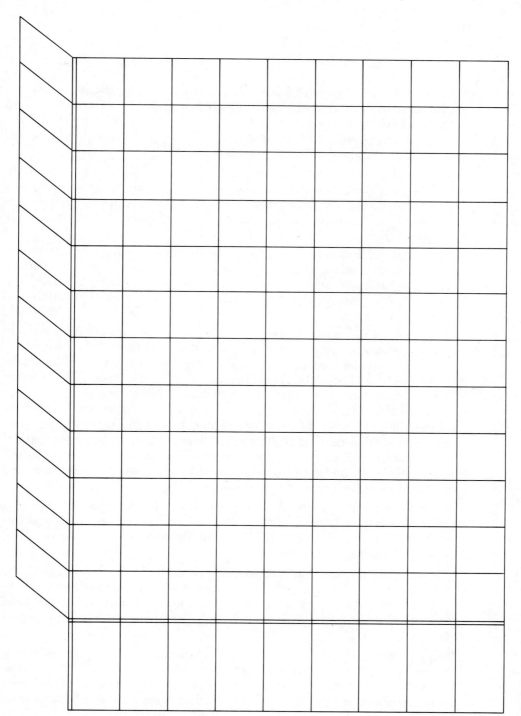

FIGURE 9-13   A blank matrix.

This is a large bank with hundreds of branches that are connected to a central computer system. Each branch has a variety of terminals and terminal controllers connected to the central system by use of the public telephone network. Some of these terminals are on dedicated leased lines, and others use the dial-up telephone network.

The security team visited six branch offices to conduct threat scenario sessions with the local branch operations staffs. The threat scenario sessions provided a good perspective of branch management, operations personnel, security, and prevailing attitudes toward embezzlement and other threats.

This network uses video terminals, teleprinter terminals, transaction terminals, local intelligent controllers, a centralized database, and a variety of other hardware devices and software programs. When possible, the bank purchases outside packages rather than developing application systems "from scratch."

Terminals at the branches and central headquarter sites are in physically secure locations. Vendors who perform maintenance are responsible for ensuring that the remote intelligent controllers, modems, and other devices operate effectively. Terminal operators use a four-digit numeric password, and each terminal is transaction coded to accept only its authorized transactions. The nine largest branches are allowed direct entry of their wire transfer business, which constitutes 92 percent of the wire transfers sent or received by the bank. There are written procedures at the local branches. Employees at several branches have started using their own or locally purchased microcomputers to increase the efficiency and throughput of the system.

Instructions for training operators are given in an extensive training manual that is updated monthly. This manual is kept in a looseleaf format. It is maintained centrally, and the updates are distributed to the various branches.

The operator terminals are transaction coded and have all the appropriate function keys; the software has various restart procedures and checkpoints that can be used to correct errors during a transaction. In some cases preprinted forms appear on the video screen so the operator can fill in the appropriate data.

During transmission, the front end communication processor performs error detection and correction and orders the retransmission of any message that is in error. A bit-oriented protocol is used. One of the good features of this system is its message switching with a store and forward capability. This gives everyone in the bank access to electronic mail. Wire transfer messages are switched from this computer to the more secure and separate wire transfer control system computer. When a user calls in, the central system calls that user back in order to control which telephone numbers are connected to its dial-up modems. Encryption currently is under consideration.

The application programs are maintained by local bank employees, although a few of the packages still use some outside consultants for maintenance. An effective program-change control procedure is operational.

The communication network control group has line monitors and other devices that are required to maintain an effective uptime ratio for this network. The network has been operating at a 98.95 percent uptime. Some of the terminal operators, however, have begun to complain about slow response time, which has been measured at an average of three seconds.

PART THREE

DESIGNING COMMUNICATION NETWORKS

Part Three of this book is devoted to describing the available communication services as well as designing data communication networks. There are chapters on . . .

- **COMMUNICATION SERVICES**
- **NETWORK DESIGN FUNDAMENTALS**
- **COMPUTERIZED NETWORK DESIGN**

10

COMMUNICATION
SERVICES

This chapter defines common carriers, tariffs, and deregulation. It also describes the primary communication services offered by common carriers, such as private circuit (lease) services, measured use services, and other special services. The costs of these communication services (tariffs) are discussed in Chapter 11.

COMMUNICATION FACILITIES

A *communication facility* is the medium that carries voice, data, or image transmissions from one point to another. These facilities (the media) may be physical, such as copper wires, coaxial cable, and optical fibers. On the other hand, the medium may be air, in which case the data communication facility transmits electromagnetic signals between microwave towers, satellites, cellular radio transmission antennas, and so on. The communication facilities may include more than just the media. End-to-end facilities actually have to include data terminal equipment (DTE), data circuit terminating equipment (DCE), specialized pieces of communication hardware, PBXs, software programs, and the like.

Even though there are separate data and voice communication facilities, technology is evolving toward a single communication facility that can handle both voice and data communications. The most prominent examples of such a facility are Integrated Services Digital Network (ISDN) and T-1 carriers, both of which were described earlier and are discussed further in this chapter.

A company or government agency that wants to develop its own voice or data communication network can select from a variety of leased communication services. If an organization chooses not to develop its own private communication network, another option is to use the public packet switched network communication facilities. In this case the organization can use dial-up to gain access to this packet network, or it can lease a private circuit between the business premise and the entry switching node to the packet network. When leasing communication circuits, the organization pays a monthly lease and has these circuits available for its private use. If the organization uses the public packet switched network, it pays charges based only on its usage of the network. Usage charges may be based either on the *time* the network is utilized or on the *volume* of data packets transmitted. Using public packet switched networks relieves the user of network design problems, most network operations tasks, maintenance and troubleshooting, and other technical operations that are required when private leased circuits are used.

Selecting a communication facility poses many problems. These problems have been compounded by the deregulation of the suppliers (common carriers) of these communication facilities, as well as by the many different rate structures known as *tariffs*. This chapter defines common carriers and tariffs and describes the major communication facilities (transmission circuits) that are offered. The costs of some of the more important communication facilities (circuits) are discussed in Chapter 11.

COMMON CARRIERS AND TARIFFS

A *common carrier* is a government-regulated private company that furnishes the general public with communication services and facilities. Common carriers are profit-oriented businesses, and their primary products are communication circuits and related services for voice, data, and image transmissions. Because this marketplace has been deregulated in the United States, common carriers now can supply a much broader range of computer-based services, such as the manufacturing and marketing of microcomputers, specialized communication hardware, and computer operations offered from the telephone company's central offices (switching centers).

The best examples of common carriers are American Telephone and Telegraph (AT&T), the seven Bell Operating Companies (BOCs), Bell Canada, MCI Communications, US Sprint, and General Telephone and Electronics (GTE). There are over 1,200 common carriers in the United States. Most of them offer communication facilities to a very small segment of the population, such as a 50- or 100-subscriber voice telephone network.

Do not let the term *deregulation* confuse you. It simply means that these common carriers can enter into other types of business instead of offering only communication circuits. They still are regulated as to the prices they can charge for the communication circuit services offered. It is the other part of their business—computers, software, and other types of hardware—that is no longer price regulated.

A *tariff* is the schedule of rates (prices) and a description of the services that are to be received when a particular type of communication service is purchased or leased. The circuits are leased, but hardware may be either leased or purchased. The best example is the price structure for home telephones and the description of what is provided for the basic monthly fee. A monthly fee allows you to be connected to the dial-up telephone network; you must either buy the telephone or pay a further small monthly fee for rental of the telephone instrument.

Tariffs are filed with the appropriate regulatory agency. There are two classes of regulatory agencies, federal and state. The best-known regulatory agency is the Federal Communications Commission (FCC), a federal government agency that regulates interstate (between states) and international communications to and from the United States. The FCC has regulatory powers to compel common carriers (companies that wish to sell/lease communication services) to conform to the federal Communications Act and its recent rewrite/revisions. The federal government currently is continuing its deregulation of interstate communications to give common carriers more independence in the competitive business environment. Every common carrier engaging in interstate or international communications is under the jurisdiction of the FCC and is required to abide by its regulations.

The other regulatory bodies are the Public Utility Commissions (PUC) in each of the 50 states. The individual state Public Utility Commissions are empowered to regulate intrastate (within a state) communications. While the federal government is continuing its deregulation of common carriers, the individual state Public Utility Commissions do not appear to be following the same course of action.

A common carrier wanting to sell communication services must have its services approved. To obtain this approval, it must file basic information with either the FCC or the state Public Utilities Commission. Such a filing provides details of its offered services, the charges for these services, justification for the charges, and so on. These documents are called tariffs, and they form the basis of the contract between the common carrier and the user of that common carrier's communication service.

Although there are many similarities in the way data communication activities have evolved in the United States and in Europe, there also are many differences. One of the primary differences is that the data communication industry in the United States operates as a series of private companies that are regulated by the government, while in Europe and most other countries of the world the PTTs (Postal Telephone and Telegraph Services) are a government monopoly that owns, controls and sells all voice and data communication services.

If the industry is regulated, as in the United States, it may be more innovative, cost effective, and able to develop new services faster. On the other hand, if the industry is a monopoly as in most countries, it may be overburdened because it supports other government agencies (such as postal services) that drain its resources. Such a situation generally fosters an agency that is uneconomical and not very innovative or progressive regarding the development of new services. These statements are generalizations, however, and have notable exceptions.

Other differences might stem from the fact that government-owned monopolies may be run based on pressures created by unemployment, protection of industrial and technical markets, political considerations, national defense considerations, and cultural and social traits within a specific country.

The European environment is not a single sovereign market. It is plagued with many geographical, political, and economic differences. Their situation could be compared with trying to interconnect and achieve total compatibility in the data communication facilities between Mexico, Canada, and the United States. It is for this reason that the European PTTs are more sensitive than the United States to the need for international standards.

In all the countries of the world either the government has a regulatory agency to control communication services and costs or it is the sole supplier of communication services (which is a monopoly situation). For example, in Mexico and South America the federal governments are the sole supplier of communication services, regardless of whether the service is voice or data. Germany's Deutschen Bundespost and France's Postes Telephonique et Telegraphique are the monopoly suppliers of communication services in those countries.

Like the United States, Canada has deregulated the communication environment and regulates only communication services and prices through its Canadian Radio-Television and Telecommunications Commission (CRTC). The Canadian CRTC is similar to the FCC in the United States. In England, the British Post Office Commission provides voice and data communication services. Until recently, the British Post Office was a government department (monopoly) that handled both mail and telecommunication services. It recently became a private company (the British government sold stock in this corporation), and now the British government regulates communications rather than being the monopoly supplier. Japan also has deregulated its communication industry and now regulates only communications. In Australia, there has been a massive deregulation of Telecom Australia's monopoly, and now it only regulates communications rather than being the sole supplier. In the Netherlands, the monopoly powers have been severely restricted, but they remain halfway between being a monopoly and a deregulated communication environment.

In a monopoly situation, private businesses in a country cannot utilize communication circuits and telecommunication equipment unless they have been either manufactured or approved by the monopoly government agency. This restriction severely limits the growth of the country's businesses and reduces the country's economic growth. Growth is limited because we live in an information-based society; and any restrictions on the communication environment (whether voice or data) are bound to have a secondary effect on the entire economy.

In a country that has deregulated its communication environment, businesses can use communication circuits from competing common carriers and communication hardware from many different vendors. Regulation occurs, first, with charges for services and, second, with meeting various technical standards or specifications as to the type of modulated signals and electrical signals that can be sent over the country's communication circuits. As an example, modem manufacturers must transmit signals that are within certain specified limits if they are to be approved for use in the United States. In reality, this is not a burdensome restriction. It is only a protective device so that one manufacturer's equipment does not harm another's or, especially, the communication circuits offered by various common carriers.

DEREGULATION

In the United States, the entire telephone system used to be run by American Telephone and Telegraph (AT&T). During the last few years of deregulation by the U.S. government, the 22 telephone companies owned by AT&T and AT&T were essentially divided in half (depending on your definition of half). The primary services retained and supplied by the first half are the local telephone services to your home or business premise, and the basic services retained and supplied by the other half (AT&T) are long distance telephone services.

First, let us begin with the local telephone services. The 22 individual telephone companies originally owned by AT&T were consolidated into seven Bell Operating Companies (BOC). Figure 10-1 shows how the United States was divided into these seven Bell Operating Companies. This figure also gives the name of each of the seven companies and lists the original telephone companies that were grouped together to form each Bell Operating Company. These are the

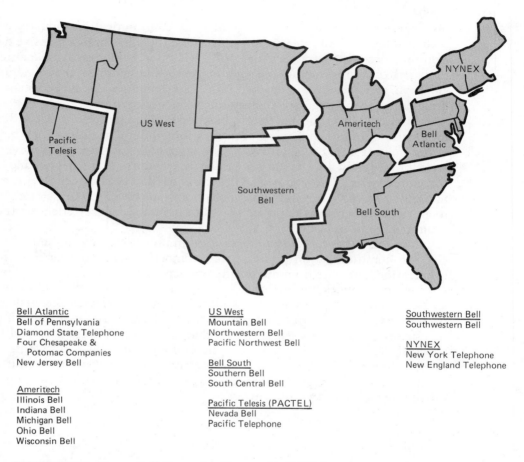

Bell Atlantic
Bell of Pennsylvania
Diamond State Telephone
Four Chesapeake &
   Potomac Companies
New Jersey Bell

Ameritech
Illinois Bell
Indiana Bell
Michigan Bell
Ohio Bell
Wisconsin Bell

US West
Mountain Bell
Northwestern Bell
Pacific Northwest Bell

Bell South
Southern Bell
South Central Bell

Pacific Telesis (PACTEL)
Nevada Bell
Pacific Telephone

Southwestern Bell
Southwestern Bell

NYNEX
New York Telephone
New England Telephone

FIGURE 10-1   Seven Bell Operating Companies (BOCs).

telephone companies that supply telephones and local loop connections between your home or office and the telephone company central offices (switching centers) for voice and data transmissions.

Because of deregulation, the Bell Operating Companies now are able to offer services other than communication circuits. Each of these seven companies has marketing agreements with manufacturers of office automation equipment, multiplexers, switchboards (PBXs), modems, cellular mobile telephone equipment, and so forth. The BOCs market equipment for these manufacturers. The next stage in the deregulation process is to allow these companies to manufacture their own equipment and computers. Of course, this will put them into direct competition with other computer manufacturers such as IBM and Digital Equipment Corporation.

To confuse matters, the Bell Operating Companies are no longer the only organizations that can supply telephone service to your home or business. There are other competing common carriers, the largest of which are MCI Communications and US Sprint. Moreover, the telephone instruments are no longer a monopoly item of these seven Bell Operating Companies. Now it is up to you to decide whether you want to purchase your own telephone from another vendor or lease it from the telephone company.

The service area of each of these seven Bell Operating Companies is broken into what is called a *local access transport area* (LATA). These areas outline the geographic area *within* which the individual Bell Operating Company can offer service and where it must turn service over to another supplier, primarily American Telephone and Telegraph (AT&T), which was the other half of the split-up during this divestiture. These LATAs define the areas in which the *Local Exchange Companies* (also known as Bell Operating Companies) can provide local exchange and exchange access services.

Local exchange service is provided when the telephone company supplies a local loop. *Exchange access service* is provided when the local telephone company interconnects through the central office so your local loop can be connected to a long distance telephone company such as AT&T. Service *between* LATAs is provided by interexchange carriers like AT&T, MCI, or US Sprint.

Equal access is a vital issue that has come into existence since deregulation. According to the terms of the deregulation agreements, local exchange companies must provide all carriers of long distance services with access to local switches equal in type, quality, and price to that which they sell to AT&T affiliates. This will allow users to dispense with the necessity of using push-button telephones and extra-digit dialing when they want to access long distance carriers such as MCI or US Sprint. While exchange access will be equal, that does not necessarily mean that all interexchange services will be equal because the investment in maintenance, servicing, and equipment will be the deciding factor in determining which carrier has the highest quality of service.

LATAs have been modeled on the concept of Standard Metropolitan Statistical Areas (SMSAs). Where possible, LATAs are based on communities of interest and conform to state boundaries. Most states are comprised of several LATAs, but some of the sparsely populated states are a single LATA. A listing of the Local Exchange Companies' LATAs is shown in Figure 10-2 where they are identified in general terms by state. There are approximately 200 LATAs in the continental United States. A number of them span state lines because they are based on SMSAs rather than political boundaries.

The Bell Operating Company Pacific Telesis (PACTEL for short) has ten LATAs for California (Figure 10-3). As you can see from the figure, some of these LATAs are quite large. For example, a call from San Francisco north to the city of Eureka stays within one service area, and so PACTEL collects the full long distance charge for it. By contrast, a call from San Francisco south to the nearby

Alabama	Champaign	Montana	Single LATA
Birmingham	Chicago	Billings	South Carolina
Huntsville	Forrest	Great Falls	Charleston
Mobile	Peoria	Nebraska	Columbia
Montgomery	Quincy	Grand Island	Florence
Arkansas	Rockford	Omaha	Greenville
Fort Smith	Springfield	Nevada	South Dakota
Little Rock	Sterling	Single LATA	Single LATA
Pine Bluff	Iowa	New Hampshire	Tennessee
Arizona	Davenport	Single LATA	Chattanooga
Phoenix	Cedar Rapids	New Jersey	Knoxville
Tucson	Des Moines	Atlantic Coastal	Memphis
California	Sioux City	Delaware Valley	Nashville
Bakersfield	Kansas	North Jersey	Texas
Chico	Topeka	New Mexico	Abilene
Fresno	Wichita	Single LATA	Amarillo
Los Angeles	Kentucky	New York	Austin
Monterey	Louisville	Albany	Beaumont
Sacramento	Owensboro	Binghamton	Brownsville
San Diego	Winchester	Buffalo	Corpus Cristi
San Francisco	Louisiana	New York Metro	Dallas
San Luis Obispo	Baton Rouge	Poughkeepsie	El Paso
Stockton	Lafayette	Syracuse	Houston
Colorado	New Orleans	North Carolina	Longview
Colorado Springs	Shreveport	Asheville	Lubbock
Denver	Maine	Charlotte	Midland
D.C.	Single LATA	Greensboro	San Antonio
Washington	Maryland	Raleigh	Waco
Florida	Baltimore	Wilmington	Wichita Falls
Daytona Beach	Hagerstown	North Dakota	Utah
Gainesville	Salisbury	Bismarck	Single LATA
Jacksonville	Massachusetts	Brainerd-Fargo	Vermont
Orlando	Eastern	Ohio	Single LATA
Panama City	Western	Akron	Virginia
Pensacola	Michigan	Cleveland	Culpeper
Southeast	Detroit	Columbus	Lynchburg
Georgia	Grand Rapids	Dayton	Norfolk
Albany	Lansing	Toledo	Richmond
Atlanta	Saginaw	Youngstown	Roanoke
Augusta	Upper Peninsula	Oklahoma	Washington
Macon	Minnesota	Oklahoma City	Seattle
Savannah	Duluth	Tulsa	Spokane
Idaho	Minneapolis	Oregon	West Virginia
Single LATA	Rochester	Eugene	Charleston
Indiana	St. Cloud	Portland	Clarksburg
Auburn-Huntington	Mississippi	Pennsylvania	Wisconsin
Bloomington	Biloxi	Altoona	Northeast
Evansville	Jackson	Capital	Northwest
Indianapolis	Missouri	Northeast	Southeast
South Bend	Kansas City	Philadelphia	Southwest
Illinois	Springfield	Pittsburgh	Wyoming
Cairo	St. Louis	Rhode Island	Single LATA

FIGURE **10-2** List of LATAs by state.

Eureka

2. Chico

3. Sacramento

9. Stockton

1. San Francisco

8. Monterey

4. Fresno

10. San Luis Obispo

7. Bakersfield

5. Los Angeles

6. San Diego

LATA's
1. San Francisco
2. Chico
3. Sacramento
4. Fresno
5. Los Angeles
6. San Diego
7. Bakersfield
8. Monterey
9. Stockton
10. San Luis Obispo

N

FIGURE 10-3 Local access transport area (LATA) for California.

city of Monterey crosses the LATA boundary, and so the charges for the call are collected by a long distance interexchange carrier. This carrier can be AT&T, MCI, US Sprint, or any other supplier of long distance service that has been approved by the California Public Utility Commission. As you can guess, each state has its own unique set of LATAs. Now let us move on to the other half of the AT&T breakup.

The *second* half of this divestiture, American Telephone and Telegraph (AT&T), retained its old "long lines department," which handled long distance communication services. This part now is called AT&T Communications. AT&T also retained the Bell Laboratories (research division) and its manufacturing divisions, primarily Western Electric. As a result, AT&T's primary business is long distance communication services. Integrated with the long distance and networking operations is the information systems equipment unit (equipment manufacturing). Other major divisions or business focuses for AT&T are the Bell Laboratories research unit and AT&T International, which markets communication products abroad.

In summary, after the divestiture AT&T was left as a single company (separate from the seven Bell Operating Companies) concerned primarily with long distance communication services. The other 22 telephone companies were spun off into seven regional Bell Operating Companies with the primary service responsibility of providing local telephone service. Of course, along with the AT&T divestiture came numerous other common carriers and equipment suppliers that are now in direct competition with both AT&T and the seven regional Bell Operating Companies.

COMMUNICATION SERVICES OFFERED

The communication services described in the remainder of this chapter may be supplied by AT&T, one or more of the seven regional Bell Operating Companies, MCI, US Sprint, General Telephone and Electronics, or one of the hundreds of other equipment suppliers and common carriers. Companies like MCI and US Sprint are in competition with AT&T for interLATA (between LATAs) service.

Communication services are divided into three basic areas to make them easy to understand: private circuit (leased) services, measured use services, and other special services.

Each of the following sections contains descriptions of selected services; usually, it is the most prominent service offered. These descriptions provide an insight into the types of communication services (circuits) offered and a basis with which to carry out preliminary design analysis for various types of data communication network designs.

The costs of these services (tariffs) are described in Chapter 11 under design step 12, Network (Circuit) Costs. We suggest that you cross-relate a specific service described in this chapter with its costs in Chapter 11 to gain a more complete understanding of each communication service. A thorough grasp of these communication services is a prerequisite to carrying out any of the network designs in Chapter 11.

PRIVATE CIRCUIT (LEASE) SERVICES

Private circuit services are those in which the user leases the service from the common carrier. In other words, private circuits are available for use by organizations twenty-four hours per day, seven days per week. In effect, they are for the exclusive use of the leasing organization.

It is helpful to understand the distinction between a *leased circuit* and a *dial-up* (measured time) circuit. If you have a leased circuit from San Francisco to Los Angeles, it is one continuous and unbroken circuit path. In other words, this leased circuit is wired around any switching equipment at the telephone company central offices. By contrast, a measured time dial-up circuit goes through all the switching equipment in the telephone company central office; there is not one continuous and unbroken circuit path for your use. Every time a call is placed on a dial-up circuit, a new circuit path is established.

Private lease circuits are so much at the disposal of the lessee that one might think the circuits were owned by the organization, even though they are leased by payment of a monthly rate. Some of the more prominent private circuit services are described below.

Voice Grade Channels. These used to be known as Series 2000/3000 Voice Grade Circuits. Technically, voice grade channels are for voice communication, remote operation of radio telephones, connection of private voice interconnecting systems, interconnecting remote central offices, data transmission, remote metering, supervisory control of electronic devices, and miscellaneous signaling. Everyone just refers to this as a voice grade equivalent channel. This channel or communication circuit was described in Chapter 4 in the section Voice Grade Leased Circuits.

When this type of circuit is leased, it may include copper wire pairs, bundles of wire pairs, microwave transmission, coaxial cables, or even optical fibers. Normally, users are told if the voice grade equivalent channel is a satellite channel because network designers must take into account the propagation delay times associated with satellite transmissions. In addition, pricing structures for a satellite voice grade equivalent channel differ from those of wires, cables, or microwave.

Conditioning, or *equalization* as it sometimes is called, is performed on communication circuits to improve their data transmission qualities. Because voice grade circuits most often transmit data in the voice bandwidth (300 to 3300 hertz), it is advisable to use conditioning or equalizing when transmitting at 4800 bits per second or greater, although it is not mandatory. This is because the signal suffers frequency/amplitude distortion or envelope delay distortion (changes in timing of the signal) during data transmission. An equalizer performs two functions.

First, it increases or decreases the amplitude of all the frequency components until they are the same amplitude as the reference component. This is accomplished by using the amplitude of the center frequency components of the voice grade channel as a reference (usually 1004 hertz). This action restores the frequency components of the signal to their former relative amplitude.

The *second* function of the equalizer is to delay each middle frequency component to the point where it is equal in delay to the edge components, so that after leaving the equalizer all the frequency components have the same relative delay. The greater the bandwidth of the transmitted signal, the greater the degree of equalization/conditioning that is required. The different sets of requirements for attenuation and delay distortion are referred to as *levels of conditioning* for a circuit.

The user decides whether the circuit should be conditioned. AT&T offers two levels of conditioning, which are used for point-to-point, multidrop, and switched configurations. *C level conditioning* has specific limits on attenuation distortion and envelope delay distortion. Such limits reduce line impairments so a data signal can get to its destination with less noise and distortion. Because there are fewer errors, fewer retransmissions are required. Bell's Technical Reference Publication 41004 gives the exact specifications. *D type conditioning* specifically limits noise and harmonic distortion. Again, this type of conditioning reduces line impairments, and there are fewer errors and fewer retransmissions.

Modem manufacturers usually specify whether conditioning is required when using their modem on voice grade circuits. The approach taken by the majority of modem manufacturers is to use a form of adaptive equalization within the modem circuitry. By using the proper circuits, they control the amount of delay distortion introduced to offset that which is on the voice grade transmission channel. Conditioning/equalization therefore can be performed by the telephone company at its central office locations, or it can be performed continuously and automatically by the circuitry within your modem.

In addition to conditioning, some local telephone companies such as New York Telephone (NYNEX) offer what is called a *straight copper circuit*. This is a pair of copper wires, and it is available only when the entire circuit is within the same telephone company central office. In essence, the straight copper circuit is comprised of two local loops connected at the central office. Because there are no repeater/amplifiers or loading coils in the circuit, the user can utilize less expensive limited distance modems (1 to 15 miles). Limited distance modems also may be used on standard Series 3000 voice grade unconditioned channels, depending on the length of the circuit and the data rate.

Other similar private lease services include Series 1000 channels, which are low speed signaling and teletypewriter channels that vary between 30- and 150-baud data transmission rates. Because of their slow signaling speed and their high cost (they may cost more than a voice grade data circuit), Series 1000 channels are falling into disuse and the telephone company is discouraging this offering.

Wideband Services   These channels are for data transmission or for alternate voice/data transmission. They are used for high speed data transmission such as 19,200, 40,800, 50,000, 56,000 and 230,400 bits per second. They also may be used in conjunction with a 50,000 bits per second switched service, a digital data service extension channel, or high speed facsimile transmission. This type of communication channel may be delivered to the user as a Group 48,000 hertz bandwidth or as 12 individual voice grade channels (48,000 hertz ÷ 12 = 4000 hertz). It also is available as a Supergroup 240,000 hertz bandwidth. These channels are used by organizations needing wider bandwidths to transmit greater quantities of data between facilities.

Digital Services These services include point-to-point and multipoint configurations for the transmission of data in a digital manner (analog modem conversion not required). This type of service operates at 2400, 4800, 9600, and 56,000 bits per second. It spans both the voice grade communication channels (4000 hertz bandwidth) and the wideband communication channels (greater than 4000 hertz). The advantage of using digital transmission is that digital modems are much less costly (although increased local loop costs may account for the lower modem cost) and the transmission error rate is far less than with analog circuits.

The designer of a network first ascertains whether digital service is available in the area in which the organization wants to transmit and then compares the cost of digital service with the appropriate analog service such as the voice grade transmission channel.

Satellite Services Several common carriers offer satellite channels for voice, data, facsimile, and various wideband applications. Basically, a *satellite channel* is defined as a four-wire voice grade circuit; therefore, users get a 4000 hertz bandwidth and a four-wire equivalent circuit. Available transmission rates range from 300 to 1,540,000 bits per second.

As was mentioned in the earlier discussion of voice grade circuits, users are notified when transmission is going over a satellite circuit because it may affect their protocols. A Binary Synchronous Communications (BSC) protocol cannot be used successfully with satellite transmission. Instead, one of the newer protocols such as X.25, SDLC, or HDLC must be used. Satellite channels usually are less costly than voice grade channels that are on the ground such as microwave transmission, wire pairs, and coaxial cables. If a greater capacity is needed, it is possible to lease a bundle or group of voice grade satellite channels to increase bandwidth beyond the standard 4000 hertz for a single voice grade channel.

The FCC has approved the use of a Direct Broadcast Satellite service to private homes. This means you can install a small (2.5-foot) dish antenna on your roof and receive television signals directly from medium-powered communication satellites orbiting some 22,000 miles above.

ISDN *Integrated Services Digital Network* is a leased service in which the common carrier offers a communication circuit with two 64,000 bits per second data transmission channels and one 16,000 bits per second signaling channel (2B + D). This is a high speed digital transmission service over which you can combine simultaneous voice conversations, data transmissions, and image (video) transmissions. It was described in the Chapter 4 section on ISDN.

T-1 Circuits   These leased circuits are digital circuits with a transmission capacity of 1.544 million bits per second. The most common method of dividing them into subchannels is to divide the entire T-1 circuit into 24 voice or data transmission channels, although multiplexing equipment can divide these circuits into 48 or 96 voice grade telephone circuits. The bandwidth available by multiplexing a T-1 circuit is so great that only larger organizations with a high volume of data and voice traffic can use this capacity to its fullest. It is important to understand that T-1 service is not dependent on any one type of medium. A T-1 channel can be provided through fiber optics, satellite, microwave, and so forth. These circuits were discussed in Chapter 2 under T-1 Carriers and in Chapter 3 under T-1 Multiplexing.

Software Defined Networks   To make traditional leased voice grade circuits more competitive, it now is possible to lease a *software defined network* (SDN) communication circuit. SDNs are built on public switched networks to provide a private leased circuit for the leasing organization. The user leases a virtual circuit from point A to point B rather than a hardwired physical circuit as is normal with a voice grade leased circuit. Needless to say, a software defined network circuit costs less than a "physical" voice grade leased circuit.

AT&T's *Software Defined Network Service* (SDNS) gives users control of their own special call-routing programs that are stored in AT&T's network. Users can access the database containing the description of their network configuration and change the configuration as their network grows or as their business needs change. SDNS also provides network management features such as restricted calling privileges, detailed reports about calls, and some security features. Users pay for the SDNS time and distance, whereas they pay a flat monthly charge for a private line leased service. Software defined networks also are discussed in Chapter 4, under Voice Grade Leased Circuits. Because SDNS includes both monthly fees and time-based usage charges, it is half way between a private circuit and measured use services.

In summary, the above private circuit (lease) services are the basic circuits/ channels used to design and develop a private network. They are the basic building blocks to which the various pieces of hardware, switchboards, protocols, and software are connected to create a working communication network.

MEASURED USE SERVICES

Measured use services are communication services in which charges are based on how much the system is used. For example, a call from your home telephone is based on the length (time) of the conversation and how far (miles) the other person is from you. For a packet switched service, charges are based on how many individual packets of information are transmitted regardless of where the other party is located. With Wide Area Telecommunications Service (WATS), a fixed monthly fee is charged for a fixed number of hours of circuit usage. If the fixed time is exceeded, another per hour rate is charged. In other words, measured use services are those in which payment is based on utilization. Some of the primary services in this group are Direct Distance Dialing (DDD), WATS, and packet switching.

Direct Distance Dialing (DDD) With direct distance dialing, the normal telephone network is used for data transmission. The user dials the host computer telephone number, receives appropriate control signaling, enters passwords/authorizations, and connects to the host computer system.

In direct distance dialing an entirely different circuit path between telephone company central offices is used each time a number is dialed. Charges are based on the distance between the two telephones (in miles) and the time the connection is held open (the data transmission). The data communication user pays the same rate as the individual who uses the telephone for voice communication. Dial-up voice grade channels have more noise and distortion than a private leased voice grade channel because the signals go through the telephone company's central office switching equipment; although, the newer digital switches bring DDD very close to a private leased circuit when they are coupled with fiber optics.

The telephone company does not make conditioning/equalization of the circuit available for DDD because each dialed call gets a different circuit path or routing. Equalization, however, can be obtained by using a more expensive modem that performs automatic equalization after the call has been routed and connected. In addition, the rate of transmission (bits per second) may be a little less than can be achieved on a private leased voice grade circuit because there is more noise and distortion on the circuit path. The point is that the user may have to transmit at a slightly lower bits per second rate to reduce the number of message blocks retransmitted because of errors. Direct distance dialing offers only two-wire connections; therefore, if full duplex transmission is required, a special modem is needed to transmit simultaneously in both directions. This special modem transmits in one direction by using one pair of frequencies and in the other direction by using a nonconflicting and different pair of frequencies.

Wide Area Telecommunications Service (WATS) *WATS* is a special bulk rate service that allows direct dial station-to-station telephone calls, although it may be eliminated in the future. It can be used for both voice communications and data transmission. WATS uses the (800) area code series in the United States.

The 48 contiguous states are divided into 58 different service areas. The geographical coverage of WATS from any one of these service areas is determined by the *band of service* to which the customer subscribes. For example, interstate service from California uses the following six bands:

- **Band 1:**   Arizona, Idaho, Nevada, Oregon, Utah, and Washington
- **Band 2:**   Colorado, Montana, Nebraska, New Mexico, and Wyoming
- **Band 3:**   Iowa, Kansas, Minnesota, Missouri, North Dakota, Oklahoma, South Dakota, and Texas
- **Band 4:**   Alabama, Arkansas, Illinois, Indiana, Kentucky, Louisiana, Michigan, Mississippi, Tennessee, and Wisconsin
- **Band 5:**   Connecticut, Delaware, Florida, Georgia, Maine, Maryland, Massachusetts, New Hampshire, New Jersey, New York, North Carolina, Ohio, Pennsylvania, Rhode Island, South Carolina, Vermont, Virginia, Washington, D.C., West Virginia, Puerto Rico, and the U.S. Virgin Islands
- **Band 6:**   Alaska and Hawaii

The state of California has two service areas, Northern California and Southern California, each of which is a different WATS band. The list of states served in bands 1 to 5 differs, depending on the state. As might be guessed, the first five bands out of New York are almost the direct opposite of bands 1 to 5 out of California. Band 5 out of California is similar to band 1 out of New York. When subscribing to a band, such as band 4, service is automatic to all lower bands (in this case bands 1 to 3).

The WATS bands described above for *interstate service* have no relationship to *intrastate WATS service*. In California, for example, there is a northern service area and a southern service area; therefore, WATS intrastate service can be for northern California only, southern California only, or statewide. Interstate WATS service does not include your home state; therefore, it is necessary to lease both interstate and intrastate WATS services if access is needed to your home state.

Interstate WATS service is available on the basis of the first 25 hours of usage, the next 75 hours, and over 100 hours of usage per month. For either intrastate or interstate, charges are a flat fee for a specified number of usage hours.

If WATS service is used for data communications and the call holding time is less than 60 seconds, billing is for one minute of usage (one minute average call holding time). WATS service also is limited to one direction only; it is either "outward dialing" or "incoming" calls only. Inward and outward capability cannot be combined into a single WATS circuit. The user has to subscribe to two circuits to do both.

AT&T Megacom™ Megacom gives users a new simplified pricing structure that allows them to call anywhere in the United States, Puerto Rico, and the U.S. Virgin Islands with potential cost savings. This new pricing is based on the mileage and duration of each call, rather than on the geographic areas associated with WATS rates. Megacom eliminates the banded geographic pricing structure under WATS, and users are billed at the same rates for calls within the same area, regardless of distance. It appears that this service is intended to replace or eliminate outward dialing WATS.

AT&T Megacom 800 This service introduces a slightly different toll-free calling that enables businesses to conduct telemarketing efforts in a more cost-effective manner. With AT&T Megacom 800, users are able to receive calls from any location or selected groups of locations in the United States, Puerto Rico, and the U.S. Virgin Islands. Again, billing is based on the duration of each call and the distance between the caller's area code and the area code of whomever is being called. Megacom and Megacom 800 are intended to be replacements for outbound and inbound WATS; Megacom is outbound and Megacom 800 is inbound.

Public Packet Switched Services Packet switched communication services offer transmission speeds up to 56,000 bits per second. The network compensates for differences in transmission speed and different protocols between various switching nodes (SN). The packet network also provides code conversion from one code to another. Data is segmented into 128-character (1024-bit) blocks called packets. Users may access the service via private communication circuits, public dial-up circuits, or other packet switched networks. Figure 10-4 shows a packet switched network that has both ground-based circuits and satellite circuits.

In packet switching the user does not design or maintain the network. The user may be, in fact, only a network user. This can be an advantage because it relieves the user of the many technical burdens related to designing and maintaining a private network.

As an example of a public packet switched service, American Telephone and Telegraph offers a service called ACCUNET™. This packet service currently is offered in 99 locations throughout the United States and offers digital transmission speeds of 4800, 9600, and 56,000 bits per second. The service conforms to the Consultative Committee on International Telephone and Telegraph (CCITT) recommendations for the X.25 packet interface protocols.

It also conforms to the X.75 gateway protocol. This X.75 interface establishes a data communication gateway between ACCUNET and other existing public packet networks, local area networks, and international packet switched networks originating from other countries.

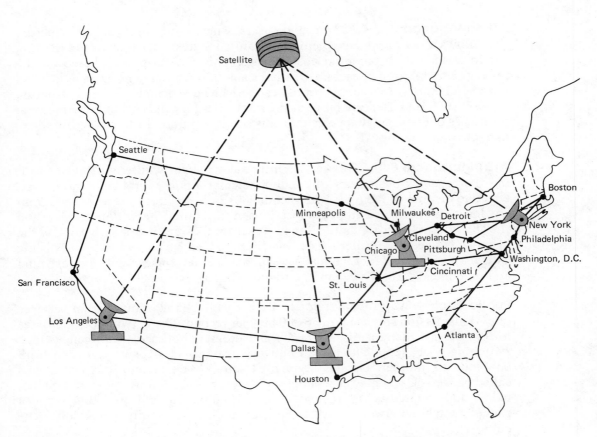

FIGURE 10-4 Packet switching network using satellite and terrestrial-based links.

DIAL-IT™ 900 Service This service allows many different users to call simultaneously a 900 exchange telephone number (900-xxxx) because it can handle 7,000 simultaneous incoming calls. These services have "sponsors" who arrange with AT&T to have their telephone number established. The sponsors have the responsibility of advertising this service to encourage people to call. Each caller is charged for the telephone call, although a sponsor may elect to pay the caller's charges, thereby making it a "free" call.

Two different arrangements are possible with DIAL-IT 900. The first is *prerecorded* information such as a national sports report or a dial-a-joke service which allows callers to hear a recording. The recording originates at the sponsor's premises and so can be changed as often as the sponsor wishes. The second arrangement is a *polling* or *call-counting service*. It allows callers to dial the 900 number to express preferences or vote on some question that is put to them through the mail, over the radio, or through a television broadcast. This polling service requires two numbers, one for each opinion. It can require more numbers if callers are asked to express several opinions.

To explain how this service works, we must describe its cost. When callers use the 900 service number, they are billed $0.50 for the first minute and $0.35 for each additional minute or fraction thereof, although the sponsor can opt to pay for the call. Once the call volume exceeds 2,000 calls or call minutes per day, the sponsor earns

$0.02 for every call minute between 2,001 and 5,000

$0.03 for every call minute between 5,001 and 10,000

$0.04 for every call minute between 10,001 and 20,000

$0.05 for every call minute over 20,000

AT&T also levies certain charges on the sponsors. With the prerecorded information arrangement there is a $250 flat fee, and the number must generate a weekly average of 2,000 calls per day per week. With the polling or voting arrangement, sponsors pay a $25 per day charge for a minimum of 500 calls per day. If the call totals for either service fall below the minimum, sponsors are charged $0.25 for each call under the required total. You may dial 1-900-555-1212 for an updated report of all currently available DIAL-IT 900 programs. There is no charge for this call!

The DIAL-IT 900 service has become controversial because numerous telephone subscribers have accidentally dialed, or their children have dialed, these numbers, and they have been billed with charges they considered unfair. Legislation is pending that may make it mandatory to lock people's telephone numbers out of the 900 exchanges if they request it. Another method of controlling these telephone calls was discussed in Chapter 9 under Voice Telephone Security.

Discount Voice Services   These services are offered by MCI, US Sprint, and many other companies. Even though these organizations offer a full line of communication services, they are best known for their discount voice services. They offer voice telephone transmission for less than the same service provided by American Telephone and Telegraph (AT&T).

When users sign up for such a service, they are given two choices. They may be given a local access number or the discount provider is registered as the primary service. If one of the discount voice services is the primary server, long distance calls can be dialed directly and the charges are billed directly to the user by the provider. The method of use in this case is identical to using AT&T; only the provider is different.

If AT&T is retained as the primary communication service provider, then a special five-digit number has to be dialed to gain access to the discount voice service communication provider. After dialing the five-digit number, the called telephone number can be dialed. Dialing this sequence of numbers can be fast if an automatic dialer telephone is used. This type of call is routed over the local loops to the discount service provider computer, from there it is routed on the long distance (IXC) circuits, and then it is routed over the local loop to the called telephone number.

These services cost the voice or data telephone user an average 10 to 20 percent less than if the same call were dialed over the Bell system DDD network. Do not be discouraged when faced with making the choice between AT&T and a discount service because it may be to your advantage to use both. All of them have dropped their minimum charges and monthly fees, and you can get some benefits in cost savings by using both types of services.

Telex/TWX Teletypewriter exchange service is a data transmission exchange between two terminals (it excludes voice communications). Each subscriber has a terminal and can contact any other subscriber in the telex or TWX network. This is nothing more than an alternative to the voice telephone network, but it has the ability to transmit hard copy between subscribers. Subscribers pay a monthly fee plus so many cents per minute of connect time. The connect time charged is only for the time it takes to transmit a message from one subscriber's terminal to another. This is a very popular service for businesses and government, especially in the area of purchasing/procurement. Western Union, the common carrier that offers both of these services, is in the process of upgrading the services to increase the speed of transmission.

Western Union Telegraph Company is developing a teletext service. They will join it with the international teletext service. Teletext, a new form of telex, is a 2400 bits per second service that provides advanced functions such as buffered CRTs, compatibility with data transmission protocols, and the ability to interface with existing business communication networks. Standards have been in development for several years by the CCITT. The basic service is designed to provide word processing terminals with local screen editing and printer capability. In the United States, Western Union will provide 2400 bits per second word processing communications to commercial and personal users. The basic goal is to deliver documents online at a cost that is competitive with airmail postage rates.

OTHER SPECIAL SERVICES

Foreign Exchange Service (FX) This service allows a user to call another central office via the dial-up telephone network, but without incurring any charges other than that of a local call.

If an organization is located in the suburbs of a major city but most of its customers are in the downtown area, then it might lease a FX circuit. The telephones at the company in the suburbs are connected directly to the telephone company's central office in the downtown metropolitan area. In other words, a FX circuit is one that runs from your telephone instrument to the telephone company's end office in another area. These circuits allow suburban subscribers to have the same "free" dialing privileges as telephone subscribers in the downtown area. When the telephone instrument is picked up, the dial tone is directly from the distant city's end office. If you list your telephone number in that city's telephone book, people may call you at a local number. In reality they reach you over the FX circuit.

The cost of FX service is the same as a voice grade private leased line plus the cost of a single telephone at the distant central office. While this service may appear to be similar to intrastate WATS, it is not because the user does not have to dial the 800 area code and it accepts both incoming and outgoing calls.

Common Control Switching Arrangement (CCSA)   This service is a private long distance dialed network. A switching arrangement is provided to allow interconnection of channels terminated in the switching equipment provided by the common carrier. The service is offered for large corporations and government agencies to interconnect several or hundreds of business operations, thus saving on telephone costs. A flat fee is paid for which the common carrier establishes a private telephone system interconnection for voice telephones, data hardware, and PBXs.

The largest CCSA is probably the U.S. government's Federal Telecommunications System (FTS). Many states also have set up private telephone systems that are designed, implemented, and maintained by the common carrier providing the CCSA service.

Hotline   This service directly connects two telephones in distant cities. When either of the two receivers is lifted, the telephone rings at the other end of the connection. This is a point-to-point service and is available only in selected cities.

SELECTED REFERENCES

1. *Computerworld: Newsweekly for the Computer Community.* Published weekly by CW Communications, 375 Cochituate Road, Box 9171, Framingham, Mass. 01701-9171, 1967– .
2. *Network World.* Published weekly by CW Communications, Box 9171, 375 Cochituate Road, Framingham, Mass. 01701-9171, 1983– .

3. *Planning Guide 1: InterLATA Telecommunications Rates and Services.* Published monthly by CCMI/McGraw-Hill, 50 South Franklin Turnpike, Ramsey, N.J. 07446.

4. *Planning Guide 2: InterLATA Telecommunications Rates and Services.* Published monthly by CCMI/McGraw-Hill, 50 South Franklin Turnpike, Ramsey, N.J. 07446.

5. *Planning Guide 3: Value Added Network and Data Private Line Tele-communications Rates and Services.* Published monthly by CCMI/McGraw-Hill, 50 South Franklin Turnpike, Ramsey, N.J. 07446.

6. Rutkowski, Anthony M. *Integrated Services Digital Networks.* Norwood, Mass.: Artech House, 1985.

QUESTIONS/PROBLEMS

1. What would be an ideal block length when using synchronous transmission over a packet switching network?

2. If Western Union tells you that they offer a 100 words per minute TWX service that uses characters consisting of seven bits plus parity, one start bit, and two stop bits (a word is six characters), what is the transmission rate in bits per second?

3. What is a common carrier?

4. What is a tariff?

5. Identify and describe some of the private circuit (lease) services.

6. Identify various measured use services.

7. Identify some of the special services that are available.

8. What is the difference between interstate/intrastate data communications and interLATA/intraLATA?

9. What is a communication facility?

10. What is the difference between the FCC and a state PUC?

11. What are the names of the seven BOCs (also referred to as local exchange carriers or local telephone companies)?

12. Which state has the most LATAs?

13. If you make a call from San Francisco to San Diego (California), is it intraLATA or interLATA?

14. What is the difference between local exchange service and exchange access service?

15. Why do you want to condition a line or circuit?

16. Over which circuit is 64,000 bits per second common?

17. Which circuit uses a digital 1.544 million bits per second transmission rate?

18. Suppose that you contracted with AT&T for a DIAL-IT 900 number. On it you offered advice on a certain instructor's exam. Assume that you averaged 3,761 call minutes per day for the ten days prior to final exams and then canceled the DIAL-IT 900 number. Did you make any money?

11

NETWORK
DESIGN
FUNDAMENTALS

This chapter presents the fundamentals of designing data communication networks. The systems approach to design is used, and 13 detailed steps are enumerated. An understanding of network design methodologies is a prerequisite to computerized design, which is described in the next chapter. Numerous design exercises are presented at the end of this chapter.

INTRODUCTION

This chapter is the first of two chapters on network design. It should be read before going on to Chapter 12, which introduces the design of networks using a sophisticated software package on a microcomputer. Before proceeding to computerized network design, you first need to understand the basics of how a communication network is conceived and designed.

It is not always possible to appreciate what a computer is doing when using computerized network design programs. For example, suppose you have a computer program that calculates the airline mileage distances between two telephone company central offices. The technique used determines the vertical and horizontal coordinates of each end office and then uses those figures in a mathematical formula to calculate the mileage. This mileage then is multiplied by the data circuit's cost per mile (the *tariff*) to determine its monthly leased cost. When the computer program says that the vertical and horizontal coordinates are used to determine the mileage, you may not understand what is meant unless you have read this chapter. We describe here the method of utilizing vertical and horizontal coordinates and give examples. In addition, important items such as needs assessment factors, feasibility factors, and evaluation criteria are discussed.

This chapter also includes basic line cost/network design problems. We use the word basic because tariffs have become very complex since the United States deregulated the communication industry. The tariffs now apply to very small local access transport areas (LATAs), and there are hundreds of different tariffs. As a result, it has become more difficult to use manual design methodologies when the number of nodes in a network exceeds 20 or 30. By reading this chapter and calculating the cost of a circuit link or designing a small three- to six-node network, you will gain an in-depth understanding of network design.

THE SYSTEMS APPROACH TO DESIGN

You should use the systems approach when planning a completely new data communication network, when enhancing a current network, or when planning for the use of public networks. Whether the network achieves success or just marginal utilization may be determined before a single piece of software or hardware is ordered. The key ingredient for success lies in planning based on the system's interface with the users. Far too often, data processing-oriented personnel take an equipment-oriented approach or a technical software-oriented approach. In today's world of data communications the designer must take a user systems application approach.

For example, there are two major classes of users for a data communication network: the organization's management and its user personnel.

Managers must accept the system and believe in it, or they will not trust the data/information/reports they receive from the system. Recall the word CATER that was discussed in Chapter 8. If the information received by management is not Consistent, Accurate, Timely, Economically feasible, and Relevant, then management will reject the system.

The *users* who work with the system on a day-to-day basis must be able to accept the system, or their productivity will fall drastically. When productivity decreases, the cost of carrying out basic office functions may increase the cost of the final product or service by 10 to 50 percent. Office productivity recently has taken on added importance because we are moving from a predominantly manufacturing society to an information (service-oriented) society. In other words, proportionately more people are involved in information-related work than in manufacturing/assembly work. We now need to have the industrial engineers from the factory environment move into the automated business office.

The above-mentioned changes are the reason why this book promulgates the systems approach to designing communication networks. Actually, it does not matter whether you are using manual or computerized design methodologies; you always should begin network design by using the systems approach.

THIRTEEN STEPS FOR NETWORK DESIGN

The following 13 steps should be used when designing a new communication network. Perhaps some steps can be omitted if a current network is being enhanced. For example, you might begin with Step 6, Analyze the Messages, or perhaps Step 7, Calculate the Traffic/Circuit Loading. It might be that your redesign involves nothing more than setting up a gateway to connect to public packet switching networks. If so, you might go directly to steps 10 and 11 (Software Considerations and Hardware Considerations). The exact sequence of steps and the number of steps used are determined by the scope of the network design project. Even so, serious consideration should be given to all 13 steps. A detailed explanation of how to carry out the step is given for each.

1. CONDUCT A FEASIBILITY STUDY

The first point that must be made about a feasibility study is that it may not be necessary to conduct one. It already may have been performed by management to identify the problem or the purpose/objectives of the proposed system. Perhaps the scope of the proposed system already has been defined. Furthermore, it is entirely possible that either management or the realities of the economic/ business environment have dictated that an online data communication network must be developed. If that is the case, it is no longer a question of whether to design a new system, but how it shall be done.

For example, can you imagine any major airline deciding that a network costs too much or does not meet its objectives? If an airline decided against a network, it would cease to be competitive with the other airlines; therefore, the feasibility of a "go/no go" network decision is made even before the airline can start to think about a feasibility study.

Of course, we now are talking about a feasibility study that helps determine whether or not to proceed with a network, rather than a more elaborate feasibility study to identify which specific network should be set up. The decision about which network is to be adopted is covered in the next 12 steps.

A primary responsibility in proceeding with a feasibility study is to define the problem clearly and put it in writing. Problem definition involves identifying all the problems that may indicate the need for a data communication network. Any of the following *needs assessment factors* may be analyzed to determine if they contribute to the need for this new network.

- Increased volume of inputs/outputs
- Inadequate data processing
- Obsolete hardware/software
- Inadequate file structures (database)
- Unsatisfactory movement of data/information throughout the organization
- Inadequate interfacing between application systems and other staff
- Documentation not available in a timely manner
- Unreliability of current systems
- Inability to maintain current systems
- Inadequate security/privacy
- Decreasing productivity
- Inadequate training
- Future growth that requires new methods
- Competition that forces change
- Negative effect of old system on employee morale
- New network viewed as having a positive effect on investments, cash flow, and so forth
- Inadequate floor space for personnel/files
- Avoidance of future costs
- Need for more timely access to information for improved decision making
- Increasing flow of information/paperwork
- Need to expand capacity for business functions/manufacturing
- Need to increase level of service quality or performance
- Need for new methods and better exchange of information in conducting international operations
- Reduction of inventories
- Need for a paperless office
- Desire to take advantage of future technology

Once the problem has been defined in this way, the purpose and objectives of the new data communication network are identified, the scope or boundaries the system will encompass are established, and perhaps some preliminary "magnitudes" of cost can be identified.

The feasibility study might include some preliminary work on the geographical scope of the network, or the physical areas of the organization that will be interconnected by it. It may be appropriate to develop a rough-draft geographical map of the intended network.

At the completion of this data gathering, a short feasibility study written report should be generated. This report is the medium by which you tell management what the problem is, what you have found its causes to be, and what you have to offer in the way of a solution. The feasibility study results might be presented verbally as well. This type of presentation provides management with an opportunity to ask questions or discuss issues that may have a bearing on whether to proceed.

Your study should help management decide whether to start a full program for the design and development of a new data communication system. It usually results in a go/no go decision.

2. PREPARE A PLAN

At this point the feasibility study has been completed, and management has given its approval to proceed with the design and development of a data communication network. Be sure to note that in this chapter we are designing a totally new network. Some of the procedures that are discussed in the remainder of this chapter may be eliminated if you are merely enhancing a current network.

In developing the plan, remember that a successful plan always takes into account the following three *feasibility factors*.

- *Technical feasibility* of the network
- *Operational feasibility* for the users who conduct their daily business using the network, and for management who has to rely on its reports
- *Economic feasibility* to keep it within budgetary limitations

The first step is to take the statement on the purpose/objectives of the network and divide it into three distinct goals. The *major* goal is the reason that the data communication network is being built. The object is to ensure that the network meets these requirements. Next, *intermediate* goals are other gains the system can make while serving its major purpose, hopefully with little or no extra expense. Finally, *minor* goals are the functions that a communication network, along with data processing applications, can perform for the organization but for which it is not quite ready (future requirements). The major goals are mandatory; the intermediate goals are desirable; and the minor goals are "wish list" items.

There is no way to outline the exact steps the plan should follow because the network must be customized for the organization and application systems it serves. The goals the network is to achieve should provide the framework for the plan. For example, the major goal might be to speed up order entry and improve cash flow through better collections. The intermediate goal might be to interface all of the accounting applications with the order entry operations. A minor goal might be to set up a voice mail/electronic mail system in the future. All too often network designers forget their priorities and concentrate on minor goals because of personal interest. When committed to writing, the goals serve as a constant reminder to avoid this trap.

The first step in developing a customized plan might be to identify the various sources of information, the types of information to be collected, and a schedule for performing various activities. It is likely that the designer will emulate the 13 key steps already listed in this chapter.

Finally, as the design begins, develop some *evaluation criteria*. If evaluation criteria are developed at the beginning, then there is a yardstick at completion for measuring the success of the data communication network/system design, development, and implementation. The following *evaluation criteria* should be considered. Insofar as possible, each criterion should be given a numerical value at the beginning of the project. This value is used at project ending to provide the means of comparison that management needs to evaluate the project properly.

- **Time** Are elapsed time, transaction time, overall processing time, response time, or other operational times reduced?

- **Cost** Are annual system cost, per unit cost, maintenance cost, or others, such as operational, investment, and implementation costs, reduced?

- **Quality** Is a better product or service being produced? Is there less rework because of the system? Has the quality of data/information improved? How can these quality factors be measured?

- **Capacity** Does the system have the capacity to handle workloads, peak loads, and average loads, as well as the long-term future capacity to meet the organization's needs in the next decade?

- **Efficiency** Is the system more efficient than the previous one?

- **Productivity** Has productivity of the user (information provider) and management (information user) improved? Is decision-making faster and more accurate because of the information provided by this system?

- **Accuracy** Are there fewer errors? Can management rely more on this system than the old one?

- **Flexibility**   Can the new system perform diverse operations that were not possible before?
- **Reliability**   Are there fewer breakdowns of this system compared with the previous one? Is uptime very high with this system? The reliability/uptime of an online network is probably the number one criterion by which to judge its design and development.
- **Acceptance**   Have both the information providers and the information users accepted the system?
- **Controls**   Are adequate security and control mechanisms in place to prevent threats to the system such as errors and omissions, fraud and defalcation, lost data, breaches of privacy, disastrous events, and the like?
- **Documentation**   Does the system have adequate written/pictorial descriptions documenting all its hardware, protocols, software, circuits, and user manuals?
- **Training**   Are training courses adequate and are they offered on a continuous basis, especially for terminal operators? Are training manuals adequate and updated on a regular basis?
- **System Life**   Is the life of the system adequate? When two to five years are spent designing and implementing a system, the system life should be of adequate duration to take advantage of the economies of scale.

The above evaluation criteria can be used to evaluate the new data communication network after it has been developed. It also may be advisable to evaluate your own performance during the design and development of this new network. In that case, examine such items as whether development time schedules were on target. Were development costs within budget or was there a large cost overrun? Were any deviations from the original purpose/objectives and scope documented? Consider interactions with those affected by the system: Do they feel they were treated fairly, and are they satisfied with you and your design? Was there a lot of turnover on the project team during the design and development?

In summary, as the plan is prepared (step-by-step approach), also develop evaluation criteria for the data communication network as well as the evaluation criteria by which to judge your own design efforts. If you ignore this step, someone else may do it and you may be judged by a set of criteria that do not relate well to your effort.

3. UNDERSTAND THE CURRENT SYSTEM

The object of this design effort step is to gain a complete understanding of the current operations (application system/messages) and any network that is functioning. This step provides a benchmark against which future design requirements can be gauged. It should provide a clear picture of the present sequence of operations, processing times, work volumes, current communication systems, existing costs, and user/management needs.

In order to be successful at this stage, begin by gathering general information or unique characteristics of the environment in which the system must operate. Next, identify the specific applications that use the data communication network and any proposed future applications.

Learn something about the background of the industry in which the network will function (what competitors are doing in this regard), as well as about your individual company and the departments that are responsible for the applications.

Determine whether there are any legal requirements, such as local, state, federal, or international laws, that might affect the network.

Consider the people in different departments who will be affected by the system. Do not overlook the fact that there are formal organizations as shown on the organization chart, and there are also informal organizations within a specific department.

It is important to be aware that company politics might affect the design effort; people may tell you what they want for their personal interests rather than what is in the best interests of the organization.

Develop a level 0 data flow diagram or an input, processing, output model[1] for *each* system that will utilize the data communication network. Your task is to identify each generic input to the application system, the typical processing steps that are performed, and each generic output. Describe and list each input/process/output.

Identify the *file formats* so database planners can start to design the database and database access methodologies. You should be aware that transmission volumes increase dramatically when the network is used for database retrieval transactions and file transfers from microcomputers.

Techniques used to complete this step might include interviewing user personnel, searching a variety of current records for message format and volumes, estimating and sampling for timings and volumes, and possibly comparing current systems with similar ones that have been put on a data communication network.

The documentation gathered during each of the above tasks can serve as a future summary of the existing system. A written summary also should be developed. This summary should include everything of importance learned during this step of the design. It is your written understanding of the existing systems. It

---

[1] *Fundamentals of Systems Analysis: Using Structured Analysis and Design Techniques*, 3rd ed., by Jerry FitzGerald and Ardra F. FitzGerald (New York: John Wiley & Sons, 1987). See Chapter 2 for data flow diagrams and page 290 for input, processing, output models.

should include any design ideas, notes on whether currently used forms or trans-mittal documents are adequate or inadequate, who was helpful or hindered prog-ress, and any other impressions gained from interviews, meetings, data flow diagrams, flowcharts, sampling, and the like. In general, the written summary should contain information that can be referred to during the detailed design steps for development of the data communication network. It is the benchmark to be used for later comparisons.

4. DESIGN THE NETWORK

By the time the network design begins, certain items already should be estab-lished, such as definition of the problem, purpose/objectives, scope of the net-work, general background information about the application systems that will use the network, and a thorough written understanding of the current system. With these items in hand, a list of general system requirements can be developed.

The object of defining the general system requirements is to assemble an over-view of the functions to be performed by the proposed network. At this point the input, processing, and output models for each of the application systems might be of great value.

During the early stages of defining the general system requirements, a review of the organization's long-range and short-range plans is advised. This review helps provide the proper perspective in which to design a system that will not be obsolete in a few years and that will meet the future requirements of the organi-zation. These long- and short-range plans indicate such information as changes in company goals, strategic plans, development plans for new products or services, projections of changing sales, research and development projects, major capital expenditures, possible changes in product mix, new offices that must be served by the communication network, emphasis on security, and future commitments to technology.

Once the system requirements have been identified, they should be prioritized. That is, they should be divided into mandatory system requirements, desirable system requirements, and wish list requirements. This information enables you to develop a minimum level of mandatory requirements and a negotiable list of desirable requirements that are dependent on cost and availability. Match these against your major, intermediate, and minor goals mentioned earlier.

System requirements should be as precise as possible. For example, rather than stating "a large quantity of characters," state requirements in more precise figures such as "50 characters per minute plus or minus 10 percent."

At this point, avoid presenting solutions; only requirements are needed. For example, a requirement might state that circuit capacity should be great enough to handle 5000 characters per minute which will triple by 1990. It would be a mistake to state this as a solution by saying that a 9600 bits per second voice grade circuit is required. Solutions should be left for later, during development of network configurations when software and hardware considerations must be interrelated with those configurations.

By definition, to design means to map out, plan, or arrange the parts into a whole which satisfies the objectives involved. The final nine steps of designing a data communication network are described in the remainder of this chapter.

5. IDENTIFY THE GEOGRAPHIC SCOPE

The scope of the application systems that are to be included on the network have been identified by now. The rough-draft geographic map that was developed during the feasibility study should be examined at this point, and a more detailed and accurate version should be prepared.

A data communication network has four basic levels of geographic scope.

- International (worldwide network)
- Country (within the boundaries and laws of a single country)
- City (within the boundaries of a specific city, state/province, or local governmental jurisdiction)
- Local facility (within a specific building or confined to a series of buildings located on the same contiguous property)

Usually it is easiest to start with the highest level, international. Begin by drawing a network map with all the international locations that must be connected. At this level it is necessary only to interconnect the major countries and/or cities around the world. It is sufficient to have a map that shows lines going between the countries/cities. Details such as the type of circuit, multiplex, multidrop, concentrators, and the like have not been decided yet. If the network does not cross international boundaries, then obviously this step can be omitted.

The next map you might prepare is the country map for each country. Interconnections should be drawn between all cities within the country or countries that require interconnection. Again, a single line drawn between the cities is quite adequate because the type of configuration has not yet been decided upon. Figure 11-1 is a typical example of a country map intermixed with an international map because of the closeness of the two countries.

The next map to prepare is one of the city or state/province. This map can be divided into two levels. The first level uses a state map with lines drawn showing the interconnection among various cities within the state. City maps are used at the second level. They show interconnection of various "local facility" locations

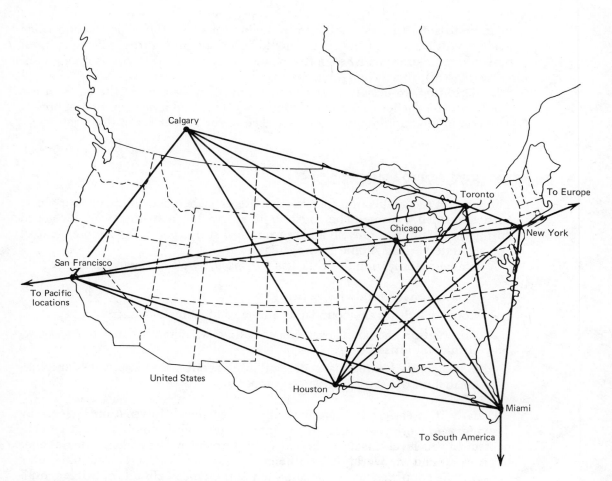

FIGURE **11-1** Country map (point to point).

within the city. When two maps are used, it does add another level. The advantage, however, is that it also decreases the complexity of simultaneously trying to design both intrastate circuits and intracity circuits. If either of these levels is omitted, the state-level maps may be the less vital. The city-level maps are needed to identify concentrator sites and/or multidrop locations, as well as individual terminal locations. At this point, lines are drawn only between the various interconnect points because configurations have not been decided upon.

The local facility "maps" are really pictorial diagrams because usually blueprints or drawings of the building floor layouts are used. Specific terminal locations can be identified on these pictorial diagrams. It is too early to identify concentrator/multiplexer sites, so this should be left until a later time. It is appropriate at this point to identify the location of current telephone equipment rooms for incoming communication circuits (voice and data).

By the end of this task there are tentative locations for individual terminals and circuit paths for the local facility, intracity, intrastate, country and international needs. To date, little is known about the volume of data that must be transmitted; nor is anything known about the type of hardware and software that might be utilized by this system.

The next step is to analyze the specific messages, although this task can be done simultaneously with the development of the geographic maps and pictorial diagrams.

6. ANALYZE THE MESSAGES

This step may be combined with the previous step on identifying the geographic scope, but it is more often combined with the following step, Calculate Traffic/ Circuit Loading. It is identified as a separate step here so that you clearly understand the level of detail that must be obtained during this very important step.

In this step each *message type* that will be transmitted or received from each application system at each terminal location is identified. Also, each *message field* (data item/attribute) is identified, along with the average number of characters for each field. Furthermore, it is necessary to identify *message length* and the *volumes* of messages transmitted per day or per hour. It probably will be necessary to visit each location where there is a system that will utilize the data communication network. These site visits are required to identify clearly each and every message type that will be transmitted or received.

If the system is a manual one, these messages might be forms in the current system, although they already might be electronically generated messages or video screen formats on terminals. Each message should be described by a short title, and a sample of the message should be attached if there is a current equivalent. If there is not an equivalent, *all* of the fields that will make up the message must be identified. *Message analysis* sometimes reveals that the system will have to handle a greater volume of data than previously was thought.

Before proceeding to the next step, we should discuss further the various fields/ data items of the message. After each message is described and samples of messages in the current system are collected, this data must be recorded. A simple form should be used to record this data, such as the one shown in Figure 11-2. The name of each individual message type is listed along with the name of each field/ data item that makes up that message. For each field in the message, list the average number of characters in each field/data item. Most data items have only an average number of characters per message; few have a peak number. It always

TELLER INQUIRY SYSTEM			
Message Name	Message Fields (data item)	Average Characters/ Field	Peak Characters Field
Passbook savings inquiry	Password	4	4
	Customer account number	9	9
	Dollar amount	6	12
	Transaction code	3	3
	Total	22	28
Loan balance inquiry	Password	4	4
	Customer loan number	16	16

FIGURE 11-2   Message contents.

is worth the effort to determine whether some of the individual data items have a peak number of characters per message. Peaks may occur during certain days of the year or hours of the day or any other time that is unique to the business situation.

Once the messages have been described and recorded, the average/peak number of characters for each message can be calculated for each application system. The most important figure is the average number of characters per message, although sometimes peak numbers of characters per message must be taken into account.

It should be noted here that most systems are built using the average number of characters for their basis, because few organizations can afford the cost of a system built on the basis of the peak number of characters. The use of averages is even more prevalent when the choice is between average number of messages per day and peak number of messages per day.

The system designer should note that a pure character count may be misleading with regard to the number of characters contained in the transmitted message. Header characters (identifying overhead-type characters within the message) and the data communication network control characters must be taken into account. The control characters can be items such as a consecutive message number, the synchronization characters, carriage returns or tabulation characters when appropriate, and line control characters for the protocol utilized (although the protocol may be unknown at this point). As a rule of thumb, 20 or 30 line control characters might be added to each message transmitted, although when messages are transmitted in contiguous groups, this figure might be closer to 20 or less. There is no rule of thumb for message header characters. Probably the best way to identify message content control information is to interview the people who run the current manual or computerized application system.

Determining the volumes of messages is critical. Now that the average number of characters for each message has been determined, the next step is to learn how many messages will be transmitted per day or per hour. To accumulate this information accurately, utilize Figure 11-3, which is a network link traffic table.

The first item in the upper left-hand corner is the identifier of a network link. This is nothing more than a first cut at determining where messages will go when users transmit from their local work area. The second column shows the name of the individual message type. The third and fourth columns show the average characters per message and the peak characters per message (if appropriate). The fifth and sixth columns show the average number of messages per day and the peak number of messages per day. The seventh and eighth columns show the average number of characters transmitted per day and the peak number of characters transmitted per day; these numbers are obtained by multiplying the characters per message by the number of messages per day.

Finally, when possible these traffic statistics (characters transmitted per day) should be broken into the hourly number of characters transmitted throughout the workday. This information can be used to help spot any problems with hourly peak volumes as the design progresses. For example, if a column total of the hourly number of characters transmitted between 9 to 10 A.M. has a volume that is 50 times the capacity of a single circuit network link, then a problem exists. The problem can solve itself if you have some messages transmitted later, such as during the next several hours. Other solutions are to have some people work overtime or to design a network link that has the capacity to meet that very high one-hour volume, although cost may prohibit the latter solution.

Even though the most important figure is the average number of characters transmitted per day, there may be important factors that cause peak volumes at various times during the day, various days during the week, or various times during the month. There also may be seasonal times of peak volumes because of holidays or legal requirements. Recall that legal requirements should have been identified earlier during the Understand the Current System phase.

The designer should plan for varying volumes at different hours of the day. For example, in an online banking network, traffic volume peaks usually are in the

Network Link	Message Type	Characters/Message		Messages/Day		Characters/Day		Hourly Number of Characters Transmitted								
		Average	Peak	Average	Peak	Average	Peak	8-9	9-10	10-11	11-12	12-1	1-2	2-3	3-4	4-5
Calgary to San Francisco	Passbook Savings Inquiry	22	28	1500	1650	33000	46200		12500	4000	4000	2000	1500	9000		
	Loan Balance Inquiry															
Down Totals						330000	405000									

SUM OF THIS COLUMN →

FIGURE 11-3 Network link traffic table.

midmorning (bank opening) and just prior to closing. Airline and rental car reservation system designers look for peak volumes of messages during holiday periods or during other vacation periods. A military system designer might look for extreme peaks in volume during crisis situations.

You can calculate message volumes by counting messages in a current system or by estimating future messages. When possible, take a random sample for several weeks of traffic and actually count the number of messages handled each day at each location.

If an online system is operational, network monitors/analyzers may be able to provide an actual circuit character count of the volume transmitted per hour or per day. Take care when selecting the sample of working days to ensure that it is not an "out of normal" situation. When estimating message volumes for a system that does not currently exist, you can use conglomerate estimating, comparison estimating, or detailed estimating.

- With *conglomerate estimating*, representatives from each application system confer to develop estimates based on past experience.

- With *comparison estimating*, the network designer meets with people inside or outside the organization who have a similar system so they can supply estimates from their network.

- With *detailed estimating*, the network designer makes a detailed study of the overall application system and its future needs to develop subestimates, which then are added together to obtain the total volume of messages as we have described above.

When making estimates of volumes, be sure to take future growth into account so the system will meet the needs of the next decade. Do not worry about the accuracy of estimates at this point, although you should make them as accurate as possible. Accuracy may not be a major concern because of the stairstep nature of communication circuits. For example, assume a situation in which a voice grade circuit is used. It can be used to transmit at 19,200 bits per second, but to meet data volumes you need to transmit at 20,000 bits per second. This would require the lease of two voice grade circuits. The combined two voice grade circuits now have a maximum capacity of 38,400 bits per second, greatly exceeding the needed 20,000 bits per second. This example demonstrates that if actual message volumes are higher than estimated, there is plenty of spare capacity. On the other hand, the opposite problem may occur if estimates are too optimistic; the organization may be forced to lease two voice grade circuits when only one is needed.

Now that individual message contents and the network link traffic table have been developed, there should be some feeling for the total volume of characters per day transmitted on each link of the proposed network. These are the volumes of characters transmitted from and to each local facility (node) where terminals will be located. The next step, which usually is carried out simultaneously with this step, is to determine traffic/circuit loading.

7. CALCULATE TRAFFIC/CIRCUIT LOADING

Now that average/peak characters transmitted per day per link have been identified, work can begin on calculating the circuit capacities required to carry that traffic. They are based on the number of characters per message and the number of messages transmitted per hour or per day. They also can be augmented through the use of modeling (described in Chapter 12).

At this point return to the geographic maps and pictorial diagrams (local facilities). Do these maps or pictorial diagrams still seem reasonable in light of the vast amount of further information that has been gathered during message analysis? At this time some of the maps or pictorial diagrams might be reconfigured slightly to further solidify the geographic configuration of the network. Remember to evaluate all the geographic maps: international, country, city/state, and local facilities.

The next step is to review all the network links over which data will travel. This may have been done when the network link traffic table was completed. If so, double-check at this time to verify that each message type was cross-referenced to the proper network link (columns 1 and 2 in Figure 11-3). If the hour-to-hour variation is significant, it may be necessary to take hourly peaks into account or adjust working schedules and work flows. Match the characters per day for each network link in Figure 11-3 with each network link that was shown on the country map (Figure 11-1). It is helpful when examining alternate configurations to list the characters per day for each link shown on Figure 11-1. It is the column totals (Figure 11-3) that really count. If the total number of characters transmitted in a single day on a single link is 330,000 or 405,000, then the network link has to operate at a speed that permits transmission of the 330,000 or 405,000 characters during the normal working hours. If it cannot meet this limit, certain adjustments have to be made. Now look at Figure 11-4. It shows San Francisco/Miami, 890,000; San Francisco/Houston, 1,250,000; and Houston/Miami, 770,000 characters per day. Later in the design, if the San Francisco/Miami traffic is multidropped through Houston, the total traffic on the Houston/Miami link will be 1,660,000 characters per day (890,000 + 770,000).

To establish circuit loading (the amount of data transmitted), the designer usually starts with the total characters transmitted per day on each link, or if possible, the number of characters transmitted per hour if peaks must be met.

Starting with the total characters transmitted per day, the system designer first

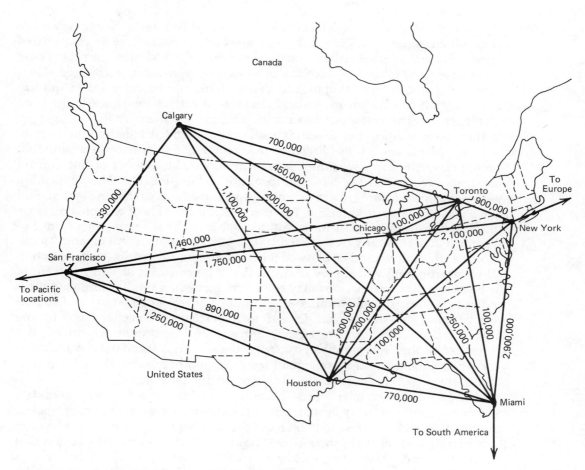

FIGURE 11-4 Link loading in characters per day.

determines whether there are any time zone differences between the various stations. This might be an international or national system that has time zone differences that must be taken into account. For example, there is a three-hour time difference between Toronto and San Francisco. This means that if a host computer in Toronto operates from 7 A.M. until 4 P.M. (Toronto time), under normal circumstances there is only a five-hour working day in San Francisco, even assuming that someone is working through the lunch hour. By the time the people arrive at work in San Francisco at 8 A.M., it is already 11 A.M. in Toronto. Then Toronto shuts down its computer at 4 P.M., and it is still only 1 P.M. in San Francisco. This leaves the San Francisco facility with a workday that extends only from 8 A.M. until 1 P.M. The practical effect of this time difference is that the 1,460,000 characters (SF/TOR link of Figure 11-4) of data must be transmitted during a five-hour period rather than the eight-hour day you might expect. These effects have to be taken into account, or work schedules must be changed. Obviously, the Toronto host computer operating hours can be extended or the San Francisco staff can start work earlier. There is no perfect solution to time zone differences, but the system designer must account for them.

Other major factors that affect circuit loading include the basic efficiency of the code utilized and TRIBs (discussed in Chapter 2). Synchronous transmission is more efficient than asynchronous transmission. The number of line control characters involved in the basic protocol affects line loading. The application systems/business future growth factor must be considered so that the system will have a reasonably useful lifetime. Forecasts should be made of expected message volumes three to seven years in the future. This growth factor may vary from 5 to 50 percent and, in some cases, exceed 100 percent for high growth organizations.

Some extra time should be allowed for transmission line errors (error detection and retransmission) which may result in the retransmission of 1 to 2 percent of the messages. Retransmissions may be even higher when small common carriers are used or when transmission is into or out of developing countries. The network designer also should consider a 10 to 20 percent contingency factor for the turnpike effect. The *turnpike effect* results when the system is utilized to a greater extent than was anticipated because the system is found to be available, is very efficient, and has electronic mail features. In other words, the system now handles message types for which it was not originally designed.

Other factors to consider when evaluating line loading might be whether to include a message priority system. High priority messages may require special identification and therefore may increase the number of characters per message. If the message mix changes and most messages become high priority over a period of time, then more characters will be transmitted during a working day. Also, a greater throughput may have to be planned to ensure that lower priority messages get through in a reasonable period of time. The learning curve of new terminal operators also may affect line loading. Operator errors and retransmissions are greater when a new system is being learned.

Another factor that might affect circuit loading is an inaccurate traffic analysis (confidence intervals). Try to account for any business operating procedures that might affect the system and the volume of data transmitted.

Other factors that must be taken into account include extra system characters transmitted (line control characters) such as polling characters, turnaround time/synchronization characters, control characters in message frames and/or packets, modem turnaround time on half duplex circuits, message propagation time subtracts from the total useful hours for transmission of data, any printer time for carriage return/tabulation/form feeding, lost time when statistical time division multiplexers are overloaded, and periods of high error rates caused by atmospheric disturbances. Use the TRIB calculations from Chapter 2.

At this point the system designer should review and establish some of the response time criteria. These are required to meet the basic needs of the application systems that will utilize the network. (Response time was covered in Chapter 2.)

Finally, begin recording on the network maps and/or pictorial diagrams some of the bits per second transmission rates that will be required for each circuit link. Sometimes it is useful to show the transmission capacity required for each link. In Figure 11-4 we show the characters per day per link. Now add the bits per second transmission rate necessary for each circuit link. This helps when alternative network configurations, software, and hardware considerations are being developed and evaluated.

Look at the San Francisco to Houston link in Figure 11-14. It shows 1,250,000 characters per day. To convert characters per day to bits per second, perform the following calculations.

$$
\begin{array}{rl}
1,250,000 & \text{characters per day} \\
\underline{\times\ 10} & \text{bits per character (asynchronous)} \\
12,500,000 & \text{bits per day} \\
\underline{\div\ 7} & \text{hours (assumes 7 working hours per day)} \\
1,785,714 & \text{bits per hour} \\
\underline{\div\ 60} & \text{minutes per hour} \\
29,762 & \text{bits per minute} \\
\underline{\div\ 60} & \text{seconds per minute} \\
496 & \text{bits per second transmission speed}
\end{array}
$$

If this example uses synchronous transmission (eight bits per character), the bits per second transmission speed is 397. Also note that the characters may not be sent uniformly over time; there may be peak periods of data bit transmission.

8. DEVELOP A CONTROL MATRIX

Because the network probably will be the "lifeline" of the entire information flow within the organization, security and control are mandatory. All of the security and control mechanisms to be included in this data communication network must be taken into consideration during the design phase. As we stated earlier, we are well into an era in which information is the single most valuable resource within an organization. For this reason, it must be protected from all types of threats such as errors and omissions, message loss or change, disasters and disruptions, breaches of privacy, security/theft, unreliability, incorrect recovery/restart, poor error handling, and lack of data validation.

At this point the network control matrix is developed. In the section Matrix of Controls in Chapter 9 we described how to develop this matrix. Review that section now to be sure you understand how to set up and continue development of this matrix through the remainder of the design project. At this time, develop the basic blank matrix, naming only the threats and components. As the design effort continues, identify controls and relate them to their threats and components by placing them in the appropriate cell of the matrix. To assist in this area, also review Appendices 1 and 2 which relate to security and control for data communication networks.

9. DETERMINE NETWORK CONFIGURATIONS

During this step of the system approach to designing a data communication network, the designer utilizes all the information collected to date. Of special value are the network maps and the traffic/circuit loading data. These are used to configure the network in such a way as to achieve the required throughput at a minimum circuit cost. Begin this step by reviewing the maps and pictorial diagrams that show the links between the station/node locations.

The object of this step is to configure the circuit paths between users and the host computer. The decision involves moving the stations/nodes about, and making judgments with regard to software and hardware. In reality, this step is performed simultaneously with the next two steps, Software Considerations and Hardware Considerations. Some goals that the network designer tries to achieve with regard to an efficient and cost-effective network include

- Minimum circuit mileage between the various stations/nodes. Computer programs/modeling can help here.
- Adequate circuit capacity to meet today's data transfer needs as well as those required three to seven years in the future.
- Reasonable response times at individual terminals. Response time needs must be met for each application.
- Reliable hardware that offers minimum cost, adequate speed and control features, a high Mean Time Between Failures (MTBF), and good diagnostic/serviceability features.
- Efficient software/protocols that can be used on a variety of circuit configurations including satellite circuits. One of the newer bit-oriented protocols that can interface with various international standards (X.25) might be used. This permits the network to interface with national/international networks as well as with electronic mail systems, utilize multivendor hardware, and connect to public packet switched networks.

- A very high level of reliability (network uptime) must be met. This may be the most important factor. The network designer always should remember that when business operations move into an online, real-time data communication network, it is as if the company has closed its doors for business when the network is down.
- Reasonable costs (not necessarily the absolute lowest).
- Acceptance of the network by both day-to-day users and managers who utilize data/information from the system.

When one is developing different network configurations, a variety of choices are available. In other words, there is a *choice set*, which is a set of all available alternatives. Each alternative is a different system or a slightly modified version of another alternative. During the deliberations, the following decisions must be considered.

- Determine the choice set, that is, all possible network configurations.
- Divide the choice set into *attainable* and *unattainable* sets. The attainable set(s) contains only those alternatives that have a reasonable chance of acceptance by management. Acceptance might be predicated on costs, software, hardware, circuit availability, or political factors within the organization.
- Review the attainable set of alternatives and place them in a ranked sequence from the most favored to the least favored, taking into account your evaluation criteria for choosing the most favored. Evaluation criteria were identified during the plan preparation phase.
- Present the most highly favored alternatives to management for review and, it is hoped, approval.

There is one other consideration in selecting the different network alternatives. The network designer must know whether the proposed alternative is going to maximize something, optimize something, or satisfice something, or if it will be a combination of the three. To *maximize* is to get the highest possible degree of use out of the system without regard to other systems. To *optimize* is to get the most favored degree of use out of the system taking into account all other systems; an optimal system does just the right amount of whatever it is supposed to do, which is not necessarily the maximum. To *satisfice* is to choose a particular level of performance for which to strive and for which management is willing to settle.

Finally, the network designer also must be aware that individual job tasks within the network may have three levels of dependence on each other.

- *Random dependence:* a job task is required because of some other job task.
- *Sequential dependence:* one particular job task must precede or follow another job task.
- *Time dependence:* a job task is required at a set time with regard to another job task.

The network designer should assess various job tasks during the development and design of network configurations. Job task interrelationships must be studied with regard to future needs and growth. Job tasks that are dependent today may not be after a new application is completed next year. Job tasks require an open-ended approach.

Now that the network maps/pictorial diagrams and traffic/circuit loading have been reviewed, line controls and modes of operation can be considered. This probably involves software and such factors as full duplex versus half duplex, whether a satellite link is used, statistical multiplexers versus pure (transparent) multiplexers, modem speeds, intelligent terminal controllers, and how different configurations operate, such as central control versus interrupt, multidrop, or point to point.

Various alternative configurations are shown in Figures 11-1, 11-5, 11-6, 11-7, and 11-8. Figure 11-1 shows a point-to-point configuration where each terminal node has its own communication circuit between all other nodes. Figure 11-5 shows the same configuration using a multidrop circuit with New York as a switching center. Figure 11-6 shows the same configuration again, using a multiplexed arrangement. The Houston site multiplexes San Francisco/Miami/Houston data to Chicago. Then Chicago multiplexes that data with the Chicago/Calgary/Toronto data and on to New York. Notice how the different configurations change overall circuit mileage. Circuits are paid for on the basis of dollars per mile per month; therefore, a minimum mileage configuration is also a minimum circuit cost configuration. Also notice that different numbers of modems are required in different configurations. For example, the point-to-point configuration requires many more modems than the multidrop configuration. Fewer modems save on modem costs.

Figure 11-7 shows a packet switching satellite network. This can be a public packet switcher or a private packet network. Figure 11-8 shows a combination of a local area network, a packet satellite network, point-to-point, multiplex, and multidrop configurations.

This step involves choosing among various network alternatives. The main constraints are the availability of software packages, hardware, and circuit links. These three factors are all interconnected and must be considered along with the performance and reliability that must be obtained. All of these factors also are interrelated with regard to cost. Therefore, when alternative network configura-

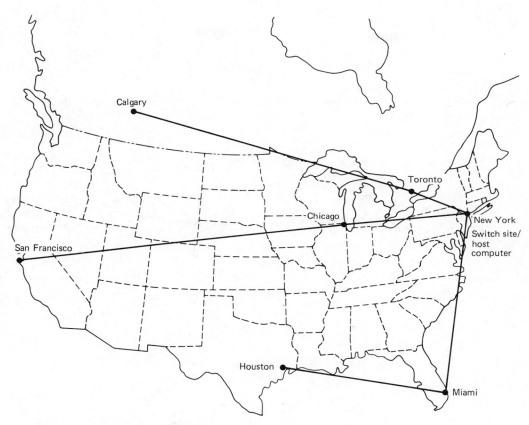

FIGURE **11-5**   Multidrop configuration.

tions are developed, consider the software, hardware, circuits, performance, and reliability, and relate these five factors to your cost/benefit analysis.

10.   SOFTWARE CONSIDERATIONS[2]

With regard to software, the type of host computer may be a major constraint. The protocols the host can handle may limit the types of terminals or other hardware that can be utilized. This limitation may be overcome through the use of protocol converters and/or the purchase of a new front end communication processor that can interface with the host and a variety of software and hardware.

[2] Ibid., Chapter 10 (page 427).

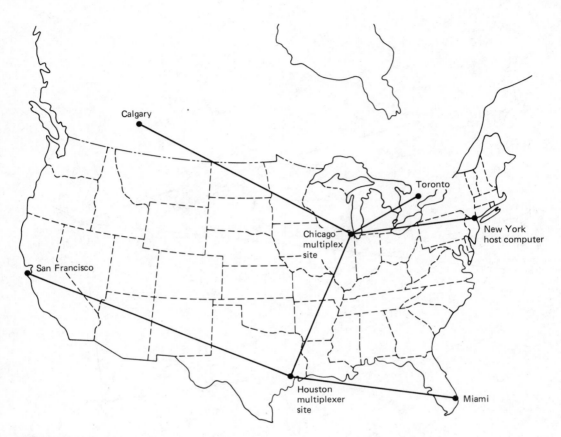

FIGURE **11-6** Multiplexed configuration.

At this point the software will determine the line control methodology/mode of operation. Decisions must be made as to whether operations will be in full duplex or half duplex, asynchronous or synchronous, and at what speeds. For a new system one of the newer bit-oriented protocols should be selected, such as X.25, SDLC, HDLC, or the like. The older byte-oriented protocols (such as Binary Synchronous Communications—BSC) probably are not a good choice because of their limitations on satellite links, slow half duplex operation, and inability to meet international standards. It is desirable to select a protocol that is compatible with the International Organization for Standardization (ISO) seven-layer model, although reality might dictate that another protocol must be utilized in order to be compatible with existing hardware.

In addition to protocols/software, other network architectures/software that

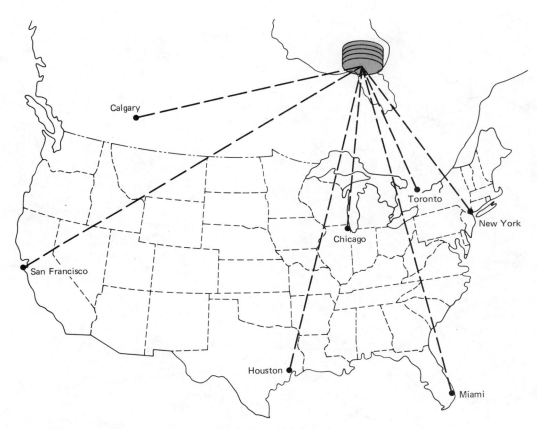

FIGURE **11-7** Public packet switching satellite configuration.

reside in the host computer and front end communication processors have to be considered. For example, telecommunication access programs and the teleprocessing monitors may affect network operations. Security software packages in the host computer also can be a constraint. Finally, the host operating system itself may be a constraint to network control and operation, as might the database management system software.

Any software programs that are located out in the network should be reviewed. These may be at remote concentrators, remote intelligent terminal control devices, statistical multiplexers, and terminals. Microcomputers also raise the issues of distributed data processing/remote application programs, micro-to-mainframe software, and local area networks.

The network designer can make a major contribution to the future by selecting

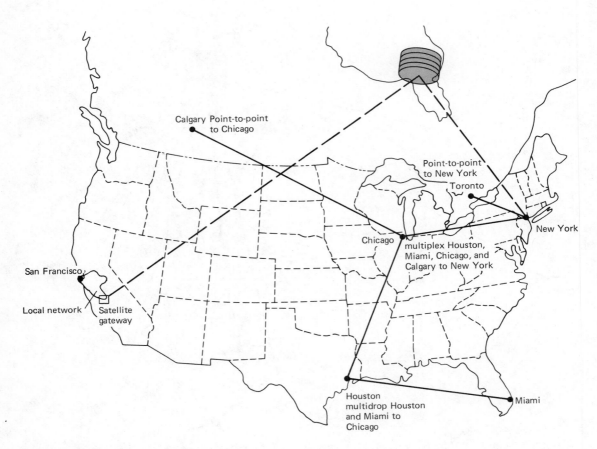

FIGURE 11-8   Multiple configurations.

a protocol that can grow, that is compatible with an internationally recognized standard, and that will not have to be changed for at least five to ten years. The protocol is crucial because the host computer network architecture must be able to interface with it. For example, the telecommunication access programs and teleprocessing monitor should be compatible with international standards. This means that the International Organization for Standardization seven-layer model should be used as the basic skeleton when protocols are interfaced to host computer/front end software packages. Also, you might want to interface with a local area network sometime in the future.

Finally, software diagnostics and maintenance must not be overlooked. Determine how quickly either in-house people or the vendor can diagnose software problems and how quickly they can fix these problems. Recall Mean Time To

Diagnose (MTTD), Mean Time To Respond (MTTR), and Mean Time To Fix (MTTF) from Chapter 8; they also apply to protocols and other software packages.

11. HARDWARE CONSIDERATIONS[3]

Hardware that interacts with the alternative network configurations is easier to handle than software because hardware is a tangible item. Some of the pieces of hardware that need to be considered are

- Terminals/microcomputers
- File servers
- Intelligent terminal controllers
- Modems (analog/digital)
- Multiplexers
- Intelligent multiplexers (STDM)/concentrators
- Line-sharing devices
- Protocol converters
- Hardware encryption boxes
- Automated switching devices
- PBX/CBX switchboards
- Data protectors
- Various communication circuit types
- Port sharing devices
- Front end communication processors
- Host computers
- Testing equipment

With this in mind, the designer uses representations of the pieces of hardware and moves them about on the various network maps and pictorial diagrams. This experimentation with configurations should take into account the protocol/software considerations. The result should be a minimum-cost network that meets the organization's data communication (throughput) requirements. This is no trivial task. Many organizations use computer simulation and modeling to carry out this task successfully.

Before ordering hardware, the design team should decide how to handle diag-

---

[3] Ibid., Chapter 10 (page 425) and Request for Proposals (Appendix 3).

nostics, troubleshooting, and repair. It should be remembered that MTTD, MTTR, and MTTF always apply to hardware because hardware usually fails more often than software. Vendor estimates of MTBF (Mean Time Between Failures) should be obtained by the design team. Issues that should be addressed include the types of test equipment that are necessary and the organizational structure of the network management group (see Chapter 8). Some hardware may have built-in diagnostic capabilities for its internal electronic circuits, as well as the ability to identify problems on the communication circuit.

Diagnostics go hand in hand with network service. The vendor's MTBF and ability to respond to service calls are essential factors that affect downtime of the network.

In summary, a network configuration that takes both hardware and software into account must be developed. Costs also are analyzed during this effort.

12. NETWORK (CIRCUIT) COSTS

Usually, it is a hindrance to propose cost limitations during the initial development of design alternatives. Of course, an effort always should be made to keep costs down; however, costs should not interfere with *preliminary* design configuration alternatives (choice sets). The point is that the various alternatives should be identified first; then costs should be related to the *attainable* design configurations. The first task is to identify the attainable/workable configurations, and the second is to identify the costs of those alternatives.

Estimating the cost of a network is much more complex than estimating the cost of a new piece of hardware. Many variables and intangibles are involved. Nevertheless, estimating the cost of a system is a necessary prerequisite to deciding whether implementation is justifiable. Some of the questions that must be considered are

- What are the major cost categories of the overall system? These may include
     Circuit costs
     Hardware costs
     Software costs
     Maintenance/network management costs
     Personnel costs
- What methods of estimating are available and what accuracy can be achieved?
- Can all costs be identified and accurately estimated?

- Can benefits be identified? Which benefits cannot be estimated in dollar terms? Can they be measured in any other way?
- What criteria will management use when evaluating these estimates? (Refer to the set of criteria developed during the "prepare a plan" step.)

Because the specific method used to gather and compile costs for presentation has to be tailored to meet the needs of the organization, several alternatives for cost analysis are presented here.[4]

Network Cost Analyzer   This is more a worksheet than a methodology. It is an excellent way to document specific network costs after a *detailed* cost analysis. Among these costs are software, hardware, and circuit costs.

Figure 11-9 shows a network cost analyzer in which each cost category is listed horizontally across the top and each circuit link is listed vertically down the left side by network link. Each individual cost is placed within a cell, such as the circuit cost for the New York City to Los Angeles circuit. By adding the costs horizontally, you can obtain a *total link* cost for each network link. Some costs, like front ends or software, may have to be allocated among all the links on a fair basis. For example, a $300,000 front end serving 20 network links might be allocated at $15,000 per link.

Totaling down a column gives the total cost of a specific category of hardware or software. Finally, the grand total in the lower right corner shows the total cost of the entire network, including its software, hardware, and circuits.

Cost/Benefit Categories   Figure 11-10 shows various cost categories associated with data communication networks, as well as the various benefit categories. The most helpful items in this figure are the direct costs, the indirect costs, and the intangible benefits. Intangibles sometimes are very difficult to identify. Other benefits, such as direct and indirect cost reductions and revenue increases, must be identified in a manner that is unique to the organization for which the network is being designed. This figure can be used to ensure that no critical cost or important benefit is overlooked. Figures 11-9 and 11-10 are used together.

Voice Grade Circuit Costs   The major factors that have to be taken into account in figuring the cost of a voice grade leased circuit are the circuit link mileage, cost (tariff) for the circuit, and any circuit termination charges at each end of the circuit link.

The common carriers use a system of *vertical and horizontal coordinates* to

[4]Ibid. Review Chapter 10 (Economic Cost Comparisons) for an in-depth presentation of 12 different cost analysis methodologies.

Network Link	Circuit Cost	Intelligent Controllers	Modems	Multi-plexer/ Concen-trators	Terminals	Front Ends	Software	Personnel and Facilities	Link Total Cost
NYC to LA	$1,420.71								
Down Totals									Grand Total

FIGURE 11-9 Network cost analyzer.

COSTS	BENEFITS
Direct costs	**Direct and indirect cost reductions**
• Computer equipment	• Elimination of clerical personnel and/or manual operations
• Communication equipment	• Reduction of inventories, manufacturing, sales, operations, and management costs
• Common carrier line charges	• Effective cost reduction, for example, less spoilage or waste, elimination of obsolete materials, and less pilferage
• Software	• Distribution of resources across demand for service
• Operations personnel costs	
• File conversion costs	**Revenue increases**
• Facilities costs (space, power, air conditioning, storage space, offices, etc.)	• Increased sales because of better responsiveness
• Spare parts costs	• Improved services
• Hardware maintenance costs	• Faster processing of operations
• Software maintenance costs	**Intangible benefits**
• Interaction with vendor and/or development group	• Smoothing of operational flows
• Development and performance of acceptance test procedures and parallel operation	• Reduced volume of paper produced and handled
• Development of documentation	• Rise in level of service quality and performance
• Costs for backup of system in case of failure	• Expansion capability
• Costs of manually performing tests during a system outage	• Improved decision process by provision of faster access to information
• Security and control	• Ability to meet the competition
	• Future cost avoidance
Indirect costs	• Positive effect on other classes of investments or resources such as better utilization of money, more efficient use of floor space or personnel, and so forth
• Personnel training	• Improved employee morale
• Transformation of operational procedures	• Keeping technical employees
• Development of support software	
• Disruption of normal activities	
• Increased system outage rate during initial operation period	
• Increase in the number of vendors (impacts fault detection and correction due to "finger pointing")	

FIGURE **11-10** Cost/benefit categories.

Location	Vertical	Horizontal
Tulsa	7707	4173
San Francisco	8492	8719

$$\text{Distance} = \sqrt{\frac{(V_1 - V_2)^2 + (H_1 - H_2)^2}{10}}$$

$$D = \sqrt{\frac{(8492 - 7707)^2 + (8719 - 4173)^2}{10}}$$

$$D = \sqrt{2,128,234.1}$$

$$D = 1,459 \text{ miles}$$

FIGURE **11-11** Vertical and horizontal coordinates formula for determining air mileage.

determine the air mileage between central offices. Figure 11-11 gives the formula for calculating the vertical and horizontal coordinates to determine air mileage. You must know the vertical and horizontal coordinates to determine the distance between any two rate centers. The V and H coordinates for Tulsa and San Francisco are shown in Figure 11-11. Next, put the V and H coordinates for the rate center into the formula in Figure 11-11 and solve it to get the air mileage. (Note that the difference always is obtained by subtracting the smaller from the larger coordinate.) Round the mileage to the nearest whole number.

The term *rate center* is a telephone company term for the central/end office where the circuits terminate. For example, one end of a local loop goes to your business or home, and the other end goes to a rate center (also called a central or end office). It is called a rate center because it is used when calculating the rates charged for the provided service.

Appendix 3 lists V and H coordinates by city only, even though they apply to individual central/end office locations throughout a city. For this reason, a given city may have many pairs of V and H coordinates. A list by city only is adequate for our design problems. You can obtain the complete set of V and H coordinates for all rate centers from FCC Tariff 10.

Each *FCC Tariff* is used for specific types of rate information. For example, Tariff 9 contains the rates and regulations for services provided among AT&T central offices and the type of "customer connection" to the central office. Tariff 10 contains the vertical and horizontal coordinates used to calculate the mileage

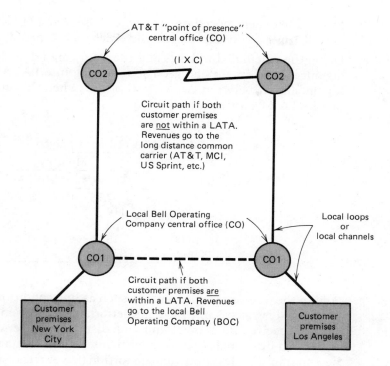

FIGURE 11-12 Voice grade circuits. The central office (CO) also is referred to as an end office or an exchange office. LATAs were described in Chapter 10; Figure 10-2 shows LATAs.

between central/end offices (rate centers). Tariff 11 contains the rates and regulations for private line local channels provided by AT&T for local distribution of traffic. These local channels usually are provided by the local Bell Operating Company.

AT&T has at least one point of presence in each local access transport area. LATAs were defined in Chapter 10 (see Figure 10-2). A *point of presence* is the end office to which your local loop is connected so that messages can go out on the long distance (IXC) circuits. These long distance IXCs are interLATA circuits; that is, they cross between LATAs.

The cost of a voice grade circuit is comprised of the following five cost elements (although number 4, conditioning, is optional).

1. Central or end office connection charge (station terminal) at each end of the IXC circuit (see Figure 11-12).

    $15.95 per month per IXC circuit end

2. Mileage between central or end offices. Use V and H coordinates to calculate the mileage. It depends on the cities (rate centers) being connected.

3. Using the calculated mileage and the following table, determine the long distance (IXC) circuit charge. The following interLATA mileage charge is for the circuit between two central offices when this circuit crosses from one LATA to another.

InterLATA Mileage Charges

Mileage	Monthly	
	Fixed	Per Mile
1–50	$ 57.75	$2.27
51–100	119.75	1.03
101–500	160.75	0.62
501–1,000	315.75	0.31
1,000+	315.75	0.31

4. Add in the cost of *channel options* such as line conditioning. Conditioning is optional because some modems perform this function. *C type conditioning* has specific limits on attenuation distortion and envelope delay distortion. Such limits reduce line impairments so that a data signal can get to its destination with less noise and distortion (fewer retransmissions caused by errors). Bell's Technical Reference Publication 41004 gives the exact specifications. *D type conditioning* specifically limits noise and harmonic distortion. Again, this type of conditioning reduces line impairments and therefore has fewer retransmissions caused by errors. There are seven levels of conditioning, but use these four levels for our design problems.

Type C
  Point to point $31.00/month
  Multidrop   52.00/month
Type D
  Point to point $22.00/month
  Multidrop   71.00/month

5. Calculate the cost of the local loop (also called the local channel). If the user's site (your office) has the same central office (Bell telephone personnel sometimes call it a *wire center*) as the AT&T point of presence central office, use zero mileage in the next table to determine the cost of the local loop. In the case of zero mileage, the cost is $66.87 for a local loop.

IntraLATA Mileage Charges

Mileage	Monthly	
	Fixed	Per Mile
zero	$ 66.87	zero
1–4	75.93	$6.34
5–8	78.24	5.77
9–25	101.31	2.88
26–50	111.62	2.48
over 50	111.62	2.48

All this complexity has been brought about by deregulation. Deregulation specifies that your local Bell Operating Company can charge for the local loops (local channels). These are *intraLATA* circuits. AT&T, along with other special common carriers such as US Sprint and MCI, can charge for the long distance interexchange channels (IXC). These are *interLATA* IXCs.

AT&T has at least one point of presence in each LATA (local access transport area). This point of presence is the end office to which your local loop must be connected when you want messages to go out on the long distance (IXC) circuits (interLATA).

Look at Figure 11-12 in which we have marked the central offices with numbers of CO1 and CO2. Central office 1 implies that it is the local Bell Operating Company central office closest to your customer premises. It is the zero mileage charge local loop in the above table. For the $66.87 listed in this table you get a circuit path from your customer premises (NYC) to CO1.

When the AT&T point of presence central office is not the same as the local Bell Operating Company central office, you must pay a mileage charge for the length of the circuit between the two central offices. In Figure 11-12 this is represented by the circuit that goes from central office 1 to central office 2. Sometimes these two central offices are the same, but usually they are not. Only a few companies that are lucky enough to be located next to the AT&T point of presence central office do not have to pay this extra (greater than zero mileage) intraLATA circuit mileage charge. These lucky companies can use zero mileage calculation for the local loop link between central office 1 (Bell Operating Company) and central office 2 (AT&T) for their intraLATA circuits.

To calculate mileage charges when the AT&T point of presence is not at CO1, you determine the mileage between CO1 and CO2 and use the above intraLATA mileage charges. For example, if the mileage between CO1 and CO2 is 7 miles, the charge is $118.63 instead of the $66.87 zero mileage charge (7 miles × $5.77 per mile + $78.24 = $118.63). For your design problems, make some estimates for this distance between the central offices (CO1 and CO2) because they are located throughout a city.

As you can see, we have given you only one intraLATA mileage charge table for this local loop mileage calculation. We have not provided all the local channel mileage tables because there are several hundred of them, they vary from state to state, and they vary between LATAs. Your local telephone company can supply them for your area. These rate tables are 30 or 40 pages long and are similar to the one above. This one is sufficient for your design problems, and it is enough to learn how to determine circuit costs. In the computerized network design software package (Chapter 12), all of the mileage charge tables are organized by area code and the first three digits of the telephone exchange number.

Once you have calculated the cost for the local channel, there may be a few additional charges, depending on the current tariff structure. For example, there may be special access surcharges, and you may be charged for access coordination functions between AT&T and the Bell Operating Company. In your design problems you can ignore these because they are in a state of flux and changing as deregulation continues. They usually are minor charges and will not affect your calculations significantly.

Now figure the cost of the circuit in Figure 11-12 if CO1 to CO2 at the New York end is 31 miles, and CO1 to CO2 at the Los Angeles end is 9 miles (assume no conditioning).

1. Central office connection charges for each end of the IXC between the CO2 in New York and the CO2 in Los Angeles: $31.90 ($2 \times \$15.95$)

2. V and H coordinates from Appendix 3.

	V	H
New York City	4997	1406
Los Angeles	9213	7878

Based on the formula in Figure 11-11, the mileage is 2,443 miles.

3. InterLATA mileage charges (NYC to LA). In Figure 11-12 this is the circuit between CO2 and CO2.

$$
\begin{array}{r}
2{,}443 \text{ miles} \\
\times\ 0.31 \text{ per mile} \\
\hline
\$757.33 \\
+\ 315.75 \text{ fixed charge} \\
\hline
\$1{,}073.08 \text{ per month}
\end{array}
$$

4. No conditioning required. Assume the modems perform conditioning.

5. Local loop cost (intraLATA mileage charges) for customer premises (NYC) to CO2 and customer premises (LA) to CO2.

NYC to CO2 (New York end)

$ 2.48 per mile

   <u>× 31</u> miles

$ 76.88 (this is the charge for CO1-CO2)

<u>+ 111.62</u> (this is the charge for NYC-CO1)

$188.50 per month

LA to CO2 (Los Angeles end)

$ 2.88 per mile

   <u>× 9</u> miles

$ 25.92 (this is the charge for CO1-CO2)

<u>+ 101.31</u> (this is the charge for LA-CO1)

$127.23 per month

And so this private voice grade lease circuit would cost

$ 31.90 central office connection

1,073.08 interLATA (CO2 to CO2)

     0.00 conditioning

  <u>315.73</u> intraLATA ($188.50 + $127.23)

$1,420.71

On the other hand, if CO1 at the Los Angeles end is the AT&T point of presence central office, what is the new cost of the local loop (it was $127.23 above)? The new cost is $66.87, which is the zero mileage amount on the previous intraLATA mileage charges table.

Dial-Up Circuit Costs   Dial-up communication circuits, just as with leased circuits, have two different rate schedules. One is for interLATA/interstate and is billed by AT&T, US Sprint, MCI, or another long distance common carrier. The other is for intraLATA/intrastate and is billed by the Bell Operating Companies (BOCs) or other local telephone company.

Figure 11-13 gives the rates for interLATA/interstate (long distance) calls as billed for AT&T, US Sprint, MCI, or another long distance common carrier. If you want to calculate the cost of an intraLATA dial-up call, we recommend that you call your local telephone company business office and ask for its dial-up rate schedule. The dial-up rate schedule looks similar to the one in Figure 11-13 except that the day rates for the initial minute and each additional minute are different. The rate mileage cutoff points also may vary because they may not go as high as 5,750.

To calculate the cost of a long distance dial-up call, determine the air mileage between the calling party and the called party (use the V and H coordinates in Appendix 3). Next, calculate the cost of the call by using the day rates in Figure

Discount Amount							
	Mon	Tues	Wed	Thurs	Fri	Sat	Sun
8:00 A.M. to 5:00 P.M.	Day Rate Period						
5:00 P.M. to 11:00 P.M.	Evening Rate Period 38% Discount from Day Rate					Evening 38% Discount	
11:00 P.M. to 8:00 A.M.	Night and Weekend Rate Period 53% Discount from Day Rate						

	Day Rates	
Rate Mileage	Initial Minute	Each Additional Minute
1–10	$0.23	$0.15
11–22	$0.28	$0.19
23–55	$0.31	$0.21
56–124	$0.33	$0.24
125–292	$0.33	$0.26
293–430	$0.34	$0.28
431–925	$0.36	$0.30
926–1,910	$0.37	$0.31
1,911–3,000	$0.40	$0.32
3,001–4,250	$0.44	$0.35
4,251–5,750	$0.46	$0.37

FIGURE **11-13** InterLATA dial-up calls (also called direct distance dialing or DDD).

11-13. Do not forget to apply the appropriate discount when the call is an evening or night call.

Let us use the circuit given in Figure 11-12 again and assume that the organization uses a dial-up call instead of a private lease line. What would its monthly telephone bill be using the interLATA rates given in Figure 11-13? Your first task is to determine how many minutes per day the telephone call lasts and whether there is one long call or many short calls each day. In this case, let us assume there is one six-hour call per day.

We already know from the previous V and H calculation that the mileage between New York City and Los Angeles is 2,443 miles. The six-hour call amounts to 360 minutes (6 hours × 60 minutes per hour). Based on the day rates in Figure 11-13, the daily cost is $115.28 for the long distance charges. To this must be added the basic monthly telephone bill from the local telephone company.

359 minutes
× 0.32 per minute
$114.88
+ 0.40 for the first minute
$115.28 per day

The monthly cost is $2,536.16 plus the basic monthly telephone bill from the local telephone company. We assumed 22 working days per month (22 days × $115.28). If you look back to the previous private lease line cost calculation, you will see that it was only $1,420.71. As you can see, this organization would do better to lease a circuit rather than use dial-up circuits. The crossover from dial-up to lease depends on how many minutes per day are used when connected via the dial-up circuits.

Wideband Circuit Costs Wideband service from AT&T provides for 19,200, 40,800, 50,000, 56,000, and 230,400 bits per second transmission speeds. The transmission speed is dependent on the modem being used. Remember, the modem is the device that determines the bits per second speed, although you do need a line/circuit that has enough bandwidth to carry the speed at which the modem transmits. Using wideband circuits, you lease either a *Group 48,000 hertz bandwidth* (equivalent to 12 voice grade circuits) or a *Supergroup 240,000 hertz bandwidth* (equivalent to 60 voice grade circuits). There is a monthly end office connection charge of $28.00 per circuit end ($28 × 2). The wideband circuit costs for both Group 48 KHz and Supergroup 240 KHz, which are shown below (use V and H coordinates to calculate the mileage), are based on the mileage between end offices.

Group 48 KHz Bandwidth (Wideband)

Mileage Band	Monthly Cost	
	Fixed	Per Mile
1–50	$ 165.28	$12.34
51–100	165.28	12.34
101–500	567.28	8.32
501–1,000	1,447.28	6.56
1,000+	3,617.28	4.39

Supergroup 240 KHz Bandwidth (Wideband)

Mileage Band	Monthly Cost	
	Fixed	Per Mile
1–50	$ 387.38	$29.18
51–100	387.38	29.18
101–500	1,342.38	19.63
501–1,000	3,407.38	15.50
1,000 +	8,577.38	10.33

Packet Switching Costs In AT&T's ACCUNET packet network, three elements can be used for calculating costs: the private leased access line, the transmission speed per port, and the number of packets transmitted.

With this service, users must assemble/disassemble their own packets and send them via a *private lease line* to the AT&T point of presence end office that handles packet switching. Therefore, you must calculate the private voice grade circuit cost as we did earlier for a voice grade circuit. This cost depends on the mileage between the user premises and the AT&T point of presence end office. For example, when calculating this cost, use the wideband circuit cost tables if transmitting at 56,000 bits per second and the voice grade mileage charge tables if transmitting at either 4800 or 9600 bits per second.

The cost element for *transmission speed per port* depends on how fast you want to transmit. This monthly charge is

- $495 for 4800 bits per second
- $695 for 9600 bits per second
- $1,225 for 56,000 bits per second.

To determine how many terminals can be connected to the port, you need to know that a 4800 bits per second port has a capacity of four virtual circuits, a 9600 bits per second port has eight, and a 56,000 bits per second port has forty.

Finally, the packet charge is based on *how many packets are transmitted* each day. To determine this monthly charge, you have to estimate or count the amount of packet traffic on the network. Remember that a packet is 128 characters long. The charges are $0.75 per kilopacket (1,000 packets) for Monday through Friday, 7:00 A.M. to 6:00 P.M. All other times are $0.30 per kilopacket. Design steps 6 and 7, described earlier in this chapter, explain how to estimate message traffic.

Satellite Circuit Costs The cost of a satellite circuit consists of three elements: the local access facility, station termination, and the satellite channel charge.

For figuring RCA Satellite tariffs, the *local access facility* is $120.00 if your

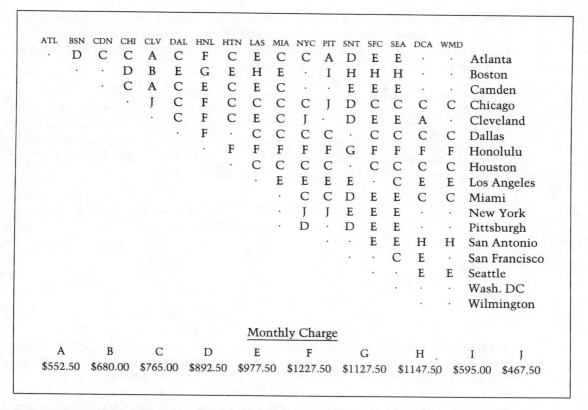

FIGURE **11-14**   Satellite channel cost for one four-wire circuit.

business location is within RCA's local distribution area. The local access facility is similar to the AT&T point of presence discussed earlier in this chapter. In this case, it is the local telephone company end office into which RCA has connected its satellite circuits. The point is that you may have to calculate the cost of a terrestrial voice grade circuit from your business premises to the RCA local access facility end office. This cost is over and above the satellite circuit cost. Use either the intraLATA or the interLATA cost tables presented earlier in this chapter, depending on whether the voice grade terrestrial circuit crosses into another LATA.

The *station termination charge* is $21.25 per satellite circuit end on the ground. (Therefore, it is 2 × $21.25.) The *satellite channel charge* is shown in Figure 11-14 for 17 major cities. This amount is a per month cost for one voice grade circuit.

There is one other optional charge for an *echo canceller:* satellite delay compensation which was discussed in Chapter 2 (see Figure 2-43). An echo canceller costs $40.00 per month. It is required when the software protocol is a half duplex, stop and wait ARQ (automatic repeat request).

Digital Circuit Costs   The cost of an all-digital transmission circuit can be calculated exactly the same way as was done earlier for voice grade circuits, with one exception. The following interLATA mileage charges should be used instead of the voice grade circuit charges given in the interLATA table in the earlier section. The tariff below is for AT&T's *DATAPHONE™Digital Service* with transmission at 2400, 4800, or 9600 bits per second.

Digital InterLATA Mileage Charges

Mileage	Monthly	
	Fixed	Per Mile
1–50	$ 79.99	$2.27
51–100	141.99	1.03
101–500	182.99	0.62
501–1,000	332.99	0.31
1,000 +	332.99	0.31

Hardware Costs   The following are costs for selected pieces of communication hardware. These costs are representative averages and therefore might not reflect the latest discounts. For this reason, you should contact hardware vendors directly or check prices in reference guides like Auerbach or Datapro. You also may request a BLACK BOX[R] hardware catalog by calling (412) 746-5530. Vendors can provide specifications for any piece of hardware you might want to use. You can call or write to vendors for the cost of specific communication software packages.

Selected Hardware (Lease) Costs

Modem (300–2400 bps)	$ 9.00 per month
Modem (4800–9600 bps)	90.00 per month
Modem (56,000 bps)	200.00 per month
Digital modem (2400–9600 bps)	20.00 per month
Video terminal	125.00 per month
Microcomputer	220.00 per month
Multiplexer (8 ports)	210.00 per month
Statistical multiplexer (16 channels with built-in 9600 bps modem)	450.00 per month
Front end processor	$2,000.00 per month

Further Design Ideas   If you want to compare more alternative circuit types than presented in this chapter, we suggest that you call AT&T for detailed tariffs

FIGURE **11-15** United States map.

for WATS (Wide Area Telecommunications Service), AT&T Megacom, and AT&T Megacom 800. These communication services were described in Chapter 10. Their costs are not presented here because they require maps of the bands emanating from specific telephone exchange local areas. You need the WATS maps and rates for the specific area of the United States in which you are located. Call AT&T at (800) 257-4636 or your local AT&T business office for this information.

The DIAL-IT 900 Service charges were included with the description of that service in Chapter 10. Perhaps you can think of an idea to make money with this service!

If you want to compare AT&T's dial-up rates listed earlier in this chapter with other discount voice service companies, call US Sprint at (800) 521-4949 or MCI Communications at (800) 624-6240 to obtain a set of their rates (tariffs).

There are design problems for you to complete at the end of this chapter. Also consider developing your own network design problem from scratch. All you need to do is identify the cities you want to connect, draw them on a map, determine the bits per second traffic volume between these cities so you know what type or number of circuits are required, and calculate the costs (circuits, modems, terminals, multiplexers, etc.), using the previous material. Put each cost element onto a network cost analyzer (Figure 11-9) as you perform these calculations. Figure 11-15 can be used to draw your network.

13. SELL AND IMPLEMENT THE NETWORK

At this point there are two more subtasks.

- Sell the system both to management and to the users who will have to work with it. This is a verbal presentation.
- Implement the system. This probably is the most difficult task of all because the various pieces of hardware, protocol/software programs, network management/test facilities, and communication circuits must be assembled into a working network.

When presenting the system to gain management/user acceptance, the designer should be prepared for objections to the proposed system. Basic objections usually follow these lines.

- The cost is too high, or it appears too low for what the system is supposed to be able to do.
- The performance is not good enough, or it is more than is required at this time.
- The new network does not meet the goals, objectives, and policies of the organization/departments that will be using it.
- The response or processing time is either too slow or too fast with respect to other operations within the organization.
- The system is not flexible enough. If changes are made in other areas, the network may collapse and the investment will be wasted.
- The quality, capacity, efficiency, accuracy, or reliability of the new network does not meet management's criteria.
- Certain management personnel may dislike or distrust the network design team's motives, personalities, or presentation methods.
- You should review the list of evaluation criteria that was prepared earlier and be ready for questions on any that were not met.

The implementation process begins *after* management has accepted the new system. Implementation consists of the installation of the new system and the removal of the old system. It involves hardware, protocols/software, communication circuits, a network management/test facility, people, written procedures that specify how each task in the network is performed, training, and complete documentation of the working system.

The steps involved in implementing a new data communication network can be very complex and demanding. A detailed implementation plan should be developed to enable implementation to proceed as smoothly as possible. The plan should specify who will do what and when they will do it. For this to be done properly, use should be made of Gantt charts, flowcharts, the Program Evaluation Review Technique (PERT), or data flow diagrams (DFD).[5]

The design/implementation team must take into account the earliest lead times that are required to order hardware, software, and circuits. In many cases these items cannot be delivered immediately. In addition, some lead time is needed for testing the protocols and software to ensure that they operate in conjunction with the hardware and circuits. Both hardware and circuits may have to be implemented in various parts of the city, a state, throughout the country, or even internationally. For this reason, it is imperative that a decision is made as to how the new system will be implemented. Four basic approaches can be used.

- All at once. All nodes and the host computer are started up at the same time (a one-for-one changeover).
- Chronologically, and in sequence, through the system. Start with the first application system, implementing those portions of the network that must be implemented with it, and then move on to the second application system.
- In predetermined phases. Similar areas within the system are started up at one time, and other areas are started up later.
- Pilot operation. Set up a pilot or test facility (this later becomes one of the working nodes) to ensure that the operation is as expected before an all-at-once or chronological cutover is made.

Once the hardware is in place, the circuits have been installed, and the protocol/software is operating, training of the users can begin, although when possible it should be started earlier. It is advantageous to obtain test terminals so terminal operators can use their particular application system in a training/test mode months before they do so in real life. Precise written procedures are required on how the terminal operators are to operate the system for data input and manipulation. Written descriptions on how to retrieve and interpret the information/data output should be provided to management.

The training should include individual operator training, extensive written training manuals, and a methodology for continual updating of these manuals. At this point, the use of Computer Assisted Instruction (CAI) should be considered. With CAI all of the training techniques and procedures are stored in the computer system; there are no written manuals. Instead, terminal operators use their terminals for training as well as for standard business operations.

---

[5] Ibid.: Gantt charts, pp. 170–173; flowcharts, pp. 551–561; PERT, pp. 536–548; DFD, pp. 61–74.

The network management/test center is a vital link in the network. This group must be in operation *before* the system is cut over to an operational status because reliability (uptime) is the single most important criterion for user acceptance.

Finally, after the system is operational, conduct follow-ups for the first six months or so to ensure that all parts of the new system actually are operating and that minor activities or operations have not been overlooked.

After the system is considered fully operational, a reevaluation should be performed. Reevaluation may come 6 or 12 months later. It is a critical review of operator/user complaints, management complaints, efficiency reports, network management trouble reports, an evaluation of statistics gathered on items such as errors during transmission and characters transmitted per link, and a review of peak load factors. Of course, it also should include a complete review of the original evaluation criteria so that you can determine the success of the design, development, and implementation of the new data communication network.

In summary, 13 steps have been completed for the design of a new data communication system. Although some steps may have been omitted because a current system was being enhanced, an orderly plan was followed. As the project is closed, pull all of the documentation together and set it up in a binder that contains 13 separate sections, one for each step that was carried out.

SELECTED REFERENCES

1. Bracker, William E., and Ray Sarch. *Case Studies in Network Design.* New York: McGraw-Hill Book Co., 1985.

2. FitzGerald, Jerry, and Ardra F. FitzGerald. *Fundamentals of Systems Analysis: Using Structured Analysis and Design Techniques*, 3rd ed. New York: John Wiley & Sons, 1987.

3. *Network Communications: Applications and Services.* Basking Ridge, N.J.: AT&T Communications Consultant Liaison Program, manual 500-936. (Contains extensive circuit cost data. For more information, call 1-800-CLIP-INFO or 1-800-432-6600.)

QUESTIONS/PROBLEMS

1. Describe the 13 steps that a designer performs when designing a new data communication network.

2. Identify two or three critical points that should appear in a feasibility study final report.

3. What are the three levels of goals that you should try to achieve when designing a network?

4. Identify and define five or six key evaluation criteria.

5. There are four levels of mapping. Can you identify and describe them?

6. When analyzing messages, which of the following peak volumes is the most important, the character per message peak or the messages per day peak?

7. Why is minimum circuit mileage between various terminal locations important?

8. The following is an excerpt from last month's progress memo from the data communication analyst Jones to Mr. Smith, his manager.

   Mr. Allen, the Vice President of Marketing, called to ask if anything could be done to improve the order entry system. I met with him and the Manager of Marketing Administration, Mrs. Johnson, and listened to their problems. We agreed on a short written definition of the problem, and determined that marketing field offices and salespeople, marketing headquarters, manufacturing, and distribution will be affected by any changes to the present system. Mr. Allen turned down my request to visit a typical field office because he feels Mrs. Johnson knows enough about their operation to fill me in. I met with the manufacturing planner, Mr. Williams, and the head of distribution, Miss Thomas. I obtained a general understanding of the current order entry system from Johnson, Williams, and Thomas. The design of the new system is under way now, and will be completed soon. I then will prepare a cost estimate of the system design and will present both to you in my report next month.

   Play the role of Mr. Smith and write a memo to Jones, commenting on his report. Be critical and try to determine any areas where Jones may not have done all he should have done.

9. Critique the following excerpts from a system requirements document:
   - The system shall be easy to operate.
   - The system shall have a mean time between failures of at least 1,000 hours.
   - The system shall transmit in half duplex mode at 2400 bits per second.
   - The system shall transmit at least 1,000 messages per hour.

10. The busiest link in a system carries 500,000 10-bit characters per 12-hour day. Allowing for a peak load equal to three times the average, for a 50 percent growth over the system life, and for a 10 percent error/retransmission factor, what is the minimum line speed in bits per second?

11. Calculate the air mileage (use V and H coordinates) from San Francisco to Denver and from Kansas City, Kansas, to Kansas City, Missouri.

12. In which step of the system design process would your plan start to identify the sources of information to be used, the types of information to be collected, the analyses to be performed, the schedules of the various activities, and the definition of the results to be produced?

13. Use Figure 11-4 and convert the characters per day on the New York to Miami link to bits per second transmission speed, assuming asynchronous transmission (10 bits per character) and a six-hour workday.

14. What are some of the items that utilize transmission time but do not transmit business data and for which the designer must account?

15. Give examples of what factors might cause peak loads and discuss how the system designer takes them into account when designing a data communication system.

16. Determine the cost of a voice grade (lease) circuit between Seattle, Washington (SEA), and Tampa, Florida (TPA). You need point-to-point type C conditioning. Use Figure 11-15 to draw the circuit.

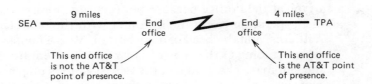

17. Assume in Question 16 that SEA uses a 16-channel statistical multiplexer transmitting at 9600 bits per second and TPA uses an 8-port multiplexer. Each end needs eight video terminals. What is the hardware cost? Use the selected hardware (lease) costs in this chapter. *HINT:* If you draw the circuit you will not forget anything.

18. Assume that the total number of transmitted characters per day from SEA to TPA in Question 16 is 400,000 and that there is a six-hour workday. If a terminal operator can type 20 words per minute (six characters per word), how many terminal operators and terminals are required?

19. Assume that the transmitted characters per day from TPA to SEA in Question 18 is 300,000. How many terminal operators and terminals are required?

20. Can the circuit handle the combined traffic of Questions 18 and 19 at 9600 bits per second?

21. Using the circuit in Question 16, calculate the monthly dial-up cost of the circuit if nine 30-minute calls per day are made. Use the month of January 1988. Which costs more, this dial-up or the leased circuit in Question 16?

22. Convert the interLATA analog circuit cost in Question 16 to an interLATA digital circuit. Does digital cost more or less?

23. Determine the cost of a multipoint circuit (private lease) from Seattle (SEA) to Salt Lake City (SLC) to Baltimore (BLT). Assume

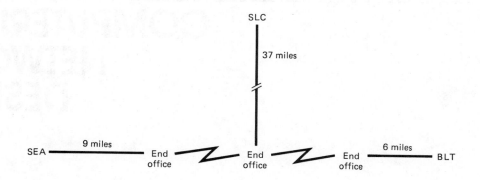

None of the three end offices is the AT&T point of presence. There is no conditioning.

24. As a group project, use Figure 11-15 and choose four or five cities to interconnect. Draw your network and cost it out. Use whichever circuits you think are best and include the hardware costs. Record your costs on a copy of Figure 11-9.

12

COMPUTERIZED
NETWORK
DESIGN

The fundamental problem in designing data communication networks is not whether the technology is available but how best to adapt its offerings to meet the changing and challenging needs of business and government organizations. This chapter introduces a microcomputer-based network optimization software package. It provides network cost analysis and response time analysis, draws the topology of the network, compares various line speeds (bits per second) for tradeoffs, and calculates mileages so you can design a data communication network using a microcomputer.

AN OVERVIEW

Many businesses today operate in multiple locations. It is not at all unusual for an organization to have 2 to 100 different sites at which business transactions take place. Managers of these organizations must have current knowledge of their geographically scattered operations to serve their customers better, to maintain close surveillance over the organization's critical activities, and especially to meet the competition. Today's information-based society calls for rapid collection, processing, and distribution of business information.

Advances in computer design, remarkable reductions in cost per computer operation, creative ideas in computer application design, and advances in both voice and data communications have brought about increased use of data communication networking. An organization's communication networks allow widely separated business locations, and the computers/terminal equipment installed at these locations can be used in a cost-effective manner to provide the response time needed for business operations. Thus, within seconds management can be apprised of the state of affairs at a branch or at any other location throughout the country, or even the world.

This chapter is the culmination of all the information in this book. Up to this point you have learned about the fundamental technical concepts of data communications (Chapters 1 to 7), network management and security (Chapters 8 to 9), and methods of designing communication networks (Chapters 10 to 11). This chapter takes the network design concepts learned in Chapter 11 and builds upon them to provide the methodology and software for computerized network design.

You must have completed Chapter 11 to use this chapter effectively because communication service offerings and the tariff environment have become increasingly complex since divestiture of the American Telephone and Telegraph (AT&T)/Bell Operating Companies (BOC). Today's network analysts need to use computerized network design systems when they want to configure a network that involves a dozen or more nodes. The network designer needs to be able to compare and evaluate service costs under a variety of scenarios, including total service from AT&T, baseline (minimal) service from AT&T with special access from other local exchange carriers (other LECs include MCI and US Sprint), and intraLATA configurations that access interexchange channels (IXC).

If you skipped over Chapter 11 (Network Design Fundamentals), it is recommended that you go back now to read that chapter. Before proceeding with computerized network design, you need to learn about planning criteria such as needs assessment factors, feasibility factors, and evaluation criteria that should be identified prior to designing any network.

Chapter 11 also shows how to analyze messages, calculate traffic/circuit loading, and design a data communication network. As you will see when you use a computerized design package, you must be able to analyze messages in order to have a computerized system calculate response time analysis. You also must understand the calculation of traffic/circuit loading to comprehend how the computerized network design program arrives at its final output calculations.

Furthermore, Chapter 11 explains the vertical and horizontal coordinate system that is used for calculating mileages between the telephone company central offices. Finally, if you have completed some of the hand-calculated circuit or network design problems in Chapter 11, you will have a better understanding of computerized network design software.

NETWORK DESIGN TOOLS

Today's sophisticated network users are turning to automated network design tools as a means of continually redesigning their voice and data networks. Such network design tools can be a major investment, ranging in price from just under $10,000 to over $100,000. On the other hand, rather than purchasing one, organizations can opt to use mainframe-based network design tools on a timesharing basis.

Network design tools should be able to perform three basic functions: cost analysis, performance analysis, and topology optimization of the network.

Cost analysis requires that the vendor from whom the network design tool is purchased continually update the databases of LATAs (local access transport areas) and tariffs (prices). These databases contain the tariff filings and LATAs for the various common carriers from whom the organization leases its data communication circuits. Depending on the sophistication of the network design tool, the databases might contain both interLATA and intraLATA tariffs.

The cost analysis portion of a network design tool should allow users to experiment with different tariffs and select from a variety of bridging arrangements. These different bridging arrangements allow users to select bridging (interconnecting) of multidrop circuits in a long-haul or intraLATA common carrier's central office. The object is to allow users to explore the consequences of changing between different communication circuits or different common carriers.

Performance analysis allows users to estimate the response times and throughput of their network. It is important to note that these network design tools provide only estimates, which may vary from the actual results. (See the next section on Modeling Networks.) Some network design tools estimate performance for a generalized class of network, such as any multidrop network, while other network design tools are built specifically for one type of network.

Topology optimization enables users to design the optimal layout of a network. Network design tools usually identify the optimal layout by interactively trying various combinations of topologies until the one that shows the lowest network cost is determined. Most modeling packages are able to optimize networks containing multidrop circuits, concentrators placed strategically throughout the network, and packet switches. The final result as to how "good" the optimization is depends on the algorithm used by the network design tool vendor.

Optimization is a CPU intensive process; therefore, large networks require a mainframe to optimize. Alternatively, microcomputers can be used for optimizing networks of less than 500 nodes. Topology optimization usually is based on establishing some given performance constraints and having the optimization algorithm optimize the network using those constraints. For example, the user might specify the maintainance of a three second response time and require that the program lay out and optimize a least cost multidrop network based on 50 nodes. Beyond the three second response time and the locations of the 50 nodes, the user also may specify other parameters such as the common carrier, transmission media, traffic volume statistics, basic characteristics of the terminals, microcomputers, host processors, front end communication processor, and the protocol that is to be used.

A sophisticated network design tool should be capable of analyzing hybrid networks that include multidrop, concentrators, and packet switching. Furthermore, the tool should allow users to modify items such as protocols, tariffs, circuit configurations, equipment characteristics, and performance objectives such as two or three second response time requirements.

Network design tools are limited by the size of the CPU on which they run. For example, large mainframes can carry out cost analysis, performance analysis, and topology optimization of networks of unlimited size. Microcomputer-based network design tools are limited to a network of about 500 nodes; although, when using the newer 80386 microprocessor chips, these tools can handle networks larger than 500 nodes.

Because the run time for one of these tools tends to increase as the square of the number of sites, users may need to reduce run times to those that can fit on a microcomputer. This is achieved by dividing the network into segments and optimizing different portions of it, such as different regions of the country or various multidrop segments.

One of these network design tools is described later in this chapter in the section Network Optimizer (MIND^SM-Data/PC). A demonstration version of this package is available (see footnote 1 later in this chapter).

MODELING NETWORKS

The complexity of designing data communication networks presents the designer with many problems. The overall goal of the designer is to construct a system that provides adequate response time to the end user while ensuring that the cost to deliver that response time is reasonable. In order to accomplish this goal, the

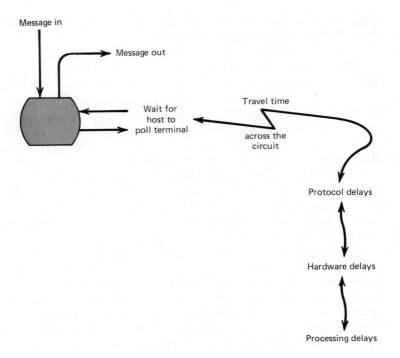

FIGURE **12-1** Message delay.

designer must understand the response time and cost tradeoff issues associated with a data communication network.

In this section we will deal with the design of multipoint polled networks. This type of network was chosen because it is the most common class of data communication network and it presents difficult issues of how user transactions are handled in a system with multiple queues.

As an example, when a user is ready to send a message to the host computer for processing, that message passes through many queues (wait times) or delay periods, as illustrated in Figure 12-1.

If there are other terminals sharing the circuit (multipoint), the queues and delays can compound or interfere with each other because usually only one terminal on a given communication circuit can communicate with the computer at a given time. Therefore, if five terminals are on a circuit and each has a message ready to send, it is easy to see that the last terminal may have to wait until the others are finished. Also, the computer may accept a message from Terminal 1 and then Terminal 2 and, before going to Terminal 3, send the response to Terminal 1's message, causing an additional delay for the others waiting for service. This example indicates some of the complexities, but understand that these are only a few of the delays encountered within data communication systems.

Fortunately, computer modeling techniques have been developed to help the designer evaluate the problem and determine a solution. These modeling programs allow the designer to examine the issues without having to get involved with all the complexities of the problem. They are, if you will, "tools of the trade" and must be looked upon as *tools*, not push-button programs. A skilled user can derive good solutions by applying the tools successfully, while an unskilled user may apply the tools incorrectly and get totally different results.

A *model* is a body of information about a system gathered for the purpose of studying the system. A *mathematical model* describes the entities of a system with the attributes being represented by mathematical variables; the activities are described by mathematical functions that interrelate the variables. Given a mathematical model of a system, it sometimes is possible to derive information about the system by *analytic means*. Where this is not possible, it is necessary to use numerical computation for solving the equations of the mathematical model. *System simulation* describes the technique of solving problems by following the changes over time using a dynamic model of a system. A *dynamic mathematical model* allows the changes of system attributes to be derived as a function of time.

The analytical models generally make several assumptions to simplify the mathematics and provide results that can be evaluated easily and cheaply. Simulation models, on the other hand, can reflect the working of a system to any finer desired level of detail. The time and cost of development and program runs are directly proportional to the level of detail incorporated.

Let us illustrate some of the analytical and simulation modeling techniques used in a data communication network design. Consider a set of terminals connected to a host computer (CPU) via a multidrop line (Figure 12-2). The CPU and the terminals bear a master/slave relationship in the sense that the transmission from the terminals are controlled by the CPU or front end. Evaluating the performance of the system under the polling discipline is the task of the system designer. The system designer uses modeling as a means to *predict* the performance of the system and answer such questions as: How does the response time of the system vary as a function of the terminal load and the number of terminals? How is the response time affected by a specific polling discipline such as giving priority to outbound traffic (CPU to terminal) over inbound traffic?

With regard to *analytical* models, the designer resorts to queuing models in order to predict analytically the response time and throughput of a polled system. A general queuing model of a polled system is shown in Figure 12-3 and may be described as queues served in cyclic order with walk times. (*Walk time* is the time to switch service from one queue to another and includes the overhead time attributable to polling messages, propagation delay, modem synchronization time, and so forth.)

Messages arrive at a terminal according to a random process, which may be

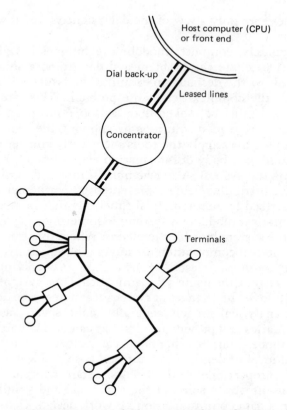

FIGURE **12-2**   Multidrop network configuration.

terminal dependent, and are queued for transmission. The server in this case is the transmission medium and is made available to each queue periodically, as defined by the polling protocol. The polling discipline also defines the amount of service received when the server arrives at a queue. Other characteristics of the system, such as poll message length and modem turnaround time, are modeled in the switchover time. Outbound response messages may be included as arrivals to an output queue at the CPU. The major difficulty in solving the above queuing model is the interrelationship between the queues at the various terminals. An exact model, therefore, has to solve an N-dimensional queuing process, which is a formidable task.

An analytical model requires a sophisticated user who is competent in mathematics for its development and use. When the user is discriminating, it can provide preliminary insights, but seldom can it yield numerical values of sufficient accuracy for the operational design of a system. To obtain a more realistic

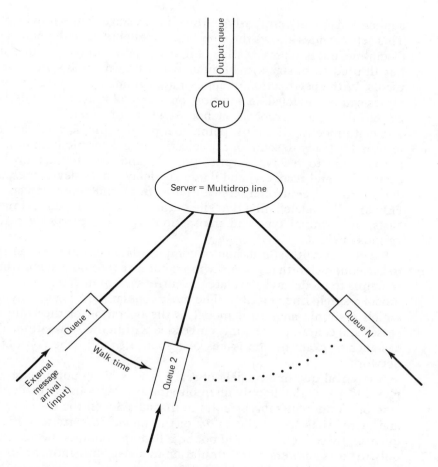

FIGURE **12-3** Queues model of polled terminals.

model capable of providing accurate numerical answers, we must resort to *simulation*. The attempt to produce a better mathematical model leads to complications. A point is reached at which a simulation model appears to be the easiest way of finding circuit utilization and response time.

Because analytical models are so complex, it often is easier to use a simulation model to answer many of the questions posed by the system designer.

Simulation models allow users to enter a model of network traffic load and obtain an analysis of waiting times and response times in various parts of the network. The model may be tailored to the user's network by entering parameter values specific to the system at hand. Alternately, the user may prefer in some applications to rely primarily on the set of average values provided by the system. This set of values allows the simulation model to operate under a specific circuit discipline and is representative of many popular protocols currently in use. It is not limited to binary synchronous communication (bisync) or SDLC-like protocols. With specification of appropriate parameters, very different circuit disciplines can be modeled. In addition, operation of most terminal-oriented systems under full duplex protocols such as SDLC can be modeled.

When modeling the multipoint line, polling (the determination of whether a terminal is ready to send input), selection (the determination of whether a terminal is ready to receive output), positive and negative acknowledgment, error occurrence and recovery, and hardware delays in all devices (e.g., terminal, modem, terminal controller, concentrator, front end communication processor or FEP) are all modeled directly. The output provides response times on an overall basis, by terminal type and transaction type, and the delay component at the various devices.

Experience with simulation of complex data communication systems has led to the conclusion that generally more harm than good is done when the designer attempts to model and simulate the entire system in detail; therefore, it is best to model a single line at a time. The most popular models use a hybrid approach in which a simple analytical model of the operation of the communication processor and central processor is embedded within the simulation. This enables the designer to examine the issues of response time in the context of a complete network.

Analytical queuing models of real systems using polling are extremely complex and are workable only if simplifying assumptions are made. This frequently (and often unpredictably) results in inaccuracies in the response times obtained and/or restricts the applicability of the model to certain traffic levels. Consequently, these models should not be relied upon unless the designer is interested only in gross answers. Pure simulation models, while more accurate, are cumbersome and costly to run.

A hybrid technique in which analysis is embedded within the simulation can yield a flexible model that provides accurate results. The input to the model can be detailed empirical data, if it is available, or it can be defined approximately by a well-known statistical process (e.g., Compound Poisson Process). A major limitation of simulation models, that of development cost, can be avoided by using the design packages developed and marketed by reputable organizations.

NETWORK OPTIMIZER (MIND^{SM}-Data/PC)

A number of software packages are available for computerized network design. Such packages consist of two major segments: (1) the software that performs the calculations; and (2) the complex database of tariffs and vertical/horizontal coordinates that are cross-referenced to area codes plus the first three digits of the telephone number.

We introduce one of these computerized network design packages, MIND^{SM}-Data/PC, in this chapter. MIND-Data, as it is known, was developed by CONTEL Business Networks and there are two versions. The first version is a full-scale computerized network design package. The second is a scaled-down version for educational use. The educational version contains only a few area codes, and it is restricted to 50 nodes (different locations); it will be described in more detail later in this chapter.

MIND-Data is an interactive tool for pricing, managing, optimizing, and designing multipoint private line data networks. It optimizes a centralized (polled) data network in terms of performance (response time and circuit utilization) and cost. The package runs on a microcomputer, and the results can interface with more powerful mainframe-based MIND (*Modular Interactive Network Designer*) software tools. For example, there are several network design packages that can do the following tasks.

- MIND-Packet addresses the design and analysis of packet switched networks.
- MIND-Voice configures the optimal mix of switched service facilities and private circuits required to carry voice data to and from a PBX (switchboard).
- MIND-Tandem addresses the dynamics of a multinode circuit switched network.
- MIND-Inventory is a PC-based telecommunication network management tool that helps telecommunication professionals document, analyze, and manage communication networks.
- MIND-Manager is a VAX-based multiuser system that provides trouble ticketing, maintenance, service order tracking, topology, and inventory capabilities.

This software package is a microcomputer-based network optimization package for telecommunication professionals who

- Have responsibility for strategic or tactical network planning
- Need cost-effective analytical tools to carry out network analysis and design
- Perform network modeling to maintain desirable cost/performance ratios
- Perform continuous reviews of circuit pricing (tariffs) and reconciliation of billings back to planned network costs

MIND-Data is capable of response time analysis and topological design, and it allows the determination of network costs versus performance tradeoffs. The complete software package offers the capability of designing a network with anywhere from 2 to 500 nodes (the educational version is limited to 50 nodes), and it is compatible with the larger mainframe-based network design tools that assist in designing a network of up to 2,000 nodes. This fact is mentioned because it is important for you to realize that in a real-life work environment a microcomputer-based network design tool still has to interface with larger network design packages. Some tasks that can be performed with this software are

- Analyzing response time for binary synchronous (bisync) protocols and for synchronous data link control (SDLC) protocols. You can vary the protocols and other design parameters to customize your design analysis.
- Mixing of line speeds (bits per second) on different circuit links in a network.
- Drawing the layout of the network, presenting the layout on the video monitor, and printing it with a graphics subsystem.
- Pricing the network using the built-in tariffs. The package uses the standard private AT&T Tariff 9 (interoffice between two different central offices) and tariffs serving various wire centers (telephone company central office) for major metropolitan areas (NECA4 and Tariff 11). Tariffs include voice grade and digital data service (DDS) ranging from 2400 to 56,000 and 1.54 million bits per second.
- Calculating the mileages between telephone company central offices through use of a database of vertical and horizontal coordinates interconnected with the LATA database.

MIND-Data covers a complete range of performance issues, along with the standard issue of designing and laying out a network. For example, it can answer a whole host of questions such as

- What is the response time for a given application profile and mix by location?
- What line speed is required?
- At what response time level is the network cost sensitive?
- What is the impact on response time if lines are upgraded from 4800 to 9600 bits per second?
- What are the cost tradeoffs between several point-to-point circuits versus a multidrop circuit?

This package operates on a CONTEL AT/4, IBM AT, or IBM-compatible computer system. It also operates on an IBM/XT or IBM/PC; however, the AT's greater processing speed makes use of it much more desirable. This computerized network optimizer functions *only* within the hard disk environment. In addition, it requires (educational version requires less disk space)

- DOS version 2.0 or higher
- At least 512,000 bytes of memory (RAM)
- 4.5 million bytes of available hard disk storage space
- A color graphics adapter card (not a monochrome)
- A color monitor or monochrome
- Printer (parallel) for reports

To load and execute, this package needs approximately 460,000 bytes of memory. Of the 512,000 bytes of RAM installed in a 512,000-byte machine, approximately 40,000 bytes are used by DOS. In addition, your memory resident background programs and print spoolers also occupy RAM. You must calculate available memory space to allow enough capacity for MIND-Data to operate. It might be desirable to remove temporarily any other memory resident RAM programs if you do not have enough RAM.

The program and data files consume over 3.5 million bytes of disk space. About 1 million bytes of disk space should be reserved for network files created during the design sessions; therefore, a total of 3 million bytes of available storage space on the hard disk is advisable.

To obtain the educational version of the MIND-Data network design optimization package, call or write to CONTEL Business Networks.[1] The educational version of this product is called "MIND[SM]-Data/PC (version 2.20)" and it contains six 5¼-inch double-sided, double-density floppy disks and the written documentation. Because six disks must be duplicated, the documentation is extensive, and because the packaging/mailing costs are high, CONTEL Business Networks must charge $50 for the educational version. Although it is an abbreviated version, its limits will not hinder you because it allows the design of networks that interconnect the area codes and telephone exchanges of several cities.

This package is designed to use a menu-driven system approach. When using the package for the first time, read the user guide before starting so that you understand the command structure. The menu-driven command line structure

[1] Orders should be sent to Mr. David Rubin, Vice President and General Manager, Network Analysis Center, CONTEL Business Networks, 130 Steamboat Road, Great Neck, New York 11024. The telephone number is (516) 829-5900.

Host: Tampa, Fla.
Circuit: 1 4800 bps voice grade

					WIRE-CNTR		VOICE GRADE	
NODE	STREET ADDRESS	CITY	ST	ACEX	V	H	CENTRAL OFFICE	
=====	=====================	==========	==	======	====	====	===================	
HOST	HEADQUARTERS	TAMPA	FL	813222	8172	1147	TAMPA	81611133
T1	OFFICE 121	TAMPA	FL	813229	8172	1147	TAMPA	81611133
T2	OFFICE 936	TAMPA	FL	813224	8172	1147	TAMPA	81611133
T5	OFFICE 994	TAMPA	FL	813221	8172	1147	TAMPA	81611133
T6	OFFICE 566	TAMPA	FL	813628	8160	1135	TAMPA	81611133

FIGURE 12-4 MIND-Data report. ACEX is a combination of the area code (813 for Tampa) and the first three digits of the telephone number. WIRE-CNTR shows the vertical and horizontal coordinates.

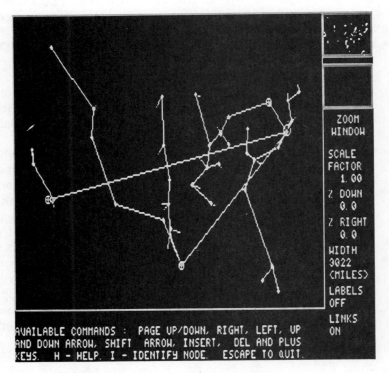

FIGURE 12-5 MIND-Data graphic presentation of a network.

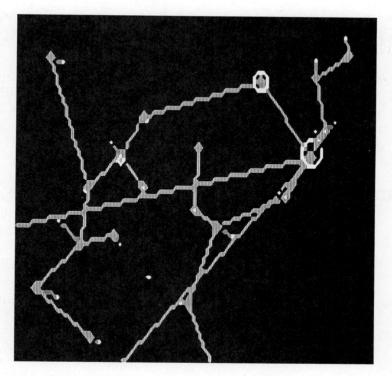

FIGURE **12-6** Magnification of a portion of the network. Compare with Figure 12-5.

allows you to step through the system and move to whatever task you want to perform, such as

- Identify a type of transaction within an application.
- Set up the nodes to design a new network.
- Add nodes to a current network.
- Save your network design.
- Optimize the network to obtain costs or a response time analysis.
- Print out the reports showing costs, response time analysis, and so forth as shown in Figure 12-4.
- Graphically display and print your network as shown in Figure 12-5.
- Magnify a portion of your network by using other software such as Pizazz (call (617) 433-5201). See Figure 12-6 and compare it with Figure 12-5.

It is recommended that you design a two-node circuit on the first attempt, and then expand that into a four- or five-node network before proceeding with something more complex.

As mentioned previously, the educational demonstration version of MIND-Data is delivered on eight diskettes. These diskettes contain the installation programs, various program files for calculations, database files for area codes, vertical and horizontal coordinates, LATA databases, and a demonstration file called DEMO.BNF. The installation procedure is described in the documentation you will receive with your educational version. (See footnote 1 in this chapter.)

SELECTED REFERENCES

1. *Call-Pricer: PC-Software for Long Distance Service Selections.* Available on five diskettes from CCMI/McGraw-Hill Book Co., 50 South Franklin Turnpike, Ramsey, N.J. 07446.

2. Ellis, Robert L. *Designing Data Networks.* Englewood Cliffs, N.J.: Prentice-Hall, 1986.

3. *MIND^SM-DATA/PC.* A software program available from CONTEL Business Networks, 130 Steamboat Road, Great Neck, N.Y. 11024, (516) 829-5900.

QUESTIONS/PROBLEMS

1. Which area code telephone exchanges are in the CONTEL educational version software design package?

2. What is the required microcomputer configuration?

3. You can compare the circuit costs that you already calculated (Chapter 11, Questions 16 and 23) using the optimizer software. Do not be surprised if they vary somewhat. The distance between the end office and your business premises may not be included. Also, the MIND-Data package you received may contain more recent cost data.

4. Run the DEMO.BNF in the optimizer software.

5. What is a model?

6. What is system simulation?

7. Use the network optimizer package to determine the cost of a circuit between Salt Lake City (801) 754 and Baltimore (301) 345. The 801 and 754 are the ACEX; 801 stands for the area code and 754 stands for the

exchange (central office), as you will learn from the user's manual. Add a circuit for Salt Lake City to Tampa (813) 254 and for Salt Lake City to Seattle (206) 345.

8. Describe the three basic functions of a network design tool.

9. Define "walk time" as described in this chapter.

10. A mathematical model describes the entities of a system with the attributes being represented by mathematical variables. Use a dictionary and define both "entity" and "attribute."

GLOSSARY

ACK An ASCII or EBCDIC code character indicating a positive acknowledgment, i.e., a message has been received correctly.

Acoustic Coupler A type of modem that permits use of a telephone handset as a connection to the public telephone network for data transmission.

ACU See **Automatic Calling Unit.**

Address

1. A coded representation of the destination of data, or of its originating terminal. For example, multiple terminals on one communication line must each have a unique address.

2. Sometimes referred to as *called number:* The group of digits that make up a telephone number. For example, an address may consist of area code, central office, and line number.

ADU See **Automatic Dialing Unit.**

ALOHA A system using a "transmit at will" access method. The name comes from a method of telecommunications whereby signals are beamed at satellites when transmission is ready to go. If it gets through, fine; if it does not, then the sender tries again. The ALOHA method of transmission was used first by Hawaiian satellite dishes beaming at communication satellites over the equator. It also was used for communicating with dishes in other Pacific Basin countries.

American Standard Code for Information Interchange See **ASCII.**

Amplifier A device used to boost the strength of a signal. Amplifiers are spaced at intervals throughout the length of a communication circuit. Also called *repeater/amplifier.*

Amplitude Modulation See **Modulation, Amplitude.**

Analog Pertaining to representation by means of continuously variable physical quantities, such as varying voltages or varying frequencies.

Analog Signal A signal in the form of a continuously varying physical quantity such as voltage, which reflects variations in the loudness of the human voice.

Analog Transmission Transmission of a continuous variable signal as opposed to a discretely variable signal. Physical quantities such as temperature are continuously variable and so are described as "analog." The normal way of transmitting a telephone or voice signal is analog.

ARQ Automatic repeat request. A system employing an error detecting code so conceived that any error initiates a repetition of the transmission of the incorrectly received message.

ASCII American Standard Code for Information Interchange. Pronounced "ask'-ee." An eight-level code for data transfer adopted by the American Standards Association to achieve compatibility among data devices.

Asynchronous Transmission Transmission in which each information character is individually synchronized, usually by the use of start and stop elements. The gap between each character is not a fixed length. Compare with **Synchronous Transmission.**

Attenuation The difference between the transmitted and received power due to loss through the equipment, communication circuits, or other devices. May be expressed in decibels.

Automatic Calling Unit (ACU) A device that permits a business machine to dial calls automatically.

Automatic Dialing Unit (ADU) A device capable of automatically dialing digits.

Automatic Equalization Equalization of a transmission channel which is adjusted while sending data signals.

Automatic Outward Identified Dialing The ability of a switching system to identify the originator of an outgoing call without operator intervention.

Automatic Repeat Request See **ARQ.**

Backbone Network A transmission facility, or arrangement of such facilities, designed to interconnect lower speed distribution channels or clusters of dispersed users or devices. See also **Long Haul Network.**

Balanced The state of impedance on a two-wire circuit when the impedance to ground as measured from one wire is equal to the impedance to ground as measured from the other wire. A balancing network is a combination of electronic components which stimulate the impedance of a uniform cable or open wire circuit over a band of frequencies.

BALUN BALanced/UNbalanced. An impedance-matching device to connect balanced twisted pair cabling with unbalanced coaxial cable.

Band Elimination Filter An electrical device that blinds a receiving unit by blocking specific frequency ranges.

Band Pass Filter An electrical device that allows a specific frequency band to enter a receiving device and attenuates all others.

Bandwidth The difference between the highest and the lowest frequencies in a band, such as 3000 hertz bandwidth in a voice grade circuit.

Baseband Signaling Transmission of a signal at its original frequencies, i.e., a signal in its original form, not changed by modulation. It can be an analog or digital signal and is usually direct electrical voltages.

Baud Unit of signaling speed. The speed in baud is the number of discrete conditions or signal elements per second. (This is applied only to the actual signals on a communication line.) If each signal event represents only one bit condition, *baud* is the same as *bits per second*. When each signal event represents other than one bit, *baud* does not equal *bits per second.*

Baudot Five-bit, 58-character alphanumeric code used in transmission of information.

BCD Binary coded decimal. Six-bit alphanumeric code.

Bell System The seven telephone operating companies known as the Bell Operating Companies (BOCs).

BERT Bit Error Rate Testing. Testing a data line with a pattern of bits that are compared before and after the transmission to detect errors.

Binary A number system using only the two symbols 0 and 1, which is especially well adapted to computer usage because 0 and 1 can be represented as "on" and "off," or as negative charges and positive charges. The binary digits appear in strings of 0s and 1s.

Binary Coded Decimal See **BCD.**

Binary Synchronous Communication (BSC/bisync) A half duplex, character-oriented synchronous data communication protocol originated by IBM in 1964.

Bipolar Coding A method of transmitting a binary stream in which binary zero is sent as a negative pulse and binary one is sent as a positive pulse.

Bit

1. An abbreviation of the term *binary digit.*
2. A single element in a binary number.
3. A single pulse in a group of pulses.
4. A unit of information capacity of a storage device.
5. Zeros and ones.

Bit Error Rate Testing See **BERT.**

Bit Rate The rate at which bits (binary digits) are transmitted over a communication path. Normally expressed in bits per second (bps). The bit rate should not be confused with the data signaling rate (*baud*), which measures the rate of signal changes being transmitted. See also **bps.**

Bit Stream A continuous series of bits being transmitted on a transmission line.

BLERT Block Error Rate Testing. Testing a data link with groups of information arranged into transmission blocks for error checking.

Block Some sets of contiguous bits and/or bytes that make up a definable quantity of information such as a message.

Block Error Rate Testing See **BLERT.**

BOC Bell Operating Company. One of the local telephone companies spun off from AT&T as a result of divestiture, now reorganized into seven regional Bell holding companies. Among the largest of the over 1,500 independent local telephone companies in the United States.

Booting The process of loading a computer's memory with essential information that allows it to function. The word comes from "pulling oneself up by one's bootstraps." Booting is of two types: cold boot and warm boot. A cold boot occurs when you initially turn your computer's power on. There is nothing in memory. When you boot, the operating system is loaded into the computer's memory. A warm boot occurs when you reload the operating system into memory. When you press the Control, Alt, and Delete keys simultaneously on an IBM or IBM-compatible, you throw the DOS operating system out of memory and reload a "new" version.

bps Bits per second. The basic unit of data communication rate measurement. Usually refers to rate of information bits transmitted. Contrast with **Baud.**

Bridge A device that provides a communication path between logically or physically separate networks using similar protocols. See also **Gateway.**

Broadband Circuit A communication circuit that has a bandwidth of greater than 4000 hertz.

Buffer A device used for temporary storage of data, primarily to compensate for differences in data flow rates (for example, between a terminal and its transmission line), but also as a security measure to allow retransmission of data if an error is detected during transmission.

Burst Error A series of consecutive errors in data transmission. Refers to the phenomenon on communication lines in which errors are highly prone to occurring in groups or clusters.

Bus A transmission path or channel. Typically an electrical connection with one or more conductors in which all attached devices receive all transmissions at the same time.

Byte A small group of bits of data that are handled as a unit. In most cases it is an 8-bit byte and it is known as a *character.*

C Conditioning A North American term for a type of conditioning that controls attenuation, distortion, and delay distortion so they lie within specified limits.

Cable Assembly of one or more conductors (usually wire) within an enveloping protective sheath.

Cable Television See **CATV.**

Camp-on A method of holding a call for a station when a busy condition is experienced, frequently signaling the station that a call is waiting, and automatically forwarding the call to the desired station when the busy condition is terminated.

Capacitance The ability to store energy such as small amounts of electricity. Capacitors can store and release energy. An inductor is another device with this ability.

Carrier An analog signal at some fixed amplitude and frequency which then is combined with an information-bearing signal in the modulation process to produce an intelligent output signal suitable for transmission of meaningful information. Also called *carrier wave* or *carrier frequency.*

Carrier Frequency The basic frequency or pulse repetition rate of a signal bearing no intelligence until it is modulated by another signal which does impart intelligence.

Carrier Wave The basic frequency or pulse repetition rate of a signal bearing no intelligence until it is modulated by another signal which does impart intelligence.

CATV Originally *Community Antenna Television.* Now also *Cable Television.* It refers to the use of coaxial cable loops to deliver television or other signals to subscribers.

CBX See **PBX.**

CCITT Consultative Committee on Internatinal Telephone and Telegraph. An international standards group.

CCSA See **Common Control Switching Arrangement.**

Central Office The telephone company switching office for the interconnection of direct dial-up calls.

Centrex A widespread telephone company switching service that uses dedicated central office switching equipment.

Channel

1. A path for transmission of electromagnetic signals. Synonym for *line* or *link.* Compare with *circuit.*

2. A data communication path. Channels may be divided into subchannels.

Channel Bank Equipment that performs multiplexing of lower speed, generally digital, channels into a higher speed composite channel. The equipment typically is in a telephone central office.

Channel Service Unit (CSU) This can be called a *digital modem*. It performs transmit and receive filtering, signal shaping, longitudinal balance, voltage isolation, equalization, and remote loopback testing. See also **Data Service Unit**.

Character A member of a set of elements upon which agreement has been reached and that is used for the organization, control, or representation of data. Characters may be letters, digits, punctuation marks, or other symbols. Also called *byte*.

Character Parity A technique of adding a redundant bit to a character code to provide error checking capability.

Checking, Echo A method of checking the accuracy of transmission data in which the received data are returned to the sending end for comparison with the original data.

Checking, Parity A check that tests whether the number of ones (or zeros) in an array of binary digits is odd (or even).

Checking, Polynomial A checking method using polynomial functions of the data transmitted to test for changes in data in transmission. Also called *cyclical redundancy check* (CRC).

CIM Computer Integrated Manufacturing. A specification that integrates computers into the manufacturing, design, and business functions of an organization.

Circuit A means of two-way communication between two data terminal installations. Compare with **Channel, Line, Link**.

Circuit Switching A method of communications whereby an electrical connection between calling and called stations is established on demand for exclusive use of the circuit until the connection is terminated.

Cladding A layer of material (usually glass) that surrounds the glass core of an optical fiber. Prevents loss of signal by reflecting light back into the core.

CLEAR TO SEND See **CTS**.

Coaxial Cable A type of electrical cable in which a solid piece of metal wire is surrounded by insulation which is surrounded by a tubular piece of metal whose axis of curvature coincides with the center of the piece of wire.

Code A transformation or representation of information in a different form according to some set of preestablished conventions. (See also **ASCII** and **EBCDIC**.

Code, Constant Ratio A code in which the ratio of ones and zeros in each character is maintained constant such as 4 zeros and 4 ones.

Code Conversion A hardware box or software that converts from one code to another, such as from ASCII to EBCDIC.

Common Carrier An organization in the business of providing regulated telephone, telegraph, telex, and data communication services. This term is applied most often to U.S. and Canadian commercial organizations, but sometimes it is used to refer to telecommunication entities (such as government-operated suppliers of communication services) in other countries. In the United States, the prices these organizations charge are regulated by the U.S. Federal Communications Commission or state public utility commissions. See also **PTT**.

Common Control Switching Arrangement (CCSA) A dedicated switched network leased by a user to handle communication requirements among various locations.

Communication Processor, Front End An auxiliary processor that is placed between a computer central processing unit and transmission facilities. This device normally handles housekeeping functions such as line management and code translation, which otherwise would interfere with efficient operation of the central processing unit. Synonym for *front end computer*.

Communication Satellite An earth satellite designed to act as a telecommunication radio relay. Most communication satellites are in geosynchronous orbit approximately 22,300 miles above the equator so they appear from earth to be stationary in space.

Communication Services The population or entire group of all transmission facilities that are available for lease or purchase.

Compaction In Systems Network Architecture, the transformation of data by packing two characters in a byte. The most frequently sent characters are compacted.

Component One of the individual pieces that, when assembled together, make up the data communication network. A component can be viewed as the item that is being reviewed or the item over which we are attempting to maintain control. Components are hardware, software, circuits, and other pieces of the network.

Compression In Systems Network Architecture, the replacement of a string of up to 64 repeated characters by an encoded control byte to reduce the length of the data stream sent to the LU–LU session partner. The encoded control byte is followed by the character that was repeated (unless that character is the prime compression character, typically the space character).

Computer, Central In data transmission, the computer that lies at the center of the network and generally performing the basic centralized functions for which the network was designed. Synonym for *host computer*.

COMSAT Communications Satellite Corporation, a private U.S. company established by statute as the exclusive international satellite carrier and representing the United States in INTELSAT.

Concentrator A device that multiplexes several low speed communication lines onto a single high speed trunk. A Remote Data Concentrator (RDC) is similar in function to a multiplexer but differs in that host computer software usually must be rewritten to accommodate the RDC. RDCs differ from statistical multiplexers because the total capacity of the high speed outgoing line, in characters per second, generally is less than or equal to the total capacity of the incoming low speed lines. Output capacity of a statistical multiplexer (stat mux), on the other hand, can exceed the total capacity of the incoming lines.

Conditioning A technique of applying electronic filtering elements to a communication line to improve the capability of that line to support higher transmission rates of data. See also **Equalization.**

Connector Cable The cable that goes between the terminal and the modem. It is usually either the RS232C or RS449 standard.

Consultative Committee for International Telephone and Telegraph See **CCITT.**

Contention A "dispute" between two or more devices over the use of a common channel at the same time.

Control Character A character whose occurrence in a particular context initiates, modifies, or stops a control operation.

Control Matrix A two-dimensional matrix that shows the relationship between all of the controls in the data communication network and the specific threats they mitigate.

cps Characters per second. A data rate unit used where circuits carry bits forming a data character.

CPU Central processing unit.

CRC Cyclical Redundancy Check. An error checking control technique utilizing a specifically binary prime divisor which results in a unique remainder.

CSMA/CD Carrier Sense Multiple Access with Collision Detection. A system used in contention networks where the network interface unit listens for the presence of a carrier before attempting to send and detects the presence of a collision by monitoring for distorted pulse.

CTS CLEAR TO SEND. A control signal between a modem and a controller used to operate over a connector cable.

Cyclical Redundancy Check See **CRC.**

D Conditioning A U.S. term for a type of conditioning that controls harmonic distortion and signal-to-noise ratio so they lie within specified limits.

DAA Data Access Arrangement. A telephone switching system protective device used to attach nontelephone company manufactured equipment to the carrier network.

Data

1. Specific individual facts or a list of such items.

2. Facts from which conclusions can be drawn.

Database A set of logically connected files that have a common access. They are the sum total of all the data items that exist for several related systems. In other words, a database might have several data items that can be assembled into many different record types.

Data Circuit Terminating Equipment See **DCE.**

Data Communications

1. The movement of encoded information by means of electrical transmission systems.

2. The transmission of data from one point to another.

Data Compression The technique that provides for the transmission of fewer data bits without the loss of information. The receiving location expands the received data bits into the original bit sequence. See also **Compression.**

Data Protectors Devices that protect the telephone company circuits from extraneous electrical signals. They limit the amount of power that can be transmitted to the telephone company central office.

Data Service Unit (DSU) The DSU is a Channel Service Unit that, in addition, provides bipolar conversion functions to ensure proper signal shaping and adequate signal strength. It, too, is a form of digital modem. See also **Channel Service Unit.**

Data Set In data communications this term generally is synonymous with modem, a device which allows a digital terminal device to transmit and receive data over an analog communication channel. The term also is used to describe a file of data, especially when referring to data processing.

Data Sink See **DTE.**

Data Terminal Equipment See **DTE.**

db See **Decibel.**

DB-25 The designation of a standard plug and jack set used in RS232C wiring. It has a 25-pin connector, with 13 pins in the top row and 12 in the bottom row.

dBm Power-level measurement unit in the telephone industry based on 600 ohms impedance and 1000 hertz frequency. Zero dBm is 1 milliwatt at 1000 hertz terminated by 600 ohms impedance.

DCE Data Circuit Terminating Equipment. The equipment installed at the user's site that provides all the functions required to establish, maintain, and terminate a connection, including the signal conversion and coding between the data terminal equipment (DTE) and the common carrier's line, e.g., data set, modem.

DDD See **Direct Distance Dialing.**

Decibel (dB) A tenth of a bel. A unit for measuring relative strength of a signal parameter such as power and voltage. The number of decibels is ten times the logarithm (base 10) of the ratio of the power of two signals, or ratio of the power of one signal to a reference level. The reference level always must be indicated, such as 1 milliwatt for power ratio.

Dedicated Circuits A leased communication circuit that goes from your site to some other location. It is a clear unbroken communication path that is yours to use 24 hours per day, seven days per week.

Delay Distortion A distortion on communication lines that is due to the different propagation speeds of signals at different frequencies. Some frequencies travel more slowly than others in a given transmission medium and therefore arrive at the destination at slightly different times. Delay distortion is measured in microseconds of delay relative to the delay at 1700 Hz. This type of distortion does not affect voice but can have a serious effect on data transmissions.

Delay Equalizer A corrective network designed to make the phase delay or envelope delay of a circuit or system substantially constant over a desired frequency range. See also **Equalizer.**

Delphi Group A small group of experts (three to five people) who meet to develop a consensus in an area where it may be impossible or too expensive to collect more accurate data. For example, a Delphi Group of communication experts might assemble to reach a consensus on the various threats to a communication network, the potential dollar losses for each occurrence of each threat, and the estimated frequency of occurrence for each threat.

Dial Tone A 90 Hz signal (the difference between 350 Hz and 440 Hz) sent to an operator or subscriber indicating that the receiving end is ready to receive dial pulses.

Dial-Up Telephone Network See **Direct Distance Dialing.**

Dibit A group of two bits. In four-phase modulation each possible dibit is encoded as one of four unique carrier phase shifts. The four possible states for dibit are 00, 01, 10, and 11.

Digital Modem See **Channel Service Unit.**

Digital PBX A PBX (switchboard) designed to switch digital signals. Telephones used in a digital PBX must digitize the voice signals. Computers and terminals can communicate directly through the digital PBX which functions as a point-to-point local area network.

Digital Signal A discrete or discontinuous signal. A signal whose various states are discrete intervals apart such as $+15$ volts and -15 volts.

Digital Termination Systems See **DTS.**

Direct Distance Dialing (DDD) A telephone exchange service that enables the telephone user to call other subscribers outside the local area without operator assistance. In the United Kingdom and some other countries, this is called *subscriber trunk dialing* (STD).

Discrete Files A set of data items and record types for one specific application. A discrete file is a separate, individual file for one application.

Diskette A disk for storing microcomputer files made of thin, flexible plastic and enclosed in a protective jacket. The most popular sizes are $5\frac{1}{4}$ inches and $3\frac{1}{2}$ inches. Also called *floppy disk.*

Distortion The unwanted modification or change of signals from their true form by some characteristic of the communication line or equipment being used for transmission, e.g., delay distortion, amplitude distortion.

Distortion Types

1. *Bias:* a type of distortion resulting when the intervals of modulation do not all have exactly their normal durations.

2. *Characteristic:* distortion caused by transients (disturbances) which, as a result of modulation, are present in the transmission circuit.

3. *Delay:* distortion occurring when the envelope delay of a circuit or system is not consistent over the frequency range required for transmission.

4. *End:* distortion of start-stop signals. The shifting of the end of all marking pulses from their proper positions in relation to the beginning of the start pulse.

5. *Jitter:* a type of distortion that results in the intermittent shortening or lengthening of the signals. This distortion is entirely random in nature and can be caused by hits on the line.

6. *Harmonic:* the resultant process of harmonic frequencies (due to nonlinear characteristics of a transmission circuit) in the response when a sinusoidal stimulus is applied.

DOS Abbreviation for Disk Operating System. A program or set of programs that instructs a disk-based computing system to schedule/supervise work, manage computer resources, and operate/control peripheral devices, including disk drives, keyboards, video monitors/screens, and printers. Different types are produced by different vendors. The most popular operating systems for PCs are MS-DOS from Microsoft (Bellevue, Washington) and PC-DOS from IBM.

Download The process of loading software/data into the nodes of a network from one node or device over the network media. Downloading usually refers to the movement of data from a host mainframe computer to a remote terminal/microcomputer.

DTE Data Terminal Equipment. Equipment comprising the data source, the data sink, or both that provides for the communication control function (protocol). Data termination equipment is actually any piece of equipment at which a communication path begins or ends, such as a terminal. The data sink is the receiving device.

DTS Digital Termination System. A form of local loop. It connects private homes and/or business locations to the common carrier switching facility.

Dual Cable A two-cable system in broadband local area networks in which the coaxial cable provides two physical paths for transmission, one for transmitting and one for receiving, instead of dividing the capacity of a single cable.

EBCDIC An acronym for Extended Binary Coded Decimal Interchange Code. A standard code consisting of a character set of 8-bit characters used for information representation and interchange among data processing and communication systems. Very common in IBM equipment.

Echo Checking See **Checking, Echo.**

Echo Suppressor A device for use in a two-way telephone channel (especially international circuits) to attenuate echo currents in one direction caused by telephone currents in the other direction. A device that suppresses echoes also suppresses data; therefore, when full duplex transmission is used the echo suppressors must be disabled. This is done by sending an appropriate disabling tone to the circuit.

Electron An elementary particle made of a tiny charge of negative electricity.

Electronic Switching System (ESS) A type of telephone switching system that uses a special purpose stored program digital computer to direct and control the switching operation. ESS permits the provision of custom calling services such as speed dialing, call transfer, and three-way calling.

Emulate When changing from one computer to a "new generation" computer, existing programs frequently cannot run on the new equipment. To save the cost of reprogramming, computer vendors provide software and hardware that accepts old programs and translates them to the new computer's machine language. In other words, the new computer equals the old computer's ability to run the programs.

Encryption The technique of modifying a known bit stream on a transmission line so it appears to an unauthorized observer to be a random sequence of bits.

End Office The telephone company switching office for the interconnection of direct dial-up calls.

Equalization The process of reducing frequency and/or phase distortion of a circuit by the introduction of networks to compensate for the difference in attenuation and/or time delay at the various frequencies in the transmission band.

Equalizer Any combination (usually adjustable) of coils, capacitors, and/or resistors inserted in the transmission line or amplifier circuit to improve its frequency response.

Error Control An arrangement that detects the presence of errors. In some systems, refinements are added that correct the detected errors, either by operations on the received data or by retransmission from the source.

ESS See **Electronic Switching System.**

Ethernet™ A baseband local area network marketed by Xerox and developed jointly by Xerox, Digital Equipment Corporation, and Intel.

Exchange Office The telephone company switching office for the interconnection of direct dial-up calls.

Extended Binary Coded Decimal Interchange Code See **EBCDIC.**

Facsimile (FAX) A system for the transmission of images. The image is scanned at the transmitter, reconstructed at the receiving station, and printed on $8\frac{1}{2} \times 11$ (or larger) paper.

FCC See **Federal Communications Commission.**

FDM See **Multiplexer.**

FDX See **Full Duplex.**

Feasibility Study A study undertaken to determine the possibility or probability of improving the existing system within a reasonable cost. Determines what the problem is and what its causes are, and makes recommendations for solving the problem.

FEC See **Forward Error Correction.**

Federal Communications Commission (FCC) A board of seven commissioners appointed by the U.S. President under the Communication Act of 1934, having the power to regulate all interstate and foreign electrical communication systems originating in the United States.

Fiber Optics Hair-thin plastic or glass fibers that carry visible light which contains information.

Firmware A set of software instructions set permanently or semipermanently into a read only memory (ROM).

Floppy Disk See **Diskette.**

Flow Control The capability of the network nodes to manage buffering schemes that allow devices of different data transmission speeds to communicate with each other.

Foreign Exchange Service (FX) A service that connects a customer's telephone to a remote exchange. This service provides the equivalent of local telephone service to and from the distant exchange.

Forward Channel A channel in which the direction of data transmission coincides with the direction of the information transfer.

Forward Error Correction (FEC) A technique that identifies errors at the received station and automatically corrects those errors without retransmitting the message.

Four-Wire Circuit A circuit using two pairs of conductors, one pair for the "go" channel and the other pair for the "return" channel. A telephone circuit carries voice signals both ways. In local loops this two-way transmission is achieved over only two wires because the waveforms traveling in each direction can be distinguished. In trunk networks, where amplifiers are needed at intervals and multiplexing is common, it is easier to separate the two directions of transmission and effectively use a pair of wires for each direction. At this point it is a four-wire circuit.

Frequency The rate at which a current alternates, measured in hertz, kilohertz, megahertz, etc. Other units of measure are cycles, kilocycles, or megacycles; hertz and cycles are synonymous.

Frequency Division Multiplexing See **Multiplexer.**

Frequency Modulation See **Modulation, Frequency.**

Frequency Shift Keying (FSK) A method of transmission whereby the carrier frequency is shifted up and down from a mean value in accordance with the binary signal; one frequency represents a binary one, and the other represents a binary zero.

Front End See **Communication Processor, Front End.**

FSK See **Frequency Shift Keying.**

Full Duplex (FDX) The capability of transmission in both directions at one time. Contrast with **Half Duplex.**

FX See **Foreign Exchange Service.**

Gateway A device that connects two systems, especially if the systems use different protocols. See also **Bridge.**

Gaussian Noise See **Noise, Gaussian.**

Geosynchronous Orbit A satellite's orbit that is over the equator and traveling in the same direction as the earth's surface, so that the satellite appears to be stationary over a point on the earth.

Gigabyte One billion bytes.

Guard Frequency A small bandwidth of frequency that separates two voice grade circuits. Also the frequencies between subchannels in FDM systems used to guard against subchannel interference.

Hacker A person who sleuths for passwords to gain illegal access to important computer files. They may rummage through corporate trash cans looking for carelessly discarded printouts.

Half Duplex (HDX) A circuit that permits transmission of a signal in two directions but not at the same time. Contrast with **Full Duplex.**

Hamming Code A forward error correction (FEC) technique named for its inventor.

Handshaking Exchange of predetermined signals when a connection is established between two data set devices. This is used to establish the circuit/message path.

HDLC High-Level Data Link Control. A Consultative Committee on International Telephone and Telegraph data communication line protocol standard.

HDX See **Half Duplex.**

Hertz (Hz) Same as cycles per second; e.g., 3000 hertz is 3000 cycles per second.

High-Level Data Link Control See **HDLC.**

Host Computer See **Computer, Central.**

Hot Line A service offered by Western Union that provides direct connection between customers in various cities using a dedicated line.

Hybrid Network A local area network or wide area backbone network with a mixture of topologies and access methods.

Idle Character A transmitted character indicating "no information" that does not manifest itself as part of a message at the destination point.

Impedance A measure of the electrical property of resistance; the amount of opposition offered by a circuit to the flow of a carrier wave. In other words, some of the energy is reflected back, which shows as a loss of power when the impedance is mismatched between two cables. Common impedances are 50, 75, 93, and 300 ohms. The resistance, inductance, and capacitance of a transmission circuit combine to give the value of its impedance.

Impulse Noise See **Noise, Impulse.**

In-Band Signaling The transmission signaling information at some frequency or frequencies that lie within a carrier channel normally used for information transmission.

Inductance The amount of stored and released electrical energy. An inductance filter provides filtering action by opposing changes in the electrical current. It helps to average out the carrier wave by keeping it from reaching either its maximum or minimum power values. Capacitors also have this ability.

Information A meaningful aggregation of data. Contrast with **Data.**

Intelligent Terminal Controller A microprocessor-based intelligent device that controls a group of terminals.

INTELSAT The International Telecommunications Satellite Consortium established in 1964 to establish a global communication satellite system.

Interexchange Channel (IXC) A channel or circuit between exchanges (central offices).

InterLATA Circuits that cross from one LATA (local access and transport area) into another.

Intermodulation Distortion An analog line impairment where two frequencies create a third erroneous frequency which in turn distorts the data signal representation.

IntraLATA Circuits that are totally within one LATA (local access and transport area).

ISDN Integrated Services Digital Network. A hierarchy of digital switching and transmission systems. The ISDN provides voice, data, and video in a unified manner. It is synchronized so that all digital elements speak the same "language" at the same speed.

ISO International Organization for Standardization, Geneva, Switzerland. The initials ISO stand for its French name. This international standards-making body is best known in data communications for developing the internationally recognized seven-layer network model called the Open System Interconnection (OSI) Reference Model.

IXC See **Interexchange Channel.**

Jack The physical connecting device at the interface which mates with a compatible receptacle—a plug. See also **RJ-11** and **RJ-45.**

Jitter Type of analog communication line distortion caused by the variation of a signal from its reference timing positions, which can cause data transmission errors, particularly at high speeds. This variation can be amplitude, time, frequency, or phase.

Jumbo Group Six U.S. master groups frequency division multiplexed together in the Bell System. A jumbo group can carry 3,600 telephone calls on one channel.

K A standard quantity measurement of computer storage. A K is defined loosely as 1,000 bytes. In fact, it is 1,024 bytes, which is the equivalent of two raised to the tenth power.

Kbps Kilo bits per second. A data rate equal to 10^3 bps (1000 bps).

Key Management   The process of controlling the secret keys used in encryption.

Key Set   A telephone instrument designed to provide push-button or switch selection of a specific channel from several possible incoming circuits.

Key System   A group of associated key sets connected to allow channel selection of a specific channel from several possible incoming circuits.

Kilo Bits per Second   See **Kbps.**

LAN   See **Local Area Network.**

Large-Scale Integration   See **LSI.**

Laser   Light amplification by stimulated emission of radiation. A device that transmits an extremely narrow and coherent beam of electromagnetic energy in the visible light spectrum. (Coherent means that the separate waves are in phase with one another rather than jumbled as in normal light.)

LATA   Local Access and Transport Area. One of approximately 160 local telephone serving areas in the United States, generally encompassing the largest Standard Metropolitan Statistical Areas (SMSA). Subdivisions established as a result of the AT&T/Bell divestiture that now distinguish local from long distance service. Circuits with both end points within the LATA (intraLATA) generally are the sole responsibility of the local telephone company. Circuits that cross outside the LATA (interLATA) are passed on to an interexchange carrier such as AT&T, MCI, or US Sprint.

Leased Circuits   Leased communication circuits that go from your site to some other location. It is a clear, unbroken communication path that is yours to use 24 hours per day, seven days per week.

Line   A circuit, channel, or link. It carries the data communication signals.

Line Control Codes/Characters   The 8-bit characters that control messages. They appear at both the beginning and end of the message to show its beginning and end. They also might appear within the message to show such things as the beginning and end of the text.

Line Loading   The total amount of transmission traffic carried by a line, usually expressed as a percentage of the total theoretical capacity of that line.

Line Protocol   A control program used to perform data communication functions over network lines. Consists of both handshaking and line control functions that move the data between transmit and receive locations.

Link   A channel or a line, normally refers to a point to point line.

Local Area Network (LAN)   A network that is located in a small geographic area, such as an office, building, complex of buildings, or a campus, and whose communication technology provides a high bandwidth, low cost medium to which many nodes can be connected.

Local Exchange Carrier   The local telephone company, such as one of the seven Bell Operating Companies (BOC).

Local Loop   That part of a communication circuit between the subscriber's equipment and the equipment in the local central office.

Log

1.   A record of everything pertinent to a system function.

2.   A collection of messages that provides a history of message traffic.

Logical Unit (LU)   In Systems Network Architecture (SNA), a port through which an end user accesses the SNA network to communicate with another end user and through which the end user accesses the functions provided by system services control points (SSCPs). An LU can support at least two sessions—one with an SSCP, and one with another logical unit—and may be capable of supporting many sessions with other logical units.

Logical Unit (LU) Services   In Systems Network Architecture (SNA), capabilities in a logical unit to

1. Receive requests from an end user and, in turn, issue requests to the system services control point (SSCP) to perform the requested functions, typically for session initiation.

2. Receive requests from the SSCP, for example, to activate LU-LU sessions via bind session requests.

3. Provide session presentation and other services for LU-LU sessions.

Long Haul Network   A network most frequently used to transfer data over distances from several thousand feet to several thousand miles. These networks can use the international telephone network to transport messages over most or part of these distances. Also called a *backbone network*.

Longitudinal Redundancy Check (LRC)   A system of error control based on the formation of a block check following preset rules. The check formation rule is applied in the same manner to each character. In a simple case, the LRC is created by forming a parity check on each bit position of all characters in the block (e.g., the first bit of the LRC character creates odd parity among the 1-bit positions of the characters in the block).

Loopback   Type of diagnostic test in which the transmitted signal is returned to the sending device after passing through a data communication link or network. This test allows a technician or diagnostic circuit to compare the returned signal with the transmitted signal to get some sense of what is wrong. Loopbacks often are done by excluding one piece of equipment after another. This allows you to figure out logically what is wrong.

LRC   See **Longitudinal Redundancy Check.**

LSI   Large-Scale Integration. A type of electronic device comprising many logic elements in one very small package (integrated circuit) to be used for data handling, storage, and processing.

LU   See **Logical Unit.**

mA   Milliampere. Measurement unit for electric current.

MAP   Manufacturing Automation Protocol. A six-layer protocol model that endorses the IEEE 802.4 token passing broadband bus local area network designed to transmit at 1, 5, or 10 million bits per second. Its special use is in factory settings. When MAP is combined with TOP (Technical Office Protocol), office functions can be integrated. When MAP is combined with CIM (Computer Integrated Manufacturing), manufacturing, design, and business functions can be integrated.

Master Group   An assembly of ten Super Groups occupying adjacent bands in the transmission spectrum for the purposes of simultaneous modulation and demodulation.

Master Number Hunting   A PBX feature that allows a station to seek an open terminal point in a predetermined sequence. In master number hunting, this "station hunting" option is activated by dialing a preset digit.

Matrix of Controls   A two-dimensional matrix that shows the relationship between all the controls in the data communication network and the specific threats they mitigate.

Mbps   A data rate equal to 10^6 bps. Sometimes called megabits per second (1,000,000 bps).

Megabyte   One million bytes.

Message A communication of information from a source to one or more destinations, usually in code. A message usually is composed of three parts.

1. A heading, containing a suitable indicator of the beginning of the message together with some of the following information: source, destination, date, time, routing.

2. A body containing information to be communicated.

3. An ending containing a suitable indicator of the end of the message.

Message Switching In this operation the entire message being transmitted is switched to the other location without regard to whether the circuits actually are interconnected at the time of your call. This usually involves a message store and forward facility.

Metered Service A combination of offerings of bulk long distance service under various terms involving flat-rate charges per hour of usage.

MHz A unit of analog frequency equal to 10^6 hertz. Also referred to as megahertz.

Microcomputer A complete computer on a microprocessor chip. Also referred to as a PC. It has input/output/processing capabilities and can support printers, disk storage, and modems. Today it is used as a workstation on various networks.

Microprocessor A single or multiple chip set that makes up a microcomputer or is the intelligence contained within a modem, multiplexer, intelligent controller, or so forth. Examples are Intel 80386 and Motorola 68030 chips.

Microsecond One-millionth of a second.

Milliampere See **mA**.

Millisecond One-thousandth of a second.

Minicomputer A computing system with capabilities generally between those of the most powerful computers and small personal and business microcomputers. Often used for specialized tasks.

Modem A contraction of the words modulator-demodulator. A modem is a device for performing necessary signal transformation between terminal devices and communication lines. They are used in pairs, one at either end of the communication line.

Modem Eliminator A small short haul device that can replace a modem when the distance of the data link is short. It does not need any external electrical power. It takes some power out of the line.

Modulation, Amplitude The form of modulation in which the amplitude of the carrier is varied in accordance with the instantaneous value of the modulating signal.

Modulation, Frequency A form of modulation in which the frequency of the carrier is varied in accordance with the instantaneous value of the modulating signal.

Modulation, Phase A form of modulation in which the phase of the carrier is varied in accordance with the instantaneous value of the modulating signal.

Modulation, Pulse See **Pulse.**

Multidrop (Multipoint) A line or circuit interconnecting several stations/nodes.

Multiplexer A device that combines data traffic from several low speed communication lines onto a single high speed line. The two popular types of multiplexing are FDM (frequency division multiplexing) and TDM (time division multiplexing). In FDM, the voice grade link is divided into subchannels, each covering a different frequency range in such a manner that each subchannel can be employed as if it were an individual line. In TDM, separate time segments are assigned to each terminal. During these times, data may be sent without conflicting with data sent from another terminal.

Multiplexing The subdivision of a transmission channel into two or more separate channels. This can be achieved by splitting the frequency range of the channel into narrower frequency bands (*frequency division multiplexing*) or by assigning a given channel successively to several different users at different times (*time division multiplexing*).

Multiprocessing Strictly, this term refers to the simultaneous application of more than one processor in a multi-CPU computer system to the execution of a single "user job," which is possible only if the job can be effectively defined in terms of a number of independently executable components. The term is used more often to denote multiprogramming operation of multi-CPU computer systems.

Multiprogramming A method of operation of a computer system whereby a number of independent jobs are processed together. Rather than allow each job to run to completion in turn, the computer switches between them to improve the utilization of the system hardware components.

Multithreading Concurrent processing of more than one message (or similar service requested) by an application program.

NAK See **Negative Acknowledgment.**

Nanosecond One-billionth of a second.

NAU See **SNA Network.**

NCP See **Network Control Program.**

Negative Acknowledgment (NAK) In the method of error control which relies on repeating any message received with (detectable) errors, the return signal that reports an error is a NAK (the opposite of *ACK*, or acknowledgment).

Network

1. A series of points connected by communication channels.

2. The switched telephone network is the network of telephone lines normally used for dialed telephone calls.

3. A private network is a network of communication channels confined to the use of one customer.

Network Control Program (NCP) The program within the software system for a data processing system that deals with control of the network. Normally it manages the allocation, use, and diagnosis of performance of all lines in the network and the availability of terminals at the ends of the network. NCP also is used as a specific term referring to a component of Systems Network Architecture (SNA).

Network Interface Controller A communication device that allows interconnection of information processing devices to a network.

Network Service An application available on a network, e.g., file transfer.

Node In a topological description of a network, a point of junction of the links. The word node also has come to mean a switching center in the context of data networks, particularly in the context of packet switching. Also called a *station*.

Noise The unwanted change in waveform that occurs between two points in a transmission circuit.

Noise, Cross Talk Noise resulting from the interchange of signals on two adjacent channels; manifests itself when you hear other people's conversations.

Noise, Echo The "hollow" or echoing characteristic that results on voice grade lines with improper echo suppression.

Noise, Gaussian Noise that is characterized statistically by a Gaussian, or random distribution.

Noise, Impulse Noise caused by individual impulses on the channel.

Noise, Intermodulation Noise resulting from the intermodulation products of two signals. This is a result of harmonic reinforcements and cancellation of frequencies.

Nonblocking Describing a switch where a through traffic path always exists for each attached station. Generically, a switch or switching environment designed never to experience a busy condition because of call volume.

Nonprinting Character A control character that is transmitted as part of the information, but not reproduced on the hard copy.

NRZ Nonreturn to zero. A binary encoding and transmission scheme in which ones and zeros are represented by opposite and alternating high and low voltages, and where there is no return to a reference (zero) voltage between encoded bits.

NRZI Nonreturn to zero inverted. A binary encoding scheme that inverts the signal on a one and leaves the signal unchanged for a zero, and where a change in the voltage state signals a one bit and the absence of a change denotes a zero-bit value.

Null Character A control character that can be inserted into or withdrawn from a sequence of characters without altering the message. Generally used to delete errors on punched paper tape.

Null Modem Cable An RS232 cable in which the two microcomputers connected at each end think they are talking through modems. Should not be confused with modem eliminators, which are short haul modemlike devices.

Octet A group of eight binary digits operated upon as an entity. Also called a *byte*.

Off Hook Activated (in regard to a telephone set). By extension, a data set automatically answering on a public switched system is said to go "off hook." The off-hook condition indicates a "busy" condition to incoming calls.

Office, Central (or End) The common carrier switching office closest to the subscriber.

Office, Tandem End A switching office that terminates a tandem trunk.

Office, Toll A switching office that terminates a toll trunk.

Ohm A unit of resistance, such that 1 ampere through it produces a potential difference of 1 volt. Ohm's law is applicable to electric components carrying direct current, and it states that the resistance is independent of the current in metallic conductors at a constant temperature and zero magnetic field.

On Hook Deactivated (in regard to a telephone set). A telephone not in use is "on hook."

Online

1. Pertaining to equipment or devices under the direct control of a central processing unit.

2. Pertaining to a user's ability to interact with a computer.

3. Pertaining to a user's access to a computer via a terminal.

Online System A system in which the input data enters the computer directly from the point of origin or in which output data is transmitted directly to where it is used.

Open Wire Communication lines that are not insulated and formed into cables, but mounted on aerial cross arms on utility poles.

Optical Fibers Hair-thin strands of very pure glass (sometimes plastic) over which light waves travel. They are used as a medium over which information is transmitted.

Out-of-Band Signaling A method of signaling which uses a frequency that is within the pass band of the transmission facility but outside of a carrier channel normally used for data transmission.

Overhead Computer time used to keep track of or run the system as compared with computer time used to process data.

PABX See **PBX.**

Packet A group of binary digits, including data and control signals, which is switched as a composite whole. The data, control signals, and possibly error control information are arranged in a specific format (128 characters).

Packet Assembler/Disassembler See **PAD.**

Packet Switching Process whereby messages are broken into finite-size packets that always are accepted by the network. The message packets are forwarded to the other party over a multitude of different circuit paths. At the other end of the circuit, the packets are reassembled into the message, which is then passed on to the receiving terminal.

Packet Switching Network A network designed to carry data in the form of packets. The packet and its format are internal to that network. The external interfaces may handle data in different formats, and conversion is done by an interface computer.

PAD Packet Assembler/Disassembler. Equipment providing packet assembly and disassembly facilities.

PAM See **Pulse Amplitude Modulation.**

Parity Bit A binary bit appended to an array of bits to make the sum of all the bits always be odd or even for an individual character.

Parity Check Addition of noninformation bits to a message in order to detect any changes in the original bit structure from the time it leaves the sending device to the time it is received.

Pass Band A range of frequencies that can be transmitted to a terminal at low attenuation.

Path Control (PC) Network In Systems Network Architecture (SNA), the part of the SNA network that includes the data link control and path control layers.

PBX Private Branch Exchange. Telephone switch located at a customer's site that primarily establishes voice communications over tie lines or circuits as well as between individual users and the switched telephone network. Typically also provides switching within a customer site and usually offers numerous other enhanced features, such as least-cost routing and call detail recording; also called PABX for private automatic branch exchange and CBX for computerized branch exchange.

PC An acronym commonly used for Personal Computer. Microcomputers often are referred to as PCs. Should not be confused with Path Control (PC) Network.

PCM See **Pulse Code Modulation.**

PDM See **Pulse Duration Modulation.**

Permissive connection Relates to the sending of *nonvoice* information (low speed data and facsimile) over the *voice* communications network. Consequently, the quality of transmission is good as long as the transmission is within the normal voice range of frequencies and conditioning. See also **Programmable Connection** or **RJ-45.**

Phase Modulation See **Modulation, Phase.**

Photon The fundamental unit of light and other forms of electromagnetic energy. Photons are to optical fibers what electrons are to copper wires; like electrons, they have a wave motion.

Phreaker A person who wants to circumvent the telephone system to make free calls.

Physical Unit (PU) In Systems Network Architecture (SNA), the component that manages and monitors the resources (such as attached links and adjacent link stations) of a node, as requested by a SSCP via a SSCP-PU session. Each node of a SNA network contains a physical unit.

Physical Unit Control Point (PUCP) In Systems Network Architecture, a component that provides a subset of system services control point (SSCP) functions for activating the physical unit (PU) within its node and its local link resources. Each peripheral node and each subarea node without a SSCP contains a PUCP.

Picosecond One-trillionth of a second.

Pirate A person who obtains the latest software programs without paying for them. A skilled software pirate is able to break the protection scheme that is designed to prevent copying.

Point of Presence Since divestiture, refers to the physical access location within a local access and transport area (LATA) of a long distance and/or interLATA common carrier. The point to which the local telephone company terminates subscribers' circuits for long distance dial-up or leased line communications.

Point to Point Denoting a channel or line that has only two terminals. A link.

Polling Any procedure that sequentially contacts several terminals in a network.

Polling, Hub Go-Ahead Sequential polling in which the polling device contacts a terminal, that terminal contacts the next terminal, and so on.

Polling, Roll Call Polling accomplished from a prespecified list in a fixed sequence, with polling restarted when the list is completed.

Polynomial Checking See **Checking, Polynomial.**

Port One of the circuit connection points on a front end communication processor or local intelligent terminal controller.

PPM See **Pulse Position Modulation.**

Preamble A sequence of encoded bits which is transmitted before each frame to allow synchronization of clocks and other circuitry at other sites on the channel. In the Ethernet specification the preamble is 64 bits.

Private Automatic Branch Exchange See **PBX.**

Private Branch Exchange See **PBX.**

Private leased Circuit A leased communication circuit that goes from your premises to some other location. It is a clear unbroken communication path that is yours to use 24 hours per day, seven days per week.

Programmable connection Relates to the sending of information (data bits) over data type circuits. See also **Permissive Connection** or **RJ-11.**

Propagation Delay The time necessary for a signal to travel from one point on the circuit to another, such as from a satellite dish up to a satellite.

Protocol A formal set of conventions governing the format and control of inputs and outputs between two communicating devices. This includes the rules by which these two devices communicate as well as handshaking and line discipline. Some protocols are ''bit-oriented,'' with a change of a single bit within a character conveying a new control message to the other end of the communication circuit. Some protocols are ''byte-oriented,'' with the entire eight bits of a character having to be changed in order to convey a control message to the other end of the communication network.

Protocol Converter A hardware device that changes the protocol of one vendor to the protocol of another. For example, if you want to connect an IBM data communication network to a Honeywell data communication network, the protocol converter converts the message formats so they are compatible. It is similar to a person who translates between French and English for two people who do not speak one another's language.

PTT Public telephone and telegraph. These are the common carriers owned by governments in which the government is the sole monopoly supplier of communication facilities.

PU See **Physical Unit.**

Public Data Network A network established and operated for the specific purpose of providing data transmission services to the public.

PUCP See **Physical Unit Control Point.**

Pulse A brief change of current or voltage produced in a circuit to operate a switch or relay or which can be detected by a logic circuit.

Pulse Amplitude Modulation (PAM) Amplitude modulation of a pulse carrier.

Pulse Code Modulation (PCM) Representation of a speech signal (or other analog signal) by sampling of a regular rate and converting of each sample to a binary number.

Pulse Duration Modulation (PDM) Pulse width modulation. A form of pulse modulation in which the durations of pulses are varied.

Pulse Modulation The modulation of the characteristics of a series of pulses in one of several ways to represent the information-bearing signal. Typical methods involve modifying the amplitude (PAM), width or duration (PDM), or position (PPM). The most common pulse modulation technique in telephone work is pulse code modulation (PCM). In PCM, the information signals are sampled at regular intervals and a series of pulses in coded form are transmitted, representing the amplitude of the information signal at that time.

Pulse Position Modulation (PPM) A form of pulse modulation in which the positions in time of pulses are varied, without their duration being modified.

QAM, QSAM Quadrature amplitude modulation, Quadrature sideband amplitude modulation. A sophisticated modulation technique, using variations in signal amplitude that allows data-encoded symbols to be represented as any of 16 or 32 different states.

Queue A line of items. In data communications there can be message input queues, output queues, and various other queues when the system cannot handle all the transactions that are arriving.

RAM Random access memory. A type of computer memory where information can be stored and retrieved by the computer in miscellaneous order without disturbing adjacent memory cells.

Rate Center A defined geographic point used by telephone companies in determining distance measurements for interLATA or intraLATA mileage rates. Basically, this is a central/end office.

RDC See **Concentrator.**

Real Time The entry of information into a network from a terminal and immediate processing of the task.

Redundancy The portion of the total information contained in a message that can be eliminated without loss of essential information.

Reliability The characteristic of equipment, software, or systems that relates to the integrity of the system against failure. Reliability usually is measured in terms of Mean Time Between Failures (MTBF), the statistical measure of the interval between successive failures of the system under consideration.

Remote Data Concentrator See **Concentrator.**

Remote Job Entry (RJE) Submission of jobs (e.g., computer production tasks) through an input unit (terminal) that has access to a computer through data communication facilities.

Repeater A device used to boost the strength of a signal. Repeaters are spaced at intervals throughout the length of a communication circuit. Also called *repeater/amplifier.*

REQUEST TO SEND See **RTS.**

Resistance The amount of opposition to the flow of the carrier wave or signal moving down the circuit.

Response Time The time the system takes to react to a given input. If a message is keyed into a terminal by an operator and the reply from the computer, when it comes, is typed at the same terminal, response time may be defined as the time interval between the operator pressing the last key and the terminal typing the first letter of the reply. It is the interval between an event and the system's response to the event. Response time thus defined includes (1) transmission time to the computer; (2) processing time at the computer, including access time to obtain any file records needed to answer the inquiry; and (3) transmission time back to the terminal.

Reverse Channel A feature of certain modems that allows simultaneous transmission (usually of control or parity information) from the receiver to the transmitters over a half duplex data transmission system. Generally, the reverse channel is a low speed channel ranging between 75 and 150 bits per second.

RJ-11 A modular telephone plug. It is the plug on your home telephone and is called a permissive connection (see Chapter 3, under Line Protectors) by the telephone company. Although it generally is used on two-wire circuits, it can be used on four-wire circuits.

RJ-45 A modular telephone plug similar to the RJ-11 plug, except that it is called a programmable connection and generally is used on four-wire circuits.

RJE See **Remote Job Entry.**

ROM Read only memory. A data storage device, the contents of which cannot be altered, except under certain circumstances. Storage in which writing over is prevented. Also, permanent storage.

Rotary Hunt An arrangement that allows calls placed to seek an idle circuit in a prearranged multichannel group. If the dialed line is busy, the call finds the next open line to establish a through channel.

Rotary Switching System An automatic telephone switching system which generally is characterized by the following features.

1. The selecting mechanisms are rotary switches.

2. The switching pulses are received and stored by controlling mechanisms that govern the subsequent operations necessary to establish a telephone connection.

RS232C A technical specification published by the Electronic Industries Association that specifies the mechanical and electrical characteristics of the interface for connecting data terminal equipment (DTE) and data circuit terminating equipment (DCE). It defines interface circuit functions and their corresponding connector pin assignments. RS232C is functionally compatible with Consultative Committee on International Telephone and Telegraph (CCITT) Recommendation V.24.

RS422 A standard operating in conjunction with RS449 that specifies electrical characteristics for balanced circuits (circuits with their own ground leads).

RS423 A standard operating in conjunction with RS449 that specifies electrical characteristics for unbalanced circuits (circuits using common or shared grounding techniques).

RS449 An Electronic Industries Association standard for data terminal equipment (DTE) and data circuit terminating equipment (DCE) connection which specifies interface requirements for expanded transmission speeds (up to 2 million bits per second), longer cable lengths, and ten additional functions. The physical connection between DTE and DCE is made through a 37-pin connector; a separate 9-pin connector is specified to service secondary channel interchange circuits when used.

RTS REQUEST TO SEND. An RS232 control signal between a modem and user digital equipment which initiates the data transmission sequence in a communication line.

Satellite See **Geosynchronous Orbit.**

Satellite Microwave Radio Microwave or beam radio system using geosynchronously orbiting communication satellites.

Satisfice To choose a particular level of performance for which to strive and for which management is willing to settle.

SC See **Session Control.**

SDLC See **Synchronous Data Link Control.**

Serial

1. Pertaining to transmitting bits one at a time in sequence.

2. Pertaining to the sequential or consecutive occurrence of two or more related activities in a single device or channel.

3. Pertaining to the sequential processing of the individual part of the whole, such as the bits of a character, or the characters of a word using the same facilities for successive parts.

Server A processor that provides a specific service to the network. Examples of servers are

1. Routing servers, which connect nodes and networks of like architectures.

2. Gateway servers, which connect nodes and networks of different architectures by performing protocol conversions.

3. Terminal servers, printer servers, disk servers, and file servers, which provide an interface between compatible peripheral devices on a local area network.

Session A logical connection between two terminals. This is the part of the message transmission when the two parties are exchanging messages. It takes place after the communication circuit has been set up and is functioning.

Session Control (SC) In Systems Network Architecture (SNA), one of the components of transmission control. Session control is used to purge data flowing in a session after an unrecoverable error occurs, to resynchronize the data flow after such an error, and to perform cryptographic verification.

Signal A signal is some thing that is sent over a communication circuit. It might be a message that you are transmitting, or it might be a control signal used by the system to control itself.

Signaling Supplying and interpreting the supervisory and address signals needed to perform the switching operation.

Signal-to-Noise Ratio The ratio, expressed in dB, of the usable signal to the noise signal present.

Simplex A circuit capable of transmission in one direction only. Contrast with **Half Duplex, Full Duplex.**

Single Cable A one-cable system in broadband local area networks in which a portion of the bandwidth is allocated for send signals and a portion for receive signals, with a guardband in between to provide isolation from interference.

SN See **Switching Node.**

SNA See **Systems Network Architecture.**

SNA Network In Systems Network Architecture (SNA), the part of a user application network that conforms to the formats and protocols of SNA. It enables transfer of data among end users and provides protocols for controlling the resources of various network configurations. The SNA network consists of network addressable units (NAUs), boundary function components, and the path control network.

Software A generic, somewhat slang term for a computer program, sometimes taken to include documentation and procedures associated with such programs.

Special Common Carrier An organization other than the public telephone companies, registered to sell or lease communication facilities.

SSCP In Systems Network Architecture (SNA), the system services control point.

SSCP-LU Session In Systems Network Architecture (SNA), a session between a system services control point (SSCP) and a logical unit (LU); the session enables the LU to request the SSCP to help initiate a LU-LU session.

SSCP-PU Session In Systems Network Architecture (SNA), a session between a system services control point (SSCP) and a physical unit (PU); SSCP-PU sessions allow SSCPs to send requests to and receive status information from individual nodes to control the network configuration.

SSCP Services In Systems Network Architecture (SNA), the components within a system services control point (SSCP) that provide configuration, maintenance, management, network, and session services for SSCP-LU, SSCP-PU, and SSCP-SSCP sessions.

SSCP-SSCP Session In Systems Network Architecture (SNA), a session between the system services control point (SSCP) in one domain and the SSCP in another domain. An SSCP-SSCP session is used to initiate and terminate cross-domain LU-LU sessions.

Start Bit A bit preceding the group of bits representing a character used to signal the arrival of the character in asynchronous transmission.

Start-Stop (Signaling) Signaling in which each group of code elements corresponding to a character is preceded by a start signal which serves to prepare the receiving mechanism for the reception and registration of the character, and is followed by a stop signal which serves to bring the receiving mechanism to rest in preparation for the reception of the next character (contrast with **Synchronous Transmission**). Start-stop transmission also is referred to as *asynchronous transmission.*

Station One of the input or output points on a network. Also called a *node.*

Station Terminal The plug supplied by the common carrier into which the modem plugs.

Statistical Multiplexer Or Stat Mux. A time division multiplexer (TDM) that dynamically allocates communication line time to each of the various attached terminals, according to whether a terminal is active or inactive at a particular moment. Buffering and queuing functions also are included.

Stop Bit A bit used following the group of bits representing a character, to signal the end of a character in asynchronous transmission.

Store and Forward A data communication technique that accepts messages or transactions, stores them until they are completely in the memory system, and then forwards them on to the next location as addressed in the message or transaction header.

Super Group A frequency division multiplexer (FDM) carrier multiplexing level containing 60 voice frequency channels. It is the assembly of five 12-channel groups occupying adjacent bands in the spectrum for the purpose of simultaneous modulation and demodulation.

Switchboard Equipment on which switching operations are performed by operators. See also **PBX.**

Switched Network Any network in which switching is present and which is used to direct messages from the sender to the ultimate recipient.

Switched Network, Circuit Switched A switched network in which switching is accomplished by disconnecting and reconnecting lines in different configurations to set up a continuous pathway between the sender and the recipient.

Switched Network, Store and Forward A switched network in which the store and forward principle is used to handle transmissions between senders and recipients.

Switching Identifying and connecting independent transmission links to form a continuous path from one location to another.

Switching Node (SN) The intelligent interface point where your equipment is connected to a public packet switching network. The switching node is a type of front end, but its primary purpose is packetizing, routing, and scheduling on the packet switching network.

Synchronous Data Link Control (SDLC) A discipline for managing synchronous, code-transparent, serial-by-bit information transfer over a link connection. Transmission exchanges may be full duplex or half duplex and over switched or non-switched links. The configurations of the link connection may be point to point, multipoint, or loop. SDLC conforms to subsets of the Advance Data Communication Control Procedures (ADCCP) of the American National Standards Institute and High-Level Data Link Control (HDLC) of the International Organization for Standardization.

Synchronous Transmission Form of transmission in which data is sent as a block or frame. If no legitimate data is available to be sent at a given time, "synch" or "idle" characters are sent to keep the transmitter and receiver in time synchronization.

System Services Control Point (SSCP) In Systems Network Architecture (SNA), a focal point within the SNA network for managing the configuration, coordinating network operator and problem determination requests, and providing directory support and other session services for end users of the network. Multiple SSCPs, cooperating as peers with one another, can divide the network into domains of control, with each SSCP having a hierarchical control relationship to the physical units and logical units within its own domain. Also see **SSCP** entries.

Systems Network Architecture (SNA) The term applied by IBM to the conceptual framework used in defining data communication interaction with computer systems.

T Carrier A hierarchy of digital systems designed to carry speech and other signals in digital form, designated T-1, T-2, and T-4.

T-1 Carrier A digital transmission system, developed by AT&T, which sends information at 1.544 million bits per second. A T-2 carrier combines four T-1s. See Figure 2-48 in Chapter 2.

Tap In baseband, the component or connector that attaches a transceiver to a cable. In broadband, a passive device used to remove a portion of the signal power from the distribution line and deliver it onto the drop line. Also called a *directional tap* or a *multitap*. Another use of the term is for a breach of security. When unauthorized people want to gain access to a system, they illegally "tap" into a communication circuit.

Tariff The schedule of rates and regulations pertaining to the services of a communication common carrier.

TASI Time Assisted Speech Interpolation. The process of sending two or more voice calls on the same telephone circuit simultaneously.

TDM See **Multiplexer.**

Telecommunication Access Programs The software programs located in the front end communication processor. They handle all of the tasks associated with the routing, scheduling, and movement of messages between remote terminal sites and the central host computer.

Telecommunication Monitors See **Teleprocessing Monitors.**

Teleprocessing Monitors A set of software programs (usually located in the host computer) that handle the various tasks required for incoming and outgoing messages. For example, the teleprocessing monitor builds the input/output queues of messages in the host computer and relieves the computer operating system of many of its tasks related to the data communication network.

Threat Any potentially adverse occurrence that can harm the data communication network, interrupt the systems that use the network, or cause a monetary loss to the organization. For example, lost messages are a potential threat.

Time Assisted Speech Interpolation See **TASI.**

Tip

1. The end of the plug that completes a circuit in a manual switchboard. It serves as the positive power source for the station because it is connected directly to the common battery.

2. In terminal interface processing, a term used by packet switching companies to designate the piece of equipment that accepts data from outlying terminals and reformats it into network language for transmission.

Token Passing A method of allocating network access wherein a terminal can send traffic only after it has acquired the network's token.

TOP Technical Office Protocol. A specification originated by Boeing defining industry local area network standards that make up its seven-layer protocol structure, for office use. Companion specification of MAP (manufacturing automation protocol).

Topology The configuration of network nodes and links. Description of the physical geometric arrangement of the links and nodes that make up a network, as determined by their physical connections.

TPA Transient Program Area. The area in random access memory (RAM) where application programs and data are stored in the microcomputer.

Tree A network arrangement in which the stations hang off a common "branch," or data bus, like leaves on the branch of a tree.

Trunk A communication channel between switching devices or central offices.

Turnaround Time The time required to reverse the direction of transmission from send to receive or vice versa on a half duplex circuit.

Two-Wire Circuit A circuit formed by two conductors insulated from each other. It is possible to use the two conductors as a one-way transmission path, a half duplex path, or a full duplex path.

Unbalanced The state of impedance on a two-wire circuit when both circuits are not evenly grounded.

Unix A versatile operating system developed at AT&T Bell Laboratories. May be used to run microcomputers, minicomputers, and host mainframes for a wide range of tasks including multiuser systems and local area networks.

Upload The process of loading software/data from the nodes of a network (terminals/microcomputers) over the network media and back to the host mainframe computer.

V.nn The V.nn series of Consultative Committee on International Telephone and Telegraph standards, relating to the connection of digital equipment to the analog public telephone network. See Chapter 5.

Value Added Common Carrier A corporation that sells services of a value added network. Such a network is built using the communication offerings of traditional common carriers, connected to computers that permit new types of telecommunication tariffs to be offered. The network may be a packet switching or message switching network. Services include transmission of data charged for by the packet and transmission of facsimile documents.

Vertical Redundancy Checking   See **VRC**.

Videotex   A two-way dialogue through a television set to a central site that offers various services in the home.

Virtual   Conceptual or appearing to be, rather than actually being.

Virtual Circuit   Software that connects end points as though through a physical circuit. Address information is contained in the packets that carry source data to the destination. This circumvents typical hardware problems like data speed mismatch and helps retransmission when there are errors.

Virtual Storage   A computer user may employ a computer as though it has a much larger memory than its real memory. The difference is made up by software rapidly moving pages in and out, to and from a backing store. The apparent memory which the user can employ is called *virtual memory*.

Virtual Terminal   A terminal that is defined as a standard on a network that can handle diverse terminals. Signals to and from each nonstandard terminal are converted to equivalent standard terminal signals by an interface computer. The network protocols then operate as though all terminals were the standard "virtual" terminals.

Voice Grade   A telecommunication link with a bandwidth (about 4 KHz) appropriate to an audio telephone line.

Voice Grade Circuit   The term applies to channels suitable for transmission of speech, digital or analog data, or facsimile, generally with a frequency range of about 300 to 3000 hertz contained within a 4000 hertz channel.

VRC   Vertical Redundancy Checking. A method of character parity checking.

WAK   Positive acknowledgment but stop sending. Contrast with **ACK** and **NAK**. It stands for Wait-Acknowledge.

Walk Time
1. The time required for the message to travel completely around a ring local area network.
2. The time to switch service from one queue to another when servicing messages.

WATS   Wide area telecommunications service.

Wide Area Network   A data communication network designed to serve an area of hundreds or thousands of miles. Compare with **Local Area Network**.

Wideband Circuit   The term applied to channels provided by common carriers capable of transferring data at speeds from 19,200 bits per second up to the 2 million bits per second region (19.2 KHz to 1000 KHz). Basically, any communication circuit that has a bandwidth greater than 4000 hertz.

Wire Center   Same as end office.

Wiring Closet   Also called a Main Distribution Function (MDF). A central point at which all the circuits in a system begin or end, to allow cross-connection.

Word
1. In communications, usually assumed to be six characters (five plus a space).
2. In computers, the unit of information transmitted, stored, and operated upon at one time.

Workstation   Input/output equipment at which an operator works. A station in which a user can send data to, or receive data from, a computer for the purpose of performing a job.

X.*nn*   The X.*nn* series of Consultative Committee on International Telephone and Telegraph standards, relating to the connection of digital equipment to a public data network which employs digital signaling. See Chapter 5.

DATA COMMUNICATION CONTROL MATRIX

This section outlines the control review matrix to be used when performing a controls review of the data communication network that interconnects remote terminals and the central computer system or the various portions of an online distributed network. The controls/safeguards listed in this matrix are designed specifically for review of the data communication network.

THE MATRIX APPROACH[1]

The internal control area to be reviewed using this matrix covers the data communication links between the computer and input/output terminals. These data communication-oriented controls may involve hardware controls, software controls, and personnel controls. When reviewing the data communication controls, match each resource/asset with its corresponding concern/exposure as listed in Figure A1-1, Data Communication Control Matrix. This matrix lists the resources in relation to potential exposures and cross-relates these with the various controls/safeguards that should be considered when reviewing the data communication controls. See the section Matrix of Controls in Chapter 9 for an explanation of how to build a controls matrix.

Following is a definition of each of the concerns/exposures that are listed across the top of the matrix in Figure A1-1 and each of the resources/assets that are listed down the left vertical column of the matrix. Following these definitions is a complete numerical listing and description of each of the controls/safeguards that are listed numerically in the cells of the matrix.

CONCERNS/EXPOSURES (THREATS)

The following concerns/exposures (threats) are directly applicable to the data communication network of an online system. The definitions for each of the exposures listed across the top of the matrix are

- **Errors and Omissions**　The accidental or intentional transmission of data that is in error, including the accidental or intentional omission of data that should have been entered or transmitted on the online system. This type of exposure includes, but is not limited to, inaccurate data, incomplete data, malfunctioning hardware, and the like.

- **Message Loss or Change**　The loss of messages as they are transmitted through the data communication system, or the accidental/intentional changing of messages during transmission.

- **Disasters and Disruptions (natural and man-made)**　The temporary or long-term disruption of normal data communication capabilities. This exposure renders the organization's normal data communication online system inoperative.

[1] Appendix 1 is taken from Chapter 4 of the book, *Internal Controls for Computerized Systems*, by Jerry FitzGerald. This book contains eight other matrices on subjects such as software, physical security, database, and application program controls. This book is available from Jerry FitzGerald & Associates, 506 Barkentine Lane, Redwood City, Calif. 94065.

THREATS/CONCERNS/EXPOSURES

COMPONENTS/RESOURCES/ASSETS	Errors and Omissions	Message Loss or Change	Disasters and Disruptions	Privacy	Security/Theft	Reliability (Uptime)	Recovery and Restart	Error Handling	Data Validation and Checking
Central System	1–4, 7, 39, 41–43, 47, 48	1–5, 7, 37, 39, 48, 49, 89	1, 8, 11, 13, 16, 29, 40, 48, 50, 51, 54, 57, 58, 64, 65, 79, 85	6, 8, 24, 35, 53, 56, 60, 62, 68, 70, 72–74, 78–80	6, 8, 24, 35, 53, 56, 60, 62, 68, 70, 72–74, 77–80	1, 13, 16, 29, 38, 40, 50, 51, 63–65, 68, 81, 88	50, 51, 63–65, 68	48, 85, 89	6, 24, 39, 41, 47, 88
Software	1–4, 7, 39, 41–43, 46–49, 52	1–5, 7, 37, 39, 41, 42, 48, 49, 52, 54, 89	1, 8, 16, 40, 48, 50–54, 57–59, 63, 85	6, 8, 24, 35, 53, 56, 60, 62, 68, 70, 72–74, 78–80	6, 8, 24, 35, 39, 53, 56, 60, 62, 68, 70, 72–74, 78–80	1, 38, 40, 50, 51, 56–59, 61, 63, 68, 88	50–52, 61, 63, 64, 68	48, 61, 85, 89	6, 24, 39, 41, 47–49, 52, 53, 55, 60, 68
Front End Communication Processor	1–4, 7, 34, 39, 41–44, 46–48	1–5, 7, 34, 37, 39, 41, 42, 49, 89	1, 8, 13, 16, 29, 40, 44, 48, 50, 51, 54, 57, 58, 64, 65, 79, 85	6, 8, 24, 35, 37, 45, 60, 62, 68, 70, 72–74, 78–80	6, 8, 24, 29, 35, 37, 39, 45, 60, 62, 68, 70, 72–74, 78–80	1, 13, 16, 29, 30, 34, 36, 40, 43, 44, 50, 51, 63–65, 81, 88	37, 50, 51, 63–65	43, 48, 85, 89	6, 24, 39, 41, 45, 47, 48, 88
Multiplexer, Concentrator Switch	1–4, 7, 37, 39, 41, 44, 46, 47	1–5, 7, 37, 39, 41, 42, 49, 89	1, 8, 13, 16, 29, 30, 32, 33, 40, 44, 48, 50, 51, 54, 57, 58, 65, 79, 85	6, 8, 24, 35, 37, 45, 60, 62, 68, 70, 72–74, 78–80	6, 8, 24, 29, 35, 37, 39, 45, 60, 62, 68, 70, 72–74, 78–80	1, 13, 16, 29, 30, 32–34, 36, 40, 44, 50, 51, 63–65, 81, 88	37, 50, 51, 63, 64	48, 85, 89	6, 24, 39, 41, 45, 47, 48, 88
Communication Circuits (lines)	12, 26	28, 70, 91	10, 15, 16, 18, 26, 63, 64, 66, 75, 76, 79, 91	25, 28, 68, 70, 75, 76, 78–80, 91	25, 28, 68, 70, 75, 76, 78–80, 91	15, 16, 20, 21, 23, 26, 27, 63, 64, 66–68, 88	63, 64, 66, 68	85	
Local Loop	12	25	25, 75, 85	25, 76	25, 29, 75, 76	68, 88	63, 64, 68	85	
Modems	12, 18	18, 24	8–11, 13–16, 18	24	24, 29	9–11, 13–18, 20, 21, 23, 36, 88	9–11, 14, 15, 63, 64	18–20, 22, 23	
People	5, 39	5, 7, 31, 39, 70	79–87	6, 8, 24, 53, 69–71, 74, 77, 79, 80	6, 8, 24, 29, 53, 69–71, 74, 77, 79, 80	81, 82, 85–87	50, 51, 86, 87	49, 86, 87, 89, 90	6, 88
Terminals/Distributed Intelligence		2		6, 8, 24, 45, 53, 56, 62, 70	6, 8, 24, 29, 45, 53, 56, 62, 70	1, 40, 88	63, 64		6, 24, 45

FIGURE A1-1 Data communication control matrix.

583

- **Privacy**   The accidental or intentional release of data about an individual, assuming that the release of this personal information is improper to the normal conduct of the business at the organization.

- **Security/Theft**   The security or theft of information that should have been kept confidential because of its proprietary nature. This is a form of privacy, but the information removed from the organization does not pertain to an individual. The information might be inadvertently (accidentally) released, or it might be the subject of an outright theft. This exposure also includes the theft of assets such as might be experienced in embezzlement, fraud, or defalcation.

- **Reliability (Uptime)**   The reliability of the data communication network and its "uptime." This includes the organization's ability to keep the data communication network operating, the Mean Time Between Failures (MTBF) at a minimum, and ability to repair equipment when it malfunctions. Reliability of hardware, software, and the maintenance of these two items are chief concerns.

- **Recovery and Restart**   The recovery and restart capabilities of the data communication network, should it fail. In other words, How does the software operate in a failure mode? How long does it take to recover from a failure? This recovery and restart concern also includes backup for key portions of the data communication network and contingency planning for backup, should there be a failure at any point of the data communication network.

- **Error Handling**   The methodologies and controls for handling errors at a remote distributed site or at the centralized computer site. This also may involve the error handling procedures of a distributed data processing system (at the distributed site). The object is to ensure that when errors are discovered they are promptly corrected and reentered into the system for processing.

- **Data Validation and Checking**   The validation of data either at the time of transmission or during transmission. The validation may take place at a remote site (intelligent terminal), at the central site (front end communication processor), or at a distributed intelligence site (concentrator or remote front end communication processor).

RESOURCES/ASSETS (COMPONENTS)

The following resources/assets (components) are those that should be reviewed during the data communication control review. The definitions for each of these assets, listed down the left vertical column of the matrix, are

- **Central System**   Most prevalent in the form of a central computer to which the data communication network transmits and from which it receives information. In a distributed system, with equal processing at each distributed node, there might not be an identifiable central system (just some other equal-sized distributed computer).

- **Software**   The software programs that operate the data communication network. These programs may reside in the central computer, a distributed system computer, the front end communication processor, a remote concentrator or statistical multiplexer, and/or a remote intelligent terminal. This software may include the telecommunication access methods, an overall teleprocessing monitor, programs that reside in the front end processors, and/or programs that reside in the intelligent terminals.

- **Front End Communication Processor**   A hardware device that connects all the data communication circuits (lines) to the central computer or distributed computers and performs a subset of the following functions: code and speed conversion, protocol, error detection and correction, format checking, authentication, data validation, statistical data gathering, polling/ addressing, insertion/deletion of line control codes, and the like.

- **Multiplexer, Concentrator, Switch**   Hardware devices that enable the data communication network to operate in the most efficient manner. The *multiplexer* is a device that combines, in one data stream, several simultaneous data signals from independent stations. The *concentrator* performs the same functions as a multiplexer except that it is intelligent and therefore can perform some of the functions of a front end communication processor. A *switch* is a device that allows the interconnection between any two circuits (lines) connected to the switch. There might be two distinct types of switch: a switch that performs message switching between stations (terminals) might be located within the data communication network facilities that are owned and operated by the organization; a circuit or line switching switch that interconnects various circuits might be located at (and owned by) the telephone company central office. For example, organizations perform message switching, and the telephone company performs circuit switching.

- **Communication Circuits (Lines)**   The common carrier facilities used as links (a *link* is the interconnection of any two stations/terminals) to interconnect the organization's stations/terminals. These communication circuits include, not to the exclusion of others, satellite facilities, public switched dial-up facilities, point-to-point private lines, multiplexed lines, multipoint or loop configured private lines, WATS services, and many others.

- **Local Loop**   The communication facility between the customer's premises and the telephone company's central office or the central office of any other special common carrier. The local loop usually is assumed to be metallic pairs of wires.

- **Modems**   A hardware device used for the conversion of data signals from terminals (digital signal) to an electrical form (analog signal) which is acceptable for transmission over the communication circuits that are owned and maintained by the telephone company or other special common carrier.

- **People**   The individuals responsible for inputting data, operating and maintaining the data communication network equipment, writing the software programs for the data communications, managing the overall data communication network, and those involved at the remote stations/terminals.

- **Terminals/Distributed Intelligence**   Any or all of the input or output devices used to interconnect with the online data communication network. This resource specifically includes, without excluding other devices, teleprinter terminals, video terminals, remote job entry terminals, transaction terminals, intelligent terminals, and any other devices used with distributed data communication networks. These may include microcomputers or minicomputers when they are input/output devices or if they are used to control portions of the data communication network.

CONTROLS/SAFEGUARDS

The following controls/safeguards should be considered when reviewing the data communication network review of an online system. This numerical listing describes each control.

It should be noted that implementation of various controls can be both costly and time-consuming. It is of great importance that a realistic and pragmatic evaluation be made concerning the probability of a specific exposure affecting a specific asset. Only then can the control for safeguarding the asset be evaluated in a cost-effective manner.

The controls, as numerically listed in the cells of the matrix, are

1. Ensure that the system can switch messages destined for a down station/terminal to an alternate station/terminal.

2. Determine whether the system can perform message switching to transmit messages between stations/terminals.

3. Provide a store and forward capability to avoid lost messages in a message switching system. A message destined for a busy station can be stored at the central switch and forwarded when the station is no longer busy.

4. Review the message or transaction logging capabilities to reduce lost messages, provide for an audit trail, restrict messages, prohibit illegal messages, and the like. These messages might be logged at the remote station (intelligent terminal), a remote concentrator/remote front end processor, or the central front end communication processor/central computer.

5. Transmit messages promptly to reduce risk of loss.

6. Identify each message by the individual user's password, the terminal, and the individual message sequence number.

7. Acknowledge the successful or unsuccessful receipt of all messages.

8. Utilize physical security controls throughout the data communication network. This includes the use of locks, guards, badges, sensors, alarms, and administrative measures to protect the physical facilities, data communication networks, and related data communication equipment. These safeguards are required for access monitoring and control to protect data communication equipment and software from damage by accident, fire, and environmental hazard either intentional or unintentional.

9. Consider using modems that have either manual or remote actuated loopback switches for fault isolation to ensure the prompt identification of malfunctioning equipment. These are extremely important in increasing uptime and identifying faults.

10. Use front panel lights on modems to indicate if the circuit/line is functioning properly (carrier signal is up). This may not be a viable alternative with organizations that have hundreds of modems.

11. Consider a modem with alternative voice capabilities for quick troubleshooting between the central site and a major remote site.

12. When feasible, use digital data transmission because it has a lower error rate than analog data transmission.

13. For data communication equipment, check the manufacturer's Mean Time Between Failures (MTBF) to ensure that the data communication equipment has the largest MTBF.

14. Consider placing unused backup modems in critical areas of the data communication network.

15. Consider using modems that have an automatic or semiautomatic dial backup capability in case the leased line fails.

16. Review the maintenance contract and Mean Time to Fix (MTTF) for all data communication equipment. Maintenance should be both fast and available. Determine from where the maintenance is dispatched, and determine if tests can be made from a remote site (for example, modems often have remote loopback capabilities).

17. Increase data transmission efficiency. The faster the modem synchronization time, the lower will be the turnaround time and thus more throughput to the system.

18. Consider modems with automatic equalization (built-in microprocessors for circuit equalization and balancing) to compensate for amplitude and phase distortions on the line. This reduces the number of transmission errors and may decrease the need for conditioned lines.

19. With regard to the efficiency of modems, make certain they have multi-speed switches so the transmission rate can be lowered when line error rates are high.

20. Utilize four-wire circuits in a pseudo-full duplex transmission mode. In other words, keep the carrier wave up in each direction on alternate pairs of wires to reduce turnaround time and gain efficiency during transmission.

21. If needed, use full duplex transmission on two-wire circuits with special modems that split the frequencies and thus achieve full duplex transmission.

22. Increase the speed of transmission. The faster the speed of transmission by the modem, the most cost effective are the data communications, but error rates may increase with speed, so more error detection and correction facilities may be needed.

23. Use a reverse channel capability for control signals (supervisory) and to keep the carrier wave up in both directions.

24. Consider the following special controls on dial-up modems when the data communication network allows incoming dial-up connections: change telephone numbers at regular intervals; keep telephone numbers confidential; remove telephone numbers from the modems in the computer operations area; require that each dial-up terminal have an electronic identification circuit chip to transmit its unique identification to the front end communication processor; do not allow automatic call receipt and connection (always have a person intercept the call to make a verbal identification); have the central site call the various terminals that are allowed connection to the system; utilize dial-out only where an incoming dialed call triggers an automatic dial-back to the caller (in this way the central system controls those telephone numbers to which it allows connection).

25. Physically trace out and, as best as possible, secure the local loop communication circuits/lines within the organization or facility. After these lines leave the facility and enter the public domain, they cannot be physically secured.

26. Consider conditioning voice grade circuits to reduce the number of errors during transmission (may be unnecessary with the newer microprocessor-based modems that perform automatic equalization and balancing).

27. Use four-wire circuits in such a way that there is little to no turnaround time. This can be done by using two wires in each direction and keeping the carrier signal up.

28. Fiber optic (laser) communication circuits can be used within a facility to totally preclude the possibility of wiretapping.

29. Ensure that there is adequate physical security at remote sites, especially for terminals, concentrators, multiplexers, and front end communication processors.

30. Determine whether the multiplexer/concentrator/remote front end hardware has redundant logic and backup power supplies with automatic fallback capabilities in case the hardware fails. This increases uptime of the many stations/terminals that might be connected to this equipment.

31. Consider logging inbound and outbound messages at the remote site.

32. Consider uninterruptible power supplies at large multiplexer/concentrator remote sites.

33. Consider multiplexer/concentrator equipment that has diagnostic lights, diagnostic capabilities, and the like.

34. If a concentrator is used, is it performing some of the controls that usually are performed by the front end communication processor, thereby increasing the efficiency and correctness of data transmissions?

35. See if the polling configuration list can be changed during the day to exclude or include specific terminals. This allows the positive exclusion of a terminal as well as allowing various terminals to come online and offline during the working day.

36. Can front ends, concentrators, modems, and the like handle automatic answering and automatic outward dialing of calls? This capability increases efficiency and accuracy when it is preprogrammed into the system.

37. Ensure that all inbound and outbound messages are logged by the central processor, front end, or remote concentrator to ensure against lost messages, keep track of message sequence numbers (identify illegal messages) and use for system restart should the entire system crash.

38. For efficiency, ensure that the central system can address a group of terminals (group address), several terminals at a time (multiple address), one terminal at a time (single address), or send a broadcast message simultaneously to all stations/terminals in the system.

39. See that each inbound and outbound message is serial numbered, as well as time and date stamped, at the time of logging.

40. Ensure that there is a "time-out" facility so the system does not get hung up trying to poll/address a station. Also, if a particular station "times out" four or five consecutive times, it should be removed from the network configuration polling list so time is not wasted on this station. This improves communication efficiency.

41. Consider having concentrators and front ends perform two levels of editing. In the first level, the front end may add items to a message, reroute the message, or rearrange the data for further transmission. It also may check a message address for accuracy and perform parity checks. In the second level of editing, the concentrator or front end is programmed to perform specific edits of the different transactions that enter the system. This editing is an application system type of editing that deals with message content rather than form and is specific to each application program being executed.

42. Have concentrators, front ends, and central computers handle the message priority system, if one exists. A priority system is used to permit a higher line utilization in certain areas of the network or ensure that certain transactions are handled before other transactions of lesser importance.

43. See that the front end collects message traffic statistics and performs correlations of traffic density and circuit availability. These analyses are mandatory for effective management of a large data communication network. Some of the items included in a traffic density report might be the number of messages handled per hour or per day on each link of the network, number of errors encountered per hour or per day, number of errors encountered per program or per program module, terminals or stations that appear to have a higher than average error record, and the like.

44. Ensure that front ends and concentrators can perform miscellaneous functions such as triggering remote alarms if certain parameters are exceeded, performing internal multiplexing operations, signaling abnormal occurrences to the central computer, slowing up input/output messages when the central computer is overburdened by heavy traffic, and the like.

45. Ensure that concentrators and front ends can validate electronic terminal identification.

46. Ensure that there is a message intercept function for inoperable terminals or invalid terminal addresses.

47. See that messages are checked for valid destination address.

48. Ensure adequate error detection and control capabilities. These might include echo checking where a message is transmitted to a remote site and the remote site echoes the message back for verification, forward error correction where special hardware boxes automatically correct some errors upon receipt of the message, or detection with retransmission. Detection with retransmission is the most common and cost effective form of error detection and correction. This may include identification of errors by reviewing the parity bit or utilizing a special code to identify errors in individual characters during transmission. A more prevalent form is to use a polynomial (mathematical algorithm) to detect errors in message blocks. Whichever way is used, when a message error is detected, it is retransmitted until it is received correctly.

49. When reviewing error detection in transmission, first determine what error rate can be tolerated, then determine the extent and pattern of errors on the communication links used by the organization, and then review the error detection and correction methodologies in use to determine if they are adequate for the application systems using the data communication network. In other words, a purely administrative message network (no critical financial data) does not require error detection and correction capabilities equal to a network that transmits critical financial data.

50. Ensure that there are adequate restart and recovery software routines to recover from such items as a trapped machine check, where instead of bringing down the entire data communication system, a quick recovery can be made and only the one transaction need be retransmitted.

51. Ensure that there are adequate restart and recovery procedures to effect both a warm start and a cold start. In other words, a data communication system should never fail completely so the user has to perform a cold start (start up as if it is a new day, all message counters cleared). The system should go into a warm start procedure, where only parts of the system are disabled and recovery can be made while the system is operating in a degraded mode.

52. Ensure that there is an audit trail logging facility to assist in the reconstruction of data files and transactions from the various stations. There should be the capability to trace back to the terminal end user.

53. Provide some tables for checking for access by terminals, people, databases, and programs. These tables should be in protected areas of memory.

54. Safe store all messages. All transactions/messages should be protected in case of a disastrous situation such as power failure.

55. Protect against concurrent file updates. If the data management software

does not provide this protection, the data communication software should.

56. For convenience, flexibility, and security, ensure that terminals can be brought up or down dynamically while the system is running.

57. Make available a systems trace capability to assist in locating problems.

58. Ensure that the system software documentation is comprehensive.

59. Provide adequate maintenance for the software programs.

60. Ensure that the system supports password protection (multilevel password protection).

61. Identify all default options in the software and their impact if they do not operate properly.

62. For entering sensitive or critical system commands, restrict these commands to one master input terminal and ensure strict physical custody over this terminal. In other words, restrict the personnel who can use this terminal.

63. Ensure that there are adequate recovery facilities and/or capabilities for a software failure, loss of key pieces of hardware, and loss of various communication circuit/lines.

64. Ensure that there are adequate backup facilities (local and remote) to back up key pieces of hardware and communication circuits/lines.

65. Consider backup power capabilities for large facilities such as the central site and various remote concentrators.

66. Consider installing capabilities to fall back to the public dial network from a lease line configuration.

67. When using multidrop or loop circuits, review uptime problems. These types of configurations are more cost effective than point-to-point configurations, but all terminals/stations downline are disconnected when there is a circuit failure close to the central site.

68. Review the physical security (local and remote) for circuits/lines (especially the local loop), hardware, software, physical facilities, storage, media, and the like.

69. For personnel who work in critical or sensitive areas, consider enforcing the following policies: insist that they take at least five consecutive days of vacation per year, check with their previous employers, perform an annual credit check, have them sign hiring agreements stating they will not sell programs, etc.

70. With regard to data security, consider encrypting all messages transmitted.

71. Develop an overall organizational security policy for the data communication network. This policy should cover specifically the security and privacy of information.

72. Ensure that all sensitive communication programs and data are stored in protected areas of memory or disk storage.

73. Ensure that all communication programs or data are stored in areas with adequate physical security when they are offline.

74. Ensure that all communication programs and data are controlled adequately when they are transferred to microfiche.

75. Lock telephone equipment rooms and install alarms on the doors of telephone equipment rooms that contain data communication circuits.

76. Do not put communication lines through the public switchboard unless it is a new electronic switchboard (ESS) and the intent is to gain verbal identification of incoming dial-up data communication calls.

77. Review the communication system's console log that shows "network supervisor terminal commands" such as disable or enable a line or station for input or output, alternately route traffic from one station to another, change the order and/or frequency of line or terminal service (polling, calling, dialing), and the like.

78. Consider packet switching networks that use alternate routes for different packets of information from the same message; this offers a form of security in case someone is intercepting messages.

79. Ensure that there is a policy for the use of test equipment. Modern test equipment may offer a new vulnerability to the organization. This test equipment is connected easily to communication lines, and all messages can be read in English. Test equipment should not be used for monitoring lines "for fun"; it should be locked (key lock or locked hood) when it is not in use and after normal working hours when it is not needed for testing and debugging; programs written for programmable test equipment should be kept locked and out of the hands of those who do not need these programs.

80. Review operational procedures, for example, administrative regulations, policies, and day-to-day activities supporting the security/safeguards of the data communication network. These procedures may include

    • Specifying the objectives of EDP security for an organization, especially as they relate to data communications.

    • Planning for contingencies of security "events," including recording of all exception conditions and activities.

- Assuring management that other safeguards are implemented, maintained, and audited, including background checks, security clearances and hiring of people with adequate security-oriented characteristics; separation of duties; mandatory vacations.
- Developing effective safeguards for deterring, detecting, preventing, and correcting undesirable security events.
- Reviewing the cost effectiveness of the system and related benefits such as better efficiency, improved reliability, and economy.
- Looking for the existence of current administrative regulations, security plans, contingency plans, risk analysis, personnel understanding of management objectives, and then reviewing the adequacy and timeliness of the specified procedures in satisfying these.

81. Review preventive maintenance and scheduled diagnostic testing such as cleaning, replacement, and inspection of equipment to evaluate its accuracy, reliability, and integrity. This may include schedules for testing and repair, adequate testing of software program changes submitted by the vendor, inventories of replacement parts (circuit boards), past maintenance records, and the like.

82. Determine if there is a central site for reporting all problems encountered in the data communication network. This usually results in faster repair time.

83. Review the financial protection afforded by insurance for various hardware, software, and data stored on magnetic media.

84. Review legal contracts with regard to agreements for performing a specific service and a specific costing basis for the data communication network hardware and software. These might include bonding of employees, conflict of interest agreements, clearances, nondisclosure agreements, agreements establishing liability for specific security events by vendors, agreements by vendors not to perform certain acts that would incur a penalty, and the like.

85. Review the organization's fault isolation/diagnostics, including techniques used to ascertain the integrity of the various hardware/software components comprising the total data communication entity. These techniques are used to audit, review, and control the total data communication environment and to isolate the offending elements either on a periodic basis or upon detection of a failure. These techniques may include diagnostic software routines, electrical loopback, test message generation, administrative and personnel procedures, and the like.

86. Review the training and education of employees with regard to the data communication network. Employees must be trained adequately in this

because of the high technical competence required for data communication networks.

87. Ensure that there is adequate documentation, including a precise description of programs, hardware, system configurations, and procedures intended to assist in the prevention, identification, and recovery from problems. The documentation should be detailed enough to assist in reconstructing the system from its parts.

88. Review techniques used for testing to validate hardware and software operation to ensure integrity. Testing, including that of personnel, should reveal departures from the specified operation.

89. Review error recording to reduce lost messages. All message transmission errors should be logged and this log should include the type of error, time and date, terminal operator, and number of times the message was retransmitted before being received correctly.

90. Review the error correction procedures. A user's manual should specify a cross-reference of error messages to the appropriate error code generated by the system. These messages help the user interpret the error that has occurred and suggest corrective action to be taken. Ensure that the errors are, in fact, corrected and the correct data reentered into the system.

91. Consider backing up key circuits/lines. This circuit backup may take the form of a second leased line, modems that have the ability to go to the public dial-up network when a leased line fails, or manual procedures where the remote stations can transmit verbal messages using the public dial-up network.

APPENDIX TWO

CONTROL LISTS FOR DATA COMMUNICATION NETWORKS

The following 14 lists of controls were excerpted from the book, *Designing Controls into Computerized Systems*, by Jerry Fitz-Gerald. This book contains 101 different lists enumerating 2,500 controls for computerized systems. It is available from Jerry FitzGerald & Associates, 506 Barkentine Lane, Redwood City, Calif. 94065.

SOFTWARE CONTROLS, DATA COMMUNICATION

The software programs that operate the data communication network portion of the system may reside in the central computer, a distributed system computer, the front end communication processor, a remote concentrator or statistical multiplexer, and/or a remote intelligent terminal. This software is concerned specifically with telecommunication access methods, telecommunication monitors that may oversee the entire data communication function, any front end communication processor software, or programs that reside in the intelligent terminals. Front end software might be located remotely with regard to the central communication center. This software review also may involve data communication software located at remote concentrator sites or the data communication software located at remote intelligent terminal devices.

1. Ensure that the system can switch properly any messages that are destined for a down station/terminal to an alternate station/terminal.

2. To avoid lost messages in a message switching system, provide a store and forward capability, where a message destined for a busy station is stored at the central switch and then forwarded at a later time when the station is no longer busy.

3. Review message or transaction logging capabilities to reduce lost messages, provide for an audit trail, restrict messages, prohibit illegal messages, and the like. These messages might be logged at the remote station (intelligent terminal), at a remote concentrator/remote front end processor, or at the central front end communication processor/central computer.

4. Identify each message by the individual user's password, the terminal, and the individual message sequence number.

5. Acknowledge the successful or unsuccessful receipt of all messages.

6. See if the polling configuration list can be changed during the day to exclude or include specific terminals. This allows the positive exclusion of a terminal as well as allowing various terminals to come online and offline during the working day.

7. Consider having concentrators and front ends perform two levels of editing. In the first level, the front end may add to a message, reroute the message, or rearrange the data for further transmission. It also may check a message address for accuracy and perform parity checks. In the second level of editing, the concentrator or front end is programmed to perform specific edits to different transactions that enter the system. This editing is an application system type of editing that deals with message content rather than form and is specific to each application program being executed.

8. See that messages are checked for valid destination address.

9. Ensure adequate error detection and control capabilities. These might include echo checking where a message is transmitted to a remote site and the remote site echoes the message back for verification, it might include forward error correction where special hardware boxes automatically correct some errors upon receipt of the message, or it might include detection with retransmission. Detection with retransmission is the most common and cost effective form of error detection and correction. This

may include identification of errors by reviewing the parity bit or utilizing a special code to identify errors in individual characters during transmission. A more prevalent form is to use a polynomial (mathematical algorithm) to detect errors in message blocks. Whichever way is used, when a message error is detected, it is retransmitted until it is received correctly.

10. Ensure that there are adequate restart and recovery software routines to recover from such items as a trapped machine check, where instead of bringing down the entire data communication system, a quick recovery can be made and only one transaction needs to be retransmitted.

11. Ensure that there are adequate restart and recovery procedures to effect both a warm start and a cold start. In other words, a data communication system never should fail completely so the user has to perform a cold start (start up as if it is a new day, all message counters cleared). The system should go into a warm start procedure, where only parts of the system are disabled and recovery can be made while the system is operating in a degraded mode.

12. Ensure that there is an audit trail logging facility to assist in the reconstruction of data files and transactions from the various stations. There should be the capability to trace back to the terminal end user.

13. Provide some tables for checking access by terminals, people, databases, and programs. These tables should be in protected areas of memory.

14. Provide adequate maintenance for the software programs.

15. Identify all default options in the software and their impact if they do not operate properly.

16. Ensure that all sensitive communication programs and data are stored in protected areas of memory or disk storage.

17. Review techniques used for testing to validate the hardware and software operation to ensure integrity. Testing, including that of personnel, should reveal departures from the specified operation.

18. Review error recording to reduce lost messages. All errors in message transmission should be logged. This log should include the type of error, time and date, terminal, circuit, terminal operator, and the number of times the message was retransmitted before it was received correctly.

19. Maintain a checksum count of the bits in the software packages. This permits a quick check to see if there are the same number of bits. If it tallies, the organization probably can rest assured that there have been no software program modifications.

20. When feasible, conduct either source code comparisons or object code

comparisons (some organizations have conducted source-to-object code comparisons). This is to determine if there have been any changes since the last source or object code comparison. This control is very time consuming and expensive because it involves the validation of a specific program on a line-by-line basis. That same program is compared, at some future time, to the validated version.

21. When sensitive software is used at distributed sites, consider downline loading that software from the central site. This provides the assurance that no illegal program changes have been made at the remote site. Also, new programs can be downline loaded every time a vendor conducts maintenance.

22. Use generalized audit software to review various functions of the system's software packages. Distribute these generalized audit software packages to personnel at remote sites. At the central site the auditors or system designers conduct this function.

23. Regularly review the logs of system restarts and accountings of rerun time caused by system malfunctions.

24. Ensure that there is a trouble log regarding software. It should contain the diagnosis of each problem and the person, software component, or device that caused the malfunction. Consider developing statistical reports from these logs and initiate appropriate actions if patterns emerge. Each malfunction should be isolated.

25. Ensure that all security features built into any of the system software packages have been considered. If they are not being used, determine the reason or reasons why.

26. Determine whether there are cleanly programmed and well-defined interfaces between any system software packages such as between operating systems, data communication software, distributed intelligence software, database management systems, and the like.

27. Determine whether the system software programmers have enumerated all the known loopholes in any of the software. They must ascertain the degree of exposure attributable to each loophole and make possible corrections.

28. If the system is running any type of queuing system, such as paging or data communication input/output transactions, review the queues, space management, and other dynamic allocation spaces to ensure that a user cannot get out of his or her address space and violate another user space or the operating system.

29. Force the queuing system to fail to determine if it leaves sensitive information spread throughout the computer system.

30. Following a system catastrophe, ensure that a terminal not logged on before the catastrophe cannot get logged on following the catastrophe without the full authentication sequence.

31. Should a communication circuit fail, ensure that the communication software does not give that open port to the next terminal signing onto the system.

32. Ensure that there is adequate maintenance and vendor support for all system software.

33. Check remote access. Terminal access to a system introduces a new dimension to system software security problems. While these may not be related directly to software security, they do pose the problem that software has the potential of being attacked from a long distance away from the physical area where the software resides. A variety of possible penetration paths may exist, so the software might check the various circuits to ensure that they are the proper circuits and that the proper input/output transactions are being entered on these circuits; consider encryption of the data; limit the use of dial-up modems when they can be connected to highly sensitive and secure software systems; ensure that, when a communication circuit drops, there is an absolute dropping of the computer software so the next person dialing into the system cannot be connected to the software or to another user's program; utilize passwords and various log-in codes, etc.

34. Examine checkpoint/restart modifications. In this case, the penetrator takes a checkpoint of his or her program during its execution and subsequently operates upon the checkpoint file as if it were a regular data file. Because the system must include all status and system registers in the checkpoint file for restart purposes, the penetrator can change effectively the contents of status or system registers (which could not otherwise be modified) by appropriately manipulating the checkpoint file. Then by restarting the program from the modified checkpoint file, the penetrator can execute in a supervisory mode or whatever status desired. Log when a restart is executed or status/systems registers are dumped.

35. Check removal or addition of software code. However achieved, the removal or addition of code from the software can pose a security threat. The use of a checksum bit count with regard to sensitive software packages might be an appropriate measure to ensure against unauthorized program changes.

36. Look for exploitable logic errors. In any major software system there are, at any point in time, some "bugs" or errors. Some of these may be documented but not yet corrected. A logic error may be exploited by a pene-

trator to compromise the integrity of the software. Logic errors should be evaluated as to their potential security risk if they must remain uncorrected for any period of time.

37. Check software generation options. Vendor software packages usually contain many options that can be called upon when generating the software system. All security options must be reviewed, evaluated, and a positive decision made when one of them is not to be used to its fullest extent. Also, any options that are left to the default state should be tested. In other words, the default mechanism should be tested to ensure that it is operating correctly and not branching to some protected or restricted area of memory or someone else's program space.

38. Check interrupt handling. During the handling of interrupt, various parameters are stored. Ensure that these parameters are stored in protected memory space, protected registers, or other protected areas.

DISASTERS AND DISRUPTIONS, DATA COMMUNICATION

Disasters or disruptions refer to either natural or human-created disruption of normal data communication capabilities. They can be either temporary or long term. They render the organization's normal online data communication system inoperable.

1. Ensure that the system can switch messages destined for a down station/terminal to an alternate station/terminal.

2. Use physical security controls throughout the data communication network. This includes the use of locks, guards, badges, sensors, alarms, and administrative measures to protect the physical facilities, data communication networks, and related data communication equipment. These safeguards are required for access monitoring and control to protect data communication equipment and software from damage by accident, fire, and environmental hazard, either intentional or unintentional.

3. Consider using modems that have either manual or remote actuated loopback switches for fault isolation to ensure the prompt identification of malfunctioning equipment. These are extremely important for increasing uptime and identifying faults.

4. Use front panel lights on modems to indicate if the circuit/line is functioning properly (carrier signal is up). This may not be a viable alternative with organizations that have hundreds of modems.

5. Consider a modem with alternate voice capabilities for quick troubleshooting between the central site and a major remote site.

6. For data communication equipment, check the manufacturer's Mean Time Between Failures (MTBF) to ensure that the data communication equipment has the largest MTBF.

7. Consider placing unused backup modems in critical areas of the data communication network.

8. Consider using modems that have an automatic or semiautomatic dial backup capability in case the leased line fails.

9. Review the maintenance contract and Mean Time to Fix (MTTF) for all data communication equipment. Maintenance should be both fast and available. Determine from where the maintenance is dispatched, and determine if tests can be made from a remote site (for example, in many cases modems have remote loopback capabilities).

10. Consider modems with automatic equalization (built-in microprocessors for circuit equalization and balancing) to compensate for amplitude and phase distortions on the line. This reduces the number of transmission errors and may decrease the need for conditioned lines.

11. Physically trace out and, as best as possible, secure the local loop communication circuits/lines within the organization or facility. After these lines leave the facility and enter the public domain, they cannot be physically secured.

12. Consider conditioning voice grade circuits to reduce the number of transmission errors (this may be unnecessary with the newer microprocessor-based modems that perform automatic equalization and balancing).

13. Ensure that there is adequate physical security at remote sites, especially for terminals, concentrators, multiplexers, and front end communication processors.

14. Determine whether the multiplexer/concentrator/remote front end hardware has redundant logic and backup power supplies with automatic fallback capabilities in case the hardware fails. This increases the uptime of the many stations/terminals that might be connected to this equipment.

15. Consider uninterruptible power supplies at large multiplexer/concentrator remote sites.

16. Consider multiplexer/concentrator equipment that has diagnostic lights, diagnostic capabilities, and the like.

17. Ensure that front ends and concentrators can perform miscellaneous functions such as triggering remote alarms if certain parameters are exceeded, performing multiplexing operations internally, signaling abnormal occurrences to the central computer, and slowing up input/output messages when the central computer is overburdened because of heavy traffic.

18. Review the organization's fault isolation/diagnostics, including the techniques used to ascertain the integrity of the various hardware/software components comprising the total data communication entity. These techniques are used to audit, review, and control the total data communication environment and to isolate the offending elements either on a periodic basis or upon detection of a failure. These techniques may include diagnostic software routines, electrical loopback, test message generation, administrative and personnel procedures, and the like.

19. Review the training and education of employees with regard to the data communication network. Employees must be trained adequately in this area because of the high technical competence required for data communication networks.

20. Ensure that there is adequate documentation, including a precise description of programs, hardware, system configuration, and procedures intended to assist in the prevention of problems, identification of problems, and recovery from problems. The documentation should be detailed enough to assist in reconstructing the system from its parts.

21. Ensure that there are adequate restart and recovery software routines to recover from such items as a trapped machine check, where instead of bringing down the entire data communication system, a quick recovery can be made and only the one transaction needs to be retransmitted.

22. Ensure that there are adequate restart and recovery procedures to effect both a warm start and a cold start. In other words, a data communication system should never fail completely so the user has to perform a cold start (start up as if it is a new day, all message counters cleared). The system should go into a warm start procedure, where only parts of the system are disabled and recovery can be made while the system is operating in a degraded mode.

23. Ensure that there is an audit trail logging facility to assist in the reconstruction of data files and transactions from the various stations. There should be the capability to trace back to the terminal end user.

24. Safe store all messages. All transactions/messages should be protected in case of a disastrous situation such as power failure.

25. Make available a systems trace capability to assist in locating problems.

26. Ensure that there are adequate recovery facilities and/or capabilities for a software failure, loss of key pieces of hardware, and loss of various communication circuits/lines.

27. Ensure that there are adequate backup facilities (local and remote) to back up key pieces of hardware and communication circuits/lines.

28. Consider backup power capabilities for large facilities such as the central site and various remote concentrators.

29. Consider installing the capability of falling back to the public dial-up network from a leased-line configuration.

30. Lock telephone equipment rooms and install alarms on the doors of telephone equipment rooms that contain data communication circuits.

31. Do not put communication lines through the public switchboard unless it is a new electronic switchboard (EES) and the intent is to gain verbal identification of incoming dial-up data communication calls.

32. Protect all electrical circuits from malicious vandalism where someone might open the circuits and cut the power. This means providing locking circuit control boxes and locating circuit control boxes in locked rooms.

33. Consider backing up key circuits/lines. This circuit backup may take the form of a second leased line, modems that have the ability to go to the public dial-up network when a leased line fails, or manual procedures whereby remote stations can transmit verbal messages using the public dial-up network.

34. Review operational procedures, for example, the administrative regulations, policies, and day-to-day activities supporting the security/safeguards of the data communication network. These procedures may include

    • Specifying the objectives of EDP security for an organization, especially as they relate to data communications.

    • Planning for contingencies of security "events," including recording of all exception conditions and activities.

    • Assuring management that other safeguards are implemented, maintained, and audited, including background checks, security clearances and hiring people with adequate security-oriented characteristics; separation of duties; mandatory vacations.

    • Developing effective safeguards for deterring, detecting, preventing, and correcting undesirable security events.

    • Reviewing the cost effectiveness of the system and the related benefits such as better efficiency, improved reliability, and economy.

    • Looking for the existence of current administrative regulations, security plans, contingency plans, risk analysis, and personnel understanding of management objectives, and then reviewing the adequacy and timeliness of the specified procedures in satisfying these.

35. Review preventive maintenance and scheduled diagnostic testing such as

cleaning, replacement, and inspection of equipment to evaluate its accuracy, reliability, and integrity. This may include schedules for testing and repair, adequate testing of software program changes submitted by the vendor, inventories of replacement parts (circuit boards), past maintenance records, and the like.

36. Determine whether there is a central site for reporting all problems encountered in the data communication network. This usually results in faster repair time.

MODEMS

A modem is a hardware device used for the conversion of data signals from terminals (digital signal) to a form that is acceptable for transmission over communication circuits that are owned and maintained by the telephone company or other special common carrier.

1. Consider using modems that have either manual or remote actuated loop-back switches for fault isolation to ensure the prompt identification of malfunctioning equipment. These are extremely important to increase uptime and identify faults.

2. Use front panel lights on modems to indicate if the circuit/line is functioning properly (carrier signal is up). This may not be a viable alternative with organizations that have hundreds of modems.

3. Consider a modem with alternate voice capabilities for quick troubleshooting between the central site and a major remote site.

4. When feasible, use digital data transmission because it has a lower error rate than analog data transmission.

5. For data communication equipment, check the manufacturer's Mean Time Between Failures (MTBF) to ensure that the data communication equipment has the largest MTBF.

6. Consider placing unused backup modems in critical areas of the data communication network.

7. Consider using modems that have an automatic or semiautomatic dial backup capability in case the leased line fails.

8. Review maintenance contracts and Mean Time to Fix (MTTF) for all data communication equipment. Maintenance should be both fast and available. Determine from where the maintenance is dispatched, and if tests can be made from a remote site (for example, modems have remote loop-back capabilities).

9. Increase data transmission efficiency. The faster the modem synchronization time, the lower the turnaround time and, thus, more throughput to the system.

10. Consider modems with automatic equalization (built-in microprocessors for circuit equalization and balancing) to compensate for amplitude and phase distortion on the line. They reduce the number of transmission errors and may decrease the need for conditioned lines.

11. With regard to the efficiency of modems, review to see if they have multispeed switches so the transmission rate can be lowered when line error rates are high.

12. Utilize four-wire circuits in a pseudo-full duplex transmission mode. In other words, keep the carrier wave up in each direction on alternate pairs of wires to reduce turnaround time and gain efficiency during transmission.

13. If needed, use full duplex transmission on two-wire circuits with special modems that split the frequencies and thus achieve full duplex transmission.

14. Increase the speed of transmission. The faster the speed of transmission by the modem, the more cost effective the data communications. Since error rates may increase with speed, you may need more error detection and correction facilities.

15. Utilize a reverse channel capability for control signals (supervisory) and to keep the carrier wave up in both directions.

16. Consider the following special controls on dial-up modems when the data communication network allows incoming dial-up connections: change the telephone numbers at regular intervals; keep telephone numbers confidential; remove the telephone numbers from modems in the computer operations area; require that each "dial-up terminal" have an electronic identification circuit chip to transmit its unique identification to the front end communication processor; do not allow automatic call receipt and connection (always have a person intercept the call and make a verbal identification); have the central site call the various terminals that are allowed connection to the system; utilize dial-out only where an incoming dialed call triggers an automatic dial-back to the caller (in this way the central system controls those telephone numbers to which it allows connection).

17. Ensure that there is adequate physical security at remote sites.

18. Ensure that front ends, concentrators, modems, and the like can handle automatic answering and automatic outward dialing of calls. This increases efficiency and accuracy when it is preprogrammed into a system.

19. Review the techniques used for testing to validate the hardware operation to ensure integrity. Testing, including that of personnel, should reveal departures from the specified operation.

MULTIPLEXER, CONCENTRATOR, SWITCH

These three hardware devices enable the data communication network to operate in the most efficient manner. The *multiplexer* is a device that combines several simultaneous data signals from independent stations into one data stream. The *concentrator* performs the same functions as a multiplexer except it is intelligent and therefore can perform some of the functions of a front end communication processor. A *switch* is a device that allows interconnection of any two circuits (lines) connected to the switch. There might be two distinct types of switch: a switch that performs message switching between stations (terminals) might be located within the data communication network facilities that are owned and operated by the organization; a circuit or line switching switch that interconnects various circuits might be located at (and owned by) the telephone company central office. Organizations perform message switching, and the telephone company performs circuit switching.

1. Ensure that the system can switch messages destined for a down station/terminal to an alternate station/terminal.

2. Determine whether the system can perform message switching to transmit messages between stations/terminals.

3. To avoid lost messages in a message switching system, provide a store and forward capability. This is where a message destined for a busy station is stored at the central switch and forwarded when the station is no longer busy.

4. Review the message or transaction logging capabilities to reduce lost messages, provide for an audit trail, restrict messages, prohibit illegal messages, and the like. These messages might be logged at the remote station (intelligent terminal), a remote concentrator/remote front end processor, or the central front end communication processor/central computer.

5. Transmit messages promptly to reduce risk of loss.

6. Identify each message by the individual user's password, the terminal, and the individual message sequence number.

7. Acknowledge the successful or unsuccessful receipt of all messages.

8. Use physical security controls throughout the data communication network. This includes the use of locks, guards, badges, sensors, alarms, and

administrative measures to protect the physical facilities, data communication networks, and related data communication equipment. These safeguards are required for access monitoring and control to protect data communication equipment and software from damage by accident, fire, or environmental hazard, either intentional or unintentional.

9. For data communication equipment, check the manufacturer's Mean Time Between Failures (MTBF) to ensure that the data communication equipment has the largest MTBF.

10. Review the maintenance contract and Mean Time to Fix (MTTF) for all data communication equipment. Maintenance should be both fast and available. Determine from where the maintenance is dispatched, and if tests can be made from a remote site (for example, modems often have remote loopback capabilities).

11. Ensure that front ends, concentrators, modems, and the like handle the automatic answering and automatic outward dialing of calls. This increases efficiency and accuracy when it is preprogrammed into the system.

12. Ensure that all inbound and outbound messages are logged by the central processor, front end, or remote concentrator to ensure against lost messages, keep track of message sequence numbers (identify illegal messages), and use for system restart should the entire system crash.

13. Ensure that there is adequate physical security at remote sites, especially for terminals, concentrators, multiplexers, and front end communication processors.

14. Determine whether the multiplexer/concentrator/remote front end hardware has redundant logic and backup power supplies with automatic fallback capabilities in case the hardware fails. This increases uptime of the many stations/terminals that might be connected to this equipment.

15. Consider uninterruptible power supplies at large multiplexer/concentrator remote sites.

16. Consider multiplexer/concentrator equipment that has diagnostic lights, diagnostic capabilities, and the like.

17. If a concentrator is used, is it performing some of the controls that usually are performed by the front end communication processor, thereby increasing the efficiency and correctness of data transmission?

18. Consider having concentrators and front ends perform two levels of editing. In the first level, the front end may add items to a message, reroute the message, or rearrange the data for further transmission. It also may check a message address for accuracy and perform parity checks. In the

second level of editing, the concentrator or front end is programmed to perform specific edits of the different transactions that enter the system. This editing is an application system type of editing that deals with message content rather than form. It is specific to each application program being executed.

19. Have concentrators, front ends, and central computers handle the message priority system, if one exists. A priority system is used to permit a higher line utilization to certain areas of the network or to ensure that certain transactions are handled before other transactions of lesser importance.

20. Ensure that front ends and concentrators can perform miscellaneous functions such as triggering remote alarms if certain parameters are exceeded, performing multiplexing operations internally, signaling abnormal occurrences to the central computer, slowing up input/output messages when the central computer is overburdened because of heavy traffic, and the like.

21. Ensure that concentrators and front ends can validate electronic terminal identification.

22. See that messages are checked for valid destination address.

23. Ensure adequate error detection and control capabilities. These might include echo checking, where a message is transmitted to a remote site and the remote site echoes the message back for verification; it might include forward error correction, where special hardware boxes automatically correct some errors upon receipt of the message; or it might include detection with retransmission. Detection with retransmission is the most common and cost effective form of error detection and correction. This may include identification of errors by reviewing the parity bit or utilizing a special code to identify errors in individual characters during transmission. A more prevalent form is to utilize a polynomial (mathematical algorithm) to detect errors in message blocks. Whichever way is used, when a message error is detected, it is retransmitted until it is received correctly.

24. Safe store all messages. All transactions/messages should be protected in case of a disastrous situation such as power failure.

25. Ensure that there are adequate backup facilities (local and remote) to back up key pieces of hardware and communication circuits/lines.

26. Consider backup power capabilities for large facilities such as the central site and various remote concentrators.

27. Review the physical security (local and remote) for circuits/lines (espe-

cially the local loop), hardware, software, physical facilities, storage media, and the like.

28. Ensure that all sensitive communication programs and data are stored in protected areas of memory or disk storage.

29. Ensure that there is a policy for the use of test equipment. Modern test equipment may offer a new vulnerability to the organization. This test equipment is connected easily to communication lines, and all messages can be read in English. Test equipment should not be used for monitoring lines "for fun"; it should be locked (key lock or locked hood) when it is not in use and after normal working hours when it is not needed for testing and debugging. Programs written for programmable test equipment should be kept locked and out of the hands of those who do not need these programs.

30. Review operational procedures, for example, the administrative regulations, policies, and day-to-day activities supporting the security/safeguards of the remote site. These procedures may include

   • Specifying the objectives of the EDP security for an organization, especially as they relate to date communications.

   • Planning for contingencies of security "events," including recording of all exception conditions and activities.

   • Assuring management that other safeguards are implemented, maintained, and audited, including background checks, security clearances and hiring of people with adequate security-oriented characteristics; separation of duties; mandatory vacations.

   • Developing effective safeguards for deterring, detecting, preventing, and correcting undesirable security events.

   • Looking for the existence of current administrative regulations, security plans, contingency plans, risk analysis, and personnel understanding of management objectives, and then reviewing the adequacy and timeliness of the specified procedures in satisfying these requirements.

31. Review preventive maintenance and scheduled diagnostic testing such as cleaning, replacement, and inspection of equipment to evaluate its accuracy, reliability, and integrity. This may include schedules for testing and repair, adequate testing of software program changes submitted by the vendor, inventories of replacement parts (circuit boards), past maintenance records, and the like.

32. Review the organization's fault isolation/diagnostics, including techniques used to ascertain the integrity of various hardware/software components comprising the total data communication entity. These

techniques are used to audit, review, and control the total data communication environment and to isolate the offending elements either on a periodic basis or upon detection of a failure. These techniques may include diagnostic software routines, electrical loopback, test message generation, administrative and personnel procedures, and the like.

33. Review techniques used for testing to validate the hardware and software operation to ensure integrity. Testing, including that of personnel, should reveal departures from the specified operation.

34. Review error recording to reduce lost messages. All message transmission errors should be logged. This log should include the type of error, time and date, terminal, circuit, terminal operator, and number of times the message was retransmitted before it was received correctly.

COMMUNICATION CIRCUITS (LINES)

The common carrier facilities are used as links (a link is the interconnection of any two stations/terminals) to interconnect the organization's stations/terminals. These communication circuits include, not to the exclusion of others, satellite facilities, public switched dial-up facilities, point-to-point private lines, multiplexed lines, multipoint or loop configured private lines, WATS service, and many others.

1. When feasible, use digital data transmission because it has a lower error rate than analog data transmission.

2. Review the maintenance contract and Mean Time to Fix (MTTF) for all data communication equipment. Maintenance should be both fast and available. Determine from where the maintenance is dispatched. Also determine if tests can be made from a remote site (for example, modems often have remote loopback capabilities).

3. Consider backing up key circuits/lines. Circuit backup may take the form of a second leased line, modems that have the ability to go to the public dial-up network when a leased line fails, or manual procedures where remote stations can transmit verbal messages using the public dial-up network.

4. Utilize four-wire circuits in a pseudo-full duplex transmission mode. In other words, keep the carrier wave up in each direction on alternate pairs of wires to reduce turnaround time and gain efficiency during transmission.

5. If needed, use full duplex transmission on two-wire circuits with special

modems that split the frequencies, thus achieving full duplex transmission.

6. Utilize a reverse channel capability for control signals (supervisory) and to keep the carrier wave up in both directions.

7. Physically trace out and, as best as possible, secure the local loop communication circuits/lines within the organization or facility. After these lines leave the facility and enter the public domain, they cannot be physically secured.

8. Consider conditioning voice grade circuits to reduce the number of transmission errors. This may be unnecessary with newer microprocessor-based modems that perform automatic equalization and balancing.

9. Fiber optic (laser) communication circuits can be used within a facility to totally preclude the possibility of wiretapping.

10. Ensure that there are adequate recovery facilities for a loss of various communication circuit/lines.

11. Consider installing the capability of falling back to the public dial-up network from a lease-line configuration.

12. When utilizing multidrop or loop circuits, review the uptime problems. These types of configurations are more cost effective than point-to-point configurations, but when there is a circuit failure close to the central site, all terminals/stations downline are disconnected.

13. Review the physical security (local and remote) for circuits/lines (especially the local loop), hardware/software, physical facilities, storage media, and the like.

14. With regard to data security, consider encrypting all messages transmitted.

15. Lock telephone equipment rooms and install alarms on the doors of the telephone equipment rooms that contain data communication circuits.

16. Do not put communication lines through the public switchboard unless it is a new electronic switchboard (ESS) and the intent is to gain verbal identification of incoming dial-up communication calls.

17. Consider packet switching networks that use alternate routes for different packets of information from the same message. This offers a form of security in case someone is intercepting messages.

18. Ensure that there is a policy for the use of test equipment. Modern test equipment may offer a new vulnerability to the organization. This test equipment is connected easily to communication lines, and all messages can be read in English. Test equipment should not be used for monitoring

lines "for fun"; it should be locked (key lock or locked hood) when it is not in use and after normal working hours when it is not needed for testing and debugging. Programs written for programmable test equipment should be kept locked and out of the hands of those who do not need these programs.

19. Review operational procedures, for example, the administrative regulations, policies, and day-to-day activities supporting the security/safeguards of the data communication network. These procedures may include

   - Specifying the objectives of the EDP security for an organization, especially as they relate to data communications.
   - Planning for contingencies of security "events," including recording of all exception conditions and activities.
   - Assuring management that other safeguards are implemented, maintained, and audited, including background checks, security clearances and hiring of people with adequate security-oriented characteristics; separation of duties; mandatory vacations.
   - Developing effective safeguards for deterring, detecting, preventing, and correcting undesirable security events.
   - Reviewing the cost effectiveness of the system and related benefits such as better efficiency, improved reliability, and economy.
   - Looking for the existence of current administrative regulations, security plans, contingency plans, risk analysis, and personnel understanding of management objectives, and then reviewing the adequacy and timeliness of the specified procedures in satisfying these requirements.

ERROR HANDLING, DATA COMMUNICATION

Error handling refers to the methodologies and controls for handling errors at a remote distributed site or at the centralized computer site. This also may involve error handling procedures of a distributed data processing system at the distributed site. It must ensure that when errors are discovered, they are corrected promptly and reentered into the system for processing.

1. To avoid lost messages in a message switching system, provide a store and forward capability. This is where a message destined for a busy station is stored at the central switch and forwarded when the station is no longer busy.

2. Review the message- or transaction-logging capabilities to reduce lost messages, provide for an audit trail, restrict messages, prohibit illegal messages, and the like. These messages might be logged at the remote station (intelligent terminal), a remote concentrator/remote front end processor, or the central front end communication processor/central computer.

3. Ensure that all inbound and outbound messages are logged by the central processor, the front end, or remote concentrator to ensure against lost messages, keep track of message sequence numbers (identify illegal messages), and use for system restart should the entire system crash.

4. See that the front end collects message traffic statistics and performs correlations of traffic density and circuit availability. These analyses are mandatory for the effective management of a large data communication network. Some of the items included in a traffic density report might be the number of messages handled per hour or per day on each link of the network, the number of errors encountered per hour or per day, the number of errors encountered per program or per program module, the terminals or stations that appear to have a higher than average error record, and the like.

5. Ensure adequate error detection and control capabilities. These might include echo checking where a message is transmitted to a remote site and the remote site echoes the message back for verification, forward error correction where special hardware boxes automatically correct some errors upon receipt of the message, or detection with retransmission. Detection with retransmission is the most common and cost effective form of error detection and correction. This may include identification of errors by reviewing the parity bit or using a special code to identify errors in individual characters during transmission. A more prevalent form is to use a polynomial (mathematical algorithm) to detect errors in message blocks. Whichever way is used, when a message error is detected, it is retransmitted until it is received correctly.

6. When reviewing error detection in transmission, first determine what error rate can be tolerated, then determine the extent and pattern of errors on the communication links used by the organization. Finally, review the error detection and correction methodologies in use and determine if they are adequate for the application systems using the data communication network. In other words, a purely administrative message network (no critical financial data) does not require error detection and correction capabilities equal to those of a network that transmits critical financial data.

7. Identify all default options in the software and their impact if they do not operate properly.

8. Review the organization's fault isolation/diagnostics, including the techniques used to ascertain the integrity of the various hardware/software components comprising the total data communication entity. These techniques are used to audit, review, and control the total data communication environment and to isolate offending elements either on a periodic basis or upon detection of a failure. These techniques may include diagnostic software routines, electrical loopback, test message generation, administrative and personnel procedures, and the like.

9. Review the training and education of employees with regard to the data communication network. Employees must be trained adequately in this area because of the high technical competence required for data communication networks.

10. Ensure that there is adequate documentation, including a precise description of programs, hardware, system configurations, and procedures intended to assist in the prevention of problems, identification of problems, and recovery from problems. The documentation should be detailed enough to assist in reconstructing the system from its parts.

11. Review error recording to reduce lost messages. All message transmission errors should be logged. This log should include the type of error, time and date, terminal, circuit, terminal operator, and number of times the message was retransmitted before it was received correctly.

12. Review error correction procedures. A user's manual should specify a cross-reference of error messages to the appropriate error code generated by the system. These messages help the user interpret the error that has occurred and suggest corrective action to be taken. Ensure that the errors are, in fact, corrected and the correct data reentered into the system.

LOCAL LOOP (LINES)

The local loop is the communication facility between the customer's premises and the telephone company's central office or the central office of any other special common carrier. It is assumed to be metallic pairs of wires.

1. When feasible, use digital data transmission because it has a lower error rate than analog data transmission.

2. Physically trace out and, as best as possible, secure the local loop com-

munication circuits/lines within the organization or facility. After these lines leave the facility and enter the public domain, they cannot be physically secured.

3. With regard to data security, consider encrypting all transmitted messages.

4. Ensure that there are adequate recovery facilities for a software failure, loss of key pieces of hardware, and loss of various communication circuits/lines.

5. Ensure that there are adequate backup facilities (local and remote) to back up key communication circuits/lines.

6. Review the physical security (local and remote) for circuits/lines (especially the local loop), hardware, software, physical facilities, storage media, and the like.

7. Lock telephone equipment rooms and install alarms on the doors of the telephone equipment rooms that contain data communication circuits.

8. Do not put communication lines through the public switchboard unless it is a new electronic switchboard (ESS) and the intent is to gain verbal identification of incoming dial-up data communication calls.

9. Review the organization's fault isolation/diagnostics, including the techniques used to ascertain the integrity of the various hardware/software components comprising the total communication entity. These techniques are used to audit, review, and control the total data communication environment and to isolate the offending elements either on a periodic basis or upon detection of a failure. These techniques may include diagnostic software routines, electrical loopback, test message generation, administrative and personnel procedures, and the like.

10. Review the techniques used for testing to validate the hardware and software operation to ensure integrity. Testing, including that of personnel, should reveal departures from the specified operation.

11. Ensure that there is a policy for the use of test equipment. Modern test equipment may offer a new vulnerability to the organization. This test equipment is connected easily to communication lines, and all messages can be read in English. Test equipment should not be used for monitoring lines "for fun"; it should be locked (key lock or locked hood) when it is not in use and after normal working hours when it is not needed for testing and debugging. Programs written for programmable test equipment should be kept locked and out of the hands of those who do not need them.

DATA ENTRY AND VALIDATION, DATA COMMUNICATION

Data entry and validation refers to the validation of data, either at the time of transmission or during transmission. The validation may take place at a remote site with an intelligent terminal, at the central site's front end communication processor, or at a distributed intelligence site's concentrator or remote front end communication processor.

1. Identify each message by the individual user's password, the terminal, and the individual message sequence number.

2. Consider the following special controls on dial-up modems when the data communication network allows incoming dial-up connections: change telephone numbers at regular intervals; keep telephone numbers confidential; remove the telephone numbers from the modems in the computer operations area; require that each "dial-up terminal" have an electronic identification circuit chip to transmit its unique identification to the front end communication processor; do not allow automatic call receipt and connection (always have a person intercept the call to make a verbal identification); have the central site call the various terminals that are allowed connection to the system; utilize dial-out only where an incoming dialed call triggers an automatic dial-back to the caller (in this way the central system controls the telephone numbers to which it allows connection).

3. See that each inbound and outbound message is serial numbered, as well as time and date stamped, at the time of logging.

4. Consider having concentrators and front ends perform two levels of editing. In the first level, the front end may add items to a message, reroute the message, or rearrange the data for further transmission. It also may check a message address for accuracy and perform parity checks. In the second level of editing, the concentrator or front end is programmed to perform specific edits of the different transactions that enter the system. This editing is an application system type of editing that deals with message content rather than form and is specific to each application program being executed.

5. Ensure that concentrators and front ends can validate electronic terminal identification.

6. See that messages are checked for valid destination address.

7. Ensure adequate error detection and control capabilities. These might include echo checking where a message is transmitted to a remote site and the remote site echoes the message back for verification, forward

error correction where special hardware boxes correct some errors automatically upon receipt of the message, or detection with retransmission. Detection with retransmission is the most common and cost effective form of error detection and correction. This may include identification of errors by reviewing the parity bit or by using a special code to identify errors in individual characters during transmission. A more prevalent form is to use a polynomial (mathematical algorithm) to detect errors in message blocks. Whichever way is used, when a message error is detected, it is retransmitted until it is received correctly.

8. When reviewing error detection in transmission, first determine what error rate can be tolerated, then determine the extent and pattern of errors on the communication links used by the organization, and then review the error detection and correction methodologies in use to determine if they are adequate for the application systems using the data communication network. In other words, a purely administrative message network (no critical financial data) does not require error detection and correction capabilities equal to those of a network that transmits critical financial data.

9. Ensure that there is an audit trail logging facility to assist in the reconstruction of data files and transactions from the various stations. There should be the capability to trace back to the terminal end user.

10. Provide some tables for checking for access by terminals, people, databases, and programs. These tables should be in protected areas of memory.

11. Protect against concurrent file updates. If the data management software does not provide this protection, the data communication software should.

12. Ensure that the system supports password protection (multilevel password protection).

13. Review the techniques used for testing to validate hardware and software operation to ensure integrity. Testing, including that of personnel, should reveal departures from the specified operation.

ERRORS AND OMISSIONS, DATA COMMUNICATION

The following controls relate to the accidental or intentional transmission of data that is in error, including the accidental or intentional loss of data or omission of data that should have been entered or transmitted via the online system. This type of threat includes, but is not limited to, inaccurate data, incomplete data, malfunctioning hardware, and the like.

1. Ensure that the system can switch messages destined for a down station/terminal to an alternate station/terminal.

2. Determine whether the system can perform message switching to transmit messages between stations/terminals.

3. To avoid lost messages in a message switching system, provide a store and forward capability. This is where a message destined for a busy station is stored at the central switch and forwarded when the station is no longer busy.

4. Review the message or transaction logging capabilities to reduce lost messages, provide for an audit trail, restrict messages, prohibit illegal messages, and the like. These messages might be logged at the remote station (intelligent terminal), remote concentrator/remote front end processor, or central front end communication processor/central computer.

5. Transmit messages promptly to reduce risk of loss.

6. Acknowledge the successful or unsuccessful receipt of all messages.

7. When feasible, use digital data transmission because it has a lower error rate than analog data transmission.

8. Consider modems with automatic equalization (built-in microprocessors for circuit equalization and balancing) to compensate for amplitude and phase distortions on the line. This reduces the number of errors in transmission and may decrease the need for conditioned lines.

9. Consider conditioning voice grade circuits to reduce the number of transmission errors. This may be unnecessary with the newer microprocessor-based modems that perform automatic equalization and balancing.

10. Ensure that all inbound and outbound messages are logged by the central processor, front end, or remote concentrator to ensure against lost messages, keep track of message sequence numbers (identify illegal messages), and use for system restart should the entire system crash.

11. See that each inbound and outbound message is serial numbered as well as time and date stamped at the time of logging.

12. Consider having concentrators and front ends perform two levels of editing. In the first level, the front end may add items to a message, reroute the message, or rearrange the data for further transmission. It also may check a message address for accuracy and perform parity checks. In the second level of editing, the concentrator or front end is programmed to perform specific edits of the different transactions that enter the system. This editing is an application system type of editing that deals with message content rather than form and is specific to each application program being executed.

13. Have concentrators, front ends, and central computers handle the message priority system, if one exists. A priority system is initiated to permit a higher line utilization to certain areas of the network or to ensure that certain types of transactions are handled before other transactions of lesser importance.

14. See that the front end collects message traffic statistics and performs correlations of traffic density and circuit availability. These analyses are mandatory for effective management of a large data communication network. Some of the items included in a traffic density report might be the number of messages handled per hour or per day on each link of the network, number of errors encountered per hour or per day, number of errors encountered per program or per program module, terminals or stations that appear to have a higher than average error record, and the like.

15. Ensure that front ends and concentrators can perform miscellaneous functions such as triggering remote alarms if certain parameters are exceeded, performing multiplexing operations internally, signaling abnormal occurrences to the central computer, slowing up input/output messages when the central computer is overburdened because of heavy traffic, and the like.

16. Ensure that there is a message intercept function for inoperable terminals or invalid terminal addresses.

17. See that messages are checked for valid destination address.

18. Ensure adequate error detection and control capabilities. These might include echo checking where a message is transmitted to a remote site and the remote site echoes the message back for verification, forward error correction where special hardware boxes automatically correct some errors upon receipt of the message, or detection with retransmission. Detection with retransmission is the most common and cost effective form of error detection and correction. This may include identification of errors by reviewing the parity bit or using a special code to identify errors in individual characters during transmission. A more prevalent form is to use a polynomial (mathematical algorithm) to detect errors in message blocks. Whichever way is used, when a message error is detected, it is retransmitted until it is received correctly.

19. When reviewing error detection in transmission, first determine what error rate can be tolerated; then determine the extent and pattern of errors on the communication links used by the organization. Finally, review the error detection and correction methodologies in use to determine if they are adequate for the application systems using the data communication network. In other words, a purely administrative message network (no

critical financial data) does not require error detection and correction capabilities equal to those of a network that transmits critical financial data.

20. Ensure that there is an audit trail logging facility to assist in the reconstruction of data files and transactions from the various stations. There should be the capability to trace back to the terminal end user.

21. With regard to the efficiency of modems, review to see if they have multispeed switches so the transmission rate can be lowered when line error rates are high.

22. Use four-wire circuits in a pseudo-full duplex transmission mode. In other words, keep the carrier wave up in each direction on alternate pairs of wires to reduce turnaround time and gain efficiency during transmission.

RESTART AND RECOVERY, DATA COMMUNICATION

Should the data communication network fail, it must have restart and recovery capabilities. In other words, how does the software operate in a failure mode? How long does it take to recover from a failure? This restart and recovery threat also includes backup for key portions of the data communication network and contingency planning for backup, should there be a failure at any point in the data communication network.

1. Consider using modems that have either manual or remote actuated loopback switches for fault isolation to ensure the prompt identification of malfunctioning equipment. These are extremely important to increase uptime and to identify faults.

2. Use front panel lights on modems to indicate if the circuit/line is functioning properly (carrier signal is up). This may not be a viable alternative with organizations that have hundreds of modems.

3. Consider a modem with alternate voice capabilities for quick troubleshooting between the central site and a major remote site.

4. Consider placing unused backup modems in critical areas of the data communication network.

5. Consider using modems that have an automatic or semiautomatic dial backup capability in case the leased line fails.

6. Ensure that all inbound and outbound messages are logged by the central processor, front end, or remote concentrator to ensure against lost mes-

sages, keep track of message sequence numbers (identify illegal messages), and use for system restart should the entire system crash.

7. Ensure that there are adequate restart and recovery software routines to recover from such items as a trapped machine check, where instead of bringing down the entire data communication system, a quick recovery can be made and only the one transaction needs to be retransmitted.

8. Ensure that there are adequate restart and recovery procedures to effect both a warm start and a cold start. In other words, a data communication system should never fail completely so the user has to perform a cold start (start up as if it is a new day, all message counters cleared). The system should go into a warm start procedure, where only parts of the system are disabled and recovery can be made while the system is operating in a degraded mode.

9. Ensure that there is an audit trail logging facility to assist in the reconstruction of data files and transactions from the various stations. There should be the capability to trace back to the terminal end user.

10. Identify all default options in the software and their impact if they do not operate properly.

11. Ensure that there are adequate recovery facilities and/or capabilities for a software failure, loss of key pieces of hardware, and loss of various communication circuit/lines.

12. Ensure that there are adequate backup facilities (local and remote) to back up key pieces of hardware and communication circuits/lines.

13. Consider backup power capabilities for large facilities such as the central site and various remote concentrators.

14. Consider installing the capability of falling back to the public dial-up network from a lease line configuration.

15. Review physical security (local and remote) for circuits/lines (especially for local loop), hardware, software, physical facilities, storage media, and the like.

16. Review the training and education of employees with regard to the data communication network. Employees must be trained adequately in this area because of the high technical competence required for data communication networks.

17. Ensure that there is adequate documentation, including a precise description of programs, hardware, system configurations, and procedures intended to assist in the prevention, identification, and recovery from problems. The documentation should be detailed enough to assist in reconstructing the system from its parts.

MESSAGE LOSS OR CHANGE, DATA COMMUNICATION

The following controls relate to the loss of messages as they are transmitted through the data communication system, or the accidental/intentional changing of messages during their transmission.

1. Ensure that the system can switch messages destined for a down station/ terminal to an alternate station/terminal.

2. Determine whether the system can perform message switching to transmit messages between stations/terminals.

3. To avoid lost messages in a message switching system, provide a store and forward capability. This is where a message destined for a busy station is stored at the central switch and forwarded when the station is no longer busy.

4. Review the message or transaction logging capabilities to reduce lost messages, provide for an audit trail, restrict messages, prohibit illegal messages, and the like. These messages might be logged at the remote station (intelligent terminal), remote concentrator/remote front end processor, or central front end communication processor/central computer.

5. Transmit messages promptly to reduce risk of loss.

6. Acknowledge the successful or unsuccessful receipt of all messages.

7. Stringently control circuit test equipment. This specifically covers network monitoring equipment with which to read data going over the communication lines. It is even more important to stringently control the programmable network monitoring equipment. This equipment is necessary, although the personnel authorized to use it should be limited. The uses to which it should be put should clearly be stated. It should be locked when not in use (this locking might involve removing it from the area, installing a keylock switch in place of the on/off switch, or covering it with a lockable metal hood so it is unusable).

8. Review error recording to reduce lost messages. All message transmission errors should be logged, and this log should include the type of error, time and date, terminal, circuit, terminal operator, and number of times the message was retransmitted before it was correctly received.

9. Fiber optic (laser) communication circuits can be used within a facility to totally preclude the possibility of wiretapping.

10. Consider logging inbound and outbound messages at the remote site.

11. Ensure that all inbound and outbound messages are logged by the central processor, front end, or remote concentrator to ensure against lost mes-

sages, keep track of message sequence numbers (identify illegal messages), and use for system restart should the entire system crash.

12. See that each inbound and outbound message is serial numbered, as well as time and date stamped, at the time of logging.

13. Consider having concentrators and front ends perform editing. The front end may check a message address for accuracy and perform parity checks.

14. Ensure adequate error detection and control capabilities. These might include echo checking where a message is transmitted to a remote site and the remote site echoes the message back for verification, forward error correction where special hardware boxes automatically correct some errors upon receipt of the message, or detection with retransmission. Detection with retransmission is the most common and cost-effective form of error detection and correction. This may include identification of errors by reviewing the parity bit or using a special code to identify errors in individual characters during transmission. A more prevalent form is to use a polynomial (mathematical algorithm) to detect errors in message blocks. Whichever way is used, when a message error is detected, it is retransmitted until it is received correctly.

15. Ensure that there is an audit trail logging facility to assist in the reconstruction of data files and transactions from the various stations. There should be the capability to trace back to the terminal end user.

16. Safe store all messages. All transactions/messages should be protected in case of a disastrous situation such as power failure.

PEOPLE CONTROLS, DATA COMMUNICATION

Individuals are responsible for managing, operating, and maintaining the data communication network and its equipment; writing software programs for the data communication system; and working at the remote stations/terminals.

1. Identify each message by the individual user's password, terminal, and individual message sequence number.

2. Use physical security controls throughout the data communication network. This includes the use of locks, guards, badges, sensors, alarms, and administrative measures to protect the physical facilities, data communication networks, and related data communication equipment. These safeguards are required for access monitoring and control to protect data communication equipment and software from damage by accident, fire, or environmental hazard, either intentional or unintentional.

3. Consider the following special controls on dial-up modems when the data communication network allows incoming dial-up connections: change telephone numbers at regular intervals; keep telephone numbers confidential; remove telephone numbers from the modems in the computer operations area; require that each "dial-up terminal" have an electronic identification circuit chip to transmit its unique identification to the front end communication processor; do not allow automatic call receipt and connection (always have a person intercept the call to make a verbal identification); have the central site call the various terminals that are allowed connection to the system; use dial-out only where an incoming dialed call triggers an automatic dial-back to the caller (in this way the central system controls the telephone numbers to which it allows connection).

4. Ensure that there is adequate physical security at remote sites, especially for terminals, concentrators, multiplexers, and front end communication processors.

5. See that each inbound and outbound message is serial numbered, as well as time and date stamped, at the time of logging.

6. Provide some tables for checking for access by terminals, people, databases, and programs. These tables should be in protected areas of memory.

7. For personnel who work in critical or sensitive areas, consider enforcing the following policies: insist that they take at least five consecutive days of vacation per year, check with their previous employers, perform an annual credit check, and have them sign hiring agreements stating that they will not sell programs, etc.

8. Review operational procedures, for example, the administrative regulations, policies, and day-to-day activities supporting the security/safeguards of the data communication network. These procedures may include:

    •   Specifying the objectives of EDP security for an organization, especially as they relate to data communications.

    •   Planning for contingencies of security "events," including recording of all exception conditions and activities.

    •   Assuring management that other safeguards are implemented, maintained, and audited, including background checks, security clearances and hiring of people with adequate security-oriented characteristics; separation of duties; mandatory vacations.

    •   Developing effective safeguards for deterring, detecting, preventing, and correcting undesirable security events.

- Reviewing the cost effectiveness of the system and related benefits such as better efficiency, improved reliability, and economy.
- Looking for the existence of current administrative regulations, security plans, contingency plans, risk analysis, and personnel understanding of management objectives, and then reviewing the adequacy and timeliness of the specified procedures in satisfying these.

9. Review the communication system's console log that shows "network supervisor terminal commands," such as disable or enable a line or station for input or output, alternately route traffic from one station to another, change the order and/or frequency of line or terminal service (polling, calling, dialing), and the like.

10. Ensure that there is a policy for the use of test equipment. Modern test equipment may offer a new vulnerability to the organization. This test equipment is connected easily to communication lines, and all messages can be read in English. Test equipment should not be used for monitoring lines "for fun"; it should be locked (key lock or locked hood) when not in use and after normal working hours when it is not needed for testing and debugging. Programs written for programmable test equipment should be kept locked and out of the hands of those who do not need these programs.

11. Review the training and education of employees with regard to the data communication network. Employees must be trained adequately in this area because of the high technical competence required for data communication networks.

12. Review error correction procedures. A user's manual should specify a cross-reference of error messages to the appropriate error code generated by the system. These messages help the user interpret the error that has occurred and suggest corrective action to be taken. Ensure that errors are, in fact, corrected and that the corrected data are reentered into the system.

13. Provide personnel with an education program in security matters so the responsibility for security is firmly fixed and clearly understood by those having such responsibilities.

FRONT END COMMUNICATION PROCESSOR

This hardware device interconnects all the data communication circuits (lines) to the central computer or distributed computer. It performs a subset of the following functions: error detection and correction, logging, message switching, store and forward, statistical data gathering, polling/addressing, insertion/deletion of line control codes, and the like.

1. Ensure that the system can switch messages destined for a down station/terminal to an alternate station/terminal.

2. Determine whether the system can perform message switching to transmit messages between stations/terminals.

3. To avoid lost messages in a message switching system, provide a store and forward capability. This is where a message destined for a busy station is stored at the central switch and forwarded when the station is no longer busy.

4. Review the message or transaction logging capabilities to reduce lost messages, provide for an audit trail, restrict messages, prohibit illegal messages, and the like. These messages might be logged at the remote station (intelligent terminal), remote concentrator/remote front end processor, or central front end communication processor/central computer.

5. Transmit messages promptly to reduce risk of loss.

6. Identify each message by the individual user's password, terminal, and individual message sequence number.

7. Acknowledge the successful or unsuccessful receipt of all messages.

8. For data communication equipment, check the manufacturer's Mean Time Between Failures (MTBF) to ensure that the data communication equipment has the largest MTBF.

9. Review the maintenance contract and Mean Time to Fix (MTTF) for all data communication equipment. Maintenance should be both fast and available. Determine from where the maintenance is dispatched and whether tests can be made from a remote site (for example, modems often have remote loopback capabilities).

10. Ensure that there is adequate physical security at remote sites, especially for terminals, concentrators, multiplexers, and front end communication processors.

11. Determine whether the multiplexer/concentrator/remote front end hardware has redundant logic and backup power supplies with automatic fallback capabilities in case the hardware fails. This increases uptime of the many stations/terminals that might be connected to this equipment.

12. If a concentrator is used, is it performing some of the controls that usually are performed by the front end communication processor, thereby increasing the efficiency and correctness of data transmissions?

13. See if the polling configuration list can be changed during the day to

exclude or include specific terminals. This allows the positive exclusion of a terminal, as well as allowing various terminals to come online and offline during the working day.

14. Can front ends, concentrators, modems, and the like handle the automatic answering and automatic outward dialing of calls? This increases efficiency and accuracy when it is preprogrammed into the system.

15. Ensure that all inbound and outbound messages are logged by the central processor, front end, or remote concentrator to ensure against lost messages, keep track of message sequence numbers (identify illegal messages), and use for system restart should the entire system crash.

16. See that each inbound and outbound message is serial numbered, as well as time and date stamped, at the time of logging.

17. Ensure that there is a "time-out" facility so the system does not get hung up trying to poll/address a station. Also, if a particular station "times out" four or five consecutive times, it should be removed from the network configuration polling list so time is not wasted on this station (improves communication efficiency).

18. Consider having concentrators and front ends perform two levels of editing. In the first level, the front end may add items to a message, reroute the message, or rearrange the data for further transmission. It also may check a message address for accuracy and perform parity checks. In the second level of editing, the concentrator or front end is programmed to perform specific edits of the different transactions that enter the system. This editing is an application system type of editing that deals with message content rather than form. It is specific to each application program being executed.

19. Have concentrators, front ends, and central computers handle the message priority system, if one exists. A priority system is used to permit a higher line utilization to certain areas of the network or to ensure that certain transactions are handled before other transactions of lesser importance.

20. See that the front end collects message traffic statistics and performs correlations of traffic density and circuit availability. These analyses are mandatory for effective management of a large data communication network. Some of the items included in a traffic density report might be the number of messages handled per hour or per day on each link of the network, number of errors encountered per hour or per day, number of errors encountered per program or per program module, terminals or stations that appear to have a higher than average error record, and the like.

21. Ensure that front ends and concentrators can perform miscellaneous func-

tions such as triggering remote alarms if certain parameters are exceeded, performing multiplexing operations internally, signaling abnormal occurrences to the central computer, slowing up input/output messages when the central computer is overburdened because of heavy traffic, and the like.

22. Ensure that concentrators and front ends can validate electronic terminal identification.

23. Ensure that there is a message intercept function for inoperable terminals or invalid terminal addresses.

24. See that messages are checked for valid destination address.

25. Ensure adequate error detection and control capabilities. These might include echo checking where a message is transmitted to a remote site and the remote site echoes the message back for verification, forward error correction where special hardware boxes automatically correct some errors upon receipt of the message, or detection with retransmission. Detection with retransmission is the most common and cost effective form of error detection and correction. This may include identification of errors by reviewing the parity bit or using a special code to identify errors in individual characters during transmission. A more prevalent form is to use a polynomial (mathematical algorithm) to detect errors in message blocks. Whichever way is used, when a message error is detected, it is retransmitted until it is received correctly.

26. When reviewing error detection in transmission, first determine what error rate can be tolerated, and then determine the extent and pattern of errors in the communication links used by the organization. Finally, review error detection and correction methodologies in use to determine if they are adequate for the application system utilizing the data communication network. In other words, a purely administrative message network (no critical financial data) does not require error detection and correction capabilities equal to those of a network that transmits critical financial data.

27. Ensure that there are adequate restart and recovery software routines to recover from such items as a trapped machine check, where instead of bringing down the entire data communication system, a quick recovery can be made and only the one transaction need be retransmitted.

28. Ensure that there are adequate restart and recovery procedures to effect both a warm start and a cold start. In other words, a data communication system should never fail completely so the user has to perform a cold start (start up as if it is a new day, all message counters cleared). The system should go into a warm start procedure, where only parts of the system are

disabled and recovery can be made while the system is operating in a degraded mode.

29. Ensure that front end documentation software is comprehensive.

30. For entering sensitive or critical system commands, restrict these commands to one master input terminal and ensure strict physical custody over this terminal. In other words, restrict the personnel who can use this terminal.

31. Ensure that there are adequate recovery facilities for loss of key pieces of hardware, and loss of various communication circuit/lines.

32. Ensure that there are adequate backup facilities (local and remote) to back up key pieces of hardware and communication circuits/lines.

33. Consider backup power capabilities for large facilities such as the central site and various remote concentrators.

34. Review preventive maintenance and scheduled diagnostic testing such as cleaning, replacement, and inspection of equipment to evaluate its accuracy, reliability, and integrity. This may include schedules for testing and repair, adequate testing of software program changes submitted by the vendor, inventories of replacement parts (circuit boards), past maintenance records, and the like.

35. Review the organization's fault isolation/diagnostics, including techniques used to ascertain the integrity of various hardware/software components comprising the total data communication entity. These techniques are used to audit, review, and control the total data communication environment and to isolate the offending elements either on a periodic basis or upon detection of a failure. These techniques may include diagnostic software routines, electric loopback, test message generation, administrative and personnel procedures, and the like.

36. Review error recording to reduce lost messages. All message transmission errors should be logged. This log should include the type of error, time and date, terminal, circuit, terminal operator, and number of times the message was retransmitted before it was received correctly.

RELIABILITY (UPTIME), DATA COMMUNICATION

The reliability of the data communication network and its uptime includes the organization's ability to keep the data communication network operating, keep Mean Time Between Failures (MTBF) at a minimum, and minimize the time to repair equipment when it malfunctions. Low reliability/breakdown of hardware, reliability of software, and the maintenance of these two items are the chief threats here.

1. Ensure that the system can switch messages destined for a down station/ terminal to an alternate station/terminal.

2. Consider using modems that have either manual or remote actuated loop-back switches for fault isolation to ensure the prompt identification of malfunctioning equipment. These are extremely important to increase uptime and identify faults.

3. Use front panel lights on modems to indicate if the circuit/line is functioning properly (carrier signal is up). This may not be a viable alternative with organizations that have hundreds of modems.

4. Consider a modem with alternate voice capabilities for quick trouble-shooting between the central site and a major remote site.

5. For data communication equipment, check the manufacturer's Mean Time Between Failures (MTBF) to ensure that the data communication equipment has the largest MTBF.

6. Consider placing unused backup modems in critical areas of the data communication network.

7. Consider using modems that have an automatic or semiautomatic dial backup capability in case the leased line fails.

8. Review the maintenance contract and Mean Time to Fix (MTTF) for all data communication equipment. Maintenance should be both fast and available. Determine from where the maintenance is dispatched, and determine if tests can be made from a remote site (for example, in many cases modems have remote loopback capabilities).

9. Increase data transmission efficiency. The faster the modem synchronization time, the lower will be the turnaround time and thus more throughput to the system.

10. Consider modems with automatic equalization (built-in microprocessors for circuit equalization and balancing) to compensate for amplitude and phase distortions on the line. This reduces the number of transmission errors and may decrease the need for conditioned lines.

11. Use four-wire circuits in a pseudo-full duplex transmission mode. In other words, keep the carrier wave up in each direction on alternate pairs of wires to reduce turnaround time and gain efficiency during transmission.

12. If needed, use full duplex transmission on two-wire circuits with special modems that split the frequencies, thereby achieving full duplex transmission.

13. Use a reverse channel capability for control signals (supervisory) and to keep the carrier wave up in both directions.

14. Consider conditioning voice grade circuits to reduce the number of errors during transmission. This may be unnecessary with the newer microprocessor-based modems that perform automatic equalization and balancing.

15. Use four-wire circuits in such a way that there is little or no turnaround time. This can be done by using two wires in each direction and keeping the carrier signal up.

16. Determine whether the multiplexer/concentrator/remote front end hardware has redundant logic and backup power supplies with automatic fallback facilities in case the hardware fails. This increases uptime of the many stations/terminals that might be connected to this equipment.

17. Consider uninterruptible power supplies at large multiplexer/concentrator remote sites.

18. Consider multiplexer/concentrator equipment that has diagnostic lights, diagnostic capabilities, and the like.

19. If a concentrator is used, is it performing some of the controls that usually are performed by the front end communication processor, thereby increasing the efficiency and correctness of data transmissions?

20. Can the front ends, concentrators, modems, and the like handle automatic answering and automatic outward handling of calls? This increases efficiency and accuracy when it is preprogrammed into the system.

21. For efficiency, ensure that the central system can address a group of terminals (group address), several terminals at a time (multiple address), or one terminal at a time (single address), or can send a broadcast message simultaneously to all stations/terminals in the system.

22. Ensure that there is a "time-out" facility so the system does not get hung up trying to poll/address a station. Also, if a particular station "times out" four or five consecutive times, it should be removed from the network configuration polling list so time is not wasted on this station (improves communication efficiency).

23. See that the front end collects message traffic statistics and performs correlations of traffic density and circuit availability. These analyses are mandatory for effective management of a large data communication network. Some of the items included in a traffic density report might be the number of messages handled per hour or per day on each link of the network, number of errors encountered per hour or per day, number of errors encountered per program or per program module, terminals or stations that appear to have a higher than average error record, and the like.

24. Ensure that front ends and concentrators can perform miscellaneous functions such as triggering remote alarms if certain parameters are exceeded, performing internal multiplexing operations, signaling abnormal occurrences to the central computer, slowing up input/output messages when the central computer is overburdened because of heavy traffic, and the like.

25. Ensure that there are adequate restart and recovery software routines to recover from such items as a trapped machine check, where instead of bringing down the entire data communication system, a quick recovery can be made and only the one transaction need be retransmitted.

26. Ensure that there are adequate restart and recovery procedures to effect both a warm start and a cold start. In other words, a data communication system should never fail completely so the user has to perform a cold start (start up as if it is a new day, all message counters cleared). The system should go into a warm start procedure, where only parts of the system are disabled and recovery can be made while the system is operating in a degraded mode.

27. Make available a systems trace capability to assist in locating problems.

28. Ensure that system software documentation is comprehensive.

29. Provide adequate maintenance for the software programs.

30. Identify all default options in the software and their impact if they do not operate properly.

31. Ensure that there are adequate recovery facilities and/or capabilities for a software failure, loss of key pieces of hardware, and loss of various communication circuits/lines.

32. Ensure that there are adequate backup facilities (local and remote) to back up key pieces of hardware and communication circuits/lines.

33. Consider backup power capabilities for large facilities such as the central site and various remote concentrators.

34. Consider installing the capability of falling back to the public dial-up network from a lease line configuration.

35. When using multidrop or loop circuits, review uptime problems. These types of configurations are more cost effective than point-to-point configurations, but all terminals/stations downline are disconnected when there is a circuit failure close to the central site.

36. Review the physical security (local and remote) for circuits/lines (especially the local loop), hardware, software, physical facilities, storage media, and the like.

37. Review preventive maintenance and scheduled diagnostic testing such as cleaning, replacement, and inspection of equipment to evaluate its accuracy, reliability, and integrity. This may include schedules for testing and repair, adequate testing of software program changes submitted by the vendor, inventories of replacement parts (circuit boards), past maintenance records, and the like.

38. Determine if there is a central site for reporting all problems encountered in the data communication network. This usually results in faster repair time.

39. Review the organization's fault isolation/diagnostics, including techniques used to ascertain the integrity of various hardware/software components comprising the total data communication entity. These techniques are used to audit, review, and control the total data communication environment and to isolate the offending elements either on a periodic basis or upon detection of a failure. These techniques may include diagnostic software routines, electrical loopback, test message generation, administrative and personnel procedures, and the like.

40. Review the training and education of employees with regard to the data communication network. Employees must be trained adequately in this area because of the high technical competence required for data communication networks.

41. Ensure that there is adequate documentation, including a precise description of programs, hardware, system configurations, and procedures intended to assist in the prevention, identification, and recovery from problems. The documentation should be detailed enough to assist in constructing the system from its parts.

42. Review the techniques used for testing to validate hardware and software operation to ensure integrity. Testing, including that of personnel, should reveal departures from the specified operation.

APPENDIX THREE

VERTICAL AND HORIZONTAL COORDINATES

This appendix contains vertical and horizontal coordinates for several hundred cities. They can be used to calculate the mileages in design problems. The calculation method was explained in Chapter 11 (see Figure 11-11). There are thousands of these *rate centers* because each central/end office in a city is a rate center. This appendix includes only one set of V and H coordinates for each city because it is not necessary to have all end offices when learning how to calculate the air mileage between end office locations (rate centers) and when learning how to design network layouts.

	V	H		V	H
Abilene, Tex.	8698 – 4513		Bridgeport, Conn.	4841 – 1360	
Akron, Ohio	5637 – 2472		Bristow, Okla.	7799 – 4216	
Albany, Ga.	7649 – 1817		Brockton, Mass.	4465 – 1205	
Albany, N.Y.	4639 – 1629		Buffalo, N.Y.	5075 – 2326	
Albuquerque, N. Mex.	8549 – 5887		Buffalo Peace		
Alexandria, La.	8409 – 3168		Bridge, N.Y.	5074 – 2334	
Allentown, Pa.	5166 – 1585		Burlington, Iowa	6449 – 3829	
Altoona, Pa.	5460 – 1972		Burlington, Vt.	4270 – 1808	
Amarillo, Tex.	8266 – 5076		Calais, Me.	3561 – 1208	
Anaheim, Calif.	9250 – 7810		Cambridge, Mass.	4425 – 1258	
Anniston, Ala.	7406 – 2304		Camden, N.J.	5249 – 1453	
Antonia, Mo.	6880 – 3507		Canton, Ohio	5676 – 2419	
Apollo, Tex.	8958 – 3482		Cape Girardeau, Mo.	7013 – 3251	
Appleton, Wis.	5589 – 3776		Carson City, Nev.	8139 – 8306	
Asheville, N.C.	6749 – 2001		Casper, Wyo.	6918 – 6297	
Atlanta, Ga.	7260 – 2083		Casselton, N. Dak.	5633 – 5241	
Atlantic City, N.J.	5284 – 1284		Cedar Rapids, Iowa	6261 – 4021	
Augusta, Ga.	7089 – 1674		Centralia, Ill.	6744 – 3311	
Austin, Tex.	9005 – 3996		Champaign-Urbana, Ill.	6371 – 3336	
Baker, Calif.	8888 – 7537		Charleston, S.C.	7021 – 1281	
Bakersfield, Calif.	8947 – 8060		Charleston, W. Va.	6152 – 2174	
Baltimore, Md.	5510 – 1575		Charlotte, N.C.	6657 – 1698	
Baton Rouge, La.	8476 – 2874		Chattanooga, Tenn.	7098 – 2366	
Beaumont, Tex.	8777 – 3344		Cheshire, Conn.	4755 – 1366	
Beckley, W. Va.	6218 – 2043		Chesterfield, Mass.	4595 – 1478	
Benton Ridge, Ohio	5847 – 2784		Cheyenne, Wyo.	7203 – 5958	
Berlin, N.J.	5257 – 1408		Chicago, Ill.	5986 – 3426	
Bethia, Va.	5957 – 1491		Chico, Calif.	8057 – 8668	
Billings, Mont.	6391 – 6790		Chipley, Fla.	7927 – 1958	
Biloxi, Miss.	8296 – 2481		Cincinnati, Ohio	6263 – 2679	
Binghampton, N.Y.	4943 – 1837		Clarksburg, W. Va.	5865 – 2095	
Birmingham, Ala.	7518 – 2446		Clarksville, Tenn.	6988 – 2837	
Bismarck, N. Dak.	5840 – 5736		Clearwater, Fla.	8203 – 1206	
Blacksburg, Va.	6247 – 1867		Cleveland, Ohio	5574 – 2543	
Bloomington, Ind.	6417 – 2984		Cocoa, Fla.	7925 – 0903	
Blue Ridge Summit, Pa.	5518 – 1746		Collinsville, Ill.	6781 – 3455	
Boise, Idaho	7096 – 7869		Colorado Springs, Colo.	7679 – 5813	
Boone, Iowa	6394 – 4355		Columbia, S.C.	6901 – 1589	
Boston, Mass.	4422 – 1249		Columbus, Ga.	7556 – 2045	
Brewton, Ala.	8001 – 2244		Columbus, Miss.	7657 – 2704	

	V – H		V – H
Columbus, Ohio	5972 – 2555	Fort Lauderdale, Fla.	8282 – 0557
Concord, N.H.	4326 – 1426	Fort Morgan, Colo.	7335 – 5739
Conyers, Ga.	7243 – 2016	Fort Myers, Fla.	8359 – 0904
Corpus Christi, Tex.	9475 – 3739	Fort Pierce, Fla.	8054 – 0737
Crestview, Fla.	8025 – 2128	Fort Walton Beach, Fla.	8097 – 2097
Crosby, N. Dak.	5495 – 6199	Fort Wayne, Ind.	5942 – 2982
Dallas, Tex.	8436 – 4034	Fort Worth, Tex.	8479 – 4122
Danville, Ky.	6558 – 2561	Framingham, Mass.	4472 – 1284
Davenport, Iowa	6273 – 3817	Frankfort, Ky.	6462 – 2634
Dayton, Ohio	6113 – 2705	Fresno, Calif.	8669 – 8239
Daytona Beach, Fla.	7791 – 1052	Gainesville, Fla.	7838 – 1310
Decatur, Ala.	7324 – 2585	Gastonia, N.C.	6683 – 1754
De Kalb, Ill.	6061 – 3591	Glenwood Springs, Colo.	7651 – 6263
Delta, Utah	7900 – 7114	Grand Forks, N. Dak.	5420 – 5300
Denver, Colo.	7501 – 5899	Grand Island, Nebr.	6901 – 4936
Des Moines, Iowa	6471 – 4275	Grand Junction, Colo.	7804 – 6438
Detroit, Mich.	5536 – 2828	Grand Rapids, Mich.	5628 – 3261
Dickinson, N. Dak.	5922 – 6024	Greeley, Colo.	7345 – 5895
Dodge City, Kans.	7640 – 4958	Green Bay, Wis.	5512 – 3747
Dodgeville, Wis.	5963 – 3890	Greensboro, N.C.	6400 – 1638
Dover, Del.	5429 – 1408	Greenville, Miss.	7888 – 3126
Duluth, Minn.	5352 – 4530	Greenville, N.C.	6250 – 1226
Eau Claire, Wis.	5698 – 4261	Greenville, S.C.	6873 – 1894
El Paso, Tex.	9231 – 5655	Greenwood, Miss.	7798 – 2993
Ennis, Tex.	8514 – 3970	Gulfport, Miss.	8317 – 2511
Eureka, Calif.	7856 – 9075	Hackensack, N.J.	4976 – 1432
Evansville, Ind.	6729 – 3019	Harlingen, Tex.	9820 – 3663
Fairmont, W. Va.	5808 – 2091	Harrisburg, Pa.	5363 – 1733
Fairview, Kans.	6956 – 4443	Hartford, Conn.	4687 – 1373
Fall River, Mass.	4543 – 1170	Hattiesburg, Miss.	8152 – 2636
Fargo, N. Dak.	5615 – 5182	Hayward, Calif.	8513 – 8660
Fayetteville, Ark.	7600 – 3872	Helena, Mont.	6336 – 7348
Fayetteville, N.C.	6501 – 1385	Herndon, Va.	5644 – 1640
Findlay, Ohio	5828 – 2766	Hinsdale, Ill.	6023 – 3461
Fitzgerald, Ga.	7539 – 1684	Hot Springs, Ark.	7827 – 3554
Flagstaff, Ariz.	8746 – 6760	Houghton, Mich.	5052 – 4088
Flint, Mich.	5461 – 2993	Houston, Tex.	8938 – 3536
Florence, S.C.	6744 – 1417	Huntington, N.Y.	4918 – 1349
Forrest City, Ark.	7555 – 3232	Huntington, W. Va.	6212 – 2299
Fort Collins, Colo.	7331 – 5965		

	V	H		V	H
Huntsville, Ala.	7267 – 2535		Lodi, Calif.	8397 – 8532	
Huron, S. Dak.	6201 – 5183		Longview, Tex.	8348 – 3660	
Indianapolis, Ind.	6272 – 2992		Logan, Utah	7367 – 7102	
Iowa City, Iowa	6313 – 3972		Los Angeles, Calif.	9213 – 7878	
Iron Mountain, Mich.	5266 – 3890		Louisville, Ky.	6529 – 2772	
Jackson, Mich.	5663 – 3009		Lubbock, Tex.	8598 – 4962	
Jackson, Miss.	8035 – 2880		Lynchburg, Va.	6093 – 1703	
Jackson, Tenn.	7282 – 2976		Lyons, Nebr.	6584 – 4732	
Jacksonville, Fla.	7649 – 1276		Macon, Ga.	7364 – 1865	
Jasper, Ala.	7497 – 2553		Madison, Wis.	5887 – 3796	
Johnson City, Tenn.	6595 – 2050		Madisonville, Ky.	6845 – 2942	
Joliet, Ill.	6088 – 3454		Manchester, N.H.	4354 – 1388	
Joplin, Mo.	7421 – 4015		Manhattan, Kans.	7143 – 4520	
Julian, Calif.	9374 – 7544		Marion, Ill.	6882 – 3202	
Kalamazoo, Mich.	5749 – 3177		Mattoon, Ill.	6502 – 3291	
Kansas City, Kans.	7028 – 4212		McComb, Miss.	8262 – 2823	
Kansas City, Mo.	7027 – 4203		Medford, Oreg.	7503 – 8892	
Kennewick, Wash.	6595 – 8391		Memphis, Tenn.	7471 – 3125	
Key West, Fla.	8745 – 0668		Meridian, Miss.	7899 – 2639	
Kingsport, Tenn.	6570 – 2107		Miami, Fla.	8351 – 0527	
Klamath Falls, Oreg.	7510 – 8711		Midland, Tex.	8934 – 4888	
Knoxville, Tenn.	6801 – 2251		Milwaukee, Wis.	5788 – 3589	
La Crosse, Wis.	5874 – 4133		Minneapolis, Minn.	5781 – 4525	
Lafayette, La.	8587 – 2996		Mobile, Ala.	8167 – 2367	
Lake Charles, La.	8679 – 3202		Mojave, Calif.	8993 – 7899	
Lake City, Fla.	7768 – 1419		Monroe, La.	8148 – 3218	
Lamar, Colo.	7720 – 5403		Montgomery, Ala.	7692 – 2247	
Lansing, Mich.	5584 – 3081		Mooers Forks, N.Y.	4215 – 1929	
La Plata, Md.	5684 – 1528		Morgantown, W. Va.	5764 – 2083	
Laredo, Tex.	9681 – 4099		Morristown, N.J.	5035 – 1478	
Laredo, Tex.	9683 – 4098		Morristown, Tenn.	6699 – 2183	
Las Cruces, N.Mex.	9132 – 5742		Muncie, Ind.	6130 – 2925	
Las Vegas, Nev.	8665 – 7411		Muskogee, Okla.	7746 – 4042	
Laurel, Miss.	8066 – 2645		Nashua, N.H.	4394 – 1356	
Laurinburg, N.C.	6610 – 1437		Nashville, Tenn.	7010 – 2710	
Lawrence, Mass.	4373 – 1311		Nassau, N.Y.	4961 – 1355	
Leesburg, Va.	5634 – 1685		Neche, N.D.	5230 – 5456	
Little Rock, Ark.	7721 – 3451		Newark, Ill.	6123 – 3527	
Littleton, Mass.	4432 – 1327		Newark, N.J.	5015 – 1430	
Locust, N.C.	6613 – 1640		New Bern, N.C.	6307 – 1119	

	V	H		V	H
New Brunswick, N.J.	5085	1434	Potsdam, N.Y.	4404	2054
New Haven, Conn.	4792	1342	Pottstown, Pa.	5246	1563
New London, Conn.	4700	1242	Poughkeepsie, N.Y.	4821	1526
New Market, Md.	5558	1676	Prescott, Ariz.	8917	6872
New Orleans, La.	8483	2638	Providence, R.I.	4550	1219
Newport News, Va.	5908	1260	Provo, Utah	7680	7006
New York City, N.Y.	4997	1406	Racine, Wis.	5837	3535
Norfolk, Va.	5918	1223	Raleigh, N.C.	6344	1436
North Bend, Nebr.	6698	4739	Reading, Pa.	5258	1612
North Bend, Wash.	6354	8815	Red Oak, Iowa	6691	4465
North Brook, Ill.	5954	3479	Redwood City, Calif.	8556	8682
Oakland, Calif.	8486	8695	Reno, Nev.	8064	8323
Ocala, Fla.	7909	1227	Richmond, Va.	5906	1472
Ogden, Utah	7480	7100	Roanoke, Va.	6196	1801
Oklahoma City, Okla.	7947	4373	Rochester, N.Y.	4913	2195
Omaha, Nebr.	6687	4595	Rockford, Ill.	6022	3675
Orangeburg, S.C.	6980	1502	Rock Island, Ill.	6276	3816
Orlando, Fla.	7954	1031	Rocky Mount, N.C.	6232	1329
Panama City, Fla.	8057	1914	Rosendale, N.Y.	4813	1564
Parkersburg, W. Va.	5976	2268	Roswell, N. Mex.	8787	5413
Pendelton, Oreg.	6707	8326	Sacramento, Calif.	8304	8580
Pensacola, Fla.	8147	2200	Saginaw, Mich.	5404	3074
Peoria, Ill.	6362	3592	Salina, Kans.	7275	4656
Petersburg, Va.	5961	1429	Salinas, Calif.	8722	8560
Petoskey, Mich.	5120	3425	Salt Lake City, Utah	7576	7065
Philadelphia, Pa.	5251	1458	San Angelo, Tex.	8944	4563
Philadelphia, Pa.	5257	1501	San Antonio, Tex.	9225	4062
Philadelphia, Pa.	5222	1493	San Bernardino, Calif.	9172	7710
Phoenix, Ariz.	9135	6748	San Diego, Calif.	9468	7629
Pine Bluff, Ark.	7803	3358	San Francisco, Calif.	8492	8719
Pittsburgh, Pa.	5621	2185	San Jose, Calif.	8583	8619
Plano, Ill.	6096	3534	San Luis Obispo, Calif.	9005	8349
Plymouth, Mich.	5562	2891	Sante Fe, N. Mex.	8389	5804
Pocatello, Idaho	7146	7250	Santa Rosa, Calif.	8354	8787
Polk City, Fla.	8067	1067	Sarasota, Fla.	8295	1094
Pontiac, Mich.	5498	2895	Scranton, Pa.	5042	1715
Port Angeles, Wash.	6206	9061	Searcy, Ark.	7581	3407
Port Huron, Mich.	5367	2813	Seattle, Wash.	6336	8896
Portland, Me.	4121	1334	Seguin, Tex.	9161	3981
Portland, Oreg.	6799	8914	Shreveport, La.	8272	3495

	V	H		V	H
Sidney, Nebr.	7112 – 5671		Tulsa, Okla.	7707 – 4173	
Sikeston, Mo.	7099 – 3221		Tupelo, Miss.	7535 – 2825	
Sioux City, Iowa	6468 – 4768		Twin Falls, Idaho	7275 – 7557	
Sioux Falls, S. Dak.	6279 – 4900		Ukiah, Calif.	8206 – 8885	
Socorro, N. Mex.	8774 – 5867		Van Nuys, Calif.	9197 – 7919	
South Bend, Ind.	5918 – 3206				
Spartanburg, S.C.	6811 – 1833		Waco, Tex.	8706 – 3993	
Spokane, Wash.	6247 – 8180		Wadena, Minn.	5606 – 4915	
Springfield, Ill.	6539 – 3513		Waldorf, Md.	5659 – 1531	
Springfield, Mass.	4620 – 1408		Warrenton, Va.	5728 – 1667	
			Washington, D.C.	5622 – 1583	
Springfield, Mo.	7310 – 3836				
Stamford, Conn.	4897 – 1388		Washington, D.C.	5603 – 1598	
Stevens Point, Wis.	5622 – 3964		Washington, D.C.	5632 – 1590	
Stockton, Cal.	8435 – 8530		Waterloo, Iowa	6208 – 4167	
St. Joseph, Mo.	6913 – 4301		Waycross, Ga.	7550 – 1485	
			Westchester, N.Y.	4921 – 1416	
St. Louis, Mo.	6807 – 3482				
St. Paul, Minn.	5776 – 4498		West Glendive, Mont.	5963 – 6322	
St. Petersburg, Fla.	8224 – 1159		West Palm Beach, Fla.	8166 – 0607	
Succasunna, N.J.	5038 – 1508		West Sweetgrass, Mont.	5829 – 7475	
Sunnyvale, Calif.	8576 – 8643		Wheeling, W. Va.	5755 – 2241	
			White River Jct., Vt.	4327 – 1585	
Superstition-Apache					
Junction, Ariz.	9123 – 6669		Wichita, Kans.	7489 – 4520	
Sweetwater, Tex.	8737 – 4632		Williamsport, Pa.	5200 – 1873	
Syracuse, N.Y.	4798 – 1990		Williamstown, Ky.	6353 – 2636	
Tallahassee, Fla.	7877 – 1716		Wilmington, Del.	5326 – 1485	
Tampa, Fla.	8173 – 1147		Winchester, Ky.	6441 – 2509	
Terre Haute, Ind.	6428 – 3145				
Thomasville, Ga.	7773 – 1709		Winston-Salem, N.C.	6440 – 1710	
Toledo, Ohio	5704 – 2820		Winter Garden, Fla.	7970 – 1069	
Topeka, Kans.	7110 – 4369		Winter Haven, Fla.	8084 – 1034	
Traverse City, Mich.	5284 – 3447		Woodstock, Ill.	5964 – 3587	
			Worcester, Mass.	4513 – 1330	
Trenton, N.J.	5164 – 1440				
Troy, Ala.	7771 – 2136		Wyoming Switch, Minn.	5686 – 4521	
Troy, N.Y.	4616 – 1633		Youngstown, Ohio	5557 – 2353	
Tucson, Ariz.	9345 – 6485		Yuma, Ariz.	9385 – 7171	
Tully, N.Y.	4838 – 1953				

APPENDIX FOUR

HOW TO USE THE LANALYZER™ DEMO DISK

The demo diskette that is available from John Wiley & Sons[1] or Excelan demonstrates the features of Excelan's LANalyzer EX 5000E Ethernet Network Analyzer. It will give you an opportunity to see a network management system in operation. All you need is an IBM PC/XT/AT or IBM-compatible running DOS 2.0 or higher.

The LANalyzer EX 5000E Ethernet Network Analyzer is a sophisticated, flexible, easy-to-use tool that allows you to

- Monitor network traffic
- Measure network performance
- Develop, debug, and maintain network protocols and applications

The EX 5000E is designed to operate on Ethernet (Versions 1.0 and 2.0) and IEEE 802.3 compliant LANs. It is available as a kit or as a package. The kit consists of hardware and software components which install into a PC. The package consists of the same hardware and software preinstalled in a COMPAQ PORTABLE 286.

---

[1] To obtain a copy of this diskette, contact Joe Dougherty at John Wiley & Sons (212) 850-6756 or Ellen Downing at Excelan (408) 434-2300.

The demonstration diskette contains a set of files that simulate the running of an EX 5000E test. The sample test is simulated because there is no network connection in this demonstration. Thus, data packets which ordinarily would originate from a network, originate from a file on the demo diskette. Similarly, where the actual LANalyzer system would store information into hardware buffers, the demonstration software writes to a demo diskette file. The speed of the demonstration therefore is not representative of actual product performance.

The sample tests and displays will show you how to

- Use the real-time network statistics and graphic displays to understand
    Distribution of activity on your LAN
    Why network performance has degraded; for example, see excessive CRC errors hidden by the protocols
    What percentage of network bandwidth is utilized
    What problems exist on which stations

- Use receive parameters, triggers, and trace data to
    Quickly isolate traffic of interest and identify problems
    Run tests unattended
    Verify protocols and application software
    Isolate time- and load-related problems
    Verify error handling

The user interface to the LANalyzer software consists of the PC's keyboard and screen. The keyboard permits user input, while the screen echoes input and displays output.

The PC keyboard function keys are redefined to correspond to LANalyzer commands. Commands are labeled on the lower portion of the screen. You enter a command simply by pressing the corresponding function key.

Output from the LANalyzer EX 5000E software is done on screens shown on the display. A typical screen identifies the task with which it is associated and contains parameter identifiers, current parameter settings, and a list of commands associated with each of the function keys. The LANalyzer EX 5000E software is organized into four main groups of screens.

- **Edit Screens.** These screens are where you create tests. On them you specify test parameters, test start and stop triggers, and files in which to save information.

- **Run Screens.** These screens display test results as they occur. On them you watch network statistics being gathered in real time.

- **Trace Screens.**   These screens display packet contents. On them you view individual packets collected during a test.
- **Statistics Screens.**   These screens display statistics collected in files. Using them you can re-create an entire test or part of a test.

Each LANalyzer screen is divided into three windows.

- **Status Window.**   This window is at the top of the screen. It displays such information as the file name and current time.
- **Data Window.**   This window is in the middle of the screen and is the largest window. It displays the screen's data.
- **Commands Window.**   This window is at the bottom of the screen. It displays the function key selections available for that screen.

As you run this demo disk, please note that your keyboard input should be limited to what is described in this document. While other keyboard input will not affect the sample test, it may cause unpredictable results. If you get to a screen position where you do not know what to do next, follow the instructions later in this section for exiting the sample test, and then start over.

To get started,

1. Boot your microcomputer.
2. Insert the demo diskette into Drive A.
3. Switch to Drive A by typing a: and pressing Return ⟨RET⟩.
4. At the system prompt, type demo and press Return.
   A> **demo** ⟨RET⟩

After a few moments, the LANalyzer EX 5000E demonstration displays a few status messages. Next you see an introduction screen. When you have read the introduction, press return to continue. If your system has a color monitor adapter, you will then be asked the following question:

Please enter monitor type (M for monochrome, C for color):

Enter **M** if your monitor is monochrome (single color); enter **C** if your monitor is color. The next thing you see is the Edit Test screen. This screen displays all the control parameters that define the sample test (see Figure A4-1).

The Edit Test screen can be divided into three portions: Receive Channel, Data Collection, and Transmit Channel. A LANalyzer EX 5000E "channel" is a set of parameters defining characteristics about data packets. A Receive Channel, for example, collects only those data packets that match its parameters.

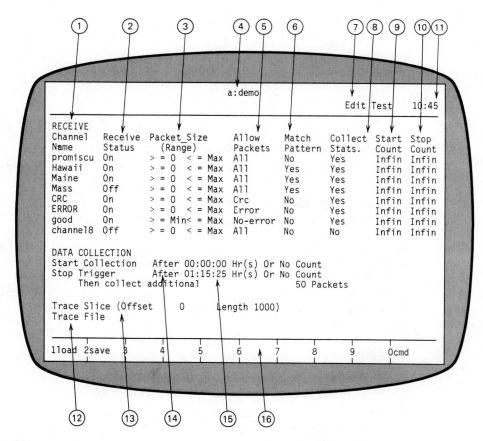

FIGURE **A4-1** The Edit Test screen (upper portion). Screen illustrations are meant to be examples only and may not match your screen display exactly.

The Receive Channel portion (consisting of the fields following the label RE-CEIVE) and the top part of the Data Collection portion (consisting of the fields following the label DATA COLLECTION) are now visible on the screen.

The following list explains the fields on the Edit Test screen shown in Figure A4-1.

① Channel Name. An alphanumeric string, up to eight characters long, that identifies a Receive Channel.

② Receive Status. Indicates whether the channel is enabled to collect packets. The parameter can have one of two values: On or Off. A Receive Channel can collect packets only if the Receive Status is On, regardless of other control parameter settings.

(3)   Packet Size (Range). Defines the acceptable packet size range in bytes. Packets within this size range will be collected on this Receive Channel. The valid range is from 0 to 9999 (bytes). For all channels in the sample test (except channel "good"), the minimum acceptable packet length is ">= 0". This means that all packets with length greater than or equal to zero (and less than the maximum length) are captured. Channel "good" captures only packets whose minimum packet length is ">= Min". "Min" refers to the minimum length of standard Ethernet packets, which is 64 bytes. All channels in the sample test have an upper length limit of "<= Max". "Max" refers to the maximum length of standard Ethernet packets, which is 1518 bytes.

(4)   Name of the file that contains the currently loaded test.

(5)   Allow Packets. Specifies the type of error(s) a packet must contain to be collected. The permissible values are All, No-error, Crc, Align, Short, and Error. Most channels in the sample test collect "All" packets, that is, packets with or without errors. Channel "CRC" captures only packets with CRC errors, channel "ERROR" captures only packets that contain errors (of any type), and channel "good" captures only packets with no errors.

(6)   Match Pattern. Indicates whether the packet must match a byte-level pattern. The value of the parameter can be either Yes or No.

(7)   The screen name.

(8)   Collect Stats. Specifies whether statistics are compiled and saved to disk for this Receive Channel. The field can have either the value Yes or the value No. If set to Yes, the name of the file in which statistics are to be stored also must be specified. The sample test does not name a statistics file; therefore, test statistics for channels requesting them are not saved.

(9)   Start Count. Used in conjunction with the Start Collection field in the Data Collection portion of the Edit Test screen. It specifies the number of packets matching this channel's parameters to ignore before packet collection begins. The Start Count parameter can have a value in the range 0 to 64,000. In the sample test, this parameter is set to "Infin" for all channels. "Infin" disables the use of this parameter.

(10)   Stop Count. Used in conjunction with the Stop Collection field in the Data Collection portion of the Edit Test screen. It specifies the number of packets that must be collected before the stop trigger fires. The Stop Count

parameter can have a value in the range 0 to 64,000. In the sample test, this parameter is set to "Infin" for all channels. "Infin" disables the use of this parameter.

(11)   The current time.

(12)   Trace File. Allows you to name a DOS file in which to save packet traces (packet information and data).

(13)   Trace Slice. Specifies the segment ("slice") of a packet to be saved. If the packets you are capturing are long, you might want to collect only a specified slice. This field has two subfields: Offset and Length. The Offset subfield specifies the packet byte at which the slice should start. The Length subfield specifies the length of the slice in bytes.

(14)   Start Collection. Defines the event causing data collection to start. The sample test specifies that packet collection should begin after zero hours (that is, start immediately) and should ignore the value in the Start Count field.

(15)   Stop Trigger. Defines the event causing data collection to stop. As defined here, an internal timer signals the end of the test after a little more than 75 minutes, although this demonstration sample test collects 511 packets and then ends. The Stop Count parameter value will be ignored. After the timed end of the test, the second portion of the stop trigger requests an additional 50 packets to be collected.

(16)   The Commands Window for this screen.

Next, let us see what a test is and how you communicate with the LANalyzer EX 5000E software. If you want to check some property of your network, you might say you want to "test" the network. All LANalyzer EX 5000E functions are realized by setting up and running tests. A test is a program in which you specify control parameters to capture packets from and/or transmit packets to the network. While a test runs, you can watch statistics being gathered. After a test completes, you can display the contents of captured packets and view tables and graphs depicting test statistics.

To capture packets from a network, you create and run a receive test. A receive test defines details (or control parameters) about the kinds of packets you want to capture. The LANalyzer system examines all packets traveling on the network and captures only the packets that match the specified parameters. When setting up a test, you also can define triggers to start and stop the test.

Once you have defined the conditions of a test, you can run it. While a test is in progress, you can monitor network status continually as it occurs. The LANalyzer EX 5000E collects traces of all packets that meet your test's defined parameters. A trace is a record of a packet—its contents and information about it. After a test finishes, you can view packet traces.

To begin running the sample test,

1. Press function key F10 (*cmd*). F10 shows as F0 on your screen.
2. Press function key F2 (*run*).

The first screen displayed is the Run Counter screen (see Figure A4-2). It shows ongoing statistics about packets being gathered on each Receive Channel. The test is running when a message in the upper left corner of the screen says, "Collecting." The test is done when the message says, "Collection stopped." Let the test run to completion. Run Counter screen information is vital to LAN administrators who might be trying, for example, to distribute workload evenly among network nodes.

The following list explains the items shown in Figure A4-2.

1. Relative time in the format *hh:mm:ss*. This is how much time has elapsed since the test began. This counter increments every second.

2. Total number of packets observed on the network since the test began.

3. Total number of packets collected on individual channels. Note that the count for channel "promiscu" is equal to the total number of packets observed on the network. This is because the "promiscu" channel is set to accept all packets.

4. Message area. "Collection stopped" indicates that the test has completed collecting network traffic data.

5. Bar graphs showing the distribution of captured packets, by channel, as a percentage of total network traffic.

6. Counts of observed packets with the indicated error.

When the test finishes, run it again and look at the Run Global screen. To do this,

1. Press function key F10 (*cmd*).
2. Press function key F2(*run*).
3. Press function key F3 (*globl*).

The screen you now see is the Run Global screen (see Figure A4-3). The Run Global screen displays ongoing test results about all observed network traffic. Let the test run to completion. The Run Global screen displays graphs depicting the

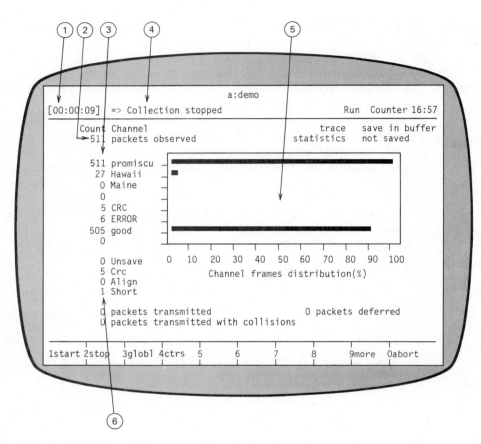

FIGURE **A4-2** The Run Counter screen.

utilization of available network bandwidth. On LANs with many terminal servers, for example, you would see high packet rates but low network utilization because of the high volume of overhead.

If the test has stopped before you display the Run Global screen, the Run Global screen displays statistics reflecting the last time period before the test ended. During the running of a test, you can switch between the Run Counter and Run Global screens as often as you wish by alternately pressing function keys F3 (*globl*) and F4 (*ctrs*).

The following list describes the items shown in Figure A4-3.

① Packet size distribution: percentage of packets in size ranges 0 to 63 bytes, 64 to 127 bytes, 128 to 255 bytes, and so on. Size ranges with less than one percent of distribution are not shown.

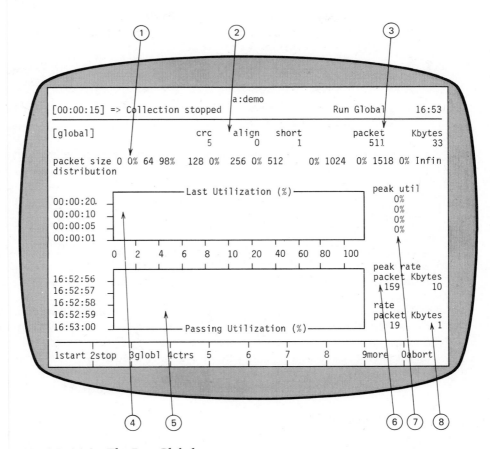

FIGURE **A4-3**   The Run Global screen.

(2) Counts of packets observed with the indicated error.

(3) Total number and cumulative size (in kilobytes) of all packets observed on the network since the test began.

(4) Last Utilization (percentage). Average network utilization (percentage) during the last 20-second, 10-second, 5-second, and 1-second time intervals. A dash (—) in this field indicates a utilization of less than 1 percent.

(5) Passing Utilization (percentage). Average network utilization (percentage) for each of the preceding five time intervals (shown as the actual clock time). The sample test time interval is one second. A dash (—) in the field indicates a utilization of less than 1 percent.

(6) Peak rate of network traffic since the start of the test, in terms of packets per second and kilobytes of packet data per second.

(7) Peak network utilization during the last 20-second, 10-second, 5-second, and 1-second time intervals.

(8) Average rate of network traffic, for the current second, in terms of packets per second and kilobytes of packet data per second.

After the test finishes, you can examine the contents of received packets using the Trace screens as follows.

1. Press function key F10 (*cmd*).
2. Press function key F3 (*trace*).

The system displays the Trace Buffer screen (see Figure A4-4). Packet trace information provides you with information needed for troubleshooting LAN problems, debugging LAN software (protocols and applications), and fine-tuning LAN performance (using the timestamp information).

The Data Window on the Trace Buffer screen is split into two discrete subwindows: the Summary Subwindow and the Packet Slice Data Subwindow. The Summary Subwindow is the upper subwindow. It reports general information about each packet. The Packet Slice Data Subwindow is the lower subwindow. It displays the data contained in the packet.

The following list explains the items shown in Figure A4-4.

(1) Number. Packets are stored in the trace buffer in the order in which they were received; they are numbered sequentially from 1.

(2) Len. Packet length in bytes.

(3) Absolut Timestmp. The absolute clock time when the packet was captured.

(4) Summary Subwindow.

(5) Elapsed time. The amount of time that has passed since the test began.

(6) Ethernet header information.

(7) Total packets. Total number of packets in the trace buffer.

(8) Channels. The channel(s) on which the packet was collected.

(9) Errs. The types of errors the packet contains, if any.

(10) Offset in the packet of the displayed slice.

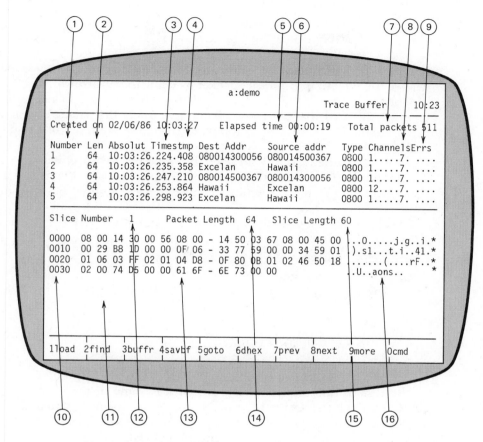

```
                                        a:demo
                                                Trace Buffer    10:23

Created on 02/06/86 10:03:27      Elapsed time 00:00:19    Total packets 511

Number Len Absolut Timestmp Dest Addr      Source addr    Type ChannelsErrs
1      64  10:03:26.224.408 080014300056 080014500367   0800 1.....7. ....
2      64  10:03:26.235.358 Excelan      Hawaii         0800 1.....7. ....
3      64  10:03:26.247.210 080014500367 080014300056   0800 1.....7. ....
4      64  10:03:26.253.864 Hawaii       Excelan        0800 12....7. ....
5      64  10:03:26.298.923 Excelan      Hawaii         0800 1.....7. ....

Slice Number    1        Packet Length   64    Slice Length 60

0000  08 00 14 30 00 56 08 00 - 14 50 03 67 08 00 45 00  ...0.....j.g..i.*
0010  00 29 B8 1D 00 00 0F 06 - 33 77 59 00 0D 34 59 01  ).s1...t.i..41.*
0020  01 06 03 FF 02 01 04 D8 - 0F 80 0B 01 02 46 50 18  .......(....rF..*
0030  02 00 74 D5 00 00 61 6F - 6E 73 00 00              .U..aons..      *



11load 2find  3buffr 4savbf 5goto  6dhex  7prev  8next  9more  0cmd
```

FIGURE **A4-4**   The Trace Buffer screen.

⑪ Packet Slice Data Subwindow.

⑫ Slice Number. The sequential number of the packet whose data slice is displayed.

⑬ Slice data in hexadecimal.

⑭ Packet Length. Length of the entire packet.

⑮ Slice Length. Length of the packet slice that has been saved in the packet trace.

⑯ Slice data in ASCII.

The Summary Subwindow on the Trace screen is a scrollable window. It provides a brief description of each packet so you can determine if its contents might be of interest. You can select a packet of interest by using the cursor control and function keys. The current selection is highlighted in reverse video. You can change the selection to another packet by using the up arrow (↑), the down arrow (↓), or the Return key. You also can change the selection to a specific packet by pressing the function key F5 (*goto*) and specifying the packet's sequential number. Pressing function key F6 (*dhex*) displays the contents of the currently selected packet.

As an experiment, scroll through the trace buffer. Specifically, select packet 214 and display its contents. Notice the errors this packet contains in the Summary Subwindow and on which channel it was captured because it contained errors.

To EXIT from the LANalyzer sample test and return to DOS, do the following.

1. Press function key F10 (*cmd*).
2. Press function key F10 (*exit*).
   The system clears the Commands Window and displays the following message:

   Exit to DOS (y/n):
3. Enter **y** and press Return.

The system returns to DOS, as indicated by the DOS prompt.

The rest of this section briefly introduces some of the other features of the LANalyzer system. Using the microcomputer's alphanumeric and symbol keys, you can enter data, control cursor movement, and scroll screen displays. Furthermore, all of the four major screen groups—Edit screens, Run screens, Trace screens, and Statistics screens—have subscreens that help define conditions and provide further information. You can display subscreens by pressing function keys, although this feature is not available on this demonstration disk.

The Edit screen, Edit Test, has already been introduced as the method for creating tests. There are several subscreens from the Edit Test screen. Here are some additional features of the Edit Test screen and its subscreens.

By scrolling down from the Receive portion of the Edit Test screen, you will find the Data Collection and Transmit portions of the screen. In the Data Collection portion, the Station Monitor field specifies whether the Run Station screen will be accessible when running a test. The Run Station screen allows you to monitor individual station (node) traffic. The Station Monitor field toggles between On and Off.

Further down on the Edit Test screen is the Transmit portion. Tests that transmit packets allow you to create a known network load and then observe node reactions. Transmit test parameters include packet definition, known packet errors, and predefined interpacket spacing, among others. A test including packet transmission injects packets into the network according to your defined parameters. A single test can both receive and transmit packets.

After creating it, you can save a test in a file with the Edit Test screen command *save*. When you want to reuse a saved test, you use the Edit Test screen command *load* with the name of the saved file.

The Match Pattern parameter on the receive portion of the Edit Test screen specifies whether incoming packets must match a pattern you have created in order to be captured. For example, you might create a pattern naming a specific source address. If the Match Pattern field for this Receive Channel is Yes, only packets originating from the specified network node (that also meet the other defined parameters) will be collected. When you move the cursor to the Match Pattern field, the command *open* appears as function key F3 in the Commands Window. If you select function key F3, you have the opportunity to create (or modify) a packet template through the Edit Test subscreen Edit Pattern.

The Edit Name subscreen allows you to assign a logical alphabetic name to each node on the network. In this way, the node's name is displayed in the address fields of various LANalyzer screens instead of its Ethernet address. You also can enter the node's name in any address field on Edit screens.

Besides the Run Counter and Run Global screens, which you see in this demonstration, the LANalyzer software has three additional Run screens that also allow you to observe network activity in real time. The Run Channel screen displays information about packets received on individual enabled channels. The Run Transmit screen displays the distribution of packets transmitted with and without collisions. The Run Station screen allows you to monitor individual station (node) traffic.

On the Edit Test screen, packet traces are placed in a buffer on the EXOS 225 Ethernet Network Analyzer board if you do not specify a trace file.

The EXOS 225 board provides a 700,000-byte circular data buffer. This large buffer allows the LANalyzer system to store somewhere between 2,700 256-byte (or smaller) packets and 450 1518-byte packets (the actual number of packets saved depends on packet size). Since the EXOS 225 board buffer is circular, new packets overwrite older ones when the buffer is full. Therefore, any packets collected beyond the first 700,000 bytes will overwrite previously collected packets.

You also can name a DOS disk file in which to save packet traces. Packet traces are stored in one or more 300,000-byte files. This length allows the files to be copied to low density $5\frac{1}{4}$-inch floppy disks for storage. While EXOS 225 onboard trace storage is temporary, traces stored to disk files are saved and can be viewed at any time after test completion.

You can view packet traces from both onboard memory and DOS files using the

Trace screens. When onboard memory overflows, a message in the Trace screen Summary Subwindow indicates that packets are "unsaved" (that is, lost because of overwriting). To search for an individual packet in the trace buffer or in a trace file, you can use the Trace screen function key, *find*. To determine the length of time between packet arrivals, you can display the Trace screen Interpacket Arrival subscreen. Also, it is possible to retransmit collected packets during subsequent tests.

While a test is running, test results are displayed as they occur on Run screens. By saving test results to a DOS disk file, you can review them at a later time on Statistics screens. On the Edit Test screen, if you name a statistics file and enable a Receive Channel's Collect Stats. parameter, test statistics will be saved to the named file. Statistics are saved in one or more 300,000-byte standard DOS files. Saved statistics can be displayed on three Statistics screens. The Global Statistics screen shows data for all traffic observed on the network during the test. The Channel Statistics screen shows test data about packets collected on enabled receive channels. The Transmit Statistics screen shows test results for all packet transmissions.

Three sample tests are provided as part of the standard LANalyzer EX 5000E software. The sample tests are of general interest, and you probably will find that you can use them to test your network without modification. The three sample tests are a default test similar to the demonstration sample test, a test to stress the network by generating an excessive load on it, and a test to collect packets sent during normal network usage between two nodes. Also included in the standard LANalyzer system software is a TCP/IP protocol decoder utility. This decoder utility displays trace files in TCP/IP format.

In this demonstration you can learn about and use the LANalyzer EX 5000E software. For more information on other LANalyzer functions, call EXCELAN at (408) 434-2300 or write to EXCELAN at 2180 Fortune Drive, San Jose, California 95131.

INDEX

Abstracted Business Information, described, 50
Acceptance, new network, 489, 532
ACCESS KEY, described, 431
Access methods:
  Carrier Sense Multiple Access (CSMA), 343
  central control, 199
  Collision Detection (CD), 343
  guaranteed, 346
  interrupt, 199
  random, 343
  telecommunication access method (TCAM), 248
  token, 346
Access time, defined, 404
Accunet:
  cost of using, 528
  packet switched, 224
  X.75 gateway to, 481
Accuracy, of information, 378
ACF (Advanced Communications Function), 251
ACK, *see* Positive acknowledgment
Acoustic couplers:
  illustrated, 162
  use of, 161

Active data line monitors, passive vs., 407
Adapters, line, 151
Adaptive differential pulse code modulation (ADPCM), defined, 120
Adaptive equalization:
  defined, 82
  modem use of, 476
Adaptive subbands excited transform, defined, 120
ADCCP (Advance Data Communications Control Procedures), 578
Address field, X.25 and SDLC frame, 265
ADPCM (adaptive differential pulse code modulation), 120
Advance Data Communication Control Procedures (ADCCP), SDLC relationship, 578
Advanced Communications Function (ACF), network control, 251
Advanced modems, 163
Advanced Program-to-Program Communications (APPC):
  described, 274–275
  impact on gateways, 348

impact on micro-to-mainframe links, 330
Air mileage, calculation of, 520
Airline reservation systems:
  data monopoly in, 22
  described, 18, 44
Alarm processing, Multi-Function Operations System (MFOS) module, 389
Alaska, WATS band, 480
Algorithms:
  asymmetric, 424
  encryption, 419, 420–427
  proprietary, 424
  public key, 424
  symmetric, 423, 424
All at once implementation, defined, 533
ALOHA system, random access, 343
Alternating current (ac), defined, 71
American Banking Association, DES encryption endorsement, 421
American National Standards Institute (ANSI):
  described, 287–288
  FDDI standard, 356
  TRIB definitions, 70

American Telephone & Telegraph (AT&T):
  CCITT membership, 288
  common carrier, 12, 198, 467
  competitors of, 474
  divestiture of, 469, 473–474, 539
  ISDN chip technology, 227–228
  local loop connection, 471
  modem certification, 166
  multiplexing by, 175–177
  T carrier capability, 121
Amplifiers:
  defined, 341
  described, 97–99
  fiber optic, 111
Amplitude deviation distortion:
  conditioning to reduce, 475
  defined, 82
Amplitude jitter, defined, 402
Amplitude modulation (AM), defined, 79, 80
Amplitude noise, defined, 130
Analog, defined, 12
Analog modulation, described, 78–82
Analog side, modem, 83, 402
Analog signals:
  illustrated, 84
  modem conversion to, 72
  modulation, 78
  see also Broadband
Analog testing:
  defined, 402
  equipment for, 403
Analog transmission:
  defined, 203
  described, 203–204
Analytical models:
  defined, 543
  polling queue assumptions in, 546
Analyzers:
  defined, 403
  network cost, 517
  response time, 406
ANSI, see American National Standards Institute
Antennas, rooftop, 20
API (Application Program Interface), 330
APPC (Advanced Program-to-Program Communications), 274–275
Application layer:
  controls for, 440
  Distributed Network Architecture (DNA) model, 282
  function of, 255, 263
  Open Systems Interconnection (OSI) model, 255, 263
Application processing time, defined, 126

Application Program Interface (API), purpose, 330
Application program software, use of, 241
Application programs, location in network, 241
Applications:
  airlines, 18, 44
  automated office, 26–34
  banking, 18, 40–43
  electronic mail, 38–40
  electronic shopping, 45
  health care, 18
  job task dependence levels in, 510
  parcel tracking, 18
  rental cars, 44
  teleconferencing, 44–45
  voice mail, 34–38
Architectures:
  Burroughs Network Architecture (BNA), 282–283
  controls to protect, 440–442
  defined, 247–248
  described, 278–286
  Digital Network Architecture (DNA), 281–282
  Distributed Communications Architecture (DCA), 282
  Distributed Network Architecture (DNA), 283
  Distributed Systems (DS), 283–284
  Distributed Systems Environment (DSE), 282
  distributed vs. hierarchical, 229
  general characteristics of, 280
  inherent broadband, 230
  Manufacturing Automation Protocol (MAP), 281
  network, 268–286
  Open Network Architecture (ONA), 284
  Systems Application Architecture (SAA), 278–279
  Systems Network Architecture (SNA), 268–278
  Transmission Control Protocol/Internet Protocol (TCP/IP), 284
  UNIX, 285–286
  XEROX Network Systems (XNS), 284
Argentina, ARPAC, 225
ARPAC network, 225
ARPANET network, 284
ARQ, 264
  see also
  Automatic Repeat Request, 104, 264
  Continuous ARQ, 105, 264, 315

  Stop and wait ARQ, 104, 264, 268, 313, 530
ASCII (United States of America Standard Code for Information Interchange), 56, 61, 63
ASCOM IV, communication software, 309
ASET (adaptive subbands excited transform), 120
Assets, protection of, 584
Asymmetric algorithms, defined, 424
Asynchronous transmission:
  defined, 60
  described, 57–58
  efficiency of, 68
  illustrated, 57
AT&T, see American Telephone & Telegraph
AT&T Accunet, 224, 481, 528
AT&T Communications, impact of divestiture, 473
AT&T DATAPHONE Digital Service, rates for, 530
AT&T DIAL-IT 900 service, 482–483, 531
AT&T 5ESS, switch management, 389
AT&T Megacom services, 481, 531
AT&T Premises Distribution System (PDS), described, 353–354
AT&T Software Defined Network (SDN), defined, 202, 478
ATM, see Automated teller machines
Attainable design configurations:
  choice set of, 509
  costs related to, 516
Attenuation:
  avoidance of, 131
  cause of distortion, 82
  defined, 97, 131
  raindrop, 107
  see also Distortion
Auditing, microcomputer, 435, 455
Auditors, network control and, 416
Austpac network, 225
Australia:
  Austpac, 225
  Telecom deregulation, 469
  Viatel, 226
AUTOEXEC.BAT:
  DOS file, 301
  Novell NetWare, 351
Automated control matrix design, 454
Automated office:
  components of, 27
  evolution of, 26–27
  functions, 30–33
  illustrated, 28, 33
  networks and the, 28

Automated teller machines (ATM):
    banking networks and, 40–41
    terminal type, 185
Automated test equipment:
    functions, 407
    types of, 408
Automatic repeat request (ARQ)
    described, 264
    satellite delays with, 104
    see also
        Continuous ARQ, 105,
            264, 315
        Stop and wait ARQ, 104,
            264, 268, 313, 530

Backbone networks:
    defined, 224, 338
    local area networks as, 366
Backup and recovery:
    controls for, 444–446
    local area network, selection
        issue, 371
Ballistic transistor, defined, 19
BALUN, defined, 355
Bands, WATS service, 480
Bandwidth:
    bits per second and, 205
    described, 203–204
    impact on throughput, 69
    voice grade circuits, 204
BANK WIRE network, 42
Banking networks:
    BANK WIRE, 42
    CHIPS, 41
    described, 18, 40–43
    encryption algorithms for,
        421
    FED WIRE, 42
    illustrated, 42, 43
    MINTS, 42
    SWIFT, 41
Baseband:
    broadband vs., 71–73, 83
    broadband vs., selection is-
        sues, 369
    defined, 12, 71, 83, 341
    disadvantages of, 341
    Ethernet, 339
    illustrated, 84, 86
    modem conversion of, 159
    signal, 71
    Token-Ring Network, 340
Basic access service (2B + D), de-
    fined, 226
Basic telecommunication access
    method (BTAM), function of,
    249–250
Baud:
    defined, 90
    setting for communications,
        306
Baudot code, described, 62, 66

BBS (bulletin board systems), see
    Newsletter, vol. 5, no. 1, Jan.
    1988 (ISBN 0-471-60129-2)
BCC (block check character), 267
BCD (Binary Coded Decimal), 62,
    65
BCM (bit compression multi-
    plexer), 176
Beginning flag, X.25 and SDLC
    frame, 265
Belgium, transborder data flow, 20
Bell, Alexander Graham, 5
Bell 201 modem, 9, 81, 82
Bell 202 modem, 205, 206
    frequencies of a, 206
Bell Canada, common carrier, 198,
    466
Bell Laboratories:
    impact of divestiture, 473–474
    UNIX development by, 197
Bell Operating Companies (BOC):
    CCITT membership, 288
    common carriers, 12, 198, 466
    competition and the, 474
    concern over teleports, 24
    consolidation into, 469–470,
        539
    geographic boundaries, illus-
        trated, 470
    modem certification, 166
BERT, see Bit-Error Rate Tester
Bias distortion, defined, 563
Bildschirimtext, 226
Binary Coded Decimal (BCD), de-
    scribed, 62, 65
Binary Synchronous Communica-
    tions (BSC):
    message format, illustrated,
        268
    Open Systems Interconnection
        (OSI) model and, 259
    protocol, 267–268, 512
    satellite circuits, problems
        with, 477, 512
Biplexers:
    described, 174, 175
    illustrated, 175
Bipolar signals, illustrated, 85–86
Bit compression multiplexers
    (BCM), described, 176
Bit-Error Rate Tester (BERT), de-
    fined, 404–405
Bit-oriented protocols:
    byte-oriented vs., 267, 512
    Distributed Communications
        Architecture, 282
Bit position, ASCII vs. EBCDIC, 65
Bit transfer rate, defined, 404
Bits:
    defined, 90
    digital transmission of, 88
    information, defined, 67
    parity, 67

patterns of, 405
    redundant, defined, 67
    serial transmission of, 56–57
    timing of, 83
Bits per second (bps):
    calculation of, 91
    bandwidth and, 205
BIU (Bus Interface Unit), 360–361
BKER (Block-Error Rate Tester),
    405
Black Box, hardware catalog, 530
BLAST, blocked asynchronous pro-
    tocol, 315
Block check characters (BCC), pur-
    pose of, 267
Block cipher, defined, 421
Block-Error Rate Tester (BKER),
    defined, 405
Block transfer time, defined, 404
BNA (Burroughs Network Archi-
    tecture), 282–283
BOC, see Bell Operating Compa-
    nies
Boeing Computer Services, 225
Books in Print, described, 50
Booting:
    defined, 300
    programs to accomplish, 302
Bose-Chaudhuri code, defined, 135
Brazil, transborder data flow, 20
Breakout box, described, 404
Bridge (device):
    defined, 152, 347
    local area network, selection
        issue, 371
    Token-Ring Network, 340
British Post Office Commission,
    role in data communications, 468
Broadband:
    baseband vs., selection issues,
        369
    defined, 12, 72, 83, 341
    disadvantages of, 342
    illustrated, 84
    Manufacturing Automation
        Protocol, 340
    modem conversion to, 159
    PC Network, 339–340
    signal, 72, 78
BRS (Bibliographic Retrieval Ser-
    vices), 50
BSC, see Binary Synchronous
    Communications
BTAM (basic telecommunication
    access method), 249–250
Bulletin board systems (BBS), see
    Newsletter, vol. 5, no. 1, Jan.
    1988 (ISBN 0-471-60129-2)
Burroughs Network Architecture
    (BNA), described, 282–283
Burst error, defined, 128
Bus Interface Unit (BIU), require-
    ments for a, 360–361

Bus topology:
    DECconnect, 354
    defined, 337
    electrical, 338
    Ethernet, 339
    illustrated, 216, 338
    Manufacturing Automation
        Protocol, 340
    PC Network, 339–340
Business operations, characteristics
    requiring data communications,
    10
Business systems, evolution of,
    14–16
Bypassing, Digital Termination
    Systems (DTS) used for, 26, 233
Byte, defined, 61
Byte-oriented protocols:
    Binary Synchronous Communi-
        nications (BSC), 267, 512
    Digital data communication
        message protocol, 282

C-band, commercial satellite use
    of, 107
C-level conditioning, defined, 476,
    522
C-message noise, defined, 402
C-notch noise, defined, 402
Cable television, role in telecom-
    muting, 25
Cables:
    coaxial, 100, 357–358
    coaxial, illustrated, 102, 358
    connector, 14, 73, 435
    dual, illustrated, 359
    dual vs. single, 359–360
    fiber optic, 356–357, 434
    fiber optic, illustrated, 114,
        357
    influence on network selec-
        tion, 352
    installation of, 358
    layout, illustrated, 361
    local area network, selection
        issue, 369
    local area network, types for,
        354–362
    null modem, 161
    single, illustrated, 360
    step-index, 113
    twisted pair, 354–355
    undersea, 112, 119
    wire, 100
CAI (Computer-Assisted Instruc-
    tion), 533
California:
    local access transport areas
        (LATA) within, illustrated,
        473
    Wide Area Telecommunica-
        tions Service (WATS) within,
        480

Call-routing programs, software
    defined networks (SDN), 478
Calls:
    conference, 208
    function, in DOS, 308
    local, defined, 200, 471
    long distance, defined, 201,
        473
    time as a cost factor, 479
    unit, defined, 200
    voice, 199–201, 211–213
Canada:
    Canadian Radio-Television &
        Telecommunications Com-
        mission (CRTC), 468
    Datapac network, 224
    satellite transmissions, 20
    Telidon videotex, 226
    transborder data flow, 20
Capacitance, cause of distortion,
    72
Capacitive loads, effect on volt-
    amps, 323
Capital, strategic resource of, 384,
    386
Captain videotex, 226
Capture buffer:
    downloading, 311
    file transfer contrasted with,
        311
Careers, telecommunication, 45–
    48
Carrier Sense Multiple Access/
    Collision Detection (CSMA/CD):
    defined, 343
    described, 343–346
    Ethernet, 339
    PC Network, 339–340
    STARLAN (local area net-
        work), 340
Carrier signals, defined, 78
Carrier waves, defined, 78
Carriers, see
    Carrier signals
    Carrier waves
    Common carriers
    T-1 carrier
CATER (acronym):
    defined, 378
    managerial information must,
        489
Cathode ray tubes (CRT), 183
CBX switchboards, defined, 229
CCEP (Commercial COMSEC En-
    dorsement Program), 423
CCITT, see Consultative Commit-
    tee on International Telephone
    and Telegraph
CCSA (Common Control Switch-
    ing Arrangement), 485
CD (Collision Detection), 343
Cells (matrix):
    empty, controls lacking,
        453

identification of, 449
Cellular radio, 234
    described, 117–188
    local loop, use of, 23
    security problems with, 433–
        434
Cellular telephones, see Cellular
    radio
Central computer, see Host com-
    puter
Central control group, failure con-
    trol by, 392
Central control mode:
    defined, 145, 199
    protocol analogy, 344
Central control system, role in
    networks, 122–123
Central office (CO):
    defined, 14, 99, 200, 520, 522
    hierarchy, illustrated, 201
    point of presence connection,
        521–525
    role in network, 99, 200
Central processing unit (CPU):
    defined, 14
    microcomputer memory, 298–
        299
    microcomputers for optimiza-
        tion vs., 541
Centralization:
    of communications, local area
        network, selection issue, 372
    of data communications, 204
    of network control, 380
CENTREX, 232–233
Certification:
    microcomputers, 303
    modems, 166, 469
Channel banks, 122
    digital, 176
    multiplexing, 173–174
Channel extenders, see Newslet-
    ter, vol. 5, no. 1, Jan. 1988 (ISBN
    0-471-60129-2)
Channel interface, defined, 147
Channel options, cost analysis of,
    522
Channel service unit (CSU), de-
    fined, 164
Channels:
    defined, 100
    digital, error free, 405
    high speed, 88
    T-1, 88, 120
    see also Circuits
Characteristic distortion, defined,
    563
Characters:
    assembly of, 149
    defined, 61
    per day average, 500
    per day peak, 500
Charter, data communications,
    written, 380

Charts:
    data flow diagrams, 533
    flowcharts, 533
    Gantt, 533
    implementation aids, 533
    network cost analyzer, 518
    organization, 385, 390
    PERT, 533
Chief Information Officer, emerging position of, 384–386, 389
CHIPS network, 42
Choice set:
    costs related to, 516
    defined, 508
Chronological implementation, defined, 533
CICS, see Customer Information Control System
CIM (computer integrated manufacturing), 281
Ciphertext, defined, 419
Circuit boards:
    add on, converters, 178
    local area network, selection issue, 370
Circuit chips, data communication use of, 25
Circuit switched digital capacity (CSDC), described, 176
Circuit switching, defined, 191, 223
Circuits:
    baseband vs. broadband, 341–342
    capacity of, 203–207, 504–507
    cellular radio, 117–118
    coaxial cable, 100–103
    conditioning of, 475–476, 522
    controls to protect, 434–435, 612–614
    copper, 476
    cost analysis of, 516–531
    dedicated, 99, 202, 228, 342–343
    defined, 13, 100, 612
    dial-up, costing of, 525–527
    dial-up, defined, 208, 479
    dial-up, signaling on, 207–208
    dial-up vs. leased, 475
    digital, 477
    digital, costs, 530
    digital, testing of, 402–403
    disaster recovery of, 444–446
    electrical protection for, 320–324
    four-wire, 96
    frequency division, illustrated, 168
    FX leasing, 485
    Group 48,000 hertz, 477
    interexchange (IXC), 14
    leased, 99, 202, 228
    loading of, 504–507
    local loop, 94

microwave, 103–104
multidrop, 154, 217–218, 477, 544
multiplexed, illustrated, 167, 168, 170, 172
multipoint, 477
open-wire pair, 100
optical, 111–117
point-to-point, 216, 477
private, 99, 202, 228, 474–478
protection of, 156–157
quality control for, 397
satellite, 104–111
satellite, costs of, 529
security of, 434–435
Series 1000, 476
Series 2000, 475
Series 3000, 475, 376
short distance radio, 119
software defined network (SDN), 202, 478
submarine, 119
Super Group 240,000 hertz, 477
20-milliamp, 160
T-1 carrier, 120–122, 466, 478
tropospheric scatter, 119
two-wire, 96
types of, 100–119
virtual, 202, 220, 478
voice grade, capacity of, 203–207
voice grade, costs of, 517, 520–525
voice grade, leased, 202–203, 475–476
voice grade, Series 2000/3000, 475
voice vs. data, 228
waveguide, 119
wideband, costs of, 527
wire cable, 100, 101
Cirrus banking network, 18
Citizens band radio, by satellite, 23
Clamp, surge protection, 322
Clear text, defined, 419
CLEAR TO SEND signal, 78, 105, 166
Clocking time, defined, 97
Cluster controllers, in micro-to-mainframe link, 327
CNA (Communication Network Architecture), 283
CO, see Central office
Coaxial cables:
    defined, 100
    Ethernet usage, 353, 358
    illustrated, 102, 358
    local area network medium, 357–358
Code book compression, defined, 156

Codes:
    conversion, 149
    defined, 61
    efficiency formula for, 68
    error correction, 136
Coding systems:
    ASCII, 61–63
    Baudot, 62, 66
    Binary Coded Decimal (BCD), 62, 65
    defined, 61
    EBCDIC, 62, 63
    4-of-8, 67
    Hollerith, 62
    M-of-N, 62, 67, 133
    self-checking, 62, 67
Coherent light, defined, 567
Collision detection (CD), defined, 343
    see also Carrier Sense Multiple Access/Collision Detection
COMMAND.COM, DOS file, 301
Commercial COMSEC Endorsement Program (CCEP), encryption algorithms, 423
Commercial databases, use with microcomputers, 306
Common carrier networks, defined, 17
Common carriers:
    defined, 12, 198, 466
    examples of, 467
    limited vs. extended, 234
Common Control Switching Arrangement (CCSA), described, 485
Communication circuits, see Circuits
Communication costs, voice vs. data, 19
Communication facility, defined, 465
Communication line control, described, 148–149
Communication modes, types of, 96–97
Communication Network Architecture (CNA), related to DNA, 283
Communication Networks, see
    Data communication networks
    Networks
    Voice communications
Communication processor, see
    Front end communication processor
Communication satellites, see Satellites
Communication services:
    described, 465–485
    measured use, 478–484
    offerings, 474–485
    private use, 474–478

Communications services (*Continued*)
  special, 484–485
Communication systems, design steps for, 490–534
Communications:
  centralization of, 372
  history of, 5
  media for, 100–119
  microcomputer usage in, 297
  security algorithms for, 423–427
  voice/data integration, 382–383
Communications Act, FCC regulatory powers granted, 467
Communications Satellite Corporation (COMSAT), teleport use by, 24–25
Compaction, data, 156
Comparison estimating, defined, 503
COMPENDEX (*Computerized Engineering Index*), 50
Components:
  computer, 411
  defined, 446
  examples of, 448, 584–586
  safeguarding of, 449
Compression, data, 156
Compromise equalization, defined, 82
CompuServe database, 224, 306
Computer, host, *see* Host computer
Computer Assisted Instruction (CAI), use in training, 533
Computer branch exchange (CBX), defined, 229
Computer crimes, examples of typical, 411–415
Computer Fraud & Abuse Act of 1986, 291
Computer integrated manufacturing (CIM), MAP/CIM, 281
Computer security, 455
Computerized design, 540–541
Computerized literature resources, 49–51
Computerized systems, controls to protect, 449, 454, 455, 597
COMSAT (Communications Satellite Corporation), 24–25
COMSEC, meaning of, 423
Concentrators:
  controls to protect, 608–612
  defined, 608
  purpose of, 174
  remote data (RDC), 560
Concerns, *see* Threats
Conditioning:
  C-level, 476, 522
  charges for, 522
  D-level, 476, 522

defined, 475
modems and, 476
seven levels of, 522
types of, 476
Conference call, defined, 208
CONFIG.SYS, DOS file, 301
Configurations:
  bus, 215–216
  cost-effective, 508
  defined, 212
  hybrid, 215–216
  local area network, 338, 368
  local area network, selection issue, 368
  multidrop, 217–218, 544
  multiplexed, 218–220
  point-to-point, 216
  ring, 215
  star, 214–215
  tree, 215
Conglomerate estimating, defined, 503
Connector cables:
  defined, 14
  role in network, 73
  security of, 435
Connectors:
  D-type, 75, 317
  fixed loss loop, 157
  permissive, 156
  port selection, 153
  programmable, 157
  universal, 157
  see also Interfaces
Consistency, of information, 378
Constant ratio codes, defined, 133
Consultative Committee on International Telephone and Telegraph (CCITT):
  error correction, 136
  Integrated Services Digital Network (ISDN), 226
  International Organization for Standardization (ISO) membership, 254
  network connection standards, 291–294
  protocol standards, 247
  role in data communications, 73, 288
  teletext standards, 485
CONTEL Business Networks, MIND-Data/PC software, 549
Continuous ARQ:
  BLAST usage, 315
  defined, 264
  satellite transmission using, 105
  see also Stop and wait ARQ
Control, centralized, 392
Control field, X.25 frame, 265
Control list, examples of, 451, 586–595, 597–635

Control matrix:
  automation of the, 454
  blank, 459
  construction of a, 446–454
  risk ranking a, 454
  use during network design, 507–508
Control points, network:
  defined, 416
  illustrated, 417
Control reviews:
  how to conduct, 446–454
  purpose of, 443
Control signals, defined, 207
Controllers:
  cluster, 327
  controls for packet switching, 428
  local intelligent, 217
  remote intelligent, 157, 429
Controlling, activities of, 379
Controls:
  adequacy of, 449, 451
  automation of, database, 454
  backup, 444–446
  CICS monitor, 441
  circuits, 434–435, 612–614
  computerized systems, 449, 454, 455, 582
  concentrators, 608–612
  data communication networks, 454–455, 581–595, 597–635
  data entry, 618–619
  data validation, 618–619
  database, 437, 442
  designing into systems, 454, 455
  disasters/disruptions, 602–606
  error handling, 614–616
  errors and omissions, 619–622
  front ends, 427–428, 627–631
  hardware, 427–434
  human error, 430
  local loops, 616–617
  matrix approach to, 446–454, 582
  message loss/change, 624–625
  microcomputers, 435–436
  modems, 428–429, 606–608
  multiplexers, 429, 608–612
  network architectures, 440–442
  network hardware, 427–434
  network management, 442–444
  Open Systems Interconnection (OSI) model, 438–440
  people, 625–627
  placement in matrix cells, 449
  protocols, 438–440
  reliability, 631–635

remote controllers, 429
restart/recovery, 442, 444–446, 622–623
software, 440–442, 597–602
switches, 608–612
switching nodes, 428
telecommunication access programs, 441–442
telephones, 432–434
teleprocessing monitors, 441, 442
terminals, 430–432
test equipment, 443
types of, 411
uptime, 631–635
Converters, protocol, 177
Coordinates:
air mileage calculation, 520
horizontal/vertical, 517, 520, 637–642
Copper circuits, NYNEX use of, 476
Corporation for Open Systems (COS), role in ISDN/OSI standards, 290
Correct, controls to, 411
COS (Corporation for Open Systems), 290
Cost analysis:
automated, features of, 540
methodologies, 517
Cost analyzers, illustrated, 399
Cost/benefit categories, 517, 519
Cost-effective network features, 508–509
Costs:
categories of, 516–517, 519
circuits, 516–531
data vs. voice, 381
direct, 519
hardware, 530
identification of, 516
indirect, 519
intangible benefits, 519
local area networks, 362–363
local area networks, cost benefits of, 372
local area networks, selection issue, 371
multiplexing, benefits of, 218
new network, 516–531
options, 522, 530
private leased/measured use, 474–485
revenue increases, 519
total local area network, 371
total link, 517
CPU, see Host computer
CRC, see Cyclical redundancy check
Cross-talk (noise), defined, 130, 355
Cross-training, defined, 384

Cross-vulnerability, encryption problem, 423
CROSSTALK XVI, operation of, 309–311
Crowding, satellite, 108
CRT (cathode ray tube), 183
CRTC (Canadian Radio-Television & Telecommunications Commission), 468
Cryption, defined, 419
Cryptography, background of, 419–420
CSDC (circuit switched digital capacity), 176
CSMA/CD (Carrier Sense Multiple Access/Collision Detection), described, 343–346
CSU (channel service unit), 164
Currents (electricity), ac vs. dc, 71
Cursor, defined, 183
Customer Information Control System (CICS):
described, 253
location in host, 252
LTERM in the, 441
Open Systems Interconnection (OSI) model and, 263
security features of, 441
sign-on table in the, 441
Cycles per second, 83, 107, 203
Cyclical parity check:
defined, 133
illustrated, 134
Cyclical redundancy check (CRC):
BLAST use of, 315
calculation of, 134
defined, 134
Ethernet use of, 345
X.PC use of, 315
Czar, Chief Information Officer as, 384

D-type conditioning, defined, 476, 522
D-type connectors, 75, 317
DAA (data access arrangements), 166
Data:
corrupted, control of, 127
defined, 378
editing, 149–150
lost, control of, 127
movement of, in networks, 54–55
storage on optical disks, 23
transmission modes for, 55–61
Data access arrangements (DAA), for modems, 166
Data bits, digital transmission, 88
Data channels:
function, 144
role in network, 137

Data circuit terminating equipment (DCE), defined, 73, 466
Data communication function, described, 377–378
Data communication networks:
basic concepts, 4–26
components of, 10, 12, 448, 583–586
control matrix review of, 581–593
controls to protect, 586–595, 597–635
defined, 9
examples of, 18
future of, 18–26
hardware in, 141
impact of technology on, 19
objections to proposed, 532
objectives of, 9, 492
role in business systems, 14–16
scope of, 492
selling proposed system, 532
threats to, 447, 582–584
types of, 16–18
written charter for, 380
see also Networks
Data communications:
applications, 18, 26–45
basic components, illustrated, 12
basic concepts of, 54
basic system, illustrated, 13
business operations needing, 10
categories of, 142–144
characteristics of problems solved by, 10
control matrix for, 452, 453, 583
controls for, 454–455
costs of, 19, 381
defined, 9, 12
design steps for, 490
error control in, 127–137, 614–616
evaluation criteria, 493–494
evolution of, 14–18
how to merge with voice, 382–384
leased circuits, defined, 228
literature on, 49–51
need to study, 4
optical disk use in, 23
risk analysis, 454
trends in, 18–26
usage modes, 11
voice vs., 380–381
see also Telecommunications
Data compaction, see Data compression
Data compression, purpose of, 156
Data editing, described, 149–150

Data Encryption Algorithm (DEA), 421
Data Encryption Standard (DES):
　hardware configuration for, 422
　importance of, 421
Data entry:
　controls for, 618–619
　defined, 618
Data flow control layer, Open Systems Interconnection (OSI) model, 262
Data flow diagrams (DFD):
　implementation use of, 533
　network, 495
Data grade twisted pair, defined, 355
Data item, analysis of, 499
Data line monitors:
　active *vs.* passive, 407
　defined, 406–407
Data link control layer:
　DNA model, 281–282
　SNA layer, 273
Data link layer:
　controls for, 438
　function of, 256, 258–259
　Open Systems Interconnection (OSI) model, 256, 258–259
Data link protocols, defined, 246
Data monopoly, defined, 22
Data-over-voice, *see* Newsletter, vol. 5, no. 1, Jan. 1988 (ISBN 0-471-60129-2)
Data PBX, defined, 231
Data processing function, traditional categories in the, 385
Data protection laws, 20
Data protectors (line protectors), 156
Data rates, high speed techniques, 81–82, 477
Data recorders, defined, 403
Data security, within organizational structure, 386
Data service unit (DSU), defined, 164
Data signaling, described, 77–78
Data synchronization, described, 77–78
Data terminal equipment (DTE):
　defined, 73, 465–466
　packetizing, 220
Data transporting network, DNA software module, 283
Data validation:
　controls for, 618–619
　defined, 618
Database management system (DBMS) software, location in network, 241
Databases:
　bibliographic, 49–51, 306

controls to protect, 437, 442
distributed, 16
local access transport area (LATA) tariffs, 540
planning formats for, 495
routing tables, 260
Databits, setting for communications, 306
DATAKIT virtual circuit switch, multiplexing, 176
DATALINK, communication software, 309
Datapac network, 224
DATAPHONE Digital Service, tariff for, 530
DB-9 connector, 75
　microcomputers and, 317
DB-25 connector, 75
　microcomputers and, 317
DCA (Distributed Communications Architecture), 282
DCE (data circuit terminating equipment), 73, 466
DDCMP (digital data communication message protocol), 282
DDD (direct distance dialing), 479
DDS (Digital Data System), 86
DEA (Data Encryption Algorithm), 421
DECconnect, Ethernet backbone, 354
Decentralization, of information, 378
DECnet, layers in, 281–282
Decoder, defined, 13
Decryption, defined, 419
Dedicated channels, *see* Dedicated circuits
Dedicated circuits, defined, 99, 202, 228, 342–343
Delay:
　caused by capacitance, 72
　caused by satellite links, 70–71, 105
　caused by statistical time division multiplexing (STDM), 171
　message, illustrated, 95
　propagation, 99, 105–106
Delay compensators, purpose of, 106
Delay distortion, defined, 131, 563
DEMO diskettes:
　control matrix design, automated, 454
　LANalyzer, network management, 367, 408, 643–656
　MIND-Data/PC, network design, 549
　risk ranking a controls matrix, 454
Demodulation, defined, 82

Dependence levels:
　random, 510
　sequential, 510
　time, 510
Deregulation:
　country examples, 468–469
　defined, 467
　impact on costing networks, 523
　United States, 469–474, 539
DES (Data Encryption Standard), 421, 422
Design and analysis:
　computerized tools for, 540–541
　network, 390–392, 400
　systems approach to, 489
Design steps:
　configuration selection, 508
　control matrix development, 507
　cost analysis, 516
　current system analysis, 494
　feasibility study, 490
　follow-up, 534
　geographic scope, 497
　hardware considerations, 515
　implementation, 532
　message analysis, 499
　network design, 496
　plan preparation, 492
　reevaluation, 534
　sell proposed network, 532
　software considerations, 511
　traffic calculation, 504
Detailed estimating, defined, 503
Detect, controls to, 411
Deutschen Bundespost, 468
Devices, defined, 199
DFT (Distributed Function Terminal), gateway, 348
Diagnostics, *see*
　Test sets
　Testing
Dial-back security, procedure for, 155
DIAL-IT 900 service, pricing, 482–483, 531
Dial-out-only facility, defined, 429
Dial-up circuits:
　cost calculation for, 515–527
　defined, 208
　direct distance dialing (DDD), 479
　distortion on, 479
　famous firsts, 209–210
　signaling on, 207–208
Dial-up modems, controls for, 429
DIALOG Information Services, 50, 224, 306
Dibits:
　defined, 90
　illustrated, 91

Differential phase shift keying (DPSK), modulation use of, 79, 81
Digital, defined, 12
Digital channel banks:
  defined, 121
  multiplexing in, 176
  use of, 121–122
Digital channels, error free guarantee, 405
Digital cross-connect switch, defined, 231
Digital cross-connect systems:
  defined, 121
  use of, 122
Digital data communication message protocol (DDCMP), 282
Digital data switches, 191–192
Digital Data System (DDS), bipolar signaling, 86
Digital Equipment Corporation, Distributed Network Architecture (DNA), 281
Digital line expanders, purpose of, 154
Digital modems, 164
Digital modulation, described, 86–89
Digital Network Architecture (DNA):
  described, 281–282
  layers of, 282
Digital PBX:
  defined, 229
  features of, 230
  illustrated, 231
Digital services:
  circuit costs, 530
  described, 477, 478
Digital side, modem, 82, 402–403
Digital signal, defined, 71
  see also Baseband
Digital switch, defined, 231
Digital switchboards, switching in, 192
Digital Termination Systems (DTS):
  bypassing and, 26, 233
  defined, 233
Digital test sets, use of, 403
Digital testing, defined, 402
Digital to analog, described, 83–85
Direct Broadcast Satellite, home use of the, 477
Direct current (dc), defined, 71
Direct distance dialing (DDD):
  cost calculation, 525–527
  described, 479
Directing, activities of, 379
Directional tap, defined, 578
Disaster plan, coverage of a, 444, 445–446

Disasters and disruptions:
  backup/recovery controls, 444–446, 602–606
  defined, 602
Discount voice services, 483–484, 531
Disengagement time, defined, 404
Disk files, local area network, selection issue, 371
Disk Operating System (DOS):
  capabilities, 306–307
  command structure, 302
  defined, 298
  files of the, 301
  importance of, 302
  MS-DOS, 563
  MS-DOS, Version 3.1, 350
  PC-DOS, 563
  typical commands in, 302
  Version 3.30 described, 304
Disk servers:
  defined, 329, 351, 369
  local area network, selection issue, 369
Disk storage, microcomputer, 298
Diskette drives, 300
Disks, optical, 23
Dispersion, cause of distortion, 82
Distortion:
  amplitude deviation, 82
  attenuation, 131
  caused by inductance, 73
  caused by leakage, 72
  caused by resistance, 73
  delay, 131
  described, 129
  digital vs. analog circuits, 98
  echoes, 96, 210–211
  envelope delay, 82, 402
  jitter, 131
  need for conditioning, 475
  noise, 129–130
  testing for, 402
  types of, 129–131, 563
  see also Noise
Distributed architecture, digital PBX, 229
Distributed Communications Architecture (DCA), described, 282
Distributed databases, 16
Distributed Function Terminal (DFT), gateway, 348
Distributed Network Architecture (DNA), described, 283
Distributed Systems (DS), network architecture, 283–284
Distributed Systems Environment (DSE), network architecture, 282
Divestiture, see Deregulation
DNA (Digital Network Architecture), 281–282

Documentation:
  local area network, selection issue, 371
  network, mandatory, 398–400
  software testing, 245
Domain, in Systems Network Architecture, defined, 272
DOS, see Disk Operating System
DOS commands:
  communication parameters with MODE, 306–307
  examples of, 301–302
  internal vs. external, 302
  MODE, 306
DOS files, use of, 301
DOS software:
  loading the, 300
  Version 3.1 for local area networks, 350
Double current loading, defined, 86
Dow-Jones News Retrieval Service, 224, 306
Downline network control, illustrated, 145
Downloading:
  capture buffer, 311
  defined, 325
  file transfer, 311
  see also Uploading
DPSK (differential phase shift keying), 79, 81
DRAM (dynamic random access memory), 19
Dropoff, defined, 217
Dropouts, defined, 402
DS (Distributed Systems) architecture, 283–284
DS-1, signaling format, 121
DSE (Distributed Systems Environment) architecture, 282
DSU (data service unit), 164
DTE see Data terminal equipment
DTS (digital termination systems), 26, 233
Dual cable:
  configuration, illustrated, 359
  single cable vs., 360
Dumb modem, smart vs., 161–162
Dumb terminal, defined, 188
Duplex, described, 96–97
Dynamic mathematical models, defined, 543
Dynamic random access memory (DRAM), described, 19

E-Mail, see Electronic mail
Eavesdropping, cellular telephones susceptibility to, 434
EBCDIC (Extended Binary Coded Decimal Interchange Code), 62, 63

Echo cancellers, cost of, 530
Echo checking, defined, 132
Echo suppressors:
    described, 210–211
    use of, 130, 210–211
Echoes:
    cause of, 210
    control of, 130
    defined, 96
    suppression of, 130, 210–211
Economic feasibility:
    of information, 378
    of network, 492
Edison, Thomas Alva, 5
Edit test, LANalyzer screen, 646
EDP auditors, network control
    and, 416
Efficiency:
    codes, formula for, 68
    network, 508–509
    transmission, defined, 67
EFS (Error Free Seconds), 405
EIA, *see* Electronic Industries As-
    sociation
Electrical bus topology, defined,
    338
Electrical fluctuations:
    local area network, selection
      issue, 371
    protection against, 320–324
Electrical ring topology, defined,
    338
Electrical signals, 83
Electrical topology, physical *vs.*,
    338
Electromagnetic interference
    (EMI), shielding against, 355
Electronic envelope, X.400 stan-
    dard, 290
Electronic funds transfer (EFT), *see*
    Banking networks
Electronic Industries Association
    (EIA):
    role in connector standards,
      73, 74
    RS232C standard, 289
Electronic mail (E-Mail):
    defined, 38
    described, 38–40
    local area network, selection
      issue, 371
    trends in, 26
Electronic shopping, 45
EMI (electromagnetic interference),
    355
Emulation board, purpose of, 326
Emulation software, purpose of,
    326
Encoding:
    adaptive Huffman, 156
    adaptive subbands excited
      transform (ASET), 120
    Huffman, 156

    purpose of, 12
    run length, 156
    see also Encryption
Encryption:
    algorithms for, 420–427
    cross-vulnerability problem in,
      423
    defined, 23, 419
    end-to-end, defined, 181
    hardware, 180–181
    hardware, defined, 180
    hardware configuration, illus-
      trated, 422
    key usage, 420
    link-to-link, defined, 181
    public key *vs.* private key,
      424
    software, defined, 180
End distortion, defined, 563
End of text (EXT), Binary Syn-
    chronous Communications (BSC),
    267
End office, defined, 14, 99, 200,
    520
End-to-end encryption, defined,
    181
End-to-end layer, Open Systems
    Interconnection (OSI) model, 261
Ending flag, X.25 frame, 265
England:
    British Post Office Commis-
      sion role, 468
    light amplifier development,
      112
    PSS network, 225
    Prestel videotex, 226
    satellite transmissions, 20
Entry devices (input), 181–189
Envelope delay distortion:
    cause of, 82
    conditioning to reduce, 475
    defined, 402
Equal access, defined, 471
Equalization:
    compromise *vs.* adaptive, 82
    defined, 82, 475
    modem feature, 165, 476
Equalizers:
    function of, 475–576
    use of, 131
Equations, *see* Formulas
Error control:
    approaches to, 131
    considerations in, 127
    front end, 150
    types of, 131–137
Error detection:
    DES *vs.* other techniques,
      422–423
    impact on network efficiency,
      506
    with retransmission, types of,
      132

Error Free Seconds (EFS), defined,
    405
Error handling:
    controls for, 614–616
    defined, 614
Error rates, circuit, 397
Errors:
    bursts of, 128
    categories of, 127
    control of, 127–137, 614–616,
      619–622
    detection by DES algorithm,
      422
    distortion as cause of, 72, 98
    human, controls to prevent,
      430
    testing for, 404–405
Errors and omissions:
    controls to prevent, 619–622
    defined, 619
Estimating:
    comparison, 503
    conglomerate, 503
    detailed, 503
Ethernet:
    baseband bus, 339
    coaxial cable, 353, 358
    DECconnect backbone, 354
    IEEE 802.3 standard, 288
    LANalyzer, 643
    packet field sizes, 345
    XEROX network, 284
Euronet, 225
Europe:
    communication environment,
      225, 468
    primary access service
      (31B + 1D ISDN), 227
    see also specific country
      names
Evaluation criteria, for network,
    493–494
Even parity, 133
Excelan, LANalyzer EX5000E, 367,
    408, 643, 656
Exchange access service, defined,
    471
Exchange Carrier Standards Asso-
    ciation, T-1 ISDN standards, 289
Exchange office, defined, 14, 99,
    200
Exchange termination box, ISDN
    use of, 227
Exit (access), defined, 441
Exposures, *see* Threats
EXT (End of text), 267
Extended Binary Coded Decimal
    Interchange Code (EBCDIC), de-
    scribed, 62, 63
Extended common carriers, de-
    fined, 234
Extended open function, interrupt
    21h, 309

External DOS commands, location of, 302
External modems, defined, 317

4-of-8 codes, 67
Facsimile (FAX) terminals, 186–188
Failure, role in logging, 137
Failure control group, function, 392–395
Failure mode analysis, defined, 243
Fast select polling, defined, 123
FAX (facsimile terminals), 186–188
FCC, see Federal Communications Commission
FCC Tariff 10, rate center coordinates, 520
FCC Tariff 11, private line rates, 521
FDDI (Fiber Distributed Data Interface), 288, 356
FDM (frequency division multiplexing), 167–168, 176
FDX (full duplex transmission), 97–98
Feasibility factors:
    economic, 378, 492
    operational, 492
    technical, 492
Feasibility study:
    conducting a, 490–492
    written report, 492
FEC (forward error correction), 23, 135
FED WIRE network, 42
Federal Communications Commission (FCC):
    digital termination systems (DTS), 234
    microcomputer certification, 303
    modem certification, 166
    role of, 12, 467
    Tariff 10, 520
    Tariff 11, 521
Federal Standard 1031, 75
Federal Telecommunications System (FTS), CCSA network, 485
Federal Wiretap Statute, 291
Femtosecond, defined, 421
FEP, see Front end communication processor
Fiber Distributed Data Interface (FDDI), local area network standard, 288, 356
Fiber optics:
    cable, illustrated, 357
    described, 111–117
    local area network medium, 356
    multiplexing, 173
    security of, 434

Fields:
    message, 499
    X.25 frame, 265
Figures, Baudot functions, 62
File formats, network design of, 495
File locking:
    defined, 351
    local area network need for, 365, 366
File redirector, defined, 307
File servers:
    defined, 307, 320, 351, 369
    local area network, selection issue, 369
File transfer:
    capture buffer vs., 311
    downloading, 311
    impact on networks, 325
File transfer program, CROSS-TALK function, 310
File transfer time, illustrated, 160
Filters (noise), amplifiers vs. repeaters, 98
Financial networks, see Banking networks
Firmware:
    defined, 247
    described, 25
Fixed loss loop connection, 157
Flags, X.25 frame, 265
Flow control:
    defined, 259, 312
    multiplexing, 171
    transport layer function, 261
Flowcharts, 533
Flywheel effect, in modems, 166
FM band, defined, 119
Follow-up, after implementation, 534
Foreign Exchange Service (FX), described 484–485
Formulas:
    air mileage with coordinates, 520
    baud, 205
    cyclical redundancy check (CRC), 134
    efficiency of codes, 68
    local area network transmission speed, 345
    network availability, 396
    number of lines (circuits), 190
    response time, 126
    TRIB, 70–71
    volt-amps, 323
Forward error correction (FEC), defined, 23, 135
Four-wire circuit, defined, 96
Frame check sequence field, X.25, 266
Frames:
    defined, 58

signaling speed calculation, 121
Synchronous Data Link Control (SDLC), 271
X.25, 265
see also Fields
France:
    laser ISL development, 108
    Postes Telephonique et Telegraphique, 468
    Teletel videotex, 226
    transborder data flow, 20
    Transpac network, 225
Freenet, radio local area network, 352
Frequency bands, satellite, 106–107
Frequency division multiplexing (FDM):
    described, 167, 176
    illustrated, 168
Frequency modulation, defined, 79, 80
Frequency shift keying (FSK), use in modulation, 79, 83, 84, 92
Frequency spectrum, illustrated, 203
Front end central computer configuration, illustrated, 144
Front end communication processor (FEP):
    controls to protect, 427–428, 627–631
    defined, 14, 146, 627
    described, 146–151
    error control, 150
    line control, 148
    message assembly, 149
    message editing, 149
    message queuing, 150
    message recording, 150
    protocol conversion, 149
    role in networks, 122
    statistical recording, 150
FSK (frequency shift keying), 79, 83, 84, 92
FTS (Federal Telecommunications System), 485
Full duplex transmission:
    defined, 97
    illustrated, 98
Function call, defined, 308
Funk and Scott Index, 50–51
Fusion splicing, described, 115
FX (Foreign Exchange Service), 484–485

Gantt, use during implementation, 533
Gateway servers, defined, 576
Gateways:
    defined, 347

Gateways (*Continued*)
   hybrid network connections, 349
   local area network, selection issue, 371
   network-to-network, 349
   Synchronous Data Link Control (SDLC), 348
   system-to-system, 349–350
   Token-Ring Network, 348
   X.75 node, 349, 481
Gaussian noise:
   defined, 129
   impact on signal-to-noise ratio, 207
GE Information Services, 225
General Telephone & Electronics (GTE):
   common carrier, 467
   competitors of, 474
Geographic boundaries, network, 492
Geographic scope, levels, 497
Geosynchronous satellites, illustrated, 104, 105, 109
Germany (West):
   Bildschirimtext, 226
   Deutschen Bundespost, 468
   laser ISL development, 108
   transborder data flow, 20
Gigahertz (GHz), defined, 107
Glossary, 555–580
Goals:
   intermediate, 492
   major, 492
   minor, 492
   network, 508–509
   network designer, 541–542
Graded index fibers, nicking of, 115
Group 48,000 hertz:
   cost of, illustrated, 527
   wideband, 476
Group multiplexing, illustrated, 214
Guardbands:
   frequency separation, 168, 204
   illustrated, 205
Guru, local area network need for a, 362, 364

H coordinates, *see* Horizontal coordinates
Hackers, example of, 414
Hagelbarger code, defined, 136
Half duplex transmission, defined, 96
Halide glasses, fiber optic use of, 111
Hamming code:
   defined, 135
   illustrated, 136

Handheld test sets, defined, 403–404
Hardware:
   controls for network, 427–434
   costs, example of, 530
   local area network, installation, 364
   network, design considerations for, 515–516
   network, selection criteria, 391–392
   protocol converter boxes, 177–178
   testing and analysis, 401–408
Hardware encryption:
   defined, 180
   illustrated, 181
Harmonic distortion, defined, 563
Hawaii, WATS band, 480
HDLC, *see* High-level Data Link Control
HDX, *see* Half duplex transmission
Health care networks, described, 18
Help desk, trouble reporting, 393
Hertz (Hz):
   defined, 83, 107
   frequency spectrum, 203
Hewlett-Packard, Distributed System (DS), 283–284
Hierarchical architecture, digital PBX, 229
High level Data Link Control (HDLC):
   AT&T ISDN chip, 228
   BLAST use of, 315
   Distributed Systems Environment (DSE) model and, 282
   new systems, recommended for use with, 512
   Open Systems Interconnection (ISO) model and, 259
   satellite channel usage, 477
High speed services:
   Accunet, 481
   Integrated Services Digital Network (ISDN), 477
   packet switched, 481
   satellite, 477
   T-1 channels, 478
   wideband, 476
   wideband, costs of, 527
Hits (signal), 402
Hollerith code, 62
Home satellite television:
   Direct Broadcast Satellite approved, 477
   future of, 20
Honeywell, Distributed Systems Environment (DSE), 282
Horizontal coordinates:
   by city, 637–642

use of, 517, 520
Host computer:
   configurations, 142–145
   defined, 14
   role in network, 137, 142–146
   trends, 146
Host services layer, Burroughs Network Architecture (BNA), 282–283
Host-to-host layer, ISO model, 260
Hotline service, described, 485
Hub (electrical), wiring, described, 338
Hub (satellite), cost of a, 107
Hub go-ahead polling, defined, 124
Hub topology, defined, 338
Huffman encoding, defined, 156
Hybrid configurations, described, 215–216
Hybrid networks:
   DECconnect, 354
   defined, 198, 349
   gateways for, 349

IBM Advanced Program-to-Program Communications (APPC):
   compared to a DECnet APPC node, 277
   described, 274–275
   Systems Network Architecture (SNA), program interface, 274
IBM Binary Synchronous Communications (BSC), 267
IBM cabling system, described, 353
IBM Disk Operating System (DOS), Version 3.30 features, 304
IBM Information Network, 224
IBM NetView, network management, 387–388
IBM Operating System/2 (OS/2), features, 304
IBM PC Network:
   broadband bus, 339–340
   NETBIOS location in the, 307
IBM Personal Computer (PC), operation, 298–302
IBM Personal System/2 (PS/2), features, 303–304
IBM Systems Network Architecture (SNA), 268–278
IBM Token-Ring Network, baseband ring, 340
ID, *see*
   Passwords
   Security locks
IEEE (Institute of Electrical and Electronics Engineers), 288–289
IEEE 802:
   Carrier Sense Multiple Access protocol, 343
   local area network standards, 288

token access method, 346
IEEE 802.3:
 Ethernet protocol, 346, 643
 Ethernet standard, 288
IEEE 802.4:
 token bus standard, 288
 token passing, Manufacturing
  Automation Protocol (MAP),
  281, 288
IEEE 802.5, token ring standard,
 288
IEEE 1003.1, portable operating
 system standard, 289
ILD (injection laser diodes), 113
Implementation:
 all at once, defined, 533
 basic approaches to, 533
 chronological, defined, 533
 follow-up after, 534
 local area network, 363–365
 new network, 532–534
 one-for-one, defined, 533
 phases, defined, 533
 pilot operation, defined, 533
 reevaluation after, 534
 tasks of, 532
 tools for use during, 533
Impulse noise, defined, 129, 402
Inductance, cause of distortion, 73
Infinet Series 90, network management, 408
Information:
 CATER to user needs, 378,
  489
 defined, 378
 strategic resource of, 4, 22,
  384, 386, 507
Information bits, defined, 67
Information control function:
 responsibilities of the, 385
 location in management hierarchy, 384
Information lag, speeding up the,
 5
Information society:
 data monopoly in the, 22
 dominated by computers, 4,
  379, 490
Information systems function, responsibilities of the, 385
Information Systems Network
 (ISN), Premises Distribution System (PDS) cabling, 353
Infrared transmission, 233–234
Inherent broadband architecture,
 defined, 230
Initialize, defined, 307
Injection laser diodes (ILD), optical
 transmission, 113
Input devices:
 digital entry, 189
 terminals as, 181–189
 voice entry, 189

Input/processing/output models,
 network design using, 495
INSPEC database, 50
Installation:
 cables, 358–359
 local area network, 363–365
 see also Implementation
Institute of Electrical and Electronics Engineers (IEEE), local
 area network standards, 288–289
Integrated packet switching, defined, 230
Integrated Services Digital Network (ISDN):
 defined, 226, 466
 described, 477–478
 illustrated, 227
 micro-to-mainframe connections, 331
 Premises Distributed System
  connection, 354
 standards for, 289, 290
 2B + D basic access service,
  226, 289, 477
 23B + 1D primary access service, North America, 226–
  227
 31B + 1D primary access service, Europe, 227
Integrated voice and data:
 defined, 230
 how to merge, 382–384
 organization of, 385
 organizational challenge, 383–
  384
 types of circuits for, 476–484
 see also Voice-over-data
Intelligent controllers, see Controllers
Intelligent Matrix 3000 Switch
 System, network management,
 408
Intelligent modems, 163
Intelligent port selectors, purpose
 of, 153
Intelligent terminal controllers,
 purpose of, 157
Intelligent terminals, defined, 188
Interexchange channels (IXC):
 controls to protect, 434
 defined, 14
 described, 99
 multiplexed, 219
 packet networks, 220
 point of presence connection,
  521, 524–525
Interfaces:
 channel, in front end, 147
 interrupt 5Ch, network interface card, 508
 line interface units, 147
 NETBIOS communications,
  307

ports, 147
 RS232C, 73–75
 RS449, 75–76
Interference, see Noise
Interlaced parity, defined, 133
InterLATA services, 471, 474
 calling rates, illustrated, 526
 digital charges, 530
 mileage charges, 522
Interleaving, data packets, 176, 222
Intermediate goals, defined, 492
Intermodulation distortion, de-
 fined, 402
Intermodulation noise, defined,
 130
Internal controls, 449, 582
Internal DOS commands, location
 of, 302
Internal modems, defined, 317
International networks, defined, 18
International Organization for
 Standardization (ISO), seven-layer
 model, 253–264, 287
Interoffice trunks, 200
Interrupt, defined, 308
Interrupt 5Ch, operation of, 308
Interrupt 21h, operation of, 308–
 309
Interrupt mode:
 defined, 145, 199
 protocol analogy, 343–344
Interrupt system, central control
 vs., 122–123
Intersatellite link (ISL), laser, 108
Interstate, defined, 467
Interstate WATS, contrasted with
 intrastate, 480
Intertoll trunks, 201
IntraLATA services, 471
 California, 473
 mileage charges, 523
Intrastate, defined, 467
Intrastate WATS, contrasted with
 interstate, 480
Inverse multiplexers, defined, 175
ISDN, see Integrated Services Digi-
 tal Network
ISL (laser intersatellite link), 108
ISN (Information Systems Net-
 work), 353
ISO (International Organization for
 Standardization), 253–264, 287
ISO model, see Open Systems In-
 terconnection Reference Model
Isochronous transmission:
 defined, 61
 described, 59–60
IXC, see Interexchange channels
IXC circuits, defined, 14

Japan:
 Captain videotex, 226

Japan (*Continued*)
  deregulation, 469
Jitter, defined, 131, 402, 563
Job tasks, impact on network design, 510

KERMIT:
  communication software, 309, 310, 313–315
  XMODEM contrasted with, 314
Key (encryption), defined, 420
Key-contents, defined, 415
Key system switchboard, defined, 228
Kilohertz (KHz), defined, 107
Ku-band, 107

LANalyzer EX5000E:
  Edit Text screen example, 646
  how to use demo diskette, 643–656
  local area network analysis, 366, 408, 643
  Run Counter screen example, 650
  Run Global screen example, 651
  Trace Buffer screen example, 653
Laser intersatellite link (ISL), development of, 108
Lasers, optical transmission, 356
  see also Fiber optics
Last mile, *see* Local loop
LATA, *see* Local access transport area
Laws:
  data protection, 20
  impact on networks, 290–291, 495
Layered approach, benefits of, 254–255
Layers:
  Burroughs Network Architecture (BNA) model, 282–283
  Distributed Network Architecture (DNA) model, 282
  Manufacturing Automation Protocol (MAP) model, 281
  Open Systems Interconnection (OSI) model, 255–264
  Systems Network Architecture (SNA) model, 273
Layout, cables in building, 361
Lead time, defined, 533
Leakage (electrical), cause of distortion, 72
Leased circuits:
  defined, 99, 202, 228
  dial-up *vs.*, 475
  types of, 474–478

Least cost routing, defined, 229
LED (light-emitting diodes), 113, 356
Legal requirements:
  impact on network design, 495
  security, 290–291
Legally enforceable standards, 290–294
Letters, Baudot function, 62
Light-emitting diodes (LED), optical transmission, 113, 356
Light pens, input by, 188
Light wave multiplexing, 176
Limited common carriers, defined, 234
Line adapters, 151
Line drivers, 160
Line interface units:
  illustrated, 152
  purpose of, 147, 151–152
Line loading, factors affecting, 506
Line outages, cause of errors, 130–131
Line protectors, purpose of, 156
Line protocol for transmission, model of a, 295
Line sharing device, defined, 14
Line splitters:
  function of, 153
  illustrated, 154
Lines:
  control of, 148–149
  defined, 100, 612
  security of, 155
  see also Circuits
Link loading, illustrated, 505
Link-to-link encryption, defined, 181
Links:
  cost of, 517
  defined, 100
  volume per, analysis, 501, 502, 504–506
  see also Circuits
Literature resources, data communications, 49–51
Local access transport area (LATA):
  defined, 471
  long distance, 201, 473
  mileage charges, 522, 523
  modeled after Standard Metropolitan Statistical Areas (SMSA), 471
  point of presence in, 521
  state-by-state, 472
  see also
    InterLATA
    IntraLATA
Local area networks (LAN):
  analysis software, 366, 643
  backup/recovery, 371
  baseband *vs.* broadband, 369
  bridges, 371

cabling decision, 369
centralization, 372
circuit boards, 370
competing methodologies, 336
computer interface, 143
configuration issues, 368
costs, 362–363, 371
defined, 198, 335
described, 334–372
disk servers, 369
documentation, 371
economic benefits of, 372
electrical protection, 320–324, 371
electronic mail, 371
file servers, 369
gateways, 371
guru for managing the, 362, 364
illustrated, 336, 338, 339
implementing, 363–365
LANalyzer for network analysis, 643–656
local disk files, 371
management tools for, 365
managing, 365–366
network servers, 370
nodes, distance between, 370
nodes, number of, 369
PBX relationship, 336–337
PC Network, 307, 308
printers, 371
protocols, 369
purchasing considerations, 363, 367–371
satellite connection, 108, 110–111
security, *see* Newsletter, vol. 5, no. 1, Jan. 1988 (ISBN 0-471-60129-2)
security features, 371
selection considerations, 367–372
software for, 350–352, 370
standards for, 339–341
surge/sag protection, 371
Token-Ring Network, 308
training, 371
users, number of, 369
voice mail, 371
wireless, 352
workstation hardware, 370
Local calls:
  defined, 200
  intraLATA, 471
Local Exchange Companies, defined, 471
Local exchange service, defined, 471
Local intelligent devices, illustrated, 217
Local loop:
  access, types of, 471

bypassing the, 26
cellular radio in, 23
connection to modem, 95
controls to protect, 616–617
defined, 14, 94, 616
local exchange service, 471
rate center connection, 520
Lock/unlock function, interrupt
21h, 308–309
Locking techniques, 351–352
Logging:
    message, 150
    purpose of, 137
Logical networks, testing, 387
Logical unit (LU):
    defined, 269
    name assignment, 270
    session types, 277–278
LOGON, communication soft-
    ware, 309
Long distance calls:
    AT&T Communications, 473
    cost calculation, 525–526
    defined, 201
Long lines, AT&T divestiture, 473
Loop checking, defined, 132
Loopback testing, modem failure,
    165, 405
Loss, signal power, 402
LTERM, terminal restriction, Cus-
    tomer Information Control Sys-
    tem (CICS), 441
LU, see Logical unit
LU 6.2:
    impact on gateways, 348
    micro-to-mainframe links, 330
    Systems Network Architec-
        ture (SNA) component, 274–
        275
    virtual telecommunication ac-
        cess method (VTAM), 251
LU-LU session, defined, 274

M-of-N codes, described, 62, 67,
    133
Machine-to-machine protocols, de-
    fined, 246
Maestro Network Management
    System, network monitoring, 408
Mailboxes, electronic, 39
Main Distribution Function
    (MDF), wiring connection, 580
Major goals, defined, 492
Management:
    controls for, 442–444
    local area networks, 365–366
Management reports, need for, 394
Managers:
    acceptance of system, 489, 532
    responsibilities of, 381–382
Manufacturing Automation Proto-
    col (MAP),
        broadband token bus, 340

described, 281
IEEE 802.4 standard, 288
layers of, 281
MAP, see Manufacturing Automa-
    tion Protocol
MAP/CIM, 281
MAP/TOP, 281
Maps:
    country, illustrated, 498
    drawing the network, 497–
        499
    network documentation, 399,
        492, 497
Mark (signaling), defined, 71, 87
Master Control Program (MCP),
    Burroughs Network Architecture
    (BNA), usage, 382
Mathematical models, defined, 543
Matrix of controls:
    automating on a PC, 454
    blank, 450, 459
    case study using, 457–461
    constructing a, 446–454
    controls placed into cells, 449
    evaluation of, example, 453
    example with controls, 452,
        583
    microcomputer use of a, 454
    network design use of, 507–
        508
    risk ranking a, 454
Maximize, defined, 509
MCI Communications:
    common carrier, 12, 198, 466,
        471
    competitors of, 474
    discount voice services, 483–
        484, 531
    local access transport access
        (LATA) services, 471
MCP (Master Control Program),
    382
MCVD (modified chemical vapor
    deposition), 116
MDF (Main Distribution Func-
    tion), 580
Mean Time Between Failures
    (MTBF), purpose, 396, 443, 516
Mean Time To Diagnose (MTTD),
    purpose, 395, 443, 516
Mean Time To Fix (MTTF), de-
    fined, 396, 443, 516
Mean Time To Repair (MTTR), de-
    fined, 396, 516
Mean Time To Respond (MTTR),
    purpose, 395, 396, 443, 516
Measured use services, described,
    478–484
Media, defined, 199, 465
Medium, defined, 10, 465
Megahertz (MHz), defined, 107
Megatrends, 5
Memory, RAM vs. ROM, 299

Message accountability, defined,
    242
Message analysis, 499–504
Message assembly, 149
Message buffering, 150
Message change:
    controls to prevent, 624–625
    defined, 624
Message contents:
    illustrated, 500
    local area network transmittal,
        339
Message controls, 624–625
Message delay, illustrated, 542
Message editing, 149–150
Message fields:
    analysis, 499
    X.25 frame, 265
Message flow:
    through hardware, 142
    through network, 54–55
    through software, 238
Message format:
    Binary Synchronous Commu-
        nications (BSC), 268
    X.25, 265
Message input time, defined, 126
Message length, analysis, 499
Message loss:
    controls to prevent, 624–625
    defined, 624
Message output time, defined, 126
Message protocols:
    Binary Synchronous Commu-
        nications (BSC) format, 268
    Open Systems Interconnection
        (OSI) model, 260
    X.25 and SDLC frame, 265
Message queuing, 150
Message recording, 150
Message security, sensitivity level,
    434
Message switching, defined, 223
Message type, analysis, 499
Message unit, defined, 271
Message volume:
    analysis, 499
    critical to design, 501
    varying, 501, 503
Messages:
    accounting for, 242–243
    integrity of, 243
    XMODEM block protocol, 313
Mexico, communications environ-
    ment, 468
MFOS (Multi-Function Operations
    System), 389
Micro-to-mainframe connections:
    described, 324–331
    establishing, 326
    factors to consider, 326–327
    IBM Cabling System connec-
        tion, 353

Micro-to-mainframe connections (*Continued*)
    illustrated, 327
    Integrated Services Digital Network (ISDN) speed, 331
    KERMIT, suitability for, 314
Micro-to-mainframe link, *see* Micro-to-mainframe connections
Micro-to-micro communications, *see* Newsletter, vol. 5, no. 1, Jan. 1988 (ISBN 0-471-60129-2)
Micro-to-modem, connectors for, 317
Microcomputers:
    auditing of, 435, 455
    booting procedure, 300–302
    business uses, 297–298
    certification of, 303
    communication software for, 306–312
    control matrix design on, 454
    controls for, 430–432, 435–436
    defined, 13, 182, 297
    described, 297–331
    electrical protection, 320–324
    internal configuration, illustrated, 300
    modems for, 316–320
    network design use of, 541, 547, 548
    protocols for, 312–316
    role in communications, 297–298
    role in data transmission, 55
    workstations *vs.*, 182
MICROLINK II, communication software, 309
Microwave tower, illustrated, 103
Microwave transmission:
    described, 103–104
    frequencies, 234
Mileage calculation, circuit costs, 520–525
MIND-Data/PC:
    described, 546–552
    educational version, 549, 552
    graphic screens, illustrated, 550, 551
    report, illustrated, 550
Minor goals, defined, 492
MINTS network, 42
MODE command, setting communication parameters, 306–307
Models:
    defined, 543
    input/processing/output, 495
    queuing of polled terminals, 544
    topology optimization, 540–541
    types of, 543–546
    see also Simulation

Modem eliminators, defined, 161
Modems:
    acoustic coupler, 161
    analog side *vs.* digital side, 82–83, 402
    baseband *vs.* broadband, 159
    Bell 201 model, 9, 81, 82
    Bell 202 model, 202, 205, 206
    certification of, 166, 469
    classification of, 159
    conditioning for, 476
    controls to protect, 428–429, 606–608
    defined, 12, 14, 82, 606
    described, 82–84
    digital, 164
    dumb, 161–162
    equalization in, 165, 476
    external, 317
    features of, 164–166
    frequency shift keying, 83
    functions, 82
    intelligent, 163
    interfaces for, 73–77
    internal, 317
    light indicators, 162
    line driver, 160
    local loop connection, 95
    microcomputer, 316–320
    multiport, 174
    null, 161, 319–320
    operation of, 84
    optical, 160
    self-testing, 405–406
    short haul, 160–161
    smart, 162
    timing in, 83
    types of, 159–164
    typical brands, 317
    ZOOM/MODEM PC, 319
Modes:
    central control, 145, 199
    interrupt, 145, 149
    transmission, 96–97
Modified chemical vapor deposition (MCVD), optical fiber fabrication, 116
Modulation:
    adaptive differential pulse code (ADPCM), 120
    amplitude (AM), 79–80
    analog, 77–82
    defined, 78, 83
    dibit schemes, 93
    differential phase shift keying (DPSK), 79, 81
    digital, 86–89
    frequency (FM), 79–80
    frequency shift keying (FSK), 79, 83
    phase, 79
    phase shift keying (PSK), 79, 81

pulse, 86
pulse amplitude (PAM), 86
pulse code (PCM), 88–89, 120
pulse duration (PDM), 86
pulse position (PPM), 86
quadrature amplitude (QAM), 81
quadrature phase shift keying (QPSK), 81
Modulator, defined, 79
Monitors:
    data line, 406
    defined, 403
    see also Test sets
Monomode fibers, defined, 357
Monopoly:
    communication services, 468–469
    data, 22
Morse, Samuel, 5
Morse code, 5
Mouse (computer), input device, 188–189
MOVE-IT, communication software, 309
MS-DOS, 316, 332, 563
MTBF (Mean Time Between Failures), 396, 443, 516
MTTD (Mean Time To Diagnose), 395, 443, 516
MTTF (Mean Time To Fix), 396, 443, 516
MTTR:
    Mean Time To Repair, 396, 516
    Mean Time To Respond, 395, 396, 443, 516
Multidrop circuits:
    defined, 154
    design problem using, 536
    digital, 477
    illustrated, 218, 511
Multidrop configuration:
    described, 217–218
    illustrated, 544
Multidrop networks, design of, 542
Multi-Function Operations System (MFOS), switch management, 389
Multimode fibers, defined, 357
Multiorganization networks, defined, 17
Multiple application networks, defined, 17
Multiple network configuration, illustrated, 514
Multiplex, defined, 167, 218
Multiplexed configuration, illustrated, 512
Multiplexers:
    biplexers, 175
    bit compression, 175–176
    circuit switched digital capacity, 176

concentrators as, 174
controls to protect, 608–612
defined, 608
digital channel banks, 176
fiber optic, 173
frequency division, 167–169
inverse, 175
light wave, 176
multiport modems, 174
pure, 169
statistical time division, 170–
173
T-1, 173–174
time compression, 176–177
time division, 169–170
time division compression,
176
types of, 167–174
virtual circuit switch, 176
Multiplexing:
AT&T, 175–177
Group, illustrated, 214
illustrated, 214
levels, illustrated, 219
voice call, 212
Multipoint circuits, see Multidrop
circuits
Multiport modems, described, 174
Multisystem communication network, case study, 457–461
Multitap, defined, 578
Multiuser system micro-to-mainframe link, defined, 328

NAK, see Negative acknowledg-
ment
Nanosecond, defined, 19
National Bureau of Standards
(NBS):
DES algorithm, 421
network protocol testing, 289
National Security Agency (NSA),
role in DES certification, 423
NBS, see National Bureau of Stan-
dards
NBSDES (NBS Data Encryption
Standard), 421
NCCF (Network Communication
Control Facility), 388
NCP (network control program),
251
NCR Corporation, Distributed
Network Architecture (DNA),
283
Needs assessment, factors in, 491
Negative acknowledgment (NAK):
satellite transmission, 104
XMODEM, 313
see also Positive acknowledg-
ment
NETBIOS:
communication interface, 307

defined, 307–308
Open Systems Interconnection
(OSI) model, function in, 307
Operating System/2 support
for, 331
PC Network, function in, 307
Synchronous Data Link Control (SDLC) gateway, 348
Netherlands, deregulation, 469
NETMAN, network management,
389
Netscan 2000 Software, network
management, 408
NetView:
programs within, 388
real-time network manage-
ment, 408
voice/data management, 387–
388
Network activity, defined, 365
Network addressable units, Sys-
tems Network Architecture
(SNA) component, 271
Network alternatives:
choice set of, 509
maximize vs. optimize vs. satisfice, 509
Network analysis, statistics, illustrated, 368
Network analyzers:
illustrated, 367
LANalyzer, 366, 408
response time, 406
Network architectures:
characteristics, 247–248
controls for, 440–442
Network availability, equation for,
396
Network changes, defined, 366
Network Communication Control
Facility (NCCF), NetView program, 388
Network configurations:
attainable vs. unattainable,
509
choice set of, 509
costs related to choice set,
516
determining, 508–511
preliminary vs. attainable, 516
Network control center, illustrated, 403
Network control program (NCP),
function, 251
Network cost analyzers:
illustrated, 518
use of, 517
Network database, controls for,
442
Network design:
basic concepts of, 488–534
computerized methods for,
538–552

evaluation criteria for, 493–
494
hardware considerations in,
515–516
manual vs. computerized, 490
software considerations in,
511–515
steps of, 490–534
strategic plans impact on, 496
system requirements, 496
tools of, 540–541, 543
Network design and analysis
group, functions, 390–392
Network error, defined, 366
Network failure control group,
functions, 392–395
Network layer:
controls for, 439
function, 256, 259–260
Open Systems Interconnection
(OSI) model, 256, 259–260
Network link traffic table, illustrated, 502
Network loading, defined, 366
Network Logical Data Manager
(NLDM), NetView program, 388
Network management:
controls for, 442–444
described, 377–408
Multi-Function Operations
System (MFOS) module, 389
tools for, 365
Network Management Productivity Facility (NMPF), NetView
function, 388
Network management systems:
defined, 404
real-time, 408
Network managers:
tasks required of, 381–382
Network operations group, functions, 392
Network optimizers:
defined, 391
MIND-Data/PC, 546–552
Network Problem Determination
Application (NPDA), NetView
program, 388
Network security:
complex environment for,
414–416
control points, illustrated,
417
defined, 366
described, 410–415
Network servers:
defined, 328, 370
local area network, selection
issue, 370
standards for, 309
Network services layer:
Burroughs Network Architecture (BNA) model, 283

Network services layer
(*Continued*)
 Distributed Network Architecture (DNA) model, 281–
  282
Network status, assessment of,
 400–401
Network switches:
 described, 192–193
 packet switching, 220–223
 switchboards, 228–233
Network termination box, Integrated Services Digital Network
 (ISDN), use, 227
Network-to-network gateways, defined, 349
Networks:
 applications, 26–45
 architectures, 268–286
 availability calculation, 396
 backbone, 224, 338
 backup and recovery, 444–
  446
 banking, 28, 40–43
 basic objections to proposed,
  532
 case study of controls, 457–
  461
 common carrier, 17
 computers for managing, 404
 configurations, 197, 212, 214–
  216, 508–511
 cost analysis, 399, 516–531
 defined, 197–198
 design and analysis, 390–392,
  400
 design fundamentals, 488
 dial-up, costs, 525–527
 digital, costs, 530
 documentation, mandatory,
  398–400
 elements of, 198
 goals of, 508–509
 hardware in, 141–194
 hardware selection criteria,
  391–392
 hybrid, 198, 349
 implementing, 532–534
 international, 18
 introduction to, 4
 local area (LAN), 198, 334
 logical *vs.* physical testing,
  387
 management of, 365–367,
  377–379, 386–397
 maps of, 399
 modeling, 541–546
 monitoring, 387–389, 393–
  397, 400–401
 monitoring, physical *vs.* logical, 387
 multidrop, design of, 542, 547
 multidrop, illustrated, 511

multiorganization, 17
multiple application, 17
multiple configuration, illustrated, 514
multiplexed, illustrated, 512
office automation and, 28
operation modes, 145, 199
optimization of, 540–541, 547
organization, 379–386
organization-wide, 17
packet switching, costs, 528
packet switching, example, 28,
 221, 222, 513
packet switching, Open Systems Interconnection (OSI)
 model, 259
PBX, illustrated, 232
point-to-point, example, 54,
 216, 498, 501
procedures for, 533
public timesharing, advantages, 223
reports, management, 394
reports, technical, 398
requirements for, 496–497
satellite circuit, costs, 528–
 530
security of, 386, 411
selling proposed system, 532
single application, 17
software for management of,
 387–389
software within, 240–242
technical concepts of, 54–137
test equipment, 401–408
total cost calculation, 517
types, 16–18
value added, 17
voice communications, 199–
 201
voice grade, costs, 517–525
volume estimation, 501–504
wide area, 198
wideband, costs, 527–528
New York Telephone (NYNEX):
 deregulation, 470
 straight copper circuits, 476
Newsletter, published 3 times per
 year to augment text. Call John
 Wiley & Sons at (212) 850-6756
 to obtain copies.
NLDM (Network Logical Data
 Manager), NetView, 388
NMPF (Network Management
 Productivity Facility), NetView,
 388
No parity, in systems, 133
Nodes:
 distance between, selection issue, 370
 number of, local area network,
  selection issue, 369
 switching, 220

Systems Network Architec-
 ture (SNA) definition of, 272
Noise (electrical):
 amplitude, 130
 analog circuit, types of, 402
 attenuation, 131
 cause of, 322
 cross-talk, 130
 delay distortion, 131
 echoes, 130, 210–211
 filtering of, 98
 Gaussian, 129
 impulse, 129
 interference, 355
 intermodulation, 130
 jitter, 131
 line outages, 130–131
 optical fibers, 115
 spikes, 129
 types, 129–131
 white, 129
Nonblocking operation, switchboard, 230
Novell NetWare, local area network operating system, 351
NPDA (Network Problem Determination Application), NetView,
 388
NSA (National Security Agency),
 DES algorithm, 423
Null modem cable, use of, 161,
 319–320
Numeric keypads, input devices,
 189
NYNEX (New York Telephone),
 470, 476
Nyquist, H., 205
Nyquist rate, 205

Objections, to new network, 532
Odd parity, in systems, 133
OECD (Organization for Economic
 Cooperation and Development),
 transborder data flow, 20
Office automation:
 data communications role in,
  27–28
 six benefits of, 32–33
 workstations in, 28–29
Ohm, defined, 322
Omissions:
 controls against, 619–622
 defined, 619
ONA (Open Network Architecture), 284
One cable configuration, illustrated, 360
One-for-one changeover, defined,
 533
Online systems:
 circuit volume estimation, 503
 microcomputer auditing of,
  435, 455

Open Network Architecture (ONA), described, 284
Open Systems Interconnection (OSI) Reference Model:
    communications with the, illustrated, 257
    controls for, 438–440
    layers of the, 253–264
    NETBIOS function in, 307
    protocols for new systems, 512
    standards for, 290
    Synchronous Data Link Control (SDLC) gateway, 348
Open wire pairs:
    defined, 100–101
    illustrated, 101
Operating system software:
    location in network, 241
    see also
        Disk Operating System
        Operating System/2
Operating System/2 (OS/2):
    Extended Edition, features, 304–305
    features, 304
    NETBIOS support, 331
    PC Network support, 331
    Systems Application Architecture (SAA) support, 331
    Token-Ring Network support, 331
Operational feasibility, network, 492
Optical circuits, described, 111–117
Optical disks, data communication use of, 23
Optical fibers:
    fabrication techniques, 116
    illustrated, 114
    monomode vs. multimode, 357
    splicing of, 115
    use of, 111
Optical modems, described, 160
Optimization:
    network, 547
    topology, 540–541
Optimize, defined, 509
Options, circuit costs, 522, 530
ORBIT Information Technologies, 50
Organization, voice/data function, 380–384, 389–390
Organization charts:
    Chief Information Officer position, 385
    information systems function, 390
Organization for Economic Cooperation and Development (OECD), transborder data flow, 20

Organization-wide networks, defined, 17
Organizing, activities of, 378–379
OSI model, see Open Systems Interconnection Reference Model
Output devices, terminals as, 181–189
Outside vapor deposition (OVD), optical fiber fabrication, 116
OVD, see Outside vapor deposition

PABX switchboards, defined, 228
Pacific Telesis:
    deregulation, 470
    local access transport arrangement (LATA) example, 471, 472
    PACTEL Spectrum Services, 388, 408
Packet assembly/disassembly (PAD), switching node, 220
Packet switching:
    controls for, 428
    defined, 220, 223
    integrated, 230
    public network use of, 23, 224–225
    service offerings, 481
Packet switching networks:
    costs of, 528
    example, 28, 221, 222, 513
    Open Systems Interconnection (OSI) model layer, 259
    penetration example, 414–415
    protocols for, 220, 264
Packetizing, defined, 220
Packets:
    defined, 58
    interleaving of, 176, 222
    size factors, 345–346
    switching of, 220
PACTEL, see Pacific Telesis
PACTEL Spectrum Services:
    network management, 388
    real-time network management, 408
PAD (packet assembly/disassembly), 220
Page scanners, described, 189
PAM (pulse amplitude modulation), 86, 87
Paperless office, 26
Parallel mode, described, 55–56
Parity:
    interlaced, 133
    setting for communications, 306
    types of, 133
Parity bits, described, 67
Parity checking:
    defined, 132
    logic, illustrated, 133

Passive data line monitors, active vs., 407
Passwords, control by use of, 431
Patch panels, defined, 403
Path control layer, Systems Network Architecture (SNA) component, 271, 273
PBX switchboards:
    data, 231
    defined, 228
    digital, 229
    IBM Cabling System connection, 353
    local area network (LAN) relationship to, 336–337
    Premises Distribution System (PDS) cabling, 354
    role in automated office, 30
    switching, 192
PC-BLAST II, data conversion protocol, 316
PC-DIAL, communication software, 309
PC-DOS, 563
PC Network:
    broadband bus, 339–340
    IBM Cabling System connection, 353
    NETBIOS location in, 307
    NETBIOS use in, 308
    Operating System/2 compatibility, 331
PCDIAL LOG, communication software, 312
PCM (Pulse code modulation), 88, 120
PCTALK III:
    communication software, 309
    XMODEM usage in, 313
PDM (pulse duration modulation), 86, 87
PDS (AT&T Premises Distribution System), 353–354
Peak-to-average ratio, defined, 402
Peer-to-peer, defined, 275, 308
Peer-to-peer protocols:
    Open Systems Interconnection (OSI) model and, 261
    NETBIOS, 308
Penetration, by hackers, 414
People:
    controls against, 625–627
    defined, 625
Performance, issue of network, 548
Performance analysis:
    automated, features, 540
    characteristics, simulation of, 126
    response time, MIND-Data/PC, 548
Peripheral node control point (PNCP), Systems Network Architecture (SNA) component, 276

Permissive connection, defined, 156
Person-to-network protocols, defined, 246
Person-to-person protocols, defined, 246
Personal Computer (PC), operation, 298–302
Personal System/2 (PS/2), features, 303
PERT (Program Evaluation Review Technique), 533
Phase implementation, defined, 533
Phase jitter, defined, 402
Phase modulation, defined, 79
Phase shift keying (PSK), use in modulation, 79, 81
Photonic switching, defined, 112
Physical layer:
    controls for, 438
    Distributed Network Architecture (DNA) model, 281–282
    function, 255, 258
    Open Systems Interconnection (OSI) model, 256, 258
Physical networks, testing, 387
Physical topology, electrical vs., 338
Physical unit (PU):
    basic, 276
    defined, 271
Pilot implementation, defined, 533
Ping-ponging, defined, 176
Pins (connector), illustrated, 74, 76, 318
Pizazz, software, 551
Plaintext, defined, 419
Planning:
    activities of, 378
    implementation tasks, 532
    new network, 492–494
Plans, long-range vs. short-range, 496
PLC (power line conditioners), 322–323
Plugs:
    RJ-11, 157, 161, 317, 321, 355, 363
    RJ-45, 157
    RJ-48, 354
    RS232, 74, 77, 161, 167, 258, 289, 317–319, 353, 362, 435
    RS449, 75–76, 258
PNCP (peripheral node control point), 276
Point of presence, defined, 521
Point-to-point circuits:
    defined, 216, 501
    digital, 477
Point-to-point network, example, 54, 216, 498

Policies, cabling responsibility, 359
Poll-to-poll time, defined, 406
Polling:
    defined, 123, 145, 148
    fast select, 123–124
    hub go-ahead, 124
    queues model, illustrated, 545
    roll call, 123
Polling discipline, purpose of, 544
Polynomial checking, defined, 134
Polynomials, error checking use of, 134
Port (hardware):
    purpose, 147
    security, 155
Port sharing devices:
    functions, 152
    illustrated, 153
Positive acknowledgment:
    ACK vs. WAK, 264
    satellite transmission of, 104, 106
    XMODEM, 313
    see also Negative acknowledgment
Postal Telephone and Telegraphs (PTT):
    CCITT membership, 288
    role of, 468
Postes Telephonique et Telegraphique, 468
Power line conditioners (PLC), purpose, 322–323
PPM (pulse position modulation), 86, 87
Predicting, circuit failures, 397
Preliminary design configurations, costs related to, 516
Premises Distribution System (PDS), ISDN connection incompatibility, 354
Presentation layer:
    controls for, 440
    function, 255, 263
    missing from Manufacturing Automation Protocol (MAP) model, 281
    Open System Interconnection (OSI) model, 255, 263
Prestel videotex, 226
Prevent, controls to, 411
Pricing structures, 474–485
Primary access service:
    Europe, 31B + 1D, defined, 227
    North America, 23B + 1D, defined, 226–227
Primitive, DOS, defined, 308
Printers, local area network, selection issue, 371
Privacy, cable television and, 25
Private automatic branch exchange, see PABX switchboards

Private branch exchange, see PBX switchboards
Private circuits:
    defined, 99, 202, 228, 474
    dial-up vs., 475
    types of, 474–478
Private key encryption, public vs., 424–425, 427
Private lines:
    defined, 99, 202
    FCC Tariff 11, 521
Private networks, 474–478
Problem incident report, contents, 394
Problem-solving methodology, defined, 393
Problem statistics, defined, 393
Problem tracking, defined, 393
Program Evaluation Review Technique (PERT), implementation use, 533
Program-to-program interface, APPC, 274–275
Programmable connection, 157
Programs, see Software
PROMT database, 50
Propagation delay:
    defined, 99
    sliding windows and, 316
Propagation time, satellite transmission, 105–106
Proprietary algorithms, defined, 424
Protocol converters:
    defined, 177
    described, 177–180
    illustrated, 178, 179
Protocol monitors, defined, 406
Protocol testing, defined, 403
Protocols:
    Binary Synchronous Communications (BSC), 267–268, 512
    bit-oriented, 267, 512
    BLAST, 315
    byte-oriented, 282, 512
    conversion, 149
    data link, 246
    data link layer, 259
    defined, 199, 246, 342
    described, 253–278
    design considerations, new systems, 512
    digital data communication message protocol (DDCMP), 282
    High level Data Link Control (HDLC) recommended for new systems, 512
    KERMIT, 313
    local area network, selection issue, 369
    machine-to-machine, 246
    message, 260, 268

microcomputer, 312–316
PC-BLAST II, 316
peer-to-peer, 261
person-to-network, 246
person-to-person, 246
polling, multidrop, 544
response time analyzers, 406
satellite channels, 477
Synchronous Data Link Control (SDLC), recommended for new systems, 512
testing of, 403
X.25, recommended for new systems, 512
X.25, illustrated, 266
X.75, gateway to Accunet, 481
X.PC, asynchronous packetizing, 315
X-ON/X-OFF, flow control, 312
XMODEM, block transmission, 313
YMODEM, *see* Newsletter, vol. 5, no. 1, Jan. 1988 (ISBN 0-471-60129-2)
Protokollon, origin of protocol, 438
PS/2 (Personal System/2), 303
PSK (phase shift keying), 79, 81
PSS network, 225
PTT (Postal Telephone and Telegraphs), 288, 468
PU (physical unit), 271, 276
PU 2.1, Systems Network Architecture (SNA) component, 274
Public directory, public key algorithm, 425–426
Public key encryption:
    defined, 424
    illustrated, 426
    private *vs.,* 424–425, 427
Public networks:
    advantages, 223
    penetration example, 414–415
Public packet switched services, 481–483
Public timesharing networks, described, 224–226
Public Utility Commissions (PUC), role of, 467
PUC, *see* Public Utility Commissions
Puerto Rico:
    AT&T Megacom service, 481
    WATS service band, 480
Pulse amplitude modulation (PAM), defined, 86, 87
Pulse code modulation (PCM), defined, 88, 120
Pulse duration modulation (PDM), defined, 86, 87
Pulse modulation, defined, 86
Pulse position modulation (PPM), defined, 86, 87

Purchasing, local area network considerations, 363, 367–371
Pure multiplexing:
    defined, 169
    statistical *vs.,* 175

Q.921 format, 289
Q.931 format, 227
QAM (quadrature amplitude modulation), 81
QPSK (quadrature phase shift keying), 81
QTAM (queued telecommunication access method), 250
Quabits, defined, 92
Quadrature amplitude modulation (QAM), 81
Quadrature phase shift keying (QPSK), use in modulation, 81
Quadrillion, relation to femtosecond, 421
Quality control, circuits, 397
Queue:
    defined, 125
    modeling of, 544
    polled terminals, illustrated, 545
Queued telecommunication access method (QTAM), function, 250

Radio frequency interference (RFI), shielding against, 355
Raindrop attenuation, KU-band, 107
RAM (random access memory), 299, 302
Random access, ALOHA system, 343
Random access memory (RAM):
    booting programs, 302
    defined, 299
Random dependence, defined, 510
Rate centers:
    coordinates, FCC Tariff 10, 520
    defined, 520, 637
Rates, *see* Tariffs
RDC (Remote Data Concentrators), 560
Read only memory (ROM):
    booting programs, 302
    defined, 299
Record locking:
    defined, 352
    local area network need for, 365, 366
Recordkeeping, network statistics, 394–395
Recovery, controls for, 442, 444–446, 622–623
Redundant bits, defined, 67
Reevaluation, after implementation, 534

Regulatory agencies, 467–469
Relevancy, of information, 378
Reliability:
    controls to safeguard, 631–635
    defined, 631
Remote Data Concentrators (RDC), defined, 560
Remote intelligent controllers:
    controls for, 429
    location in network, 157
Remote job entry (RJE) terminals:
    controls, 430–432
    described, 185
Rental car systems, described, 44
Repeaters, defined, 98, 341
Reports:
    feasibility study, 492
    management, 394
    technical, 398
Request for Proposals (RFP), 515
REQUEST TO SEND signal, 78, 166
Requirements, defining the, 496–497
Resistance (electrical), cause of distortion, 73
Resource, information as a, 4, 22, 384, 386, 507
Response time:
    analyzers for, 406
    calculation of, 126
    defined, 124
    equation, 126
Restart:
    controls for, 442, 444–446, 622–623
    defined, 622
Retrain time:
    defined, 97
    modem feature, 166
Retransmission, in error control, 132
Return loss, defined, 402
Reverse channel, modem feature, 165
RFI (radio frequency interference), 355
RFP (Request for Proposals), 515
Ring topology:
    defined, 337
    electrical, 338
    illustrated, 215, 336
    Token-Ring Network, 340
Risk ranking, network controls, 454
RJ-11 plug:
    BALUN end, 355
    IBM Cabling System, 363
    internal modem, 317
    short haul modems, 161
    surge protection, 321
    use of, 157
RJ-45 plug, use of, 157

RJ-48 plug, Premises Distribution System (PDS), 354
Roll call polling, defined, 123
ROM (read only memory), 299, 302
ROMBIOS (Read Only Memory Basic Input/Output Systems), defined, 301, 308
Routing servers, defined, 576
Routing tables, network layer, 260
RS232 connector/cable:
    described, 73–75
    Electronic Industries Association standard, 289
    external modems, 317
    IBM Cabling System, 353
    local area network use, 362
    modem control, 77
    multiplexing and the, 167
    null modems and, 319
    Open Systems Interconnection (OSI) model and, 258
    pins of the, 74, 318
    security of, 435
    short haul modems, 161
RS449 connector/cable:
    described, 75–76
    Open Systems Interconnection (OSI) model and, 258
    pins of the, 76
Rule of thumb, staff per number of terminals, 386
Run Counter, LANalyzer screen, 650
Run Global, LANalyzer screen, 651
Run length encoding, use of, 156

SAA (Systems Application Architecture), 278–280, 331
Safeguards, see Controls
Sags:
    defined, 321, 322
    local area network, selection issue, 371
    major electrical problem, 324
Satellite channel, defined, 477
Satellites:
    Binary Synchronous Control (BSC) protocols, 268
    channel costs, illustrated, 529
    citizens band and, 23
    delay, 70–71, 104–106
    impact on TRIB, 70–71
    frequencies of, 106–107
    geosynchronous, 104–110
    home use of, 20, 477
    issues concerning, 108
    local area network usage, 108, 110–111
    packet switching configuration with, illustrated, 513
    sliding windows in, links, 316

Satisfice, defined, 509
Scanners, page, 189
Schema, defined, 241
Scoop System, traffic management, 408
Scope, network, 492, 497
Screens (video), touch, 188
Scripts, defined, 316
SDLC, see Synchronous Data Link Control
SDN, see Software defined networks
Security:
    dial-back, 155
    diskless workstations, see Newsletter, vol. 5, no. 1, Jan. 1988 (ISBN 0-471-60129-2)
    fiber optic communications, 113
    local area network, 365
    local area network, selection issue, 371
    need for, in networks, 411–414
    satellite communications, 108
Security locks, 330, 351–352, 365
Security logging, 137, 150
Security software:
    impact on system, 442
    location in network, 241
Selecting (polling), defined, 123, 148
Selection, issues influencing local area network, 367–371
Self-checking code, described, 62, 67
Self-testing modems, defined, 405
Sender, defined, 12
Sequential dependence, defined, 510
Serial mode, described, 56–57
Series 1000 circuits, defined, 476
Series 2000 circuits, media for, 475
Series 3000 circuits:
    media for, 475
    voice grade, 475
Servers:
    defined, 198
    disk, defined, 329, 351, 369
    file, defined, 307, 330, 351, 369
    network, defined, 328, 370
    network, standards for, 309
    types of, 576
Session, defined, 207
Session layer:
    controls for, 439
    Distributed Network Architecture (DNA) model missing, 282
    function of, 255, 262
    Open System Interconnection (OSI) model, 255, 262

Seven-layer OSI model, see Open System Interconnection (OSI) Reference Model
Shannon, Claude, 206
Shannon's law, 207
Shielding:
    against radio frequency interference, 355
    illustrated, 356
Short distance radio, described, 119
Short distance transmission, infrared, 233
Short haul modems, described, 160–161
Shunt, surge protection, 322
Sign-on table, terminal restriction in customer information control system (CICS) by, 441
Signal-to-noise ratio, defined, 207
Signaling, dial-up circuits, 207–208
Signaling format, T-1 links, 121
Signaling rate:
    calculation of, 121
    defined, 90, 94
Signals:
    analog, 72,
    baseband, 71–72
    bipolar, 85–86
    broadband, 72–73, 78
    control, 207
    digital, 72–73, 85
    distortion of, 72
    double current, 86
    light, 111
    progression through hardware, 142
    progression through network, 54
    progression through software, 238
    synchronization of, 77–78
    unipolar, 85–86
Signature verification, 425
Simplex transmission, defined, 96
Simulation:
    defined, 125
    purpose of, 545–546
    system, defined, 543
    see also Models
Simulators, performance characteristics, 126
Sine waves, defined, 71, 72
Single application network, defined, 17
Single bit transmission, illustrated, 91
Single cable:
    dual vs., 360
    illustrated, 360
Single frequency interference, defined, 402
Sink, defined, 10

Sliding window protocol:
  BLAST usage, 316
  continuous ARQ, 264
  X.PC usage, 316
Smart modems, dumb *vs.*, 162
SMARTCOM II, communication
  software, 309
SMSA (Standard Metropolitan Statistical Areas), 471
SN (switching node), 220, 428
SNA, *see* Systems Network Architecture
Society for Worldwide Interbank
  Financial Telecommunications
  (SWIFT), 41
Software:
  basic concepts, 238–245
  basic concepts, illustrated,
    239
  booting, 302
  communication packages,
    309–312
  controls for, 440–442, 597–
    602
  data communication system,
    248–253
  defined, 147, 247, 597
  design precepts, 242–244
  diagnostics for, 514–515
  Disk Operating System (DOS),
    298, 300
  front end location, 147
  functional specification, 244
  impact on message flow, 238
  local area networks, 350–352
  local area networks, analysis,
    366
  local area networks, installa-
    tion, 364
  local area networks, selection
    issue, 370
  location in network, 239, 247
  matrix development, 454
  microcomputer communica-
    tions, 306–312
  network design considerations,
    511–515
  network management, 387–
    390
  network optimization, 546–
    552
  ROMBIOS (Read Only Mem-
    ory Basic Input/Output Sys-
    tems), 301
  testing precepts, 244–245
  types in network, 240–242
Software defined networks (SDN):
  defined, 202, 478
  quality of (AT&T, MCI, US
    Sprint), *see* Newsletter, vol.
    5, no. 1, Jan. 1988 (ISBN 0-
    471-60129-2)
Software encryption, defined, 180

Software protocol conversion pack-
  ages, 178–180
Software testing, types of, 244–245
SOH (start of heading), 267
Source, defined, 10
Source, The (organization), 224,
  306
Source-to-destination layer, Open
  Systems Interconnection (OSI)
  model, 261
Space (signaling), defined, 71, 87
Spain, transborder data flow, 20
Special services, 484–485
Specification, software testing,
  244
Spectrum Services, 388, 408
Sperry Univac, Distributed Com-
  munications Architecture (DCA),
  282
Spikes:
  defined, 321
  source of errors, 129
Splicing, optical fiber, 115
Split streaming, modem feature,
  165
Spoofing, defined, 433
SSCP (System Services Control
  Point), 271
Staffing:
  activities of, 379
  cross-training of employees,
    384
  see also People
Standalone communication config-
  uration, illustrated, 143
Standard Metropolitan Statistical
  Areas (SMSA), relationship to
  LATAs, 471
Standards:
  connectors and Open Systems
    Interconnection (OSI) model,
    258
  encryption, 421
  European environment for,
    468
  facsimile, 187
  Integrated Services Digital
    Network (ISDN), 226–228
  legally enforceable, 290–294
  local area networks (LAN),
    339–341
  network servers, 309
  Open Systems Interconnection
    (OSI) model, 253–264
  protocols, 246–247
  telecommunication, 286–294
  V.24, 73
  V.28, 73
  V.32, 136
  V.*nn* series, 293–294
  X.3, 220
  X.20, 75
  X.21, 75, 258

X.25, 259, 264, 267, 282, 283,
  315, 477
X.75 gateway, 349, 481
X.400, 290, 308
X.*nn* series, 291–292
STAR local area network, see
  STARLAN
Star topology:
  defined, 338
  IBM Cabling System, 353
  illustrated, 214, 339
STARLAN:
  Premises Distribution System
    (PDS) cabling, 353
  twisted pair CSMA/CD, 340
Start of heading (SOH), Binary
  Synchronous Communications
  (BSC), 267
Start of text (STX), Binary Syn-
  chronous Communications (BSC),
  267
Start-stop transmission, 57
Stat muxes, 171
Static (electricity), protection
  against, 324
Station terminals, role in network,
  94
Stations, defined, 199, 272
Statistical recording, front end, 150
Statistical time division multiplex-
  ing (STDM), described, 170–173
Statistics:
  network traffic, 501, 503
  normal operation, 396–397
  problems in network, 393
STD (Subscriber Trunk Dialing),
  562
STDM (statistical time division
  multiplexing), 170–173
Step-index cables, defined, 113
Stop and wait ARQ:
  Binary Synchronous Commu-
    nications (BSC), satellite link
    use, 268
  cause of delay, 104
  defined, 264
  satellite echo canceller, 530
  XMODEM usage, 313
  see also Continuous ARQ
Stopbits, setting for communications, 306
Store and forward:
  defined, 22
  switching, 191, 220
Store and forward switching, defined, 223
Straight copper circuits, defined,
  476
Strategic plans, critical to system
  design, 496
Strategic resources:
  capital *vs.* information, 384,
    386

Strategic resources (*Continued*)
   industrial *vs.* information societies, 4, 22
STX (start of text), 267
Subchannels, defined, 100
Sub-LAN, defined, 362
Submarine cable transmission, 119
Subnetwork, Open Systems Interconnection (OSI) model, 259
Subschema, defined, 241
Subscriber loop, defined, 94
Subscriber Trunk Dialing (STD), defined, 562
Success:
   evaluation criteria to measure, 493–494
   feasibility factors and, 492
   information on environment related to, 495
   key ingredient of, 489
Super Group 240,000 hertz:
   cost, illustrated, 528
   wideband, 476
Superconductivity, *see* Newsletter, vol. 5, no. 1, Jan. 1988 (ISBN 0-471-60129-2)
Surges:
   defined, 321
   protection, local area network, selection issue, 371
Sweden, transborder data flow, 20
SWIFT network, 42, 43
Switch database administration, Multi-Function Operations System (MFOS) module, 389
Switch management, Multi-Function Operations System (MFOS) module, 389
Switchboards:
   digital, 182
   trends, 229–230
   types of, 228–229
Switches:
   controls for, 608–612
   data (PBX), 231
   defined, 608
   digital cross-connect, 231
   functions of, 190–191
   management of, 389
   number of lines required, calculation, 190
   types of, 189–194
   uses for, 193–194
   virtual circuit, 176
Switching:
   circuit, described, 191
   circuit, defined, 223
   defined, 190
   digital data, 191–192
   impact of ballistic transistor on, 19
   network, 192
   photonic, 112

role in data communications, 99
store and forward, 191
store and forward, defined, 223
Switching node (SN):
   controls for, 428
   purpose of, 220
Symmetric algorithms, defined, 423
SYN characters, 78, 406
Sync characters, in transmission, 59
Synchronization, synchronous transmission and, 77
Synchronous Data Link Control (SDLC):
   BLAST use of, 315
   Distributed Network Architecture (DNA) model, 283
   gateway, 348
   new system use, 512
   Open System Interconnection (OSI) model, 256, 259
   satellite channels, 477
Synchronous transmission:
   defined, 60
   described, 58–59
   synchronization in, 77
System database, controls for, 442
System Services Control Point (SSCP), Systems Network Architecture (SNA), 271
System simulation, defined, 543
System-to-system gateways, defined, 349
Systems analysis, structured, 495, 534
Systems Application Architecture (SAA):
   components of, 279
   described, 278–279
   features, 280
   Operating System/2 support for, 331
Systems approach, network design, 489, 490
Systems Network Architecture (SNA):
   Advanced Program-to-Program Communications (APPC), 274–275
   data link control layer, 273
   described, 268–278
   file transfer, 325
   gateway, 347–348
   logical unit (LU), 269, 270, 271
   LU 6.2, 274–275
   message unit in, 271
   NETBIOS usage, 308
   network addressable units in, 271
   node in, 272

Open Systems Interconnection (OSI) model, 261
path control layer, 273
path control network in, 271, 273
peripheral node control point (PNCP), 276
physical unit (PU), 271
System Services Control Point (SSCP) in, 271
telecommunication access method (TCAM), 250

2B + D ISDN, 226, 289, 477
20-milliamp circuit, described, 160
23B + 1D ISDN, 227
31B + 1D ISDN, 227
T carrier system, defined, 120
T-1 carrier, described, 120–122, 466, 478
T-1 channels, hierarchy of, 88, 120
T-1 multiplexing, described, 173–174
T-1 Network Diagnostic System, network management, 408
T-2 channels, hierarchy of, 88
T-4 channels, hierarchy of, 88
Tandem trunks, 200
Tapping:
   directional, 578
   fiber optic cable, 113, 115, 434
   multitap, 578
Tariffs:
   defined, 467–468, 489
   FCC Tariff 10, 520
   FCC Tariff 11, 521
   satellite, 528
TASI (Time Assignment Speech Interpolation), illustrated, 213
Tasks:
   levels of dependence, 509–510
   network managerial, 381–382
TCAM (telecommunication access method), 250
TCM (time compression multiplexing), 176–177
TCP/IP (Transmission Control Protocol/Internet Protocol), 284
TDM (time division multiplexing), 169–170, 176
Technical and office protocol (TOP), MAP/TOP, 281
Technical feasibility, network, 492
Technical reports, need for, 398
Technology, impact on communications, 19
Telecom, Australian deregulation, 469
Telecommunication access method (TCAM), function of, 250
Telecommunication access programs:
   controls for, 441–442

function of, 249
location, 240, 249
Telecommunications:
    careers in, 44–48
    defined, 12
    degree programs in, 47–48
    standards, 286–294
Telecommunications function, responsibilities of the, 385
Telecommuting, defined, 25
Teleconferencing, described, 25, 44–45
Telenet network, 224
Telephone sentinel, defined, 432
Telephone system, famous firsts, 209–210
Telephones:
    basic voice network, 200
    controls for, 432–434
Teleports, defined, 24
Teleprinter terminals, 185
Teleprocessing, defined, 12
Teleprocessing monitors:
    controls for, 441, 442
    function of, 252–253, 441
    location, 240–241, 252
Teletel videotex, 226
Teletext, CCITT standards for, 485
Telex service, 484
Telidon videotex, 226
Terminal block, defined, 94
Terminal program, CROSSTALK function, 310
Terminal servers, defined, 576
Terminals:
    attributes of, 188–189
    categories of, 181–188
    controls for, 430–432
    defined, 13
    dumb, 188
    facsimile, 186–188
    intelligent, 188
    microcomputer, 182–183
    remote job entry (RJE), 185
    teleprinter, 185
    transaction, 185
    video, 183–185
Test sets:
    analog vs. digital, 403
    control of, 443
    handheld, 403
Testing:
    automated equipment for, 407–408
    equipment for networks, 401–408
    loopback, 165, 405
    software, 244–245
    test documentation, 245
    test execution, 245
    test plan, 244
Testing/problem management group, purpose of, 395–397

Threats:
    defined, 446
    examples of, 447, 582–584
    faced by components, 411
    illustrated, 412
    mitigate, 449
Throughput:
    data, 69–71
    offloading to increase, 441
Time Assignment Speech Interpolation (TASI):
    illustrated, 213
    packet interleaving, 176
    use of, 211–213
Time charges:
    software defined network (SDN) service, 478
    usage charges vs., 466
Time compression multiplexing (TCM), 176–177
Time dependence, defined, 510
Time division multiplexing (TDM):
    described, 169–170, 176
    digital channel banks, 176
Timeliness, of information, 378
Times:
    access, 404
    application processing, 126
    bit transfer rate, 404
    block transfer, 404
    clocking, 97, 166
    delay, in satellite channels, 70–71, 104
    disengagement, 404
    femtosecond, 421
    file transfer, 160
    lead, 533
    message input, 126
    message output, 126
    nanosecond, defined, 19
    poll/response, 406
    poll-to-poll, 406
    quadrillion, relationship to femtosecond, 421
    response, 124, 126–127, 242
    response, analyzers for, 406
    retrain, 97, 166
    turnaround, 97
    walk, 346, 543
    zone differences affecting design, 505–506
    see also
        Gigahertz
        Hertz
        Kilohertz
        Mean times
        Megahertz
Timesharing networks, advantages, 223
Timing:
    modem speeds, 83
    signal synchronization, 77–78

Token, defined, 346
Token access method, defined, 346
Token passing:
    IEEE 802.4 support by MAP, 281, 288
    Manufacturing Automation Protocol (MAP), 340
    Token-Ring Network, 340
Token-Ring Network:
    baseband ring, 340
    gateway, 348
    IBM Cabling System connection, 353
    NETBIOS use in, 308
    Operating System/2 (OS/2) compatibility, 331
Tools:
    implementation, 533
    matrix development, 454
    network design, 540–541, 543
TOP (technical and office protocol), 281
Topology:
    defined, 212, 337
    electrical vs. physical, 338
    types, 337–338
Topology optimization:
    automated layout for, 540–541
    MIND-Data/PC, 548
Touch screens, 188
TPA (Transient Program Area), 302
Trace Buffer, LANalyzer screen, 653
Tracking networks, described, 18
Traffic:
    capacity analysis, 504–507
    loading estimation, inaccurate, 506
    network link, table, 502
Traffic data collection and analysis, Multi-Function Operations Systems (MFOS) module, 389
Traffic statistics, network design, 501
Training:
    cross-training of staff, 384
    implementation task, 533
    local area network, selection issue, 371
Transaction terminals, 185–186
Transaction trail, 137
Transborder data flow:
    international policy, 20
    regulations by country, 20
    satellites impact on, 20
Transient (electrical), defined, 321
Transient program area (TPA), programs in the, 302
Transistor, ballistic, 19
Transmission:
    access methods, 199, 248, 343, 346
    analog, 203

Transmission (*Continued*)
  asynchronous, 57–58
  baseband *vs.* broadband, 71–73, 83
  cellular radio, 117–118
  coaxial cable, 100
  efficiency of, 67–69
  fiber optic, 111–117
  FM radio, 119
  infrared, 233–234
  isochronous, 59–60
  microwave, 103–104, 234
  open wire pairs, 100
  parallel, 55–56
  serial, 56–57
  short distance radio, 119
  start-stop, 57
  submarine cables, 119
  synchronous, 58–59, 77
  tropospheric, 110
  waveguide, 119
  wire cables, 100
Transmission Control Protocol/Internet Protocol (TCP/IP), described, 284
Transmission media, types of, 13, 100–119
Transmission methods, contrasted, 96
Transmission modes, 55–61
  comparison of, 60–61
Transmission Rate of Information Bits (TRIB):
  defined, 70
  impact on circuit loading, 506
Transmission speeds:
  fiber optic, 112–113
  local area network capacity, calculation, 344–345
  see also High speed services
Transmitters, optical, 112
Transpac network, 225
Transparent, defined, 167
Transport header, defined, 261
Transport layer:
  controls for, 439
  Distributed Network Architecture (DNA) model, 281–282
  function, 255–256, 260–262
  Open Systems Interconnection (OSI) model, 255–256, 260–262
Tree configuration, illustrated, 215
Trellis code, 136
Trends, data communications, 18–26
TRIB (Transmission Rate of Information Bits), 70, 506
Tribits, defined, 92
Tropospheric scatter, described, 119
Trouble log, defined, 395

Trouble ticketing, Multi-Function Operations System (MFOS) module, 389
Trouble tickets, defined, 393
Troubleshooting function, centralized, 393
Trunk line, defined, 100, 228
Trunks:
  defined, 100
  interoffice, 200
  intertoll, 201
  tandem, 200
Turnaround time:
  defined, 97
  TRIB calculation, 70
Turnpike effect, defined, 506
Twisted pair wiring:
  data grade, defined, 355
  DECconnect, 354
  local area network cabling, 352, 355, 362
  STARLAN, 340, 353
Two-cable configuration, illustrated, 359
Two-point circuits, defined, 216
Two-point segment, defined, 100
Two-wire circuit, defined, 96
TWX, *see* Telex
Tymnet public timesharing network, 224
  X.PC development, 315

UDLC (Universal Data Link Control), DCA model, 282
Ulrich's International Periodicals Directory, 51
Unattainable design configurations, choice set, 509
Uninet network, 225
Uninterruptible power supply (UPS), purpose, 323–324
Unipolar signals, illustrated, 85–86
Unit call, defined, 200
United States:
  communications with Mexico, 468
  deregulation in the, 469–474
  Federal Standard 1031, 75
  laser ISL development, 108
  microcomputer certification, 303
  modem certification, 166, 469
  videotex development, 226
United States of America Standard Code for Information Interchange (USASCII), 56, 61, 63
Universal connection, 157
Universal Data Link Control (UDLC), Distributed Communications Architecture model, 282
UNIX, described, 285–286
Uploading, defined, 310, 325
  see also Downloading

UPS (uninterruptible power supply), 232–324
Uptime:
  controls to safeguard, 631–635
  defined, 631
US Sprint:
  common carrier, 12, 198, 466, 471
  competitors of, 474
  discount voice service, 483–484, 531
  local access transport area (LATA) service, 471
Usage charges, time *vs.* volume basis, 466, 478
USASCII (United States of America Standard Code for Information Interchange), 56, 61, 63
Users:
  acceptance of system, 489, 532
  number of, local area network, selection issue, 369

V coordinates, *see* Vertical coordinates
V standards, 73, 291, 293–294
V.24 cable, 73
V.28 cable, 73
V.32 error correction, 136
Vacuum cleaning (telephone), defined, 433
Value added networks (VAN), defined, 17, 224
VAN, *see* Value added networks
VAX minicomputer, APPC application limitations, 277
VDT (video display terminals), 183
VDU (video display units), 183
Vertical coordinates:
  by city, 637–642
  use of, 517, 520
Very small aperture terminals (VSAT):
  Ku-band, 107
  network usage, 234
Viatel videotex, 226
Video display terminals (VDT), 183
Video display units (VDU), 183
Video teleconferencing, described, 25, 44–45
Video terminals:
  described, 183–185
  selection considerations, 183–185
Videotex, defined, 19, 225
Virgin Islands (U.S.):
  AT&T Megacom service, 481
  WATS service band, 480
Virtual circuits:
  defined, 220
  switches, 176
Virtual leased circuits, defined, 202, 478

Virtual links:
  defined, 256
  illustrated, 257
Virtual telecommunication access
  method (VTAM), function of,
  250–251
VNCA (VTAM Node Control Application), 388
Voice and data communication,
  see Newsletter, vol. 5, no. 1, Jan.
  1988 (ISBN 0-471-60129-2)
Voice call multiplexing, use of,
  212
Voice communications:
  controls for, 432–434
  costs of, 19, 381
  data communications integra-
    tion with, 380
  discount services, 483–484
  how to merge with data, 382–
    384
  trunk lines, defined, 228
Voice entry, 189
Voice grade circuits:
  bandwidth, 204
  capacity of, 203–207
  costs of, 517, 520–525
  design problem, 536
  leased, defined, 202, 475
  point of presence, illustrated,
    521
  Series 2000/3000, 475
Voice jack, defined, 94
Voice mail:
  defined, 35
  described, 34–38
  features of, 36–38
  local area network, selection
    issue, 371
  trends in, 26
Voice message systems, 22–23
Voice networks, 199–201
Volt-amps, calculation of, 323
Voltage regulators, purpose, 322
Volume locking:
  defined, 351
  local area networks, need for,
    365, 366
VSAT (very small aperture terminals), 107, 234
VTAM (virtual telecommunication
  access method), 250–251
VTAM Node Control Application
  (VNCA), NetView function, 388

WAK, defined, 264
Walk time, defined, 346, 543
WAN (wide area networks), 198
WATS, see Wide Area Telecommunications Service
WATSON, communication software, 312
Watts:
  uninterruptible power supply
    (UPS) rating, 323
  volt-amp calculation, 323
Waveforms:
  analog, 72
  carrier, 78
  sine, 71
Waveguides:
  defined, 119
  fiber optic, 356
Western Union, Telex, 484
White noise:
  defined, 129
  impact on signal-to-noise ra-
    tio, 207
Wide area networks (WAN), defined, 198
Wide Area Telecommunications
  Service (WATS):
  described, 479–480
  interstate vs. intrastate,
    480
  tariffs for, 531
Wideband services:
  cost of, 527
  described, 476–477
Wire cable:
  defined, 100
  illustrated, 101
Wire center, defined, 522
Wiring closet, defined, 353
Wiring hub, defined, 338
Wolverton, Van, MS-DOS, 332
Workstations:
  defined, 29, 182
  diskless, see Newsletter, vol.
    5, no. 1, Jan. 1988 (ISBN 0-
    471-60129-2)
  features of, 29
  impact on office automation,
    28–29
  local area network hardware,
    selection issue, 370
  microcomputers vs., 182
Written procedures, network,
  533

X standards, 291–292
X.3 packet assembly/disassembly
  (PAD), 220
X.20 cable, asynchronous, 75
X.21 cable:
  Open Systems Interconnection
    (OSI) model, 258
  synchronous, 75
X.25:
  Burroughs Network Architec-
    ture (BNA) model, 283
  Distributed Communications
    Architecture model, 282
  defined, 264
  Distributed Network Archi-
    tecture (DNA) model, 283
  Distributed Systems (DS)
    model, 284
  frame, illustrated, 265
  illustrated, 266
  new system protocols, 512
  Open Systems Interconnection
    (OSI) model, 259
  satellite channel usage, 477
  Synchronous Data Link Con-
    trol frame identical to, 274
  X.75 gateway connection, 349
  X.PC packet conversion to,
    315
X.75 gateway node, 349, 481
X.400:
  electronic mail standard, 290
  NETBIOS usage, 308
X.PC, packet protocol, 315
X-ON/X-OFF, flow control proto-
  col, 312
XEROX Network Systems (XNS),
  described, 284
XMODEM:
  CROSSTALK protocol, 310
  KERMIT contrasted with, 314
  local area network error cor-
    rection, 362
  message block protocol, 313
XNS (XEROX Network Systems),
  284

YMODEM, see Newsletter, vol. 5,
  no. 1, Jan. 1988 (ISBN 0-471-
  60129-2)

Zero slot LAN, defined, 362
ZOOM/MODEM PC, modem
  functions described, 319